Sliding Mode Control of Power Converters in Renewable Energy Systems

Sliding Mode Control of Power Converters in Renewable Energy Systems

Special Issue Editor

Luis Martinez-Salamero

MDPI • Basel • Beijing • Wuhan • Barcelona • Belgrade

Special Issue Editor
Luis Martinez-Salamero
Rovira i Virgili University
Spain

Editorial Office
MDPI
St. Alban-Anlage 66
4052 Basel, Switzerland

This is a reprint of articles from the Special Issue published online in the open access journal *Energies* (ISSN 1996-1073) in 2019 (available at: https://www.mdpi.com/journal/energies/special_issues/ sliding_mode_control).

For citation purposes, cite each article independently as indicated on the article page online and as indicated below:

LastName, A.A.; LastName, B.B.; LastName, C.C. Article Title. *Journal Name* **Year**, *Article Number*, Page Range.

ISBN 978-3-03928-098-8 (Hbk)
ISBN 978-3-03928-099-5 (PDF)

Contents

About the Special Issue Editor

Luis Martinez-Salamero taught circuit theory, analog electronics, and power processing from 1978 to 1992 at the Escuela Técnica Superior de Ingenieros de Telecomunicación de Barcelona, Polytechnic University of Catalonia, Barcelona, Spain. From 1992 to 1993, he was visiting scholar at the Center for Solid State Power Conditioning and Control, Department of Electrical Engineering, Duke University, Durham, NC. He continued his cooperation with Duke University for three months per year in 1995 and 1996. From 2003 to 2004, 2010 to 2011, and March–August of 2018, he was visiting scholar at the Laboratory of Architecture and Systems Analysis (LAAS), National Agency for Scientific Research (CNRS), Toulouse, France. Since 1995 he has been a full professor with the Department of Electrical Electronic and Automatic Control Engineering ,School of Electrical and Computer Engineering, Rovira i Virgili University (URV), Tarragona, Spain, where he managed the Research Group in Industrial Electronics and Automatic Control (GAEI) in the period 1998–2018 (http://deeea.urv.cat/gaei/index.php). His research interests include the structure and control of power conditioning systems, namely, the electrical architecture of satellites, power distribution in hybrid and electric vehicles, as well as the nonlinear control of converters and drives, and power conditioning for renewable energy. He has published a large number of papers in scientific journals and conference proceedings in the fields of modelling, simulation, and control of power converters, and holds a U.S. patent on dual-voltage electrical distribution in vehicles. He was Guest Editor of the IEEE Transactions on Circuits and Systems Special Issue on Simulation, Theory and Design of Switched-Analog Networks (Aug. 1997). In cooperation with the European Space Agency (ESA), he organized the 5th European Space Power Conference (ESPC-98) in Tarragona and served during two terms (1996–2002) as a Dean of the School of Electrical and Computer Engineering in URV. He has supervised 15 doctoral theses and given lectures in the universities of Toulouse, Bordeaux, and Tel-Aviv. He has conducted numerous seminars on the use of non-linear control in power converters in both Spanish and foreign universities, and has been invited speaker in four plenary sessions in international conferences. He was president of the Spanish Joint Chapter of the IEEE Power Electronics and Industrial Electronics Societies from 2005 to 2008, and distinguished lecturer of the IEEE Circuits and Systems Society in the period 2001–2002. He served as a president of the committee for Communications, Electronics and Computer Science for the Research Activity Evaluation (CNEAI) of the Spanish Ministry of Education in 2012, and as a national coordinator in the area of Electrical, Electronic and Automatic Control Engineering of the National Agency of Prospect and Evaluation (ANEP) of the Spanish Ministry of Science and Innovation in the period 2008–2011. He is currently a distinguished professor of Rovira i Virgili University.

Preface to "Sliding Mode Control of Power Converters in Renewable Energy Systems"

The first paper on dc-to-dc switching converters published in a scientific journal had only one page and did not mention the term power converter in the description of that emergent family of circuits (1). Almost fifty years later, around 7000 patents have been issued on dc–dc conversion in the USA alone, making us consider whether the interdisciplinary field known as Power Electronics reached maturity long time ago, and if there is still space for innovation.

Some data can help us to answer affirmatively to the second question. They are supported by the concern for the global warming, which has renewed the interest in renewable energy technologies. Underpinning a sustainable environment implies the use of more efficient devices and better control strategies in the generation, transport, and conversion of electric energy. Two vectors catalyze the new research: advances in technology and the new paradigm in power distribution. In the first case, the increasing use of wide-gap power devices and digital processors is contributing to reduce the size, weight, and internal losses of power converters in parallel with better dynamic performances. In the second case, electric vehicles and smart grids are the main industrial actors of the new research, in which the notion of a single converter has been substituted by the concept of multi-converters (i.e., an important number of converters interacting with the energy sources or the loads, using intermediate elements called buses for ac or dc power distribution).

In this context, sliding-mode control continues to become established in the control of power converters since it constitutes a reliable and efficient solution to many open problems. It offers robustness in the face of parametric uncertainty, fast response, and systematic approach in the control design, which feeds on a rigorous general theory that has been successfully applied in other engineering fields like process control and electromechanical systems. This Special Issue of Energies reflects the state of the art of the sliding-mode control of power converters, and the selected papers are representative of different applications in the field of renewable energies.

The first paper illustrates the control design of a three-phase voltage source supplying dc loads without a rectifier. In this work, Alsmadi, Chairez, and Utkin develop switching commands for the power devices of a three-phase PWM AC/DC voltage source converter in such a way that the output voltage can track a desired positive time-varying function.

In the second paper, Yang and Tan exhaustingly cover the recent development of sliding-mode control applications for renewable energy systems and examine the current trends to improve their efficiency and load protections in the face of large-signal variations. A comparative study between sliding-mode control and proportional–integral control is presented in three cases (i.e., a low-power wind energy conversion system, a series–series compensated wireless power transfer system, and a multiple energy storage system in a DC microgrid).

The next paper addresses the design of the input filter of a power converter supplying a constant power load (CPL). Anderson, Moré, and Puleston present a Lyapunov analysis to tackle the nonlinear internal dynamics of the sliding-mode controlled converter with CPL. Using a Lienard-type description, the authors establish the stability conditions and provide a secure operation region, which is eventually translated into filter design guidelines.

Tuning the parameters of a second-order sliding-mode control of the grid-side converter of a doubly-fed induction generator (DFIG) under unbalanced and harmonically distorted grid voltage is the subject of the fourth paper by Susperregui, Herrero, Martinez, Tapia-Otaegui, and Blasco. A multi-objective optimization is applied for tuning, and two versions of the control algorithm are compared. A Pareto front is eventually derived to help the designer to understand the trade-off among objectives and select the final solution.

A novel adaptive-gain second-order sliding mode direct power control strategy for a wind-turbine-driven DFIG is the subject of the fifth paper. Han and Ma present the control scheme in detail, derive the adaptive gains from a Lyapunov stability approach, and show the resulting reduction of the rotor voltage chattering. Active and reactive power regulation is attained under a two-phase stationary reference frame for both balanced and unbalanced grid voltage.

The dual-stator winding generator (DWIG) is a promising electrical machine for wind energy systems in the low/mid power range. In the sixth paper, Talpone, Puleston, Cendoya, and Barrado-Rodrigo propose a super-twisting algorithm sliding-mode control with moderate real-time computation burden to maximize power extraction during low wind regimes. The results show an extended range of operation of the machine, fast finite convergence for maximum power tracking, reduction of both chattering and mechanical stress, and robustness against parameter uncertainties.

Efficient lighting has become a must for either domestic or industrial applications, the discharge lamps being a representative example of the devices involved in the conversion of electricity into light. In the seventh paper, Valderrama-Blavi, Leon-Masich , Olalla and Cid-Pastor present a versatile ballast for discharge lamps of different type. The ballast has two stages. The first stage is a boost converter exhibiting loss-free resistor characteristics imposed by a sliding-mode control. The second stage is a resonant inverter supplying the discharge lamp at high frequencies. Successful ignition, warm-up and both nominal and dimming operations are illustrated in the paper.

An example of an inductive device with potential use in sea-wave energy harvesting is introduced in the next paper by Garriga-Castillo, Valderrama-Blavi, Barrado-Rodrigo, and Cid-Pastor. The energy obtained by a magnetic pick-up is stored in a battery and used to supply a dc load. An interface circuit for impedance matching between the magnetic pick-up and the battery is used. The interface is a dc-to-dc switching converter exhibiting loss-free resistor characteristics imposed by a sliding-mode control. Two candidates for the role of interface, i.e., a SEPIC converter and the cascade connection of a buck and a boost converters show similar performances after being compared on equal experimental basis.

The control of both output and circulating currents in a modular multilevel converter by means of sliding-mode control is described in the ninth paper. Uddin, Zeb, M.A. Khan, Ishfaq, I. Khan, S.ul Islam, Kim, Park , and Lee use a first-order switching strategy to control the output current, and a second-order switching law-based super-twisting algorithm to control the circulating current and suppress its second harmonic. The proposed control shows better performances than the conventional proportional-resonant regulation scheme.

Digital implementation of sliding-mode control is an important research area today because it has recently opened the way to implement sliding-mode controllers resulting in constant switching frequency. One of these ways is analyzed in detail in the tenth paper by Vidal-Idiarte, Restrepo, El Aroudi, Calvente, and Giral. They interpret a digital input-output linearization strategy in a buck converter under the optics of discrete-time sliding-mode control theory, and their theoretical approach is verified by means of simulations and experiments.

Designing boost inverters is a well-known problem for engineers due to the difficulty of the analytic understanding of how these circuits operate in either autonomous or grid-connected mode. Lopez-Caiza, Flores-Bahamonde, Kouro, Santana, Müller, and Chub present in the eleventh paper in a dual boost inverter a combination of a resonant control and a sliding-mode control to regulate the current injected to the grid and the balance between the inductor currents respectively. The experimental results validate the theoretical predictions and meet the requirements of power quality standards.

Micro-inverters development is an example of recent directions in power conversion for photovoltaic systems. A design based on the cascade connection of two boost converters and a full-bridge circuit is analyzed in the twelfth paper by Valderrama-Blavi, Rodriguez-Ramos, Olalla, and Genaro-Muñoz. Two sliding-mode control alternatives are analyzed and their performances measured in an experimental prototype. The first one regulates the system energy through the control of the input current while the second one enforces a self-oscillating transformer behavior with variable transformer ratio.

Rivera, Ortega-Cisneros, and Chavira address the control of a boost converter for output tracking of a DC biased sinusoidal signal in the thirteenth paper. They apply discontinuous output regulation based on the use of a sliding function made of a linear combination of tracking errors and an integral term. Experimental results show good tracking of the output voltage with THD less than the 5% standard limit, and excellent rejection of input voltage disturbances.

Multiphase converters have become a practical solution in industry to improve efficiency while reducing size of passive components. In the fourteenth paper, Pajer, Chowdhury, and Rodic propose a multiphase buck converter for battery emulation, which is regulated by a cascade control scheme. The inner loop uses a sliding-mode control for phase currents while the outer loop applies a proportional controller with output current feedforward. Disturbance observers are used in both loops for mismatch compensation. The theoretical analysis is corroborated by experimental results in a 4-phase synchronous prototype.

Power converters supplying a constant power load (CPL) are open-loop unstable, so they can only operate in closed-loop with an appropriate control scheme. El Aroudi, Martinez-Treviño, Vidal-Idiarte, and Cid-Pastor approach the problem in the fifteenth paper by proposing a digital sliding mode -based control with PWM that results in inrush current limitation. The paper covers exhaustingly the design of a cascade control that eventually yields excellent output voltage regulation and the suppression of inrush current in a boost converter experimental prototype feeding a CPL of 1 kW.

Lopez-Santos, Cabeza-Cabeza, Garcia, and Martinez-Salamero illustrate in the last paper the use of sliding-mode control to obtain high power factor in the bridgeless isolated version of the SEPIC converter, which is used as unidirectional isolated interface between an AC source and a low voltage DC distribution bus. Zero-crossing points are considered here as an additional mode, which is analyzed in detail to demonstrate how the switching surface is reached and the sliding motions ensured. The simplicity of the implementation and the low level of resulting THD show that the proposed control is comparable to the best strategies reported in the technical literature.

Finally, I want to express my gratitude to the large number of reviewers who carried out manuscript reviews in record time. The quality of this special issue is also due to their deep and thorough work.

Luis Martinez-Salamero
Special Issue Editor

Article

Design of a Continuous Signal Generator Based on Sliding Mode Control of Three-Phase AC-DC Power Converters

Yazan M. Alsmadi [1,*], **Isaac Chairez** [2] **and Vadim Utkin** [3]

[1] Department of Electrical Engineering, Jordan University of Science and Technology, Irbid 22110, Jordan
[2] Bioprocesses Department at the UPIBI, The National Polytechnic Institute, Mexico City 07340, Mexico; jchairezo@ipn.mx
[3] Department of Electrical and Computer Engineering, The Ohio State University, Columbus, OH 43210, USA; utkin.2@osu.edu
* Correspondence: ymalsmadi@just.edu.jo

Received: 11 October 2019; Accepted: 18 November 2019; Published: 23 November 2019

Abstract: In recent years, hundreds of technical papers have been published which describe the use of sliding mode control (SMC) techniques for power electronic equipment and electrical drives. SMC with discontinuous control actions has the potential to circumvent parameter variation effects with low implementation complexity. The problem of controlling time-varying DC loads has been studied in literature if three-phase input voltage sources are available. The conventional approach implies the design of a three-phase AC/DC converter with a constant output voltage. Then, an additional DC/DC converter is utilized as an additional stage in the output of the converter to generate the required voltage for the load. A controllable AC/DC converter is always used to have a high quality of the consumed power. The aim of this study is to design a controlled continuous signal generator based on the sliding mode control of a three-phase AC-DC power converter, which yields the production of continuous variations of the output DC voltage. A sliding mode current tracking system is designed with reference phase currents proportional to the source voltage. The proportionality time-varying gain is selected such that the output voltage is equal to the desired time function. The proposed new topology also offers the capability to get rid of the additional DC/DC power converter and produces the desired time-varying control function in the output of AC/DC power converter. The effectiveness of the proposed control design is demonstrated through a wide range of MATLAB/Simulink simulations.

Keywords: sliding mode control (SMC); power converter; continuous signal generator; equivalent control; AC-DC power converter

---

1. Introduction

Sliding mode control (SMC) has high order reduction property, good dynamic performance, low sensitivity to disturbances, and plant parameter variations, allowing SMC to handle nonlinear systems with uncertain dynamics and disturbances. Additionally, SMC is decoupled into independent lower dimensional subsystems, simplifying feedback control design. These properties allow SMC to be used in a wide range of applications such as automotive control, robotics, aviation, power systems, power electronics, and electric motors [1–5].

Power electronic converters are controlled by switching electrical components, which can produce two dissimilar values at the gating terminals [6,7]. Their controlled variables may take values from a two valued discrete set. Moreover, linearization is not required [1,2,8,9]. Hence, SMC is a preferred method to realize the control of power converter devices.

Within the wide diversity of available power electronic devices, the well-known three-phase AC/DC is commonly applied in energy conversion plants. Nevertheless, inherent complications appear with regards to reactive power generation, as well as the higher harmonic content in the input current. These characteristics appear as practical disadvantages that have become more relevant as the AC/DC converter capacity turn out to be larger and larger [10–12].

The idealized AC/DC converter shows up as a constant DC voltage as controlled output (or current) and a sinusoidal input set of currents at unity power factor at the AC line. Nevertheless, the current technology of thyristor phase-controlled converters has two intrinsic disadvantages: First, the larger firing angle, the smaller power factor; and second, the line current has moderately large harmonics components [13,14]. As AC/DC converters are more and more controlled using PWM switching patterns, the input as well as the output performances improve. These PWM AC/DC converters offer numerous advantages compared to some traditional rectifiers [15,16]: Unity power factor, low harmonic components in input current, bidirectional power flow, and low ripple in output voltage. These characteristics make simpler the filtering processes on both AC and DC sides of the proposed converter [1,2,10,16].

Conventional control design techniques for this type of power converter device usually solve the maintaining of the DC output voltage at a given reference level firstly, and at second place, try to seek for the minimization of high order harmonics and reactive power at the input. SMC of AC/DC power converters, presented in [1], offers the inverse sequence of actions. First, a current tracking system was designed with sinusoidal current references that are proportional to the AC input voltages with a constant gain of proportionality. This automatically sets the reactive power to zero. Second, it was proven that the output voltage will be constant and only depends on the amplitude of the reference input. However, the proposed control method requires an additional DC/DC converter to control the DC load. This introduces the following control design challenge: Is it possible to avoid using an additional DC/DC converter and generate any arbitrary desired time varying function at the output of the AC/DC converter such that the DC load can be directly controlled?

The main contributions of this study are:

(a) Sliding mode control is an appropriate tool for application for wide range of power converters. The first publications on DC/DC converters [1–4,7,8] demonstrated its efficiency. The methods of minimization heat losses and of chattering amplitude based on harmonic cancellation principle, switching frequency control for DC/DC converters can be found in [2–4]. Multidimensional sliding modes were utilized in power converters to control AC load with DC energy source [5,6]. The design methodology to control output constant voltage and power factor simultaneously for AC-DC converter was developed in [7]. We are not aware of publications with our problem statement—to have an arbitrary time function (not constant) in the output of AC/DC converter. The attempts to find time of the varying gain as an algebraic state function of state failed, because it should satisfy the algebraic equation.

(b) A SMC has been proposed to generate continuous waveforms based on a controlled switched sequence of a three-phase AC-DC converter. This achievement was a consequence of solving a trajectory tracking of estimated reference currents. The realization of such tracking enforces the production of bounded derivative DC output voltage. The tracking controller implemented a time-varying relationship between the currents and voltages on the AC side of the converter. Such given positive relation between voltage and current justified the positive power efficiency of the controlled power converter.

2. Problem Statement

The design problem considered in this study is the generation of switching commands for the power electronics-switching elements of the AC/DC converter (shown in Figure 1) in such a way that the output voltage can track the desired output of the converter $f(t)$, which should be positive. This condition agrees with the classical realization of AC/DC, buck, and boost power converters. Since the input AC voltage is bounded, the output capacitor C can be charged at a limited velocity, which means that the time derivative $\dot{f}(t)$ should be bounded. Therefore, the problem can be described as fixing the switching sequence such that:

$$\lim_{t \to \infty}\left|f(t) - v_{dc}\right| = 0 \tag{1}$$

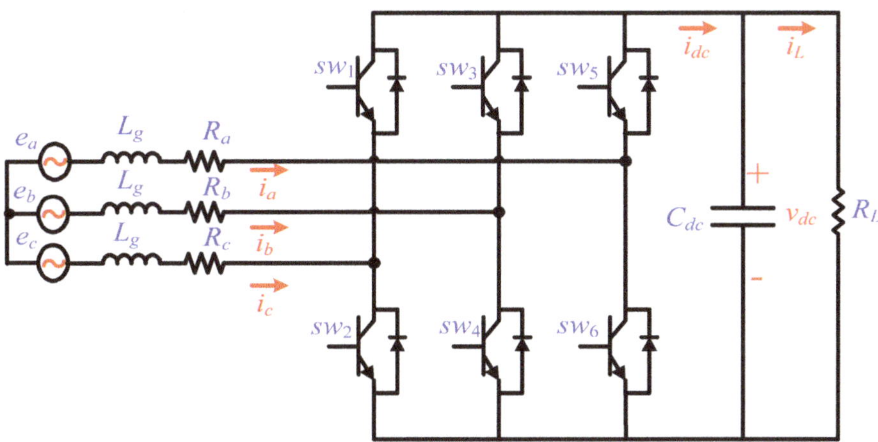

Figure 1. Scheme of the three-phase PWM AC/DC voltage source converter.

The power efficiency can be maximized if each input phase current of the power converter is proportional to the corresponding phase voltage with a positive gain. Therefore, the objective of this paper is to design a new control algorithm such that the input phase currents track preselected reference inputs and the positive gain of the proportionality is selected as a time varying function. Accordingly, the output voltage is equal to the desired function $f(t)$.

3. Circuit Model of the Three-Phase PWM AC/DC Voltage Source Converter Scheme

Figure 1 shows the three-phase PWM AC/DC voltage source converter scheme. e_a, e_b, e_c are the balanced three-phase AC input voltages; i_{dc} is dc-link current; R_L is a resistive load connected to the DC side; i_L is the load current; i_a, i_b, i_c are the three-phase AC input currents; C_{dc} is the dc-link capacitance v_{dc} is dc-link voltage; R_g and L_g represent the grid-side resistance and inductance, respectively.

The balanced three-phase AC input currents are given by:

$$L_g \frac{di_a}{dt} = e_a - R_g i_a - v_{an} \tag{2}$$

$$L_g \frac{di_b}{dt} = e_b - R_g i_a - v_{bn} \tag{3}$$

$$L_g \frac{di_c}{dt} = e_c - R_g i_c - v_{cn} \tag{4}$$

where v_{an}, v_{bn}, v_{cn} are the AC side phase voltages of the converter. The balanced three-phase AC voltages are given by:

$$e_a = E_0 \sin(\omega t) \tag{5}$$

$$e_b = E_0 \sin\left(\omega t - \frac{2\pi}{3}\right) \tag{6}$$

$$e_c = E_0 \sin\left(\omega t + \frac{2\pi}{3}\right) \tag{7}$$

Here, ω is the AC power source angular frequency and E_0 is the amplitude of the phase voltages. Assume that $i_{abc} = \begin{bmatrix} i_a \\ i_b \\ i_c \end{bmatrix}$, $e_{abc} = \begin{bmatrix} e_a \\ e_b \\ e_c \end{bmatrix}$, $v_s = \begin{bmatrix} v_{an} \\ v_{bn} \\ v_{cn} \end{bmatrix}$, then Equations (2)–(4) can be re-written in a compact form:

$$L_g \frac{di_{abc}}{dt} = e_{abc} - R_g i - v_s \tag{8}$$

Define the switching function S of each switch as:

$$S_j = \begin{cases} 1, & S_j \text{ is close} \\ -1, & S_j \text{ is open} \end{cases} \quad j = a, b, c \tag{9}$$

As a result, the voltage vector v_s can be given in terms of the switching functions $S = \begin{bmatrix} S_a \\ S_b \\ S_c \end{bmatrix}$ as:

$$v_s = \frac{1}{3} v_{dc} \begin{bmatrix} 2 & 1 & 1 \\ 1 & 2 & 1 \\ 1 & 1 & 2 \end{bmatrix} S \tag{10}$$

By substituting Equation (10) into (8), the AC input current equations can be given by:

$$L_g \frac{di_{abc}}{dt} = e_{abc} - R_g i_{abc} - \frac{1}{3} v_{dc} \begin{bmatrix} 2 & 1 & 1 \\ 1 & 2 & 1 \\ 1 & 1 & 2 \end{bmatrix} S \tag{11}$$

In conclusion, the output voltage can be given by:

$$C \frac{dv_{dc}}{dt} = -i_L + i^T S \tag{12}$$

4. Sliding Mode Current-Tracking Control

As previously indicated, a sliding mode-based current tracking system is designed such that sinusoidal reference inputs are tracked by phase currents proportional to input AC voltages. Rewriting Equation (11) as:

$$L_g \frac{di_{abc}}{dt} = e_{abc} - R_g i_{abc} - \frac{1}{3} v_{dc} \Gamma_0 S \tag{13}$$

where, matrix Γ_0 is given by:

$$\Gamma_0 = \begin{bmatrix} 2 & 1 & 1 \\ 1 & 2 & 1 \\ 1 & 1 & 2 \end{bmatrix}, det\Gamma_0 = 0 \tag{14}$$

Since the sum of the three-phase currents is zero, only three state variables should be controlled in a system with a three-dimensional control vector S: Two phase currents and output voltage. However, because matrix Γ_0 is singular, the conventional sliding mode approach cannot be directly applied. Therefore, a tracking system for two phase currents only is first designed. As a result, the sliding mode should be enforced on the intersection of two surfaces $\sigma_a = L_g\left(i_{aref} - i_a\right)$ and $\sigma_b = L_g\left(i_{bref} - i_b\right)$, or in a vector form:

$$\sigma_{ab} = L_g\left(i_{abref} - i_{ab}\right) \tag{15}$$

Excluding the phase current $i_c = -i_a - i_b$ yields:

$$L_g \frac{di_{ab}}{dt} = e_{ab} - R_g i_{ab} - \frac{1}{3} v_{dc} \Gamma S \tag{16}$$

$$C\frac{dv_{dc}}{dt} = -\frac{v_{dc}}{R_L} + (i_a, i_b, -i_a - i_b)S \tag{17}$$

where Γ is a 2×3 matrix given by:

$$\Gamma = \begin{bmatrix} 2 & 1 & 1 \\ 1 & 2 & 1 \end{bmatrix} \tag{18}$$

The ideal tracking system is based on the Lyapanov function:

$$V = \frac{1}{2} \sigma_{ab}{}^T \sigma_{ab} \tag{19}$$

$\dot{V}$ has to be negative definite and calculated on the system trajectory when selecting discontinuous control:

$$\dot{V} = \sigma_{ab}{}^T F(.) - (u_{dc}/L)(\alpha, \beta, \gamma)S \tag{20}$$

where $F(.)$ is state function, which does not depend on control. α, β, and γ are given by:

$$\begin{cases} \alpha = (2\sigma_a + \sigma_b) \\ \beta = (\sigma_a + 2\sigma_b) \\ \gamma = (\sigma_a + \sigma_b) \end{cases} \tag{21}$$

If v_{dc}/L is large enough, $F(.)$ can be suppressed with control S given by:

$$S = \begin{cases} S_a = sign\,(\alpha) \\ S_b = sign\,(\beta) \\ S_c = sign\,(\gamma) \end{cases} \tag{22}$$

such that $\dot{V} = \sigma_{ab}{}^T F(.) - (u_{dc}/L)\big(|\alpha| + |\beta| + |\gamma|\big) < 0$. As a result, σ_{ab} tends to zero and σ_{ab} becomes zero after a finite time interval [17–19]. Consequently, sliding mode occurs with $i_{ab} = i_{abref}$. Therefore, the current tracking system is developed with sinusoidal current references proportional to the input AC voltages as:

$$i_{ab} = K(t) \times e_{ab} \tag{23}$$

Calculate the equivalent control [2]:

$$(\Gamma S)_{eq} = \left(-L_g\frac{di_{abref}}{dt} + e_{ab} - R_g i_{ab}\right)\frac{3}{u_c} \tag{24}$$

The three phase input currents can also be expressed as:

$$\left(i_a,\ i_b,\ -i_a - i_b\right) = \frac{1}{3}\left(i_a - i_c,\ i_b - i_c\right)\Gamma \tag{25}$$

The sliding mode equation can be obtained by substituting Equations (24) and (25) into (17):

$$C\left(\frac{dv_{dc}}{dt}\right) = \left(-\frac{v_{dc}}{R}\right) + \frac{1}{v_{dc}}\,(i_a - i_c, i_b - i_c)\left(-L_g\frac{d}{dt}i_{abref} + e_{ab} - R_g i_{ab}\right) \tag{26}$$

After sliding mode occurs $(i_{ab} = K(t)e_{ab})$:

$$i_{ab} = i_{ab,ref}$$
$$\tfrac{d}{dt}i_{ab} = \tfrac{d}{dt}i_{ab,ref}$$

Therefore,

$$\frac{d}{dt}i_{ab} = K(t)\frac{d}{dt}e_{ab} + e_{ab}\frac{d}{dt}k(t) \tag{27}$$

where, $i_a - i_c = K(t)(e_a - e_c, e_b - e_c)$.

It can be shown that the output voltage v_{dc} is given by:

$$C\frac{dv_{dc}}{dt} = \frac{-v_{dc}}{R_L} + \frac{1}{v_{dc}}\left(-\frac{3}{2}L_g KE_0^2\frac{d}{dt}K + \frac{3}{2}KE_0^2 - \frac{3}{2}R_g K^2 E_0^2\right) \tag{28}$$

If $y = v_{dc}^2$, then the gain K should satisfy the differential equation

$$\left(\frac{1}{2}C\frac{dy}{dt} + \frac{1}{R_L}y\right)\frac{2}{3KE_0^2} = -L_g\frac{dK}{dt} + 1 - R_g K \tag{29}$$

Assume that f is the voltage reference and $\hat{y} = f^2$, then the equation

$$\left(\frac{1}{2}C\frac{d\hat{y}}{dt} + \frac{1}{R_L}\hat{y}\right)\frac{2}{3KE_0^2} = -L\frac{dK}{dt} + 1 - R_g K \tag{30}$$

can be simulated in the controller,

$$\Delta y = \hat{y} - y \tag{31}$$

$$\frac{1}{2}C\frac{d\Delta y}{dt} + \frac{1}{R_L}\Delta y = 0 \tag{32}$$

Equation (32) shows that $\Delta y \to 0$ and the output voltage tends to be the reference input $v_{dc} \to f$. Assume that

$$\left|\frac{3}{2}\left(\frac{1}{2}C\frac{d}{dt}f^2 + \frac{1}{R_L}f^2\right)\right| \leq M,$$

then:

$$L\dot{K} \geq 1 - R_g K - \frac{M}{KE_0^2} \tag{33}$$

If $R_g K = \varepsilon < 1$, then

$$\lim_{E_0^2 \to \infty}\left(1 - R_g K - \frac{M}{KE_0^2}\right) < 1 \tag{34}$$

It means that

$$L\frac{dK}{dt}\bigg|_{R_g K = \varepsilon} > 0 \tag{35}$$

and K is always positive if $R_g K(0) > \varepsilon$. This corresponds to a power factor equal to 1 ($i_{abc} = Ke_{abc}$) for a high enough amplitude of the input source voltage. It is important that $L\dot{K}$ in (29) will be negative for high enough values of K. Hence, the gain K is bounded.

Remark: The implementation of the proposed controller in embedded systems requires the online measurement of the dc voltage at the capacitor and the possibility of realizing fast enough oscillations on the switching electronic elements. This part of the problem can be solved using available fast dedicated microcontrollers devices. Notice that the accuracy of the produced signal is a function of the relative relationship between the switching devices operation frequency and the main frequency components of the desired signals. Evidently, the exact reconstruction of the desired signal cannot be acquired. Nevertheless, the studies regarding discrete implementation of sliding mode has shown that the accuracy of the sliding mode realization is proportional to the square power of the sampling period if the explicit discretization is considered. For more details on the implementation issues, please see references [20–23]. Remark: The proposed control strategy offers an alternative to some other sliding mode controllers designs considering adaptive pulse width modulation [24], sub-optimal regulation [25], and multitype restrictive [26] approaches which have been applied on DC-DC power converters to obtain arbitrary signals. However, not one of them has been tested on the AC-DC device.

5. Simulation Results

In order to evaluate the proposed sliding mode control design procedure several computer simulations have been conducted using MATLAB/Simulink software. The control algorithm is represented in the following flow diagram (Figure 2):

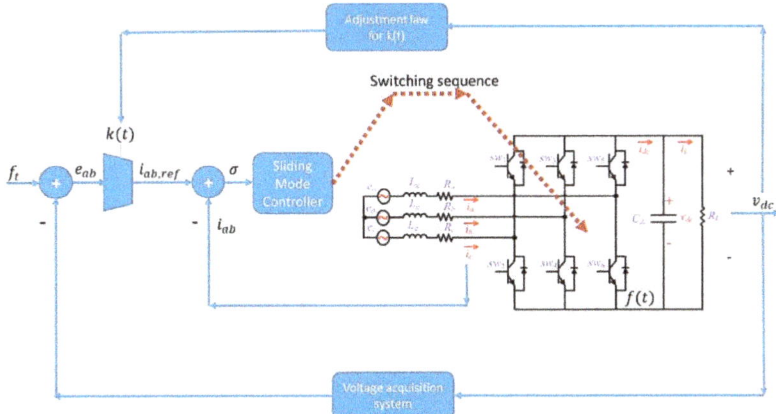

Figure 2. Flow diagram describing the sliding mode control realization.

Different generated signals have confirmed the abilities of the proposed continuous waveform generators.

The simulation was performed for power converter governed by Equation (10) with control (21) and different desired functions $f(t)$ in the converter output. The differential equation for time varying gain $K(t)$ in Equation (29).

5.1. Sinusoidal Waveform Generation

The first signal is a pure sinusoidal waveform, which corresponds to a traditional signal used in diverse signal generators. The selected reference waveform is:

$$f(t) = 250(\sin(\omega t) + 1) \geq 0 \ (35) \tag{36}$$

The results of the simulation are shown in Figure 3.

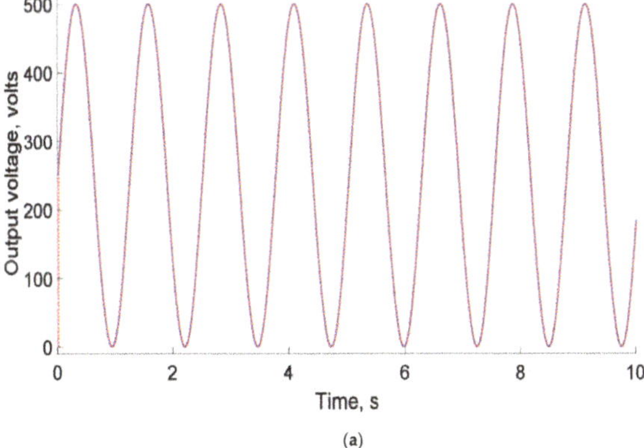

Figure 3. *Cont.*

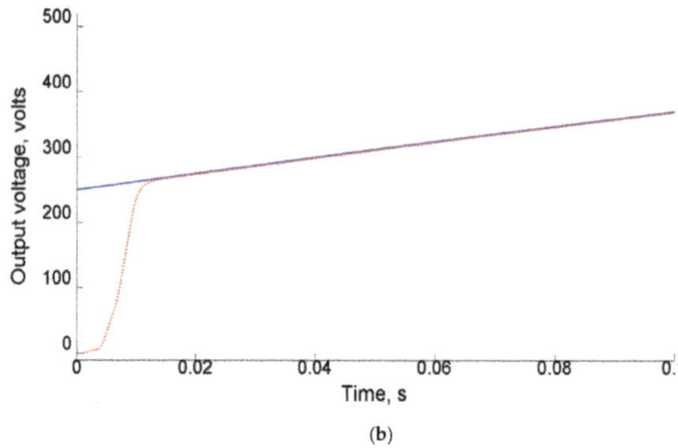

(b)

Figure 3. Comparison of the reference and the generated sinusoidal continuous waveform for a fixed period: (**a**) (0,10.0) and (**b**) (0,0.1) s.

The input currents obtained by the application of the suggested first sliding-mode controller show a modulated sinusoidal shape in all three branches (Figure 4).

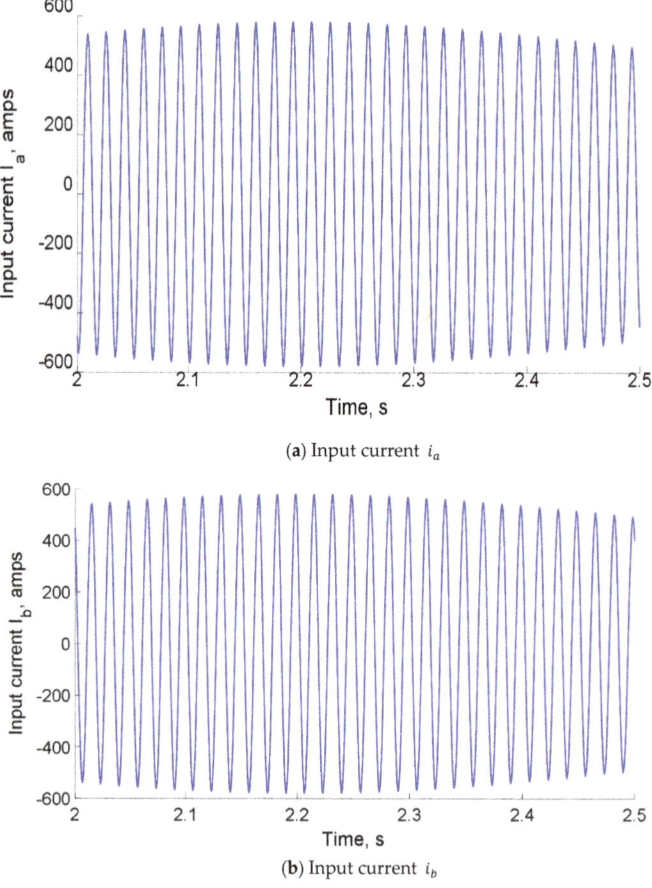

(**a**) Input current i_a

(**b**) Input current i_b

Figure 4. *Cont.*

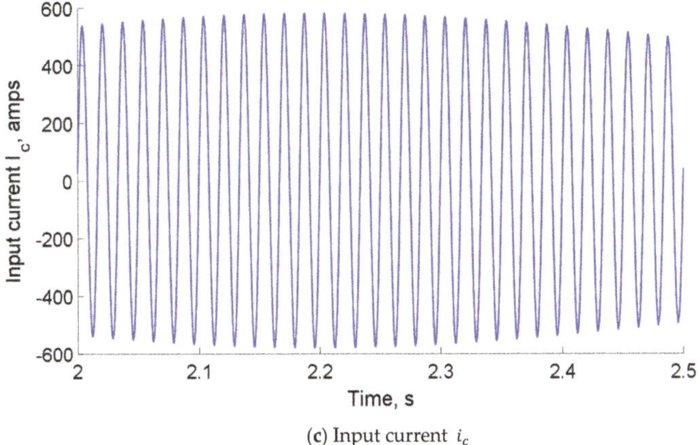

(c) Input current i_c

Figure 4. Time variation of the input currents at the three branches of the AC source on the period (2.0–2.5) s (closer view) for the three branches: (**a**) i_a, (**b**) i_b, and (**c**) i_c with the a sinusoidal reference.

The time dependence of the current $\hat{i}_s$ also shows the expected modulation with the frequency of the desired output current, which is 15 Hz (Figure 5).

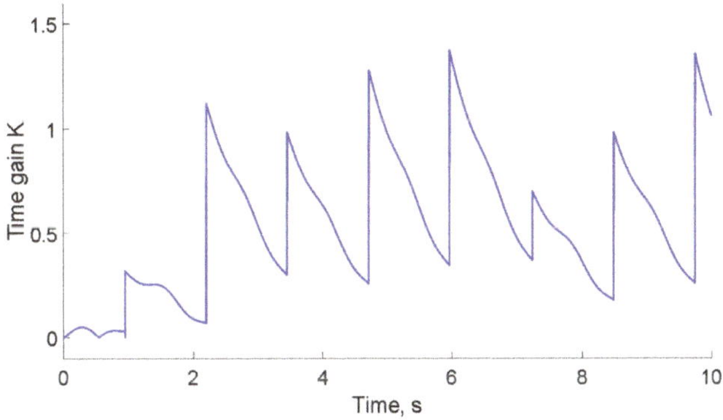

Figure 5. Time variation of the controller gain K.

The phase relations between the input currents hold both in the transient and the steady state periods. In the period between 0.0 and 0.1 s, the current decreases exponentially to the steady state, which is detected after 0.1 s. A phase shift of 2/3 is evidenced, which also confirms the efficiency of the suggested controller. Notice that the simultaneous dependence of K with respect to the reference waveform as well as its derivative does not relate it to the reference voltage form. The gain variation is continuous but not necessarily differentiable, considering the gain structure estimated in this study.

5.1.1. Variable Frequency Sinusoidal Waveform Generation

The second proposed reference signal is a composite sinusoidal waveform, which corresponds to a class of simplified chirp signal. Such waveforms can be used for testing the spectral response of diverse systems for calibration purposes. The selected composite sinusoidal reference waveform is:

$$f_2(t) = \begin{cases} 200(\sin(5t)+1) & if \quad 0 \le t < 2 \\ 100(\sin(t)+1) & if \quad 2 \le t < 8 \\ 200(\sin(5t)+1) & if \quad 8 \le t \le 10 \end{cases} \tag{37}$$

Once more, the selected bias constants mean a positive waveform. The selected transition times between the sinusoidal forms can design a continuous composite waveform with a bounded derivative. The simulation results are shown in Figure 6.

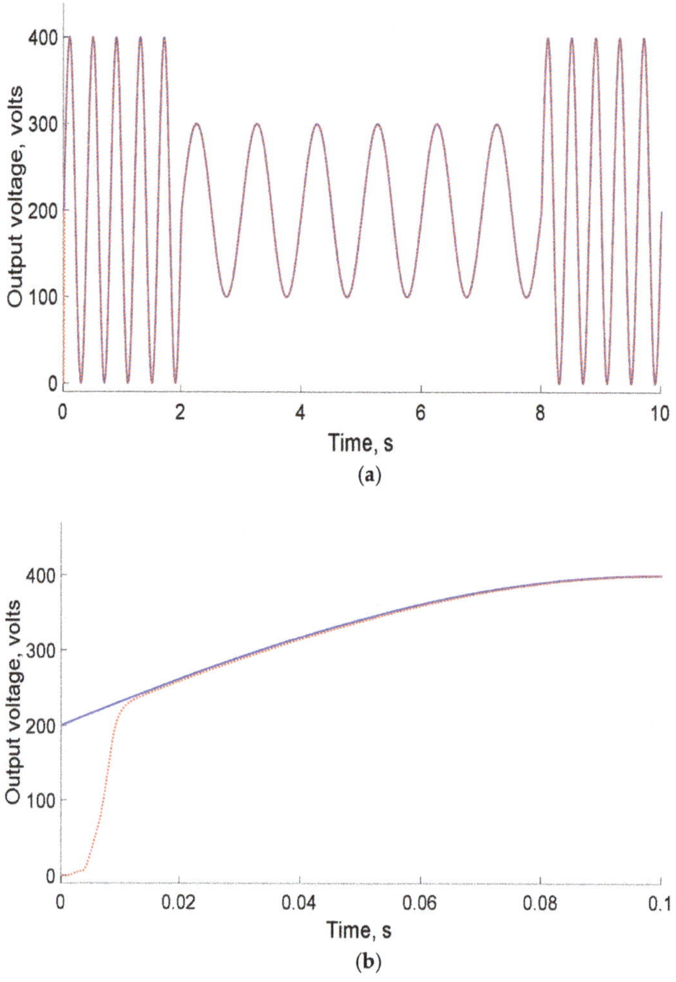

Figure 6. Comparison of the reference and the generated sinusoidal continuous waveform for a fixed period: (**a**) (0,10.0) and (**b**) (0,0.1) s.

The input currents agree with the variation of the sinusoidal frequency by the application of the suggested first sliding mode controller with the corresponding modulated sinusoidal shape in all three branches (Figure 7).

When looking at the time variation of the gain K for the controller, notice that the simultaneous dependence of K with respect to the reference waveform, as well as its derivative does not relate it to the reference voltage form. The second waveform considered in this study produces a smoother variation of the gain K. The gain variation is continuous considering the gain structure estimated in this study (Figure 8).

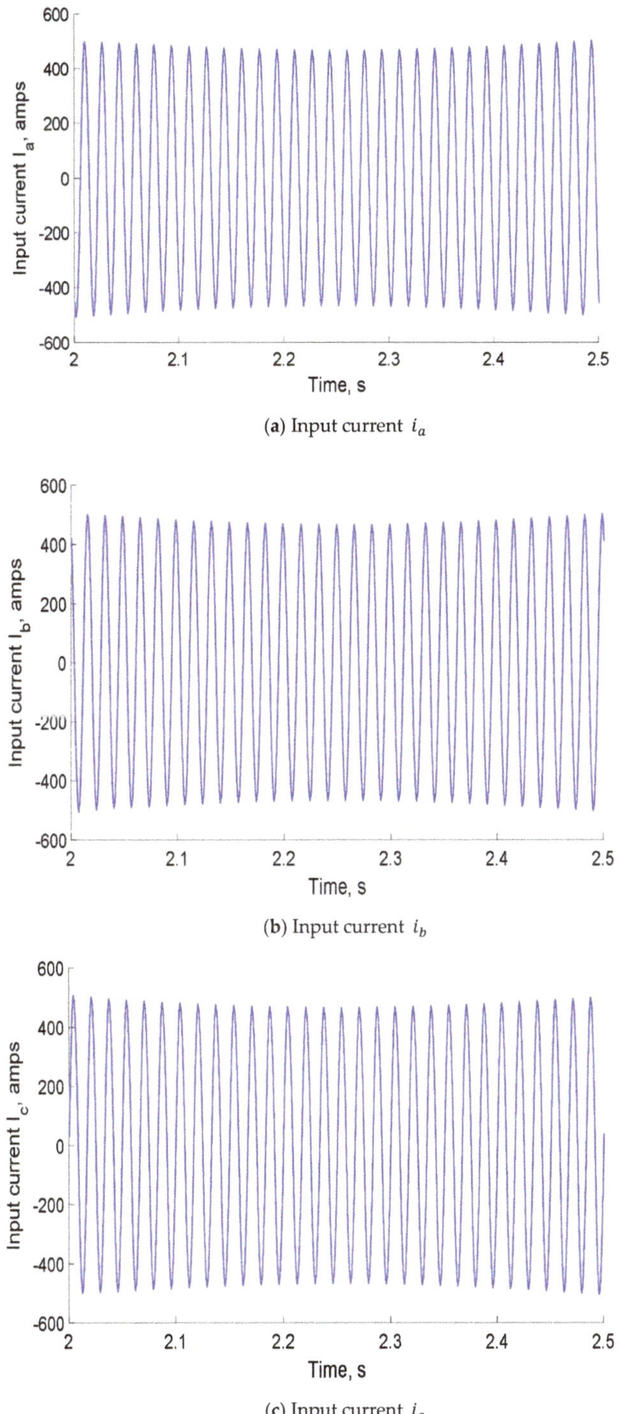

(a) Input current i_a

(b) Input current i_b

(c) Input current i_c

Figure 7. Time variation of the input currents at the three branches of the AC source on the period (2.0–2.5) s (closer view) for the three branches: (**a**) i_a, (**b**) i_b, and (**c**) i_c with the reference sinusoidal.

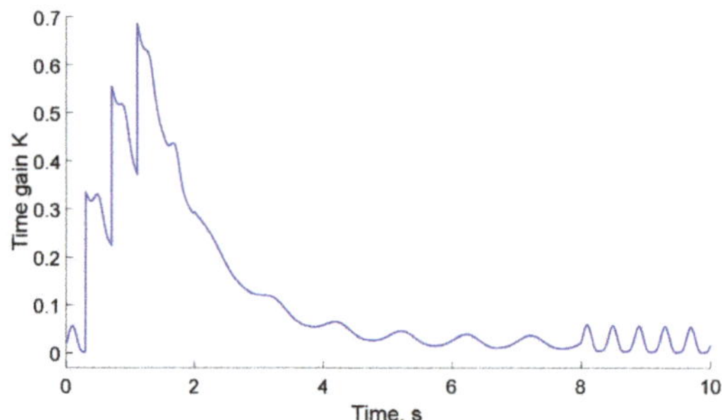

Figure 8. Time variation of the controller gain K.

5.1.2. Triangular Waveform Generation

The third suggested reference signal is a triangular signal, which is also a common signal used in the calibration of diverse devices. Notice that this signal has a bounded but not continuous derivative. Consequently, the suggested controllers are applicable (Figure 9).

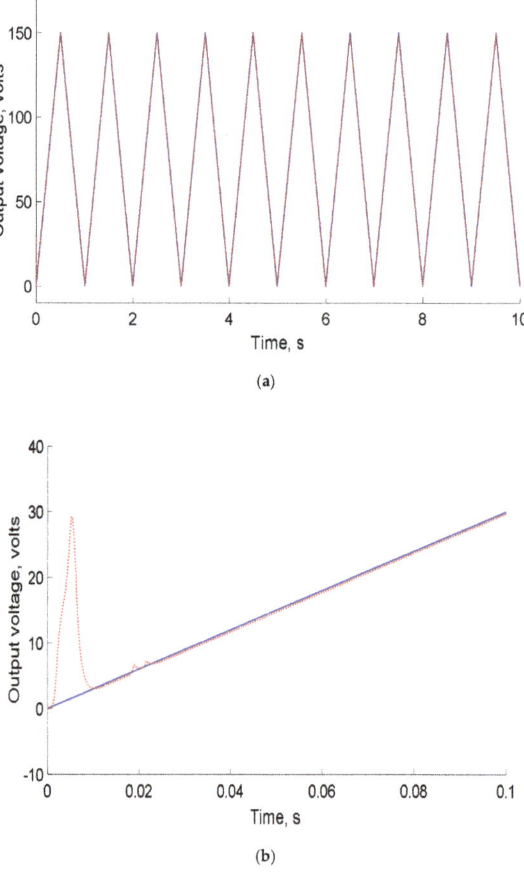

(a)

(b)

Figure 9. Comparison of the reference and the generated triangular waveform for a fixed period: (**a**) (0,10.0) and (**b**) (0,0.1) s.

For the class of triangular signal, the controller succeeded at reconstructing the suggested reference signal as shown in Figure 9a. Consequentially, the tracking error of the reference voltage is reduced to less than 0.05% over a period of 0.08 s (Figure 9b).

The time variation of the gain K for the controller with the reference triangular signal appears in Figure 10. The gain variation function is continuous.

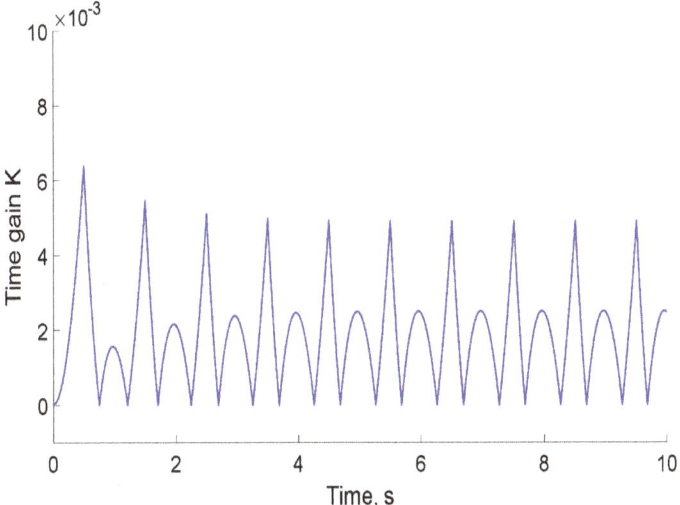

Figure 10. Time variation of the controller gain K.

Even if the exact sliding motion can be acquired if and only if the switches in the power converter oscillate at the infinite frequency, the current available technology allows to oscillate at such high frequencies ensuring the existence of the practical sliding motion. On the other hand, the required high frequency oscillations of the sliding mode may produce heath losses which could damage the switching circuit. In all the presented cases, chattering phenomenon should mentioned always when applying sliding mode control. The set of chattering suppression methods has been developed in the framework of sliding mode control theory. They are surveyed in [27]. The harmonics cancellation principle is the most efficient for power converters and can be applied for our case. The design idea consists in using several parallel converters with controlled phases such that high order harmonics can be cancelled.

6. Conclusions

This paper has presented the control design procedure to directly control DC loads using a three-phase voltage source without a rectifier. It consists of two steps. First, the current tracking problem is solved with reference currents proportional to phase voltages. Then, the time varying proportionality coefficient is selected such that the output voltage is equal to the desired time function. It has been shown that the proportionality gain should satisfy the first order differential equation, which is implemented in the controller. The behavior of the system with a positive coefficient is equivalent to having a unity power factor. Stability of the complete system (the power converter and controller dynamics) was also proved. A wide range of computer simulations were provided to demonstrate efficiency of the proposed control design for different types of sinusoidal and triangular functions as voltage reference inputs.

Author Contributions: Formal analysis, Y.M.A., I.C., and V.U.; Investigation, Y.M.A., I.C., and V.U.; Methodology, Y.M.A., I.C., and V.U.; Supervision, V.U.; Validation, V.U.; Writing—Original draft, Y.M.A., I.C., and V.U.

Funding: This research work was partially funded by the [Consejo Nacional de Ciencia y Tecnología] grant number [CB-20181457] and the [Instituto Politécnico Nacional] grant number [SIP-20191724].

Conflicts of Interest: The authors declare no conflict of interest.

References

1. Alsmadi, Y.; Utkin, V.; Haj-Ahmed, M.; Xu, L.; Abdelaziz, A. Sliding-mode control of power converters: AC/DC converters & DC/AC inverters. *Int. J. Control* **2018**, *91*, 2573–2587.
2. Utkin, V.; Guldner, J.; Shi, J. *Sliding Mode Control in Electro-Mechanical Systems*; CRC Press, Taylor Francis Group: London, UK, 2009.
3. Utkin, V. *Sliding Modes in Control and Optimization*; Springer: Berlin, Germany, 1992.
4. Utkin, V. *Sliding Modes and Their Application in Variable Structure Systems*; MIR Publishers: Moscow, Russia, 1978.
5. Bartolini, G.; Fridman, L.; Pisano, A.; Usai, E. *Modern Sliding Mode Control Theory: New Perspectives and Applications*; Springer: Berlin, Germany, 2008.
6. Mohan, N.; Undeland, T.; Robbins, W. *Power Electronics: Converters, Applications, and Designs*; John Wiley Sons, Inc: Hoboken, NJ, USA, 2003.
7. Krein, P. *Elements of Power Electronics*; Oxford University Press: Oxford, UK, 1997.
8. Sira-Ramirez, H.; Ortigoza, R.S. *Control Design Techniques in Power Electronics Devices*; Springer: Germany, Berlin, 2006.
9. Martinez-Salamero, L.; Cid-Pastor, A.; Giral, R.; Calvente, J.; Utkin, V. Why is sliding mode control methodology needed for power converters? In Proceedings of the 14th International Power Electronics and Motion Control Conference (EPE/PEMC), Ohrid, Macedonia, 6–8 September 2010; Volume 1, pp. S9-25–S9-31.
10. Chen, L.; Blaabjerg, F.; Frederiksen, P.S. An improved predictive control for three-phase PWM AC/DC converter with low sampling frequency. In Proceedings of the 20th International Conference on Industrial Electronics, Control and Automation (IECON 94), Bologna, Italy, 5–9 Septmber 1994; pp. 399–404.
11. Ooi, T.; Dixon, J.W.; Kulkarni, A.B.; Nishimoto, M. An integrated AC drive system using a controlled-current PWM rectifier/inverter link. *IEEE Trans. Ind. Appl.* **1988**, *3*, 64–71. [CrossRef]
12. Wong, C.; Mohan, N.; He, J. Adaptive phase control for three phase PWM AC-to-DC converters with constant switching frequency. In Proceedings of the Conference Record of the Power Conversion Conference, Yokomaha, Japan, 19–21 April 1993; pp. 73–78.
13. Komurcugil, H.; Kukrer, O. A novel current-control method for three-phase PWM AC/DC voltage-source converters. *IEEE Trans. Ind. Electron.* **1996**, *46*, 544–553. [CrossRef]
14. Wu, R.; Dewan, S.B.; Slemon, G.R. A PWM AC-to-DC converter with fixed switching frequency. *IEEE Trans. Ind. Appl.* **1990**, *26*, 880–885. [CrossRef]
15. Habetler, T.G. A space vector-based rectifier regulator for AC/DC/AC converters. *IEEE Trans. Power Electron.* **1993**, *8*, 30–36. [CrossRef]
16. Wu, R.; Dewan, S.B.; Slemon, G.R. Analysis of an AC-to-DC voltage source converter using PWM with phase and amplitude control. *IEEE Trans. Ind. Appl.* **1991**, *27*, 355–364. [CrossRef]
17. Alsmadi, Y.; Utkin, V.; Xu, L. Sliding Mode Control of Three-Phase Boost-Type and Single-Phase, Three-Wire AC/DC Power Converters. In Proceedings of the IEEE 13th Workshop on Variable Structure Systems (VSS), Nantes, France, 29 June–2 July 2014; pp. 1–6.
18. Alsmadi, Y.; Utkin, V.; Xu, L. Sliding Mode Control of AC/DC Power Converters. In Proceedings of the IEEE 4th International Conference on Power Engineering, Energy and Electrical Drives, Istanbul, Turkey, 20 February 2013; pp. 1229–1234.
19. DeCarlo, R.A.; Zak, S.H.; Matthews, G.P. Variable structure control of nonlinear multivariable systems: A tutorial. *Proc. IEEE* **1988**, *76*, 212–232. [CrossRef]
20. Utkin, V.I. *Sliding Modes in Control and Optimization*; Springer Science & Business Media: Berlin, Germany, 2013.
21. Brogliato, B.; Polyakov, A. Globally stable implicit Euler time-discretization of a nonlinear single-input sliding-mode control system. In Proceedings of the 2015 54th IEEE Conference on Decision and Control (CDC), Osaka, Japan, 14 December 2015; pp. 5426–5431.
22. Chakrabarty, S.; Bandyopadhyay, B. Osaka, JapaA generalized reaching law for discrete time sliding mode control. *Automatica* **2015**, *52*, 83–86. [CrossRef]

23. Huber, O.; Brogliato, B.; Acary, V.; Boubakir, A.; Plestan, F.; Wang, B. Experimental results on implicit and explicit time-discretization of equivalent-control-based sliding-mode control. In *Sliding-Mode Control*; Fridman, L., Barbot, J.P., Eds.; IET: BeiJing, China, 2016; Volume 102, pp. 207–235.
24. Baldi, S.; Papachristodoulou, A.; Kosmatopoulos, E.B. Adaptive pulse width modulation design for power converters based on affine switched systems. *Nonlinear Anal. Hybrid Syst.* **2018**, *30*, 306–322. [CrossRef]
25. Goudarzian, A.; Khosravi, A.; Raeisi, H.A. Optimized sliding mode current controller for power converters with non-minimum phase nature. *J. Frankl. Inst.* **2019**, *356*, 8569–8594. [CrossRef]
26. Wang, B.; Ma, G.; Xu, D.; Zhang, L.; Zhou, J. Switching sliding-mode control strategy based on multi-type restrictive condition for voltage control of buck converter in auxiliary energy source. *Appl. Energy* **2018**, *228*, 1373–1384. [CrossRef]
27. Lee, H.; Utkin, I.V. Chattering suppression methods in sliding mode control systems. *Annu. Rev. Control.* **2007**, *31*, 179–188. [CrossRef]

 *energies*

Article

Trends and Development of Sliding Mode Control Applications for Renewable Energy Systems

Yun Yang and Siew Chong Tan *

Department of Electrical and Electronic Engineering, The University of Hong Kong, Hong Kong, China
* Correspondence: sctan@eee.hku.hk; Tel.: +852-3917-2707

Received: 25 June 2019; Accepted: 19 July 2019; Published: 24 July 2019

Abstract: Based on the matured theoretical framework of sliding mode control for varied, nonlinear, and unpredictable systems, practical designs of sliding mode control have been developed to suit the purpose of controlling power converters under various operating conditions. These design guidelines are particularly valuable for emerging technologies with renewable energy sources. This paper presents a discussion on the recent development of sliding mode control applications for renewable energy systems, and further examines the current trends of achieving efficiency improvement of renewable energy systems and load protections against large overshoot/undershoot in transient states, by utilizing the fast-dynamic-tracking capability of the sliding mode control. Three comparative case studies between the sliding mode control and proportional-integral control involving, namely, a low-power wind energy conversion system, a series-series-compensated wireless power transfer system, and a multiple energy storage system in a direct current (DC) microgrid, are provided.

Keywords: sliding mode control; renewable energy systems; fast dynamic response; wind energy conversion system; series-series-compensated wireless power transfer system; energy harvesting

---

1. Introduction

The sliding mode (SM) control is a unique type of nonlinear control that is particularly suited for variable structure systems. Its control mechanism involves a state-feedback discontinuous control law that continuously actuates the controlled system to abruptly change its operation from one continuous state to another, at high frequency, such that the controlled system's dynamics are constricted to following a particular reference track. In so doing, superior large-signal control responses and small-signal stability of systems with significant parameter uncertainties and large operating point changes are more easily achievable with SM control compared to other control approaches [1]. Additionally, the advantages of SM control compared to other nonlinear control methods include being generally free of online system identification and ease of implementation.

Since the early development of variable structure control in the 1950s and 1960s [2–4], various branches of SM control, including discrete SM control, adaptive SM control, terminal SM control, and global SM control, etc., have been investigated for nonlinear systems, infinite-dimensional systems, multiple-input-multiple-output systems, stochastic systems, discrete systems, and variable structure systems, etc. [5–12]. V. I. Utkin surveyed a class of variable structure systems in 1977 [5]. S. Z. Sarpturk et al. demonstrated the control input must be upper and lower bounded for discrete SM control [6]. H. Sira-Ramirez et al. combined the advantages of chattering-free dynamic SM control and the adaptive backstepping technique to regulate linearizable systems [7]. G. Wheeler et al. improved the adaption law on the upper bound of uncertainties to guarantee the boundness of both states of the plant and the estimated control gains [8]. Y. Feng et al. presented a new terminal sliding mode manifold for the second-order system to resolve the singularity issue associated with the conventional terminal SM control [9]. H. S. Choi et al. proposed a global SM control with the consideration of input disturbances and uncertainties of the parameters to ensure sliding behavior throughout an entire response for motor drives. In that study, the SM control further enhanced the dynamic tracking performance by minimizing the reaching time of the sliding surface [10]. The neural network learning is achieved with an online adaption algorithm that inherits robustness and high-speed learning from the SM control. Y. S. Lu et al. proposed a self-organizing fuzzy SM control to achieve rapid and accurate tracking of a class of nonlinear systems [11]. B. Yoo et al. adopted fuzzy logic approximators to approximate the unknown system functions in designing the SM control of nonlinear systems [12]. Since then, most SM control works have been redirected toward exploring potential real-life utilizations, of which their applications in switching power converters is one area that has been most extensively studied.

The earliest publication on SM control of power converters can be traced back to 1983 by F. Bilalović, where the feasibility of using SM control for buck converters was examined [13]. This was followed by R. Venkataramanan et al., who presented comprehensive applications of equivalent SM control on basic second-order DC-DC converters to achieve constant-frequency SM control [14]. Since then, SM control for higher-order and more complex topologies and configuration of converters have been exhaustively studied [15–24], including that of L. Martínez-Salamero et al. and P. Mattavelli et al., who proposed design methods for achieving locally-stable SM control for high-order converters [18,19]. In addition, the works covering the performance evaluation of SM controlled DC-DC converters include that of G. Escobar et al. [25] and S. C. Tan et al. [1], who compared and validated the superior dynamic tracking performance of the SM control over other controls. Further, S. P. Huang et al. applied the SM control on a fourth-order Ćuk converter in 1989 [15]. E. Fossas et al. investigated the audio-susceptibility and load disturbance of a Ćuk converter controlled by SM control [16]. L. Malesani et al. summarized the SM control designs for the Ćuk converter [17]. M. Castilla et al. presented an SM control for quantum resonant converters [20]. P.F. Donoso-Garcia et al. and Y.B. Shtessel et al. proposed the use of SM control to regulate the output voltage and balance the current of modular DC-DC converters [21]. M. López et al. and R. Giral et al. applied the SM control for interleaving parallel-connected boost-type modular converters [22]. M. López et al. also designed an SM control, which is more relevant to practical power converters, for parallel-connected boost converters [23].

The pioneering works on SM control of power converters have laid a strong foundation for the subsequent works on SM control of renewable energy systems. Most renewable energy systems are linked to or contain power electronics components as part of their systems. By far, SM control has been diversely applied to the power electronics components in renewable energy systems to improve the power system's quality and dynamic responses. This is particularly appropriate since many renewable sources are intermittent, largely fluctuating, and highly uncertain in nature. Further, achieving good power quality, such as high voltage and frequency stability, high power factor, and low voltage and current harmonics, has always been the primary pursuit in such applications.

For the reasons above, most earlier investigations on SM control of renewable energy systems were focused solely on their power quality enhancement [26–32]. In [26], the stabilization of the voltage amplitude and frequency was attained via the mitigation of the power imbalance between the power supply and demand in the presence of appreciable wind power generation using SM control. In [27], the SM control was adopted in dedicated local controllers of wind energy conversion (WEC) systems, smart loads, and energy storage systems (ESS) to robustly regulate the power flow and direct current (DC) bus voltage, even in the presence of model uncertainties and external disturbances. In [28], the SM control was designed for both positive and negative sequence power control to improve the power sharing and harmonics cancellation of a hybrid AC/DC microgrid. In [29], a double SM frequency control strategy with a disturbance observer was applied to an isolated hybrid micro-grid to achieve frequency regulation. The control method can significantly reduce frequency deviation and maintain the power balance of the plant even with unmatched uncertainties and resource variations. In [30], the fixed-frequency SM control was applied to a voltage-fed quasi-Z-source inverter to ensure stable operation of the battery storage system with a variable renewable energy source. In [31], a second-order SM controller was implemented to regulate the zero-sequence injection of a four-leg three-level neutral-point-clamped inverter to control the power flow of a hybrid ESS, which comprises a lithium-ion battery and a vanadium redox flow battery. The controller manages the hybrid ESS in terms of improving the power quality and stability, and also performs the control of the renewable energy injection into a microgrid. In [32], a high-order SM controller was adopted to ensure the stability of a wind turbine in the presence of parametric uncertainties of the turbine, as well as electric grid disturbances.

In more recent works, the state-of-the-art research on SM control of renewable energy systems involve, in addition to achieving high power quality, the dynamic tracking improvement of the renewable energy systems. By far, several investigations have been carried out to showcase the advantages of SM control over linear control in dynamic power tracking of the systems with renewables [33–39]. In [35], a reduced-order SM controller was designed for a cascaded boost converter in a photovoltaic (PV) system to achieve a high conversion ratio with an efficiency of close to 95% for a wide operating range. The reduced-order SM control is derived based on the full-order switched model of the cascaded boost converter, taking into account the sliding mode constraints, the nonlinear characteristic of the PV module, and the dynamics of the maximum power point tracking (MPPT) controller. In [36], the boundary control with variable sliding surfaces was adopted in the MPPT controller to attain maximum power extraction from PV cells in dynamic conditions. It is found that significant improvement is achievable over conventional proportional-integral (PI) control. In [37], SM control was applied to both the mechanical system and the power converter of a low-power WEC system in a DC microgrid. Compared to the conventional MPPT schemes using linear controllers, the SM control regulates the dynamics of the WEC system more rapidly and achieves better energy harvesting. This is due to the fast-dynamic-tracking merit of SM control, which significantly improves dynamic energy-harvesting property as compared to linear controllers in conditions of variable sun and wind. In [38], a pulse-width-modulation (PWM)-based SM control was applied to the buck-boost regulator of the series-series (SS)-compensated wireless power transfer (WPT) system with variable power generations and consumptions. The PWM-based SM control exhibits better dynamic tracking performance of the output voltage than the conventional linear control when the input voltage, the mutual inductance, or the load condition of the WPT system is variable. In [39], an SM control-based direct power control (DPC) was designed for a dual-active-bridge (DAB) DC-DC converter, which has been widely adopted in DC and hybrid microgrids, to achieve accurate reference tracking against line and load disturbances, as well as low overshoot and undershoot in dynamics.

Clearly, the intentional exploitation of the advantage of fast dynamic tracking of SM control can further result in more energy harvesting of renewable energy sources and possibly better load protections for various renewable energy systems. This could be a trend for future SM control applications in renewable energy systems. In this paper, through the illustrations of some case studies, an examination is provided of how the emerging use of SM control is extendable to such complex renewable power-electronics systems and how their advantages can be significant in such applications.

2. Case Study 1-SM Control for Low-Power WEC Systems in DC Microgrids

In [37], the absolute MPPT of a low-power WEC system without energy storage was achieved only if the power consumption of the load fully matched the maximum power generation of the wind turbine. In practice, wind energy is intermittent. To achieve real-time MPPT of the WEC system, both the power generation and the consumption are required to be dynamically and instantaneously matched. The quicker the power generation matches the consumption, the higher the amount of energy that can be extracted for the time period. This section presents the application of SM control on both the wind turbine and its subsequent grid-connected converter in achieving fast dynamic power tracking of a low-power WEC system in a DC microgrid in maximizing its harvested energy.

The schematic diagram of the system in the case study is shown in Figure 1. The WEC system consists of a wind turbine system and a power conversion system. The wind turbine system extracts the maximum power from the wind and converters it into electrical power, which is further converted by the power conversion system to feed the DC grid with proper control. Specifically, the torque of the wind turbine is controlled by the MPPT controller-1 via the regulation of the output voltage (i.e., E_{rmf}) of the generator, thus the maximum power generation (MPG) from wind energy can be implemented. However, by only adopting the MPPT controller-1, the wind turbine system cannot guarantee all the generated power being injected into the DC grid without using any local ESS. The MPPT controller-2 needs to be used for the power conversion system to achieve maximum power injection (MPI) based on the state feedback signals of $V_{DC\text{-link}}$, i_{R2}, and v_C, such that the injected power into the DC grid matches the MPG.

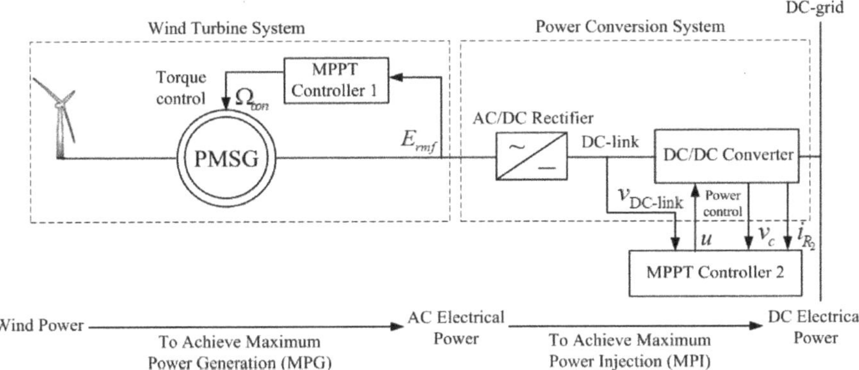

Figure 1. Schematic diagram of a low-power wind energy conversion (WEC) system in a DC microgrid.

For the wind turbine system, the mechanical equation of the shaft is:

$$J\frac{d\Omega}{dt} = T_g - T_e - f\Omega \tag{1}$$

where J is the total moment of inertia; Ω is the mechanical generator speed; T_g is the gearbox torque; T_e is the electrical torque; and f is the viscous friction coefficient. The mechanical generator speed reference is:

$$\Omega_{ref} = \frac{\lambda_{opt} v G}{R} \tag{2}$$

where R is the radius of the blade; G is the gear ratio of the gear box; v is the speed of wind; and λ_{opt} is the optimal tip ratio for the maximum conversion ratio. Obviously, the MPG of the wind turbine system can be implemented by tracking the mechanical generator speed reference. Then, the sliding surface of the SM control for the MPPT controller-1 can be selected as:

$$S = \dot{\Omega}_{ref} - \dot{\Omega} + c(\Omega_{ref} - \Omega) \tag{3}$$

where c is a tuning coefficient. By substituting (1) and (2) into (3),

$$S = \left(\frac{f^2}{J^2} + \frac{cf}{J}\right)S + \left(\frac{f}{J^2} + \frac{c}{J}\right)\left(T_e - T_g\right) + \ddot{\Omega}_{ref} + c\dot{\Omega}_{ref} - \left(\frac{f^2}{J^2} + \frac{cf}{J}\right)\Omega_{ref} \tag{4}$$

To satisfy the criteria $S \cdot \dot{S} < 0$, the control signal of the SM control for the MPPT controller-1 can thus be derived as:

$$\Omega_{\text{con}} = f\Omega_{\text{ref}} - \alpha_1 \ddot{\Omega}_{\text{ref}} - \left(J - \frac{f}{J}\alpha_1\right)\dot{\Omega}_{\text{ref}} - \alpha_2 \text{sgn}(\Omega_{\text{ref}} - \Omega) \tag{5}$$

where sgn (.) indicates the signum function; and α_1 and α_2 are the sliding coefficients.

Alternatively, the control signal of the conventional PI control for the MPPT controller-1 is:

$$\Omega_{\text{con}} = K_{\text{p}}(\Omega_{\text{ref}} - \Omega) + K_{\text{i}} \int_0^t (\Omega_{\text{ref}} - \Omega)dt \tag{6}$$

where K_{p} and K_{i} are the proportional and integral gains. In this paper, the sliding coefficient α_2 of the adopted

SM control and the tuning coefficients K_{p} and K_{i} of the PI control are tuned by the trial-and-error method. The sliding coefficient α_1 of the SM control satisfies:

$$\alpha_1 = \frac{J^2}{cJ + f} \tag{7}$$

For the power conversion system, the state-space equation of the grid-connected converter is:

$$\begin{cases} v_{\text{DC-link}} u = L_1 \frac{di_{L_1}}{dt} + R_1 i_{L_1} + v_c \\ v_{\text{DC-grid}} = -R_2 i_{L_1} + R_2 C \frac{dv_c}{dt} + v_c \end{cases} \tag{8}$$

The sliding surface of the SM control for the MPPT controller-2 is chosen as:

$$S(x,t) = \sum_{i=1}^{3} \alpha_i x_i(t) \tag{9}$$

where α_i indicates the sliding coefficients and $x_i(t) \in x(t)$. Besides,

$$x = \begin{bmatrix} x_1 \\ x_2 \\ x_3 \end{bmatrix} = \begin{bmatrix} V_{\text{ref}} - \beta v_c \\ \frac{d(V_{\text{ref}} - \beta v_c)}{dt} \\ \int (V_{\text{ref}} - \beta v_c)dt \end{bmatrix} \tag{10}$$

Then, by substituting (8) into (10), the control signal of the SM control for the MPPT controller-2 can be derived as:

$$v_{\text{con}} = K_1 i_c + K_2 v_c + K_3 \tag{11}$$

and

$$\left|v_{\text{ramp}}\right| = \beta v_{\text{DC-link}} \tag{12}$$

where $K_1 = -\frac{\alpha_1 \beta L_1}{\alpha_2} + \beta\left(\frac{L_1}{R_2 C} + R_1\right)$, $K_2 = -\frac{\alpha_3 \beta L_1 C}{\alpha_2} + \beta\left(\frac{R_1}{R_2} + 1\right)$, and $K_3 = -\frac{\beta R_1}{R_2} V_{\text{DC-grid}} + \frac{\alpha_3}{\alpha_2} L_1 C\left(\frac{P_{\text{ref}} v_{\text{DC-link}}}{V_{\text{DC-grid}}} R_2 + V_{\text{DC-grid}}\right)$, which are tuned to satisfy the hitting, existence, and stability conditions of sliding mode operation. The sliding coefficients satisfy $\alpha_1 > 0$, $\alpha_2 > 0$, and $\alpha_3 > 0$ to ensure the Hurwitz-Routh stability of the sliding surface. Alternatively, the control signal of the conventional PI control for the MPPT controller-2 is:

$$v_{\text{con}} = K_{\text{p}}(v_{\text{ref}} - \beta v_c) + K_{\text{i}} \int_0^t (v_{\text{ref}} - \beta v_c)dt \tag{13}$$

and

$$\left|v_{\text{ramp}}\right| = \text{constant} \tag{14}$$

Then, the duty cycle of the grid-connected converter for both control is:

$$d = \frac{v_{\text{con}}}{\left| v_{\text{ramp}} \right|} \tag{15}$$

The sliding coefficients α_1, α_2, and α_3 of the SM control are tuned based on the guidelines in [1]. The tuning coefficients K_p and K_i of the PI control are tuned by the trial-and-error method.

The comparisons are conducted in experiments between the SM controllers and the conventional PI controllers for the wind turbine system with MPG, the power conversion system with MPI, and the entire WEC system with MPPT. With practical considerations, the wind speed is constrained within 6.6 m/s to 7.9 m/s. For the wind turbine system with MPG, the waveforms of the output voltage, the output current, and the output power of the wind turbine controlled by the conventional PI control and SM control are presented in Figure 2a,b, respectively. Here, both the PI control and SM control are optimally tuned under this wind speed change condition. The output power over the two transient cycles is approximately 430.7 W for the PI control and 431.6 W for the SM control. The difference of the extracted power between the two controls is quite small.

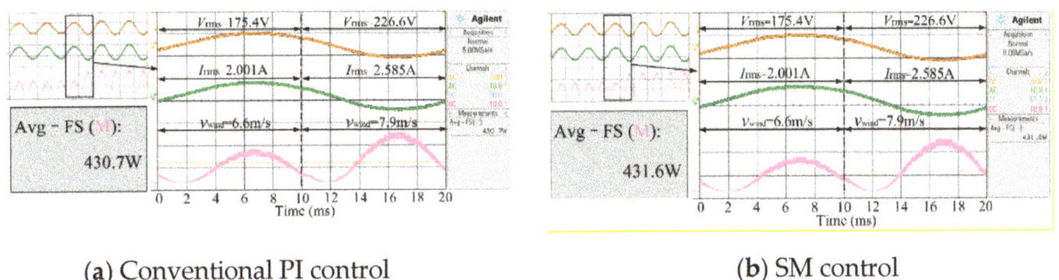

<table>
<tr><td>(a) Conventional PI control</td><td>(b) SM control</td></tr>
</table>

Figure 2. The experimental waveforms of the output voltage, the output current, and the output power of wind turbine system for wind speed altering from 6.6 to 7.9 m/s.

For the power conversion system with MPI, the root-mean-square (RMS) value of the input AC voltage is altered from 180 V to 220 V to simulate a step change in wind energy. The experimental waveforms of the DC grid voltage and the output power of the grid-connected converter controlled by the conventional PI control and SM control, are shown in Figure 3a,b. Here, both the PI control and SM control are optimally tuned under this voltage change condition. The settling time of the output power for both control are identical at 15.8 ms.

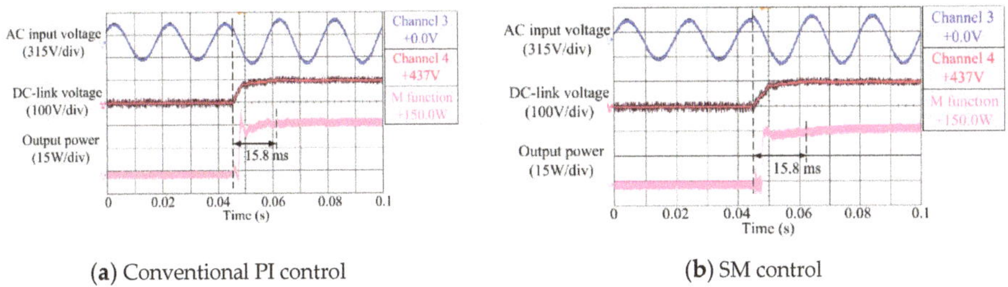

<table>
<tr><td>(a) Conventional PI control</td><td>(b) SM control</td></tr>
</table>

Figure 3. The experimental waveforms of the DC grid voltage and the output power of the power conversion system for input voltage changing from 180 to 220 V.

Then, with both the PI control and SM control settings kept in the abovementioned tuned values, the MPPT performance of the entire system is evaluated. Various experiments are performed with the wind speed first changing from 7.9 m/s to 6.6 m/s and then from 6.6 m/s back to 7.9 m/s. The corresponding waveforms are shown in Figure 4. From Figure 4a,b, it can be seen that the SM control takes a shorter time of 11.1 ms to reach the steady state MPPT position as compared to the PI control, which takes approximately 16.8 ms (about 34% more time), when there is a step change of wind speed from 6.6 m/s to 7.9 m/s. For the case of stepping down the wind speed from 7.9 m/s to 6.6 m/s, the SM control is also quicker in responding to the change with settling time of 30.1 ms, as compared to that of PI control at 43.7 ms (about 45% more time). The experiments were also conducted for the case of the wind speed changing from 7.3 m/s to 6.6 m/s and then from 6.6 m/s back to 7.3 m/s (results not included in paper). All results validate that the WEC system with SM control tracks faster the MPPT position, and thus presumably harvests more energy than conventional linear controllers on account of faster dynamics.

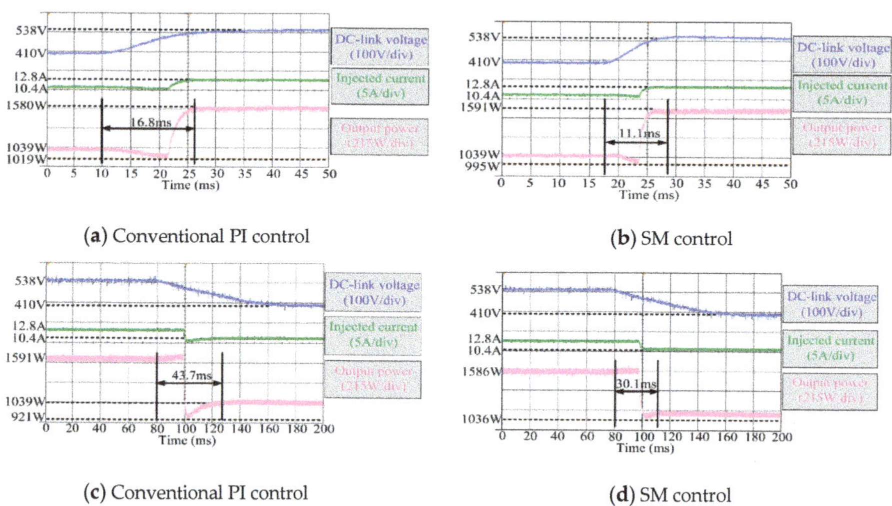

(**a**) Conventional PI control   (**b**) SM control

(**c**) Conventional PI control   (**d**) SM control

Figure 4. The experimental waveforms of the DC grid voltage, the injected current, and the output power of the entire WEC system for wind speeds altering from 6.6 to 7.9 m/s and vice versa.

The experiments were also carried out with a random wind profile within the wind speed between 6.6 m/s and 7.9 m/s. The results are depicted in Figure 5. A comparison of the energy harvested over 50 s of the WEC system controlled by the two control schemes exhibit the SM control for both MPPT controllers which can harvest approximately 0.44% more energy than the conventional PI control, as shown in Figure 6.

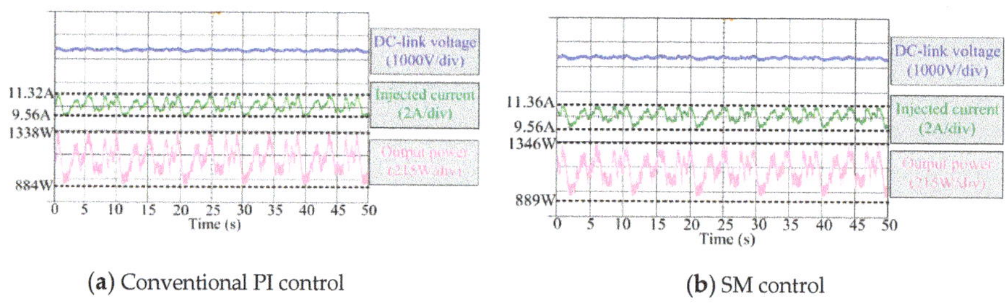

(**a**) Conventional PI control   (**b**) SM control

Figure 5. The experimental waveforms of the DC grid voltage, injected current, and output power of the entire WEC system for random wind profile under (**a**) proportional-integral (PI) and (**b**) sliding mode (SM) control.

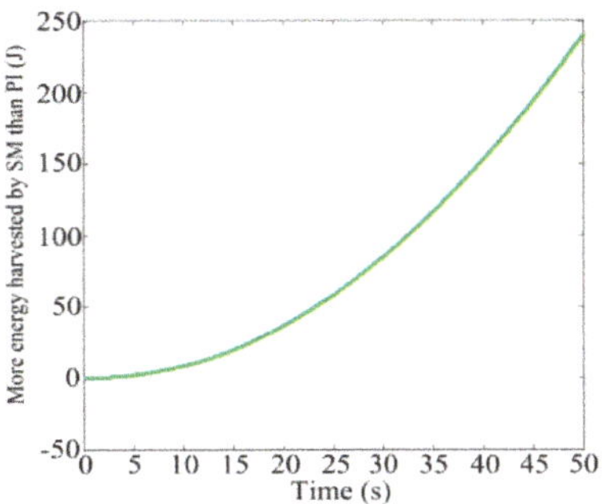

Figure 6. Additional energy harvested of the WEC system with a random wind profile by the SM control than the conventional PI control.

In conclusion, the experimental results from this study validate that the energy harvesting property of renewable systems that have a fluctuating source, can be better improved using SM control compared to using conventional optimally-tuned PI control. SM control provides better dynamic tracking of the MPPT position for varying input and load conditions than linear control.

3. Case Study 2-SM Control of SS-Compensated WPT Systems with Renewable Energy Sources

In [38], SM control was designed for an SS-compensated WPT system to achieve maximum energy efficiency as well as output voltage regulation. The transmitter of the SS-compensated WPT consists of a full-bridge high-frequency DC/AC resonant inverter (S_1 to S_4) being controlled by a perturb-and-observation (P and O) phase-displacement control to minimize the input current from the renewable source, thereby achieving minimum input power. The receiver of the SS-compensated WPT consists of a buck-boost converter being controlled by a PI control or a SM control to dynamically match the optimal load condition. This section presents a brief discussion on the investigation.

The schematic diagram of a typical two-stage SS-compensated WPT system is plotted in Figure 7. The renewable source provides a time-varying DC power supply that may be fluctuating in amplitude with respect to solar irradiance, temperature, and load conditions. L_1 and L_2 represent the wireless power transmitter and receiver coil, respectively; C_1 and C_2 are the series compensation capacitors; and R_{p1} and R_{p2} are the equivalent parasitic resistance of the transmitter and receiver coil, respectively. The output of the receiver coil is connected to a full-bridge high-frequency rectifier diode circuit comprising D_1 to D_4 with storage capacitor C_b. Controller 1 is a P and O phase-displacement control for achieving minimum input power of the variable renewable source via the regulation of the zero-state phase angle of the output voltage of the DC/AC inverter V_{in} (schematics of Controller 1 can be found in [38]). Controller 2 is for regulating the output voltage the buck-boost converter and in this study, both the conventional PI control and the SM control are adopted for comparison.

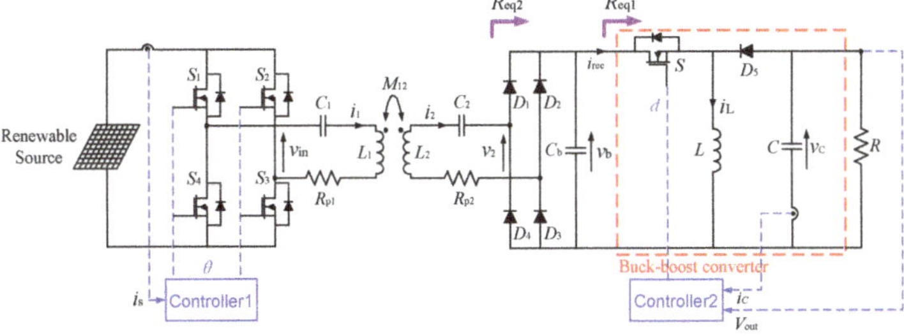

Figure 7. Schematic diagram of a typical two-stage SS-compensated wireless power transfer (WPT) system.

Conventionally, the PI control is adopted to regulate the buck-boost converter of the receiver,

$$\Delta u(k) \;=\; K_{\mathrm{p}}\Delta e(k) + K_{\mathrm{i}}e(k) \tag{16}$$

where $\Delta u(k) = u(k) - u(k-1)$ is the control signal increment; $e(k) = V_{\mathrm{ref}} - V_{\mathrm{out}}$ is the voltage error between the reference and the output voltage; and $\Delta e(k) = e(k) - e(k-1)$ is the voltage erroring increment. The control signal $u(k)$ is obtained by adding the control signal increment $\Delta u(k)$ to the control signal $u(k-1)$. The tuning coefficients K_{p} and K_{i} are tuned by the trial-and-error method.

However, the output voltage of the WPT system controlled by the conventional PI control (a typical linear control) can suffer from high overshoot/undershoot and long settling time during the dynamics, such as load-point change and P and O searching process of the phase-displacement control for the full-bridge inverter. To overcome such issues, SM control can be adopted for the buck-boost converter. The control variables $x(k)$ of the SM control can be expressed in the following form:

$$x(k) \;=\; \begin{bmatrix} x_1(k) \\ x_2(k) \\ x_3(k) \end{bmatrix} \;=\; \begin{bmatrix} V_{\mathrm{ref}} - V_{\mathrm{out}}(k) \\ \frac{1}{T_{\mathrm{s1}}}[V_{\mathrm{out}}(k) - V_{\mathrm{out}}(k+1)] \\ T_{\mathrm{s1}} \sum\limits_{j\,=\,0}^{k-1} [V'_{\mathrm{ref}} - V_{\mathrm{out}}(j)] \end{bmatrix} \tag{17}$$

The state-space model for the SM control can be further derived as:

$$x(k+1) \;=\; \begin{bmatrix} 1 & T_{\mathrm{s1}} & 0 \\ 0 & 1-\frac{T_{\mathrm{s1}}}{RC} & 0 \\ T_{\mathrm{s1}} & 0 & 1 \end{bmatrix} \begin{bmatrix} x_1(k) \\ x_2(k) \\ x_3(k) \end{bmatrix} + \begin{bmatrix} 0 \\ \frac{T_{\mathrm{s1}}V_{\mathrm{out}}(k)}{LC} - \frac{T_{\mathrm{s1}}V_{\mathrm{out}}(k)}{LC}u(k) \\ 0 \end{bmatrix} \tag{18}$$

The sliding surface of the SM control is given as:

$$S(k) \;=\; \alpha_1 x_1(k) + \alpha_2 x_2(k) + \alpha_3 x_3(k) \tag{19}$$

where α_1, α_2, and α_3 are the sliding coefficients (i.e., $\alpha_1 > 0$, $\alpha_2 > 0$ and $\alpha_3 > 0$). Then, the equivalent control signal $u_{\mathrm{eq}}(k)$ can be obtained by letting $S(k+1) = S(k)$,

$$u_{\mathrm{eq}}(k) \;=\; 1 - \frac{K_1}{V_{\mathrm{ref}}} + \frac{K_1}{V_{\mathrm{out}}(k)} + \frac{K_2 i_c(k)}{V_{\mathrm{out}}(k)} \tag{20}$$

where $K_1 = \dfrac{\alpha_3 LCV'_{\mathrm{ref}}}{\alpha_2}$ and $K_2 = \dfrac{L}{RC} - \dfrac{\alpha_1 L}{\alpha_2}$ are the tuning coefficients of the SM control. The equivalent control signal satisfies $0 \le u_{\mathrm{eq}}(k) < 1$, which equalizes the duty ratio $d(k)$ of the pulse-width modulation. The sliding coefficients α_1, α_2, and α_3 are optimally tuned based on the guidelines in [1].

The results of the dynamic performance of the system between the conventional PI control and the SM control for the SS-compensated WPT system are shown in Figures 8 and 9. The PI control is designed under the operating condition when V_{in} has no zero-state in its output voltage waveform, i.e., zero-state phase angle is 0. From Figure 8a, it is shown that in the event that the zero-state phase angle of V_{in} is step-changed from 120° to 100°, the overshoot of the output voltage V_{out} of the buck-boost converter controlled by the conventional PI control is approximately 2.5 V and the settling time is approximately 151.8 ms during the transient state. For the same conditions, the overshoot of V_{out} of the buck-boost converter controlled by the SM control is approximately 2 V and the settling time is approximately 48 ms, as shown in Figure 8b. The overshoot is reduced by approximately 20% and the settling time is reduced by approximately 68.4%.

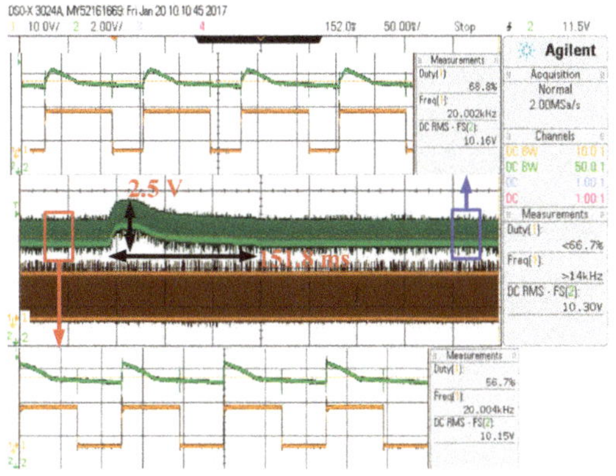

(**a**) Conventional PI control

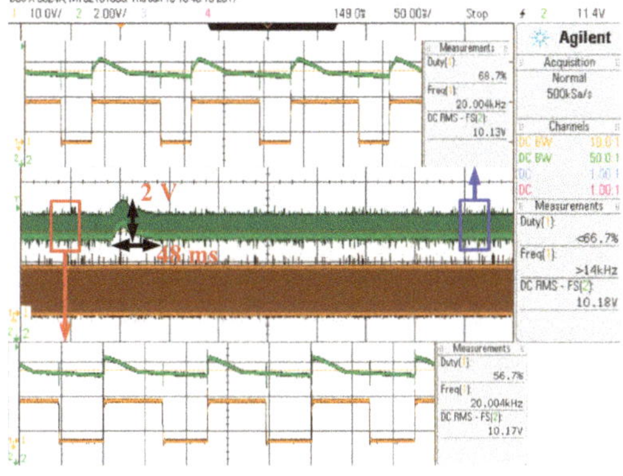

(**b**) SM control

Figure 8. The waveforms of V_{out} and the switching signals when the phase displacement is altered from 120° to 100°.

For the case of step-changing, the zero-state phase angle of V_{in} from 100° to 110°, an undershoot of V_{out} of approximately 2 V is present and the settling time is approximately 164.3 ms with the PI control (see Figure 9a), while it is approximately 1.9 V and approximately 39.3 ms with the SM control. The same experiment is applied for other step change conditions and the results indicate that with the SM control, the output voltage can always have a lower overshoot/undershoot and settling time as compared to PI control.

In conclusion, this study shows and validates that SM control performs better dynamic regulation of the output voltage than PI control in systems, in terms of lowering overshoots and having faster settling time in the event of the variation of the renewable source. This is particularly important in modern clean energy environments such as that using WPT for electric vehicle charging. The improvement can prevent the load suffering from undesirable overshoot/undershoot during dynamics and gives better over-voltage and under-voltage protection to the loads and complementing circuitries.

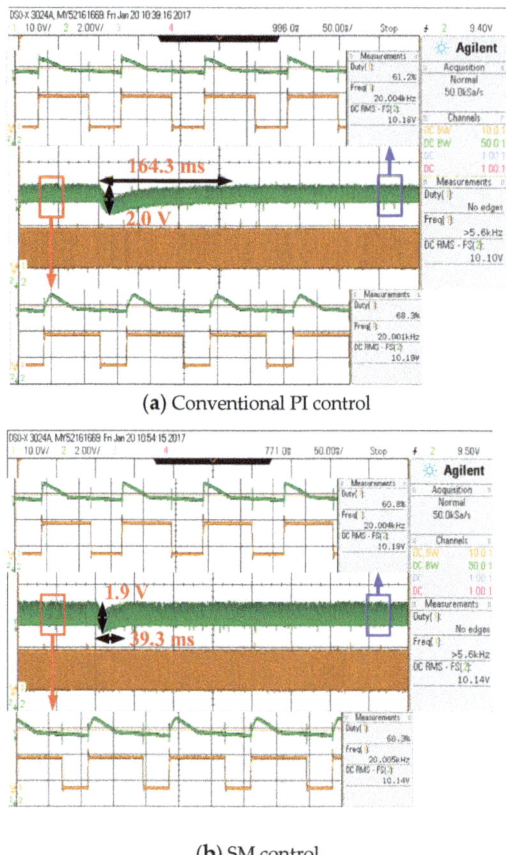

(**a**) Conventional PI control

(**b**) SM control

Figure 9. The waveforms of V_{out} and the switching signals when the phase displacement is altered from 100° to 110°.

4. Case Study 3-SM Control of Multiple Energy Storage Systems in DC Microgrids

In [40], a two-layer hierarchical control scheme was adopted in a system with multiple ESS to mitigate the distribution power loss on the transmission power cables of standalone DC microgrids. The hierarchical control comprises a centralized model predictive control (CMPC) with adaptive weighting factors and multiple local PI controls. The CMPC is a secondary control that generates bus voltage references for the local PI control of each ESS based on the optimization of the flow power in the standalone DC microgrid. The multiple local PI controllers are primary controllers that track the references provided by the CMPC to achieve the overall functions. The two-layer hierarchical control scheme coordinates the various ESS to reduce the power loss on the transmission power cables of the standalone DC microgrid, while simultaneously regulating the DC bus voltages within the tolerances (i.e., ±5% tolerance). However, the hierarchical control scheme reported in [40] only guaranteed the bus voltages to be regulated within a ±5% tolerance in a steady state, while the overshoot and undershoot during transient states may exceed the tolerance. This may cause unwanted electrical tripping of the microgrids, especially when voltage level of the renewable energy sources (RES) fluctuate widely and randomly on certain occasions. In this section, the simulation results and discussions based on the use of SM control in replacement of the conventional PI control adopted in the hierarchical control scheme reported in [40], are presented. The SM control is adopted as a local control for the grid-connected converters of the ESS. The structure of the investigated 380 V five-bus standalone DC microgrid is depicted in Figure 10. The RES 1 and RES 2 are integrated at Bus 1 and Bus 5, respectively. Three ESS are installed at Buses 1, 3, and 4, respectively. The grid-connected converters of the ESS are non-isolated boost converters shown in Figure 11. The overall control block diagram of multiple ESS in the five-bus standalone DC microgrid is shown in Figure 12. For fair comparison, both the conventional PI control and the SM control are adopted as local controls to regulate the bus voltage and to track the bus voltage reference generated by the CMPC (schematics of the CMPC can be found in [40]).

For the conventional PI control, the control signal is

$$u_{\mathrm{PI}} = K_{\mathrm{pI}}(i_{\mathrm{Lref}} - i_{\mathrm{L}}) + K_{\mathrm{iI}} \int_0^t (i_{\mathrm{Lref}} - i_{\mathrm{L}})dt \tag{21}$$

where the inductor current reference is generated based on

$$i_{\mathrm{Lref}} = K_{\mathrm{pV}}(V_{\mathrm{busref}} - V_{\mathrm{bus}}) + K_{\mathrm{iV}} \int_0^t (V_{\mathrm{busref}} - V_{\mathrm{bus}})dt \tag{22}$$

The tuning coefficients of the conventional PI control (i.e., K_{pI}, K_{iI}, K_{pV}, and K_{iV}) are optimized offline by the genetic algorithm (GA). The schematic diagram of the parameter tuning of the conventional PI control is depicted in Figure 13. Here, V_{err} is the difference between the bus voltage reference and the bus voltage (i.e., $V_{\mathrm{err}} = V_{\mathrm{busref}} - V_{\mathrm{bus}}$).

For the SM control, the control variables are:

$$x(k) = \begin{bmatrix} x_1 \\ x_2 \\ x_3 \end{bmatrix} = \begin{bmatrix} i_{\mathrm{Lref}} - i_{\mathrm{L}} \\ V_{\mathrm{busref}} - V_{\mathrm{bus}} \\ \int(i_{\mathrm{Lref}} - i_{\mathrm{L}})dt + \int(V_{\mathrm{busref}} - V_{\mathrm{bus}})dt \end{bmatrix} \tag{23}$$

The sliding surface of the SM control is given as:

$$s = \alpha_1 x_1 + \alpha_2 x_2 + \alpha_3 x_3 \tag{24}$$

where α_1, α_2, and α_3 are the sliding coefficients (i.e., $\alpha_1 > 0$, $\alpha_2 > 0$ and $\alpha_3 > 0$). Then, the equivalent control signal u_{SM} can be obtained by letting $S \cdot \dot{S} < 0$ to give:

$$u_{\mathrm{SM}} = 1 - \frac{K_2 \frac{V_{\mathrm{bus}}}{R_{\mathrm{L}}} - V_{\mathrm{bat}} + K_3(V_{\mathrm{busref}} - V_{\mathrm{bus}}) - K_3(i_{\mathrm{Lref}} - i_{\mathrm{L}})}{K_2 i_{\mathrm{L}} - V_{\mathrm{bus}}} \tag{25}$$

where $K_1 = \frac{\alpha_3 L(K+1)}{\alpha_1}$, $K_2 = \frac{L}{C}\left(K + \frac{\alpha_2}{\alpha_1}\right)$, and $K_3 = \frac{\alpha_3 L}{\alpha_1}$. The sliding coefficients α_1, α_2, α_3, and K are optimally tuned based on the guidelines in [1].

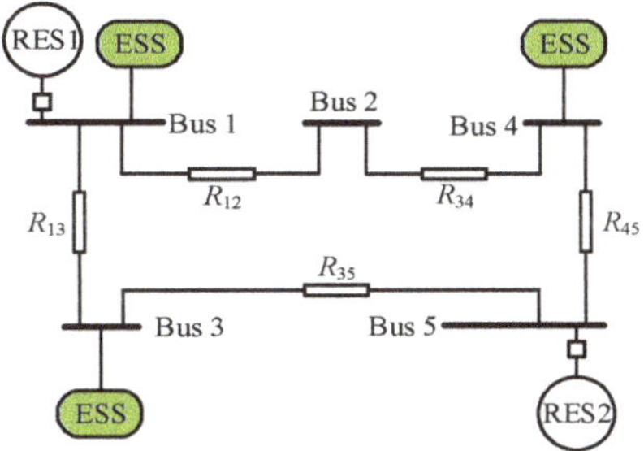

Figure 10. The structure of the investigated five-bus standalone DC microgrid.

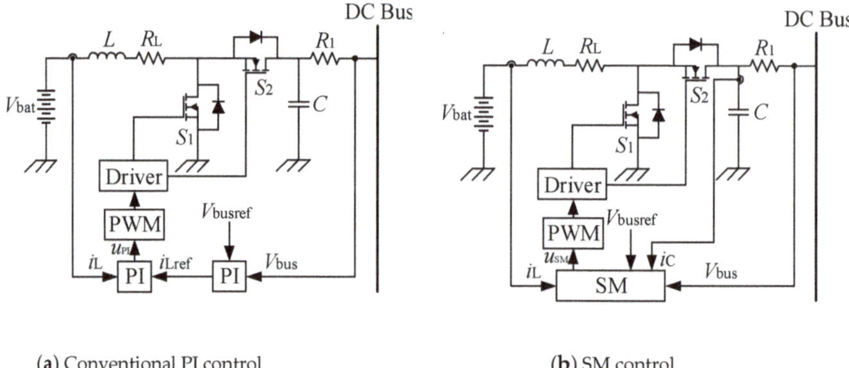

(**a**) Conventional PI control   (**b**) SM control

Figure 11. The circuitry and control of the grid-connected converters for energy storage systems (ESS).

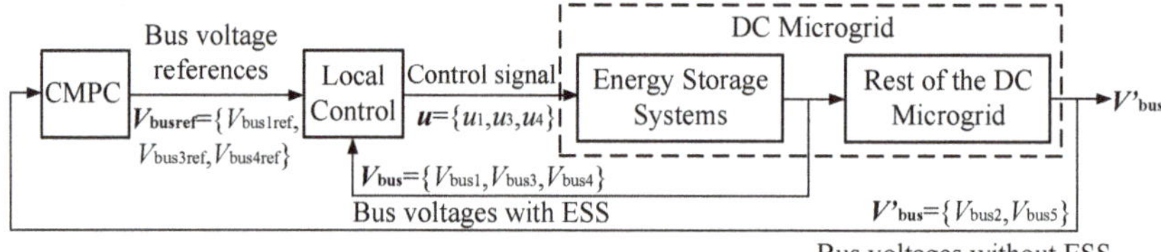

Figure 12. Overall control block diagram of multiple ESS in the five-bus standalone DC microgrid.

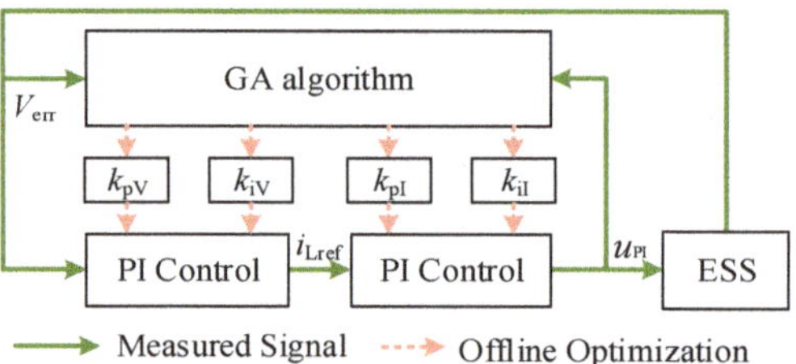

Figure 13. Schematic diagram of the parameter tuning of the conventional PI control.

Figure 14 shows the comparison of the energy loss reduction of the five-bus standalone DC microgrid over 5 s (1–6 s) based on the conventional PI control and the SM control. The difference between the two control schemes are negligible, which means that both the conventional PI control and the SM control are tracking the references generated by CMPC rapidly and have similar effects in reducing the distribution power loss. The waveforms of the bus voltages over a time period of 5 s under both control schemes are shown in Figure 15. Apparently, the voltages of Bus 1 during the intervals of 1–2 s, 4–5 s, and 5–6 s, Bus 2 during the intervals of 2–3 s and 5–6 s, Bus 3 during the interval of 4–5 s, Bus 4 during the intervals of 2–3 s and 5–6 s, and Bus 5 during the interval of 4–5 s are all out of the ±5% tolerance during transient when the grid-connected converters of ESS are using the conventional PI control. However, all the bus voltages are within the tolerance during transient when they are controlled using the SM control. This is important if grid protection is of concern and when achieving stringent voltage regulation is valued. It is apparent that SM control exhibits strength in achieving both features in this case.

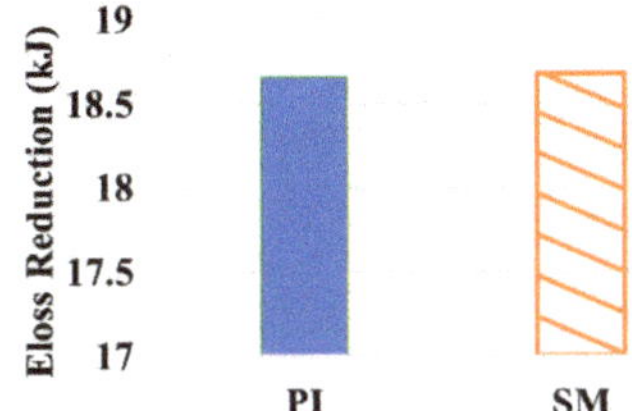

Figure 14. The energy loss reduction of the DC microgrid with PI and SM control.

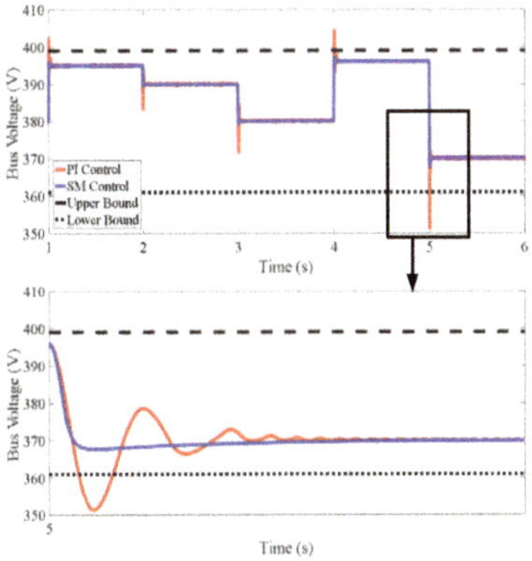

(**a**) Bus 1

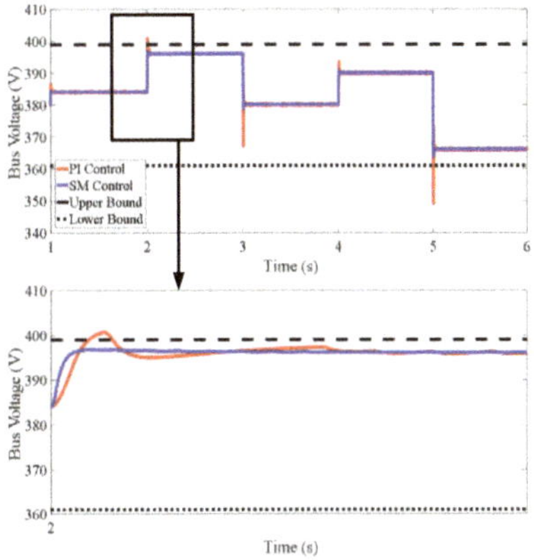

(**b**) Bus 2

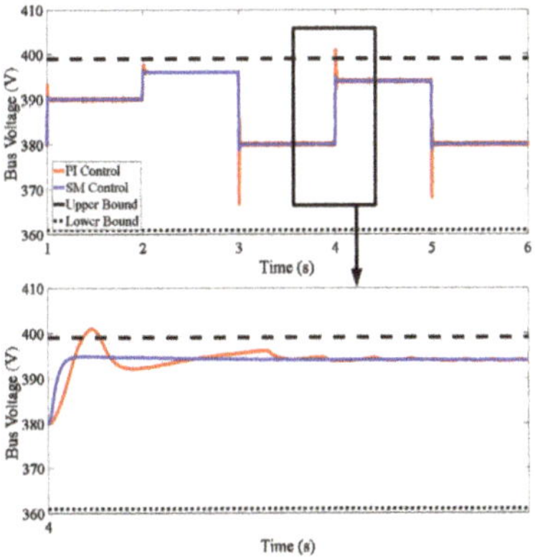

(**c**) Bus 3

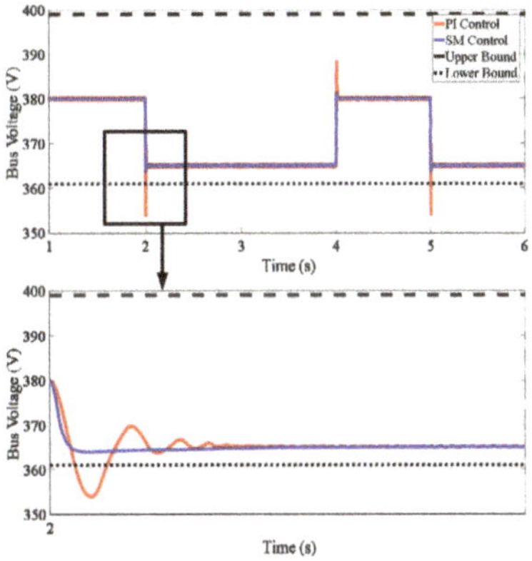

(**d**) Bus 4

Figure 15. *Cont.*

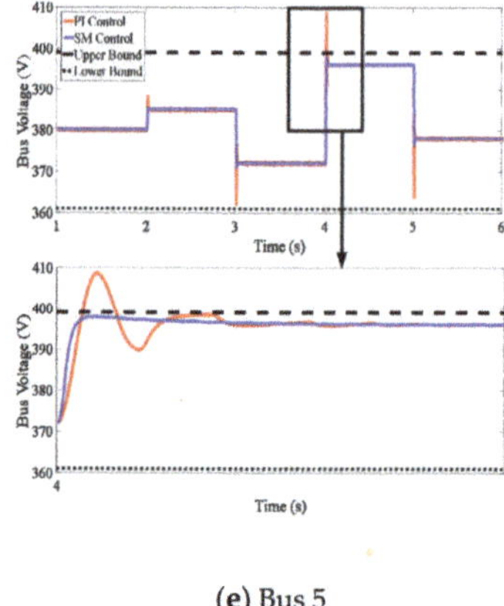

(**e**) Bus 5

Figure 15. Bus voltages of the DC microgrid over a simulation period of 5 s with both PI and SM control.

5. Conclusions and Future Discussions

Sliding mode control is a well-established control method that has been widely applied in power converters for its properties of high robustness against parameter uncertainties, fast dynamic responses, high flexibility in the design, and easy implementations as compared to other nonlinear control techniques. This paper further presented the recent developments and trends of sliding mode control applications to power-electronics based renewable energy systems. Notably, in addition to achieving high power quality, such as high voltage and frequency stability, high power factor, and low voltage and current harmonics, the applications of sliding mode control can be extended to include efficiency improvement and better load protections of renewable energy systems, via the exploitation of the fast-dynamic-response property of sliding mode control. The case studies provided in the paper have shown that sliding mode control can increase the energy harvested from renewable sources and achieves better dynamic regulation of output and bus voltages. This, thereby, protects the load and the system from possible large overshoot/undershoot that may arise as a result of poor control of the system or that of any constituting component.

Most of the present applications of sliding mode control for power converters in renewable energy systems focus on the single-input-single-output pulse-width-modulation-based converters. The sliding mode control may be further applied for multiple-input-multiple-output pulse-width-modulation-based converters, which include single multiple-input-multiple-output converters, e.g., multi-port converters for high voltage step-up/down applications, and multiple single-input-single-output converters distributed and installed in a network. The single multiple-input-multiple-output converters are widely used in data centers and other isolated grids, e.g., grids in electric ships. The multiple single-input-single-output converters are widely adopted for distributed renewable energy sources and energy storage systems in microgrids. The sliding mode control designs for multiple-input-multiple-output pulse-width -modulation-based converters may consider hierarchical structures, which comprises two or more layers of sliding mode control.

Author Contributions: Conceptualization, Y.Y. and S.C.T.; methodology, Y.Y. and S.C.T.; software, Y.Y.; validation, Y.Y. and S.C.T.; formal analysis, Y.Y. and S.C.T.; investigation, Y.Y. and S.C.T.; resources, S.C.T.; data curation, Y.Y.; writing—original draft preparation, Y.Y.; writing—review and editing, S.C.T.; supervision, S.C.T.; project administration, S.C.T.

Funding: This research received no external funding.

Conflicts of Interest: The authors declare no conflict of interest.

References

1. Tan, S.C.; Lai, Y.M.; Tse, C.K. *Sliding Mode Control of Switching Power Converters-Techniques and Implementation*; CRC Press: Boca Raton, FL, USA, 2012.
2. Emelyanov, S.V. *Variable Structure Control Systems*; Nauka: Moscow, Russia, 1967. (In Russian)
3. Itkis, Y. *Control Systems of Variable Structure*; Wiley: New York, NY, USA, 1976.
4. Utkin, V.I. *Sliding Mode and Their Application in Variable Structure Systems*; Mir Publisher: Moscow, Russia, 1978.
5. Utkin, V.I. Variable structure systems with sliding modes. *IEEE Trans. Autom. Control* **1977**, *22*, 212–222. [CrossRef]
6. Sarpturk, S.Z.; Istefanopulos, Y.; Kaynak, O. On the stability of discrete-time sliding mode control system. *IEEE Trans. Autom. Control* **1987**, *32*, 930–932. [CrossRef]
7. Sira-Ramirez, H.; Llanes, S.O. Adaptive dynamical sliding mode control via backstepping. In Proceedings of the 32nd IEEE Conference on Decision and Control, San Antonio, TX, USA, 15–17 December 1993; pp. 1422–1427.
8. Wheeler, G.; Su, C.H.; Stepanenko, Y. A sliding mode controller with improved adaption laws for the upper bounds on the norm of uncertainties. *Automatica* **1998**, *34*, 1657–1661. [CrossRef]
9. Feng, Y.; Yu, X.H.; Man, Z.H. Non-singular terminal sliding mode control of rigid manipulator. *Automatica* **2002**, *38*, 281–288. [CrossRef]
10. Choi, H.S.; Park, Y.H.; Cho, Y.S.; Lee, M.H. Global sliding mode control. *IEEE Control Syst. Mag.* **2001**, *21*, 27–35.
11. Lu, Y.S.; Chen, J.S. A self-organizing fuzzy sliding-mode controller design for a class of nonlinear servo systems. *IEEE Trans. Ind. Electron.* **1994**, *41*, 492–502.
12. Yoo, B.; Ham, W. Adaptive fuzzy sliding mode control of nonlinear system. *IEEE Trans. Fuzzy Syst.* **1998**, *6*, 315–321.
13. Bilalović, F.; Mušić, O.; Šabanović, A. Buck converter regulator operating in the sliding mode. In Proceedings of the Seventh International Conference on Power Conversion, Dallas, TX, USA, 7–8 February 2007; pp. 331–340.
14. Venkataramanan, R.; Šabanović, A.; Ćuk, S. Sliding mode control of DC-to-DC converters. In Proceedings of the IEEE Conference on Industrial Electronics, Control and Instrumentations, Dubrovnik, Croatia, 15–18 September 2002; pp. 251–258.
15. Huang, S.P.; Xu, H.Q.; Liu, Y.F. Sliding-mode controlled of DC/DC converters. In Proceedings of the IEEE Power Electronics Specialists Conference Record, Milwaukee, WI, USA, 26–29 June 1989; pp. 124–129.
16. Fossas, E.; Martínez, L.; Ordinas, J. Sliding mode control reduces audiosusceptibility and load perturbation in the Ćuk converter. *IEEE Trans. Circuits Syst. Part I* **1992**, *39*, 847–849. [CrossRef]
17. Malesani, L.; Rossetto, L.; Spiazzi, G.; Tenti, P. Performance optimization of Ćuk converters by sliding-mode control. *IEEE Trans. Power Electron.* **1995**, *10*, 302–309. [CrossRef]
18. Martínez-Salamero, L.; Calvente, J.; Giral, R.; Poveda, A.; Fossas, E. Analysis of a bidirectional coupled-inductor Ćuk converter operating in sliding mode. *IEEE Trans. Circuits Syst. Part I* **1998**, *45*, 355–363. [CrossRef]
19. Mattavelli, P.; Rossetto, L.; Spiazzi, G. Small-signal analysis of DC-DC converters with sliding mode control. *IEEE Trans. Power Electron.* **1997**, *12*, 96–102. [CrossRef]
20. Castilla, M.; De Vicuna, L.C.; Lopez, O.; Matas, J. On the design of sliding mode control schemes for quantum resonant converters. *IEEE Trans. Power Electron.* **2000**, *15*, 960–973. [CrossRef]
21. Donoso-Garcia, P.F.; Cortizo, P.C.; De Menezes, B.R.; Severo Mendes, M.A. Sliding mode control for current distribution on DC-to-DC converters connected in parallel. In Proceedings of the IEEE Power Electronics Specialists Conference Record, Baveno, Italy, 23–27 June 1996; pp. 1513–1518.
22. Giral, R.; Martínez-Salamero, L.; Leyva, R.; Maixe, J. Sliding-mode control of interleaved boost converters. *IEEE Trans. Circuits Syst. Part I* **1998**, *47*, 1330–1339.
23. López, M.; De-Vicuňa, L.G.; Castilla, M.; Gaya, P.; López, O. Current distribution control design for paralleled DC/DC converters using sliding-mode control. *IEEE Trans. Ind. Electron.* **2004**, *45*, 1091–1100. [CrossRef]
24. Sira-Ramirez, H.; Rios-Bolivar, M. Sliding mode control of DC-to-DC power converters via extended linearization. *IEEE Trans. Circuits Syst. Part I* **1994**, *41*, 652–661. [CrossRef]
25. Escobar, G.; Ortega, R.; Sira-Ramirez, H.; Vilain, J.P.; Zein, I. An experimental comparison of several nonlinear controllers for power converters. *IEEE Control Syst. Mag.* **1999**, *19*, 66–82.

26. Bashash, S.; Fathy, H.K. Modeling and control of aggregate air conditioning loads for robust renewable power management. *IEEE Trans. Control Syst. Technol.* **2013**, *21*, 1318–1327. [CrossRef]

27. Moré, J.J.; Puleston, P.F.; Kunusch, C.; Fantova, M.A. Development and implementation of a supervisor strategy and sliding mode control setup for fuel-cell-based hybrid generation systems. *IEEE Trans. Energy Convers.* **2015**, *30*, 218–225. [CrossRef]

28. Baghaee, H.R.; Mirsalim, M.; Gharegpetian, G.B.; Talebi, H.A. A decentralized power management and sliding mode control strategy for hybrid AC/DC microgrids including renewable energy resources. *IEEE Trans. Ind. Informat.* **2017**, in press. [CrossRef]

29. Wang, C.; Mi, Y.; Fu, Y.; Wang, P. Frequency control of an isolated micro-grid using double sliding mode controllers and disturbance observer. *IEEE Trans. Smart Grid* **2018**, *9*, 923–930. [CrossRef]

30. Liu, J.; Jiang, S.; Cao, D.; Lu, X.; Peng, F.Z. Sliding mode control of quasi-z-source inverter with battery for renewable energy system. In Proceedings of the IEEE Energy Conversion Congress and Exposition, Phoenix, AZ, USA, 16–21 September 2011; pp. 3665–3671.

31. Tabart, Q.; Vechiu, I.; Etxeberria, A.; Bacha, S. Hybrid energy storage system microgrids integration for power quality improvement using four-leg three-level NPC inverter and second-order sliding mode control. *IEEE Trans. Ind. Electron.* **2018**, *65*, 424–435. [CrossRef]

32. Beltran, B.; Tarek, A.A.; Benbouzid, M.E.H. High-order sliding-mode control of variable-speed wind turbines. *IEEE Trans. Ind. Electron.* **2009**, *56*, 3314–3321. [CrossRef]

33. Hoseini, S.K.; Pouresmaeil, E.; Hosseinnia, S.H.; Catalão, J.P.S. A control approach for the operation of DG units under variations of interfacing impedance in grid-connected mode. *Int. J. Electr. Power Energy Syst.* **2016**, *74*, 1–8. [CrossRef]

34. Hoseini, S.K.; Pouresmaeil, E.; Adabi, J.; Catalão, J.P.S. Stable integration of power electronics-based DG links to the utility grid with interfacing impedance uncertainties. In Proceedings of the Doctoral Conference on Computing, Electrical and Industrial Systems, Costa de Caparica, Portugal, 28 March 2015; pp. 502–511.

35. Haroun, R.; Aroudi, A.E.; Cid-Pastor, A.; Garcia, G.; Olalla, C.; Martínez-Salamero, L. Impedance matching in photovoltaic systems using cascaded boost converters and sliding-mode control. *IEEE Trans. Power Electron.* **2015**, *30*, 3185–3199. [CrossRef]

36. Yang, Z.; Ho, C.N.M.; Siu, K.K.M. A fast and accurate MPP control technique using boundary controller for PV applications. In Proceedings of the IEEE Applied Power Electronics Conference and Exposition, Anaheim, FL, USA, 26–30 March 2017; pp. 2822–2829.

37. Yang, Y.; Mok, K.T.; Tan, S.C.; Hui, S.Y.R. Nonlinear dynamic power tracking of low-power wind energy conversion system. *IEEE Trans. Power Electron.* **2015**, *30*, 5223–5236. [CrossRef]

38. Yang, Y.; Zhong, W.X.; Kiratipongvoot, S.; Tan, S.C.; Hui, S.Y.R. Dynamic improvement of series-series compensated wireless power transfer systems using discrete sliding mode control. *IEEE Trans. Power Electron.* **2018**, *33*, 6351–6360. [CrossRef]

39. Li, K.; Yang, Y.; Tan, S.C.; Hui, S.Y.R. Sliding-mode-based direct power control of dual-active-bridge DC-DC converters. In Proceedings of the IEEE Applied Power Electronics Conference and Exposition, Anaheim, CA, USA, 17–21 March 2019; pp. 188–192.

40. Yang, Y.; Tan, S.C.; Hui, S.Y.R. Mitigating distribution power loss of DC microgrids with DC electric spring. *IEEE Trans. Smart Grid* **2018**, *9*, 5897–5906. [CrossRef]

Article

Stability Criteria for Input Filter Design in Converters with CPL: Applications in Sliding Mode Controlled Power Systems

Jorge Luis Anderson Azzano * , Jerónimo J. Moré and Paul F. Puleston

Instituto LEICI, Facultad de Ingeniería UNLP—CONICET, B1900 La Plata, Argentina; jmore@ing.unlp.edu.ar (J.J.M.); puleston@ing.unlp.edu.ar (P.F.P.)
* Correspondence: anderson.jorgeluis@gmail.com

Received: 30 September 2019; Accepted: 21 October 2019; Published: 24 October 2019

Abstract: Microgrids are versatile systems for integration of renewable energy sources and non-conventional storage devices. Sliding Mode techniques grant excellent features of robustness controlling power conditioning systems, making them highly suitable for microgrid applications. However, problems may arise when a converter is set to behave as a Constant Power Load (CPL). These issues manifest in the stability of internal dynamics (or Zero Dynamics), which is determined by the input filter of the power module. In this paper, a special Lyapunov analysis is conducted to address the nonlinear internal dynamics of SM controlled power modules with CPL. It takes advantage of a Liérnad-type description, establishing stability conditions and providing a secure operation region. These conditions are translated into conductance and invariant region diagrams, turning them into tools for the design of power module filters.

Keywords: Lyapunov-based filter design; constant power load; Sliding Mode controlled power module; zero dynamics stability

---

1. Introduction

Microgrids can be succinctly understood as an interactive hybrid system that combines distributed generation modules, storage modules and loads. The number and nature of these modules, as well as the microgrid topologies, vary greatly [1–3]. Presently, the modules interconnection is predominantly done using AC buses, but DC microgrids are expected to increase in the coming years [4]. In particular, DC microgrids are very versatile for integration of renewable energy sources and non conventional storage devices, such as photovoltaic panels, wind turbines, lithium batteries, flow batteries, supercapacitors, fuel cells and many others (see Figure 1). The increasing development of power electronics and novel control techniques have favoured their implementation, making it possible to obtain higher efficiency and reliability. Furthermore, distribution power systems have been widespread in several applications due to their flexibility features and reduced size and cost.

In this growing scenario, where a wide range of new topologies is being developed, technological challenges are continuously arising, many of them at the level of the modules power converters and their related controllers. The nonlinear behaviour of power converters in these systems often makes conventional linear control techniques not sufficiently effective. In addition, the existence of uncertainties and system disturbances require the development of robust controllers to ensure system stability and efficiency. In this context, Sliding Mode (SM) control have been increasingly accepted for controlling power conditioning systems [5–9]. These control techniques proved to be particularly suitable for electronic converters, granting excellent features of robustness and finite-time convergence, which make them highly appropriate for microgrid applications [10–14].

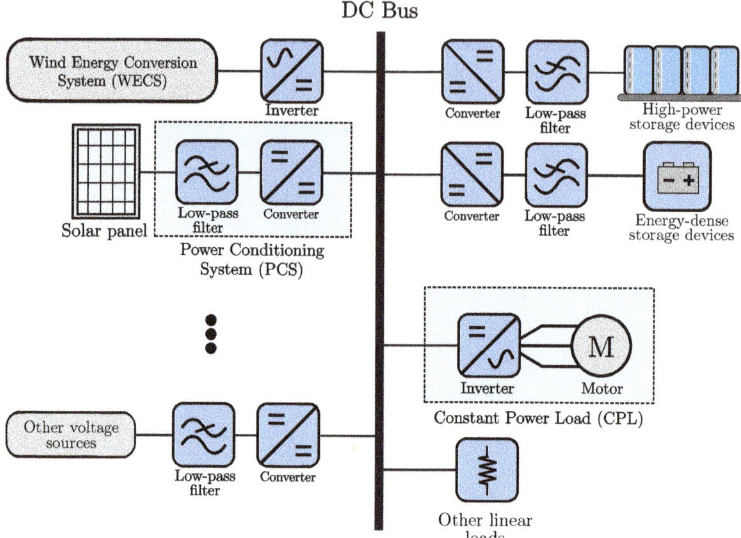

Figure 1. Illustrative representation of a DC microgrid combining renewable modules, non conventional storage modules and constant power load.

However, there are certain issues to be aware of at the moment of designing and implementing those power modules. Effectively, it has been extensively reported in the literature that, when a controlled power system tightly regulates energy supplied to a DC bus, the converter behaves as a Constant Power Load (CPL) for the upstream source [15–17]. Such CPL acts as a negative impedance, i.e., an increase in its terminal voltage results in a decrease in its current (and vice versa), which undermines the stability margins of the system.

In the case of SM controlled power systems, these issues are manifested in the stability of the Zero Dynamics (ZD). This internal dynamics is determined by the input filter of the power module, which exists to protect the source or the upstream device against switching harmonics propagation. If ZD is not carefully analysed and its stability is not thoroughly addressed during the design of the controlled power system, then performance and robustness could be severely compromised, causing undesired oscillations or even system failures.

Diverse criteria have been proposed for the analysis of input filters dynamics with CPLs in DC microgrids. The most widely found in the literature is the small-signal criterion by eigenvalue analysis through system linearization [17,18]. Also, several methods have been presented using Nyquist approaches, for instance, Gain margin and Phase margin criterion [19], Middlebrook's criterion [20] and the opposing argument criterion [21] among the most important ones. Moreover, in [22] a passivity-based criterion is proposed for a LC filter and in [23] stability conditions are found taking into account parameters variations through bifurcation analysis.

Those methods do not guarantee large-signal stability and give no precise information about the delimitation of a stable region for secure operation of the power module [21]. So as a contribution to this field, in this paper, a special Lyapunov analysis is conducted to address the nonlinear internal dynamics of SM controlled power modules with CPL. This study is performed taking advantage of a transformation to a Liérnad-type description, establishing sufficient conditions to define the stability of the nonlinear system and providing a secure operation region. These conditions are, subsequently, condensed into conductance and invariant region diagrams, which are proposed as versatile tools for the design of power module filters.

The paper is organised as follows. In Section 2, the proposed input filter design method for power converters with CPL is developed. It is divided into two main parts. The first one, Section 2.1, deals with the search of the stability conditions via Liénard-based nonlinear Lyapunov approach. Then, the second part, Section 2.2, translates those conditions into appropriate diagrams, based on which the filter design procedure will be grounded. In Section 3, simulation results are presented and analysed, considering as application case a SM controlled Boost convert with second-order filter. Finally, in Section 4, conclusions and future lines of research are discussed.

2. Development of the Proposed Input Filter Design Method for Power Converters with CPL

Two steps are required to develop the proposed method for input filter design. The first one is to obtain sufficient conditions for the zero dynamics (ZD) stability, over the full nonlinear operation range of SM controlled converters with CPL. Based on those conditions, the second step consists of elaborating a filter design criteria with the help of ad hoc conductance diagram description.

In Figure 2, a schematic diagram showing the topology of the power module under study is presented.

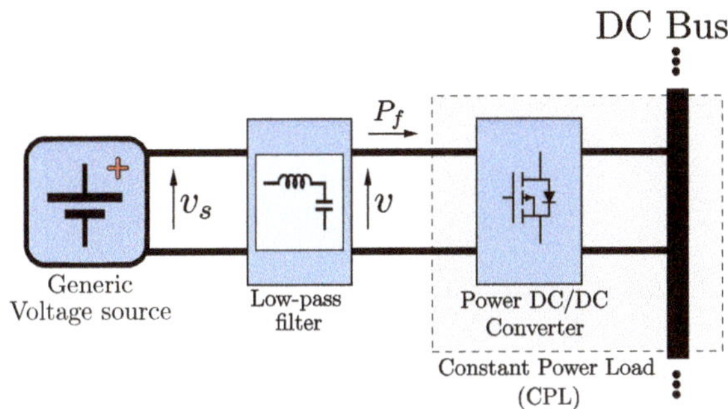

Figure 2. Schematic diagram of the Power Conditioning System (PCS) employed for the ZD analysis.

In the following subsections, both aforementioned steps are treated in detail.

2.1. Search of Stability Conditions via Liénard Based Nonlinear Lyapunov Approach

In this subsection, the ZD is thoroughly studied and sufficient conditions are obtained by mean of nonlinear Lyapunov analysis. For the sake of clarity, a path of progressive complexity is followed. Initially, a nonlinear Lyapunov approach is applied to a simple topology based on a first-order capacitive filter, and a sufficient condition for its stability is obtained. Then, the study is broadened to a widely-used second-order LC filter, describing the system in a Liénard form and analyzing its stability through a special energy-like Lyapunov function.

2.1.1. First-Order Capacitive Filter Topology

The electrical circuit of the capacitive filter supplying a constant power P_f is shown in Figure 3. Its electrical model is defined by the first order differential equation:

$$C_f \dot{v} = \frac{1}{R_s} V_{oc} - \frac{1}{R_s} v - \frac{P_f}{v}, \quad \forall v > 0 \tag{1}$$

where C_f is the filter capacitance, V_{oc} is the open circuit source voltage, R_s is the internal source resistance and v the voltage across the capacitor. The nonlinear term $\frac{P_f}{v}$, corresponding to the CPL current, acts as a negative impedance, when the voltage v is increased the filter output current is decreased. This negative impedance reduces the equivalent resistance of the system, leading to a reduction of the stability margin.

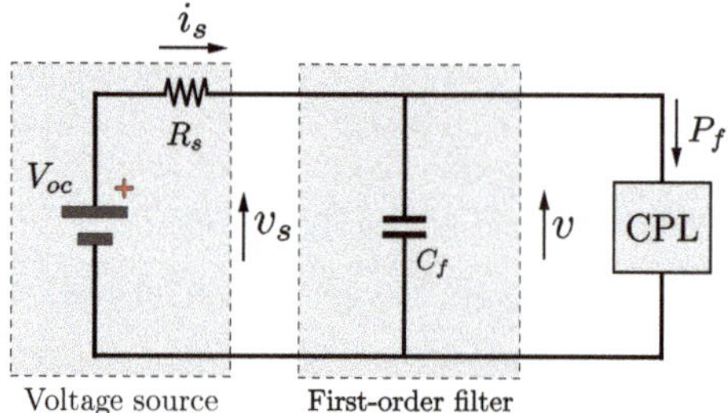

Figure 3. Scheme of the equivalent electrical circuit of the first-order input filter with CPL.

From (1), the well-known condition for the existence of equilibrium in CPL is [24]:

$$P_f \leq \frac{V_{oc}^2}{4R_s} = P_{fmax}.$$

(2)

which if strictly satisfied, results in two solutions. The first one,

$$v_0 = \frac{V_{oc}}{2} + \frac{1}{2}\sqrt{V_{oc}^2 - 4P_f R_s},$$

(3)

is the equilibrium capacitor voltage v_0, that defines the equilibrium point of operation of the power module. The second one,

$$v_{lim} = \frac{V_{oc}}{2} - \frac{1}{2}\sqrt{V_{oc}^2 - 4P_f R_s},$$

(4)

has no physical meaning, and represents a voltage limit for the system stability. Additionally, by multiplying (3) by (4), it is straightforward to obtain $v_{lim} = \dfrac{P_f R_s}{v_0}$ (the operation equilibrium, given by v_0, and the stability limit, determined by v_{lim}, can be visualised further in Figure 4, at the end of this subsection).

As previously discussed, traditional stability study of the system described by (1) has been addressed by linearization, determining the local stability near the equilibrium voltage v_0 through eigenvalue analysis. In accordance with this conventional analysis, if P_f satisfies:

$$P_f < \frac{v_0^2}{R_s}$$

(5)

then the eigenvalues of the linearised system around the equilibrium point have negative real part and so the trajectories of the system converge to the equilibrium voltage v_0, for close enough initial conditions.

It is well known that this linear analysis only ensures the stability for trajectories sufficiently close to v_0, but gives no information about the attraction region. In fact, it will be shown that a power system whose eigenvalues analysis results are stable, would in practice become unstable if voltage v drops below the stability limit voltage v_{lim}.

Effectively, through the nonlinear Lyapunov analysis presented in the sequel, it is possible to completely define the system stability, allowing establishing the actual range of v where secure operation is guaranteed.

To this end, in this simple first case study, the equilibrium point is translated to the origin through the state transformation:

$$z = v - v_0 \tag{6}$$

it leads to:

$$\dot{z} = -a(z) \tag{7}$$

with

$$a(z) = \frac{1}{C_f R_s} z + \frac{P_f}{C_f}\left(\frac{1}{z+v_0} - \frac{1}{v_0}\right). \tag{8}$$

Now, the proposed energy-like Lyapunov function

$$V(z) = \int_0^z a(\xi)d\xi \tag{9}$$

is locally positive-definite if condition

$$za(z) > 0 \tag{10}$$

is satisfied for some interval that includes the origin. Therefore, it follows that the first derivative of V is equal to:

$$\dot{V}(z) = -a(z)^2 \tag{11}$$

which is definite-negative for $a(z) \neq 0$.

The Lyapunov function, defined in (9), is shown in Figure 4, with $z_0 = 0$ and $z_{lim} = v_{lim} - v_0$ obtained from (3) and (4), respectively, in combination with (6). As can be appreciated, the local maximum and minimum of the function $V(z)$ are at z_{lim} and z_0 (corresponding to capacitor voltages $v = v_{lim}$ and $v = v_0$, respectively). These values of z are zeros of the function $\dot{V}(z)$, i.e., zeros of $a(z)$, and thus zeros of $\dot{z}$ (see (7)).

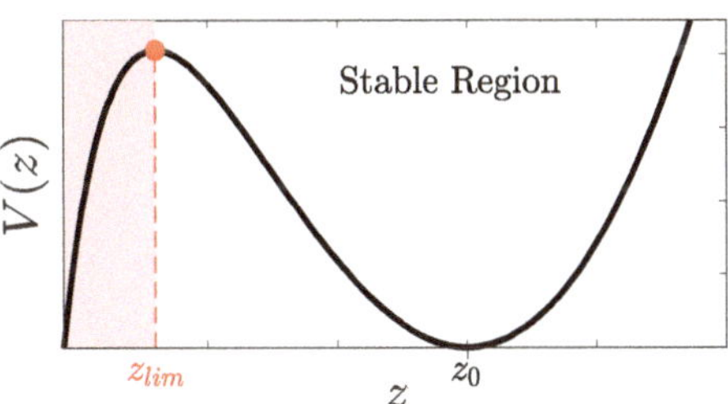

Figure 4. Energy-like Lyapunov function for first order filter study.

Stability condition in terms of conductances. It can straightforwardly obtained by computing inequality (10) as a function of the voltage v:

$$g_0(P_f, v) = \frac{P_f}{v_0}\frac{1}{v} < \frac{1}{R_s} = g_s \tag{12}$$

where g_0 is an equivalent output conductance viewed from the filter and g_s is the internal conductance of the voltage source.

Stable operation region $\mathcal{S}$ for first-order filter. It should be remarked that, unlike condition (5) obtained via eigenvalue linear analysis, the condition given by (12) defines the stable region of the first-order filter, fully determining the admissible range for the capacitor voltage v. It establishes that if the voltage v satisfies:

$$v > v_{lim} = \frac{P_f R_s}{v_0} \text{ with } P_f \leq P_{fmax} \tag{13}$$

all system trajectories converge to the equilibrium voltage v_0. Otherwise, the system trajectories become unstable.

2.1.2. Second-Order LC Filter Topology

Founded on the previous approach, in this subsection, the Zero Dynamics stability of the power module is studied for widespread second-order input filters. To this end, nonlinear Lyapunov analysis is therefore employed to obtain stability conditions in terms of equivalent conductances. Then, these conditions will help to determine an invariant region where secure operation of power modules will be completely guaranteed.

Regretfully, this is not as direct as it is for the first-order capacitive filter. To be able to perform the analysis, in this work, the ZD is firstly rewritten into a special Liénard-type form [25], by means of a linear transformation. This allows a suitable energy-like Lyapunov function capable of dealing with this nonlinear topology.

Figure 5 shows a schematic model of the second-order LC filter connected to the converter acting as a constant power load.

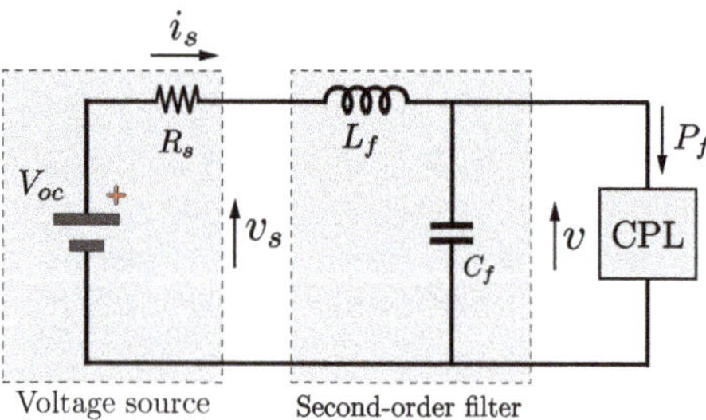

Figure 5. Scheme of the equivalent electrical circuit of the second-order input filter with CPL.

The system is described by:

$$\begin{cases} L_f \cdot \dot{i}_s &= -v - R_s \cdot i_s + V_{oc} \\ C_f \cdot \dot{v} &= i_s - \dfrac{P_f}{v}. \end{cases} \tag{14}$$

The above system presents an operation equilibrium point, defined by both the equilibrium source current i_{s0} and (if (2) holds) the equilibrium filter voltage v_0:

$$i_{s0} = \frac{P_f}{v_0} \tag{15}$$

$$v_0 = \frac{V_{oc}}{2} + \frac{1}{2}\sqrt{V_{oc}^2 - 4P_f R_s} \tag{16}$$

where the latter is the solution with physical meaning for the power system operation, whereas there exists a second solution

$$v_{lim} = \frac{V_{oc}}{2} - \frac{1}{2}\sqrt{V_{oc}^2 - 4P_f R_s} = \frac{P_f R_s}{v_0} \tag{17}$$

which again represents a voltage limit for the system stability.

Analogously to the previous case, the equilibrium point is shifted to the origin by redefining the system states as:

$$x_1 = i_s - i_{s0} \tag{18}$$

$$x_2 = v - v_0, \tag{19}$$

leading to the transformed system:

$$\begin{cases} \dot{x}_1 & = -\dfrac{R_s}{L_f}x_1 - \dfrac{1}{L_f}x_2 \\[2mm] \dot{x}_2 & = \dfrac{1}{C_f}x_1 + \dfrac{P_f}{C_f}\left(\dfrac{1}{x_2 + v_0} - \dfrac{1}{v_0}\right). \end{cases} \tag{20}$$

To obtain a ZD representation convenient for the Lyapunov analysis, the following state linear transformation is proposed:

$$z_1 = \frac{1}{C_f}x_1 + \frac{R_s}{L_f}x_2 \tag{21}$$

$$z_2 = x_2, \tag{22}$$

and consequently, the transformed state system acquires the Liénard-type form [25]:

$$\dot{z}_1 = -a(z_2) \tag{23}$$

$$\dot{z}_2 = z_1 - b(z_2) \tag{24}$$

with

$$a(z_2) = \frac{1}{L_f C_f}\cdot z_2 + \frac{R_s P_f}{L_f C_f}\left(\frac{1}{z_2 + v_0} - \frac{1}{v_0}\right) \tag{25}$$

$$b(z_2) = \frac{R_s}{L_f}\cdot z_2 + \frac{P_f}{C_f}\left(\frac{1}{z_2 + v_0} - \frac{1}{v_0}\right). \tag{26}$$

Now, it can be proposed the appropriate energy-like Lyapunov function, depending on new states z_1 and z_2:

$$V(z_1, z_2) = \frac{1}{2}\cdot z_1^2 + \int_0^{z_2} a(\zeta)d\zeta \tag{27}$$

with $V(0,0) = 0$. Then, similar to condition (10), if z_2 satisfies:

$$z_2 a(z_2) > 0 \tag{28}$$

the integral term will be positive, and in consequence, $V(z_1, z_2)$ will be positive-definite for some region around the origin.

Next, the first derivative of V is straightly computed as:

$$\dot{V}(z_1, z_2) = z_1 \dot{z}_1 + a(z_2)\dot{z}_2 = -z_1 a(z_2) + z_1 a(z_2) - a(z_2)b(z_2) \tag{29}$$

$$\implies \dot{V}(z_1, z_2) = -a(z_2)b(z_2) \tag{30}$$

which means that $\dot{V}$ will be locally negative-definite if it satisfies $a(z_2)b(z_2) > 0$ for some interval around the origin. Therefore, taking into account condition (28), it leads to:

$$a(z_2)b(z_2) > 0 \implies a(z_2)^2 \cdot z_2 \cdot b(z_2) > 0$$

$$\Longleftrightarrow z_2 \cdot b(z_2) > 0. \tag{31}$$

Hereunder, the Lyapunov conditions (28) and (31) will be rewritten in terms of equivalent conductances, preparing the ground for the completion of the proposed design method in the following subsection.

First Lyapunov condition in terms of conductances. It is obtained combining condition (28) with (25), giving:

$$\frac{1}{L_f C_f} z_2^2 \left(\frac{1}{R_s} - \frac{P_f}{(z_2 + v_0)v_0} \right) > 0 \tag{32}$$

$$\implies g_0(P_f, v) = \frac{P_f}{vv_0} < \frac{1}{R_s} = g_s \tag{33}$$

where similarly to (12), g_0 is the equivalent output conductance viewed from the filter and g_s is the internal conductance of the source.

Second Lyapunov condition in terms of conductances. It is computed from condition (26) and (31), resulting in:

$$z_2^2 \left(\frac{R_s}{L_f} - \frac{1}{C_f} \frac{P_f}{(z_2 + v_0)v_0} \right) > 0 \tag{34}$$

$$\implies g_0(P_f, v) = \frac{P_f}{vv_0} < \frac{R_s C_f}{L_f} = g_{lc} \tag{35}$$

where the new equivalent conductance g_{lc} is called filter conductance.

Furthermore, both conditions can be summarised in one inequality as:

$$g_0(P_f, v) < \min(g_s; g_{lc}) \tag{36}$$

or, similarly to the first-order filter, in terms of voltage v, as:

$$v > v_{min} = \max\left(v_{lim} = \frac{P_f R_s}{v_0}; \frac{P_f L_f}{R_s C_f v_0} \right) \text{ with } P_f \leq P_{fmax} \tag{37}$$

However, unlike the first-order case, the above condition (alternatively (36) or (33)–(35)) is not sufficient to ensure the ZD stable behaviour, because not all trajectories will necessarily converge to (i_{s0}, v_0) for any initial conditions $v > v_{min}$. In fact, with a second-order filter, the stable region also depends on the inductor current i_s.

Nevertheless, (37) does guarantee the existence of an invariant region around the equilibrium point, within which all system trajectories would converge to the desired equilibrium. Moreover, in contrast to the eigenvalues approach, following the subsequent analysis this stable region can be fully determined and eventually used as a helping tool for the filter design.

Stable operation region S for second-order filter. Condition (37) is therefore used to define such invariant region in the ZD state plane, where performance and secure operation of the SM controlled power system can be ensured.

For better interpretation, Figure 6 schematically depicts the Lyapunov function $V(z_1, z_2)$, including both equilibrium solutions, these are the the desired equilibrium point of operation, $p_0 = (i_{s0}, v_0)$ in green, and the limit point $p_{lim} = (i_{s0}, v_{lim})$ in red. The rationale behind the proposed procedure is quite simple. A basin of attraction targeting to p_0 must be devised by intersecting $V(z_1, z_2)$ with an appropriate horizontal plane, $Plane_{invar}$.

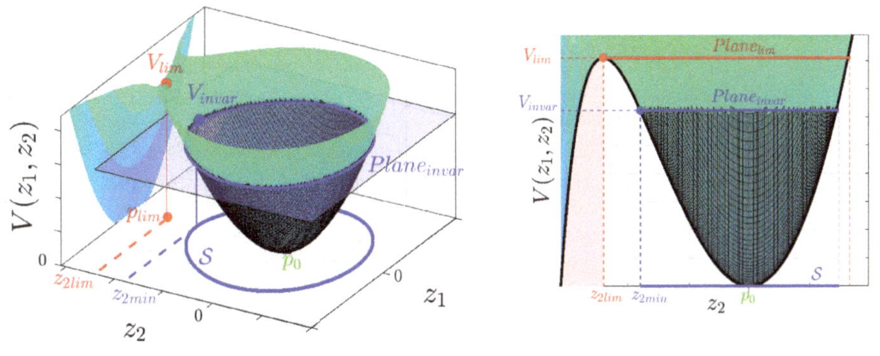

(a) Lyapunov function $V(z_1, z_2)$.

(b) Lyapunov function $V(z_1, z_2)$. Lateral view.

Figure 6. Illustration of the invariant region obtained through Lyapunov stability conditions.

It can be easily proved that the limit point p_{lim} is an unstable equilibrium, hence it defines a plane, $Plane_{lim}$ (given by $V(z_1, z_2) = V_{lim}$, with constant $V_{lim} = V(p_{lim})$), which naturally is an upper bound for $Plane_{invar}$. Moreover, $Plane_{lim}$ delimited a basin-shaped, yet not necessarily attractive, Lyapunov function.

Therefore, to ensure attractiveness to the desired equilibrium, the lower (or equal) $Plane_{invar}$ should be obtained such that, below it, Lyapunov conditions hold (or, equivalently, condition (37) holds). Then, the proposed invariant region (in blue solid line) is delimited by the intersection of $V(z_1, z_2)$ with $Plane_{invar}$, defined by $V(z_1, z_2) = V_{invar}$ (see Figure 6a). The constant value V_{invar} will be computed below using v_{min}, such that $\dot{V}$ results negative for all $(z_1, z_2) : V(z_1, z_2) < V_{invar}$, thus the basin of attraction is attained and all ZD trajectories which start below $Plane_{invar}$ converge to p_0. Therefore, the stable invariant region can be formally defined

$$\forall (z_1, z_2) \in S : V(z_1, z_2) < V_{invar} \tag{38}$$

where the constant V_{invar} is computed according to (27):

$$V_{invar} = \min \left(V(z_1, z_{2,min}) \right) = \min \left(\frac{1}{2} z_1^2 + \int_0^{z_{2,min}} a(\xi) d\xi \right) \tag{39}$$

$$\implies V_{invar} = \int_0^{z_{2,min}} a(\xi) d\xi \tag{40}$$

with

$$z_{2,min} = v_{min} - v_0. \tag{41}$$

Please note that the way $v_{min} = \max\left(v_{lim}; \dfrac{P_f L_f}{R_s C_f v_0}\right)$ is defined in (37), effectively implies that $Plane_{lim}$ is the upper limit plane for the $Plane_{invar}$.

2.2. Proposed Filter Design Criteria through Conductance Diagram Description

The second step to conclude the method for input filter design is developed in this subsection. It is based on translating the previously obtained stability conditions into appropriate conductance diagrams. These diagrams, complemented with those of the stable operation region of the power module, are then proposed as practical tools for systematic design.

2.2.1. Design Criterion for the First-Order Input Filter

The case of the power module with a first-order input filter is relatively simple and can be straightforwardly designed with the assistance of condition (12) (or, equivalently, (13)). Therefore, the main contribution of this case to the paper is not the design itself, but the introduction and interpretation of the conductance diagrams, familiarising the reader with their use in the framework of a filter design procedure, aiming to the more interesting second-order topology to come.

For the case under study, the condition arrived in (12) provides a relation between the output conductance g_0 and the source conductance g_s for the stability of the ZD system described in (1). This relation can be compared for different values of the output power P_f in a conductance diagram, where a stable operation region can be established.

Figure 7 presents the conductance diagram for a voltage source with open-circuit voltage $V_{oc} = 24$ V and internal resistance $R_s = 0.144\ \Omega$. The diagram is constructed using the conditions (12)–(13), obtained from the procedure described in Section 2.1.1. Each hyperbola (dashed black lines) represents an output conductance g_0 for a specific value of power P_f. For the sake of illustration, the conductance g_0 for a given power $P_f = P_{fnom}$ is highlighted in thick black line. The upper limit curve corresponds to the maximum allowed power P_{fmax}, computed through (2).

In addition, the conductance g_0 at each equilibrium point (green line) is obtained as:

$$g_0(P_f, v_0) = \frac{P_f}{v_0^2}. \tag{42}$$

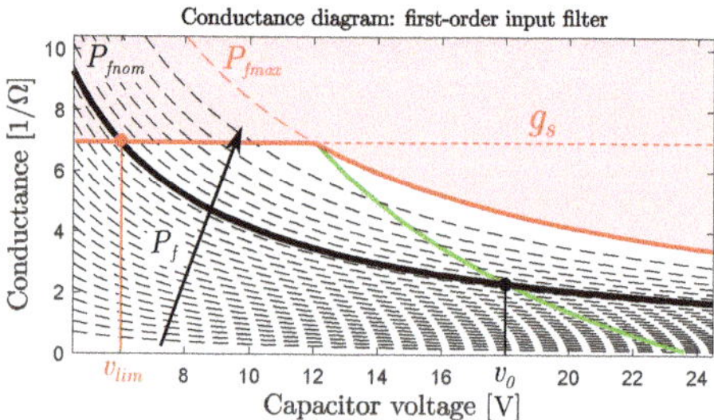

Figure 7. First-order filter conductance diagram for variable filter power P_f.

The intersection between the source conductance g_s (horizontal red line) and each g_0 defines the limit voltage v_{lim}, which establishes the secure operation of the power system (the locus constituted by these points for different P_f corresponds to the solid red segment). Then, the region of stable operation is delimited by the boundary conformed by the latter and the maximum allowed power, depicted in solid red line. All trajectories that start inside this region (white background in the figure) will converge to the equilibrium voltage v_0 (green line). Otherwise, they will become unstable.

Please note that for the first order filter case the conductance diagram is defined by the source parameters R_s and V_s, thus the stability region will not be altered by the selection of the filter capacitance C_f.

Therefore, from the above, the design procedure for this case study is direct. Firstly, knowing the desired cut-off frequency f_c for the filter and R_s, capacitance C_f is computed.

Then, from the stability region of the conductance diagram, for a desired power $P_f = P_{fnom}$ specified in the design, the admissible range of v can be obtained. Alternatively, for a known admissible range of variation of the capacitor voltage v, an upper limit for the P_{fnom} of the power module can be set. It can be appreciated in the conductance diagram that exists a trade-off between the selection of the operation power P_{fnom} and the admissible variation range of v.

2.2.2. Design Criterion for Second-Order Input Filter

Building on the previous approach, in this subsection the design criterion for more prevalent second-order filter in converters with CPL is presented. As it was mentioned in Section 2.1.2, unlike the first-order case, the condition $v > v_{min}$ given by (37) is necessary but not sufficient for the design. Therefore, the conductance diagram remains as a helpful design tool to determine and easily visualise the viable stable operation points. However, to ensure a design that provides wholly secure operation of the power module, it must be complemented with an invariant operation region diagram. This invariant region is constructed through the proposed procedure presented in Section 2.1.2 by means of (38)–(41). The value of the constant V_{invar}, which defines the stable region, is computed from (39), using diverse values of filter power P_f and filter parameters.

To facilitate the understanding of those diagrams for this topology, a descriptive analysis is presented below, together with illustrative diagrams of conductance and of the invariant region for different values of P_f, assuming fixed cut-off frequency f_c and filter capacitor C_f.

As with the first-order case, the conductance diagram shown in Figure 8a is constructed through the set of Equations (32)–(37). However, now, the extra condition defined by (35), which strongly depends on the filter parameters, also needs to be considered.

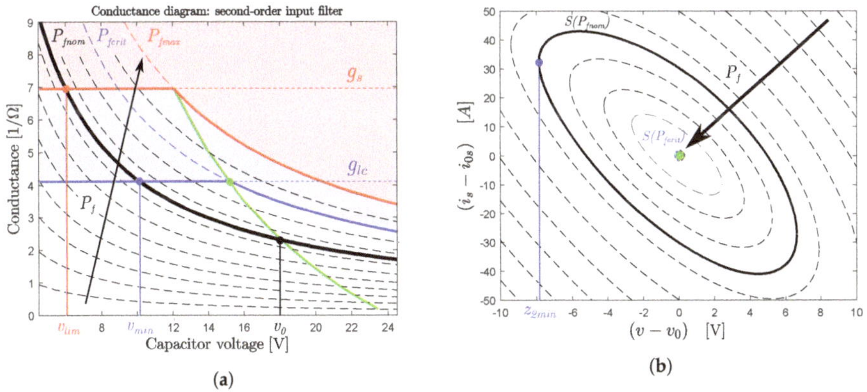

Figure 8. llustrative diagrams (**a**) of conductance and (**b**) of invariant region for second-order filter with variable power P_f and fixed f_c and C_f.

As in the previous analysis, the green line depicts the equilibrium locus, constituted by the conductance g_0 at each equilibrium point $g_0(P_f, v_0) = \dfrac{P_f}{v_0^2}$, for different P_f. However, in this case, the red solid line defined by the source conductance g_s and P_{fmax} is not a stability boundary. In fact, the conductance condition in this topology, set by inequality (36), depends not only on g_s, but also on the equivalent conductance g_{lc} (dashed blue line). So, a more restrictive boundary (in thick solid blue line) arises, conformed by $min(g_s; g_{lc})$ and a critical power P_{fcrit}. The latter is an upper bound for P_f, beyond which the equilibrium points of the green locus are unstable. It can be computed by

$$P_{fcrit} = \frac{v_0^2 R_s C_f}{L_f}. \tag{43}$$

Please note that the white-background region delimited by the blue boundary, does define the stable equilibria, hence it allows the visualisation of the admissible values of P_f that can be selected as operation filter power.

Nevertheless, it does not provide per se a region that secures stable operation, as it was for the first-order filter. Thus, to take into account in the design procedure the admissible variation ranges of v and i_s, the assistance of the invariant region diagram is required. In Figure 8b invariant regions $\mathcal{S}$ are plotted, parametrised for different values of filter power P_f, with their boundaries computed as $V(z_1, z_2) = V_{invar}$ in accordance with (38)–(41).

The admissible ranges of v and i_s strongly depend on the selected operation power P_f. Note the existing trade-off for the design. The higher the selected P_f, getting closer to the P_{fcrit}, the lower the invariant region, reducing the stability tolerance to admissible variations of v and i_s with respect to the nominal operation point.

In this framework, a sequential design procedure will be succinctly presented below, as an illustrative example of application.

- Firstly, the desired cut-off frequency f_c is chosen to protect the source against switching harmonics propagation. It should be carefully selected, considering that a small value of f_c could lead to large components sizing, which would imply a higher cost and weight. On the other hand, there is an upper bound for f_c given by the switching converter frequency, whose practical higher limit is typically selected one decade below such switching frequency. In this case study, a f_c value equals to 1 kHz is taken.
- Next, the operation power P_f is selected, in accordance with the power module requirements. In this example, the second-order input filter is designed for supplying an electrical power up to $P_f = 750$ W (solid black line in Figure 9a).
- The goal of the following step is to obtain the range of selectable values for capacitor C_f. These values are those which guarantee stable operation points for the f_c and P_f previously chosen. This step is easily fulfilled using the conductance diagram in Figure 9a, where g_{lc} is plotted for different values of capacitor C_f (blue horizontal lines).

It can be observed in Figure 9a that there exists a minimum value of capacitor C_f, equals to 0.65 mF, for which the equilibrium point of $P_f = 750$ W (black dot) would become unstable. So, designing C_f greater than this so-called C_{fmin} ensures that the stable equilibria region (white background zone) will contain the desired equilibrium point.

Therefore, from the conductance diagram, the range of selectable values for the filter capacitor can be set as $C_f > 0.65$ mF $= C_{fmin}$. If required, this value can be analytically computed through (35), as a function of the cut-off frequency:

$$C_{fmin} = \frac{1}{2\pi f_c} \frac{1}{v_0} \sqrt{\frac{P_f}{R_s}}. \tag{44}$$

- Then, from those admissible values of C_f, one should be chosen, according to the required variation ranges of v and i_s, which are contemplated in the specifications of the power module. This design step is conducted with the help of the invariant region diagram in Figure 9b, parametrised in terms of C_f.

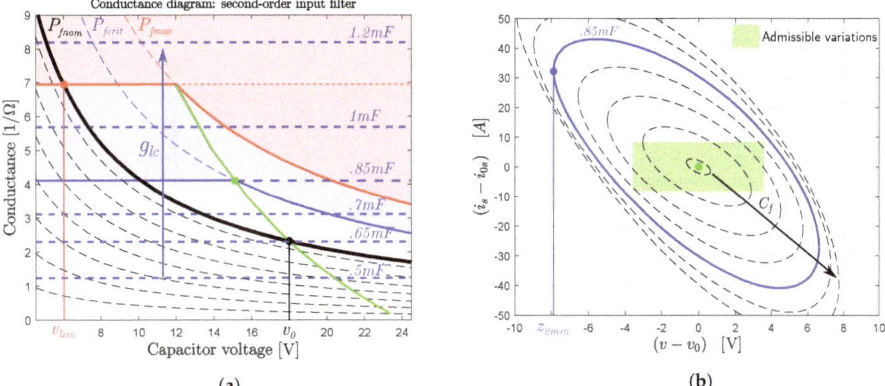

Figure 9. Design example: diagrams (**a**) of conductance and (**b**) of invariant region for second-order filter with variable filter capacitor C_f and fixed $f_c = 1$ kHz and $P_f = 750$ W.

To illustrate the selection procedure, variations up to $\pm 20\%$ for both voltage v and current i_s, are considered admissible for the normal operation of the power module (represented by the green area Figure 9b).

Now, the proposed design criterion is simple. The smallest C_f should be selected, such that its corresponding stable operation invariant region includes the admissible variation green area. Therefore, for this case study, $C_f = 850$ µF is chosen (see curve in blue line).

Please note that there is a trade-off between the capacitor sizing and the allowed voltage v and current i_s variations. In this particular case, all capacitances larger than 850 µF would provide secure stable operation to the power module.

- Finally, once the required C_f was selected and knowing the desired f_c, the inductance L_f can be easily obtained from:

$$L_f = \left(\frac{1}{2\pi f_c}\right)^2 \cdot \frac{1}{C_f},\tag{45}$$

which for this illustrative example, results $L_f = 30$ µH.

To conclude, it should be remarked that the previous procedure is just one of diverse possible ways to perform the filter design, which may differ depending on each particular case, available data and specifications.

3. Application Case: SM Controlled Boost Converter with Second-Order Filter

In this section, following the above proposed criteria, the ZD stability of a second-order input filter for a SM controlled boost converter-based power module is evaluated. An illustrative electrical diagram of the considered system is presented in Figure 10. The power module is intended to operate as a part of a microgrid, as the one presented in Figure 1, connected to a common DC Bus, in this case of 48 V. In addition, the power module is fed by the power source $v_s(i_s)$, characterized by its open circuit voltage and internal resistance, which may represent different kinds of sources (i.e., Li-Ion batteries, Flow batteries, Fuel Cells, Supercapacitors, etc.).

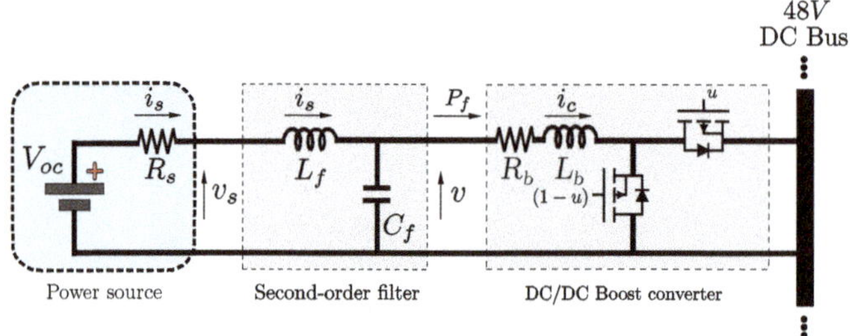

Figure 10. Electrical diagram of the PCS based on boost converter with second-order input filter.

In this particular case study, the power converter is assumed to be controlled by means of a Second Order Sliding Mode (SOSM) algorithm, designed to regulate a constant power to the DC bus (see details in Appendix A).

Under this CPL assumption, the filter is design following the criterion explained in Section 2.2.2. Then, in the sequel, to validate the proposed design method, the power system is evaluated under different initial conditions and filter power P_f.

Recapitulating, a cut-off frequency f_c = 1 kHz and an operation filter power P_f = 750 W were selected. Next, the proposed design provided a filter capacitor of C_f = 850 μF, considering admissible voltage and current variations up to ±20%. Lastly, L_f was computed using (45), giving L_f = 30 μH.

The complete set of system parameters employed for the subsequent simulations results are displayed in Table 1. It is important to remark that with this set of parameters, the resulting critical power P_{fcrit} is of about 930 W. If the system exceeds this power value, the equilibrium points become unstable (see Figure 9a).

Table 1. Parameters of the simulated power module.

System Parameters				STA Parameters	
V_{oc}	24 V	f_c	1 kHz	β	200
R_s	0.144 Ω	C_f	0.85 mF	α	5×10^{-3}
P_f	750 W	L_f	30 μH		
L_b	50 μH	R_b	1 mΩ		
f_s	50 kHz	V_{dc}	48 V		

Then, to evaluate the system behaviour, several in-silico test were conducted. In the first place, the system is forced to operate under different values of constant power, until the critical power P_{fcrit} is reached. This results are shown in Figure 11.

Figure 11a presents the filter power evolution for such stepped power reference. It can be seen that the delivered power results equal to the reference, until the critical power P_{fcrit} is surpassed (~0.75 s). At this moment, the ZD becomes unstable and the filter power collapses to zero.

Figure 11b depicts the corresponding filter voltage and current. Please note that as the the filter power increases, the v and i_s variations during transients become bigger. Once the critical power is surpassed, the voltage drops to zero and the current saturates at its maximum ($i_s = V_{oc}/R_s$).

The following set of tests aims to show that region $\mathcal{S}$, obtained via the proposed design, is an invariant region and that the inside trajectories converge to the desired equilibrium point. To this end, Figure 12 displays the trajectories in the $i_s - v$ plane for several initial conditions (red dots) placed near the boundary of $\mathcal{S}$ (750 W) (blue curve). Effectively, as was previously discussed, it can be appreciated that this region ensures ZD stability of the SM controlled module, guaranteeing its secure operation.

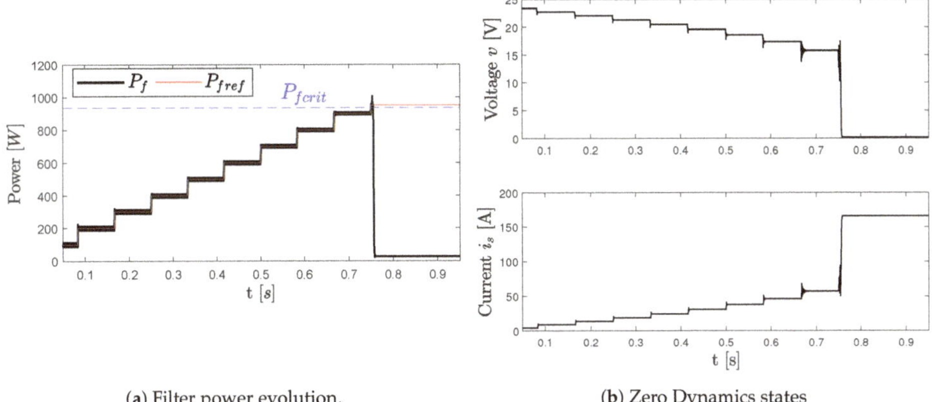

(**a**) Filter power evolution.   (**b**) Zero Dynamics states

Figure 11. Time response of the controlled power module for multiple reference steps.

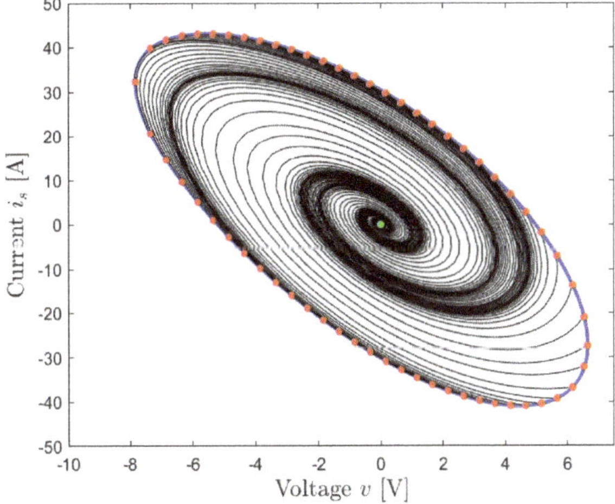

Figure 12. $i_s - v$ plane for stable trajectories with initial conditions inside invariant region $\mathcal{S}(750\text{ W})$.

The final test presented in Figure 13 is intended to illustrate the unstable behaviour for an initial condition outside the invariant region $\mathcal{S}(750\text{ W})$. The latter is plotted in blue line in the $i_s - v$ plane (top left figure in Figure 13). It can be observed in this example, as the ZD stability is not guaranteed, the system trajectory is not converging to equilibrium. It is important to stress that not all possible initial conditions outside this invariant region $\mathcal{S}$ result in unstable trajectories.

It is worth noticing that the SOSM-STA manages to reach the power reference of 750 W only for about 0.5 ms (see top right figure in Figure 13. Power reference in red line). As the ZD under this conditions results unstable, the controller is not able to remain over the sliding surface for a long time. Hence, the filter voltage v drops to zero while the filter current saturates at its maximum ($i_s = V_{oc}/R_s$), leading to a system failure.

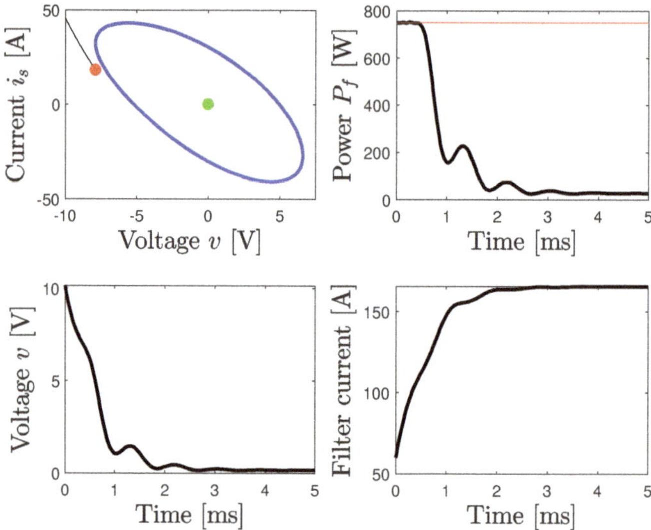

Figure 13. Unstable behaviour for initial condition outside invariant region $\mathcal{S}$ (750 W).

4. Conclusions

This paper proposed design criteria for input filters operating in DC microgrid applications with CPL. The presented methods were aimed to ensure the stable operation of power modules controlled by Sliding Mode Control techniques. To this end, conductance and invariant region diagrams were proposed as practical tools for filter design. The use of these diagrams represents a handy visual assistance for the selection of the filter elements.

To develop this procedure, firstly, the Zero Dynamics of the SM controlled module were brought into a particular Liérnad-type form, through a special transformation. Then, stability conditions, in term of equivalent conductances, were obtained using a suitable energy-like Lyapunov function capable to deal with the ZD nonlinear structure. Finally, the proposed conductance diagram and invariant region were presented, allowing establishing a sequential filter design procedure.

With the purpose of validating the proposed methodology, a SM controlled Boost converter with a second-order filter was assessed by simulation. The output power of the module was assumed regulated by means of a Second Order Sliding Mode Super-Twisting Control. The module tests consisted on leading the system to different power condition until reaching critical power condition, where the system becomes unstable. Also, it was proved that, for initial conditions outside the proposed stable region, the operation of the SM controlled system was not guaranteed.

Future studies will be focused on extending the proposed methodology to nonlinear input power source. In addition, in the next stage of this research, to evaluate the design criteria against real unmodelled dynamics and perturbation, the implementation of experimental tests will be undertaken.

Author Contributions: Conceptualization, P.F.P.; formal analysis, J.L.A.A. and J.J.M.; investigation, J.L.A.A., J.J.M. and P.F.P.; methodology, J.L.A.A. and P.F.P.; project administration, P.F.P.; supervision, P.F.P.; validation, J.L.A.A. and J.J.M.

Funding: This research was funded by ANPCyT PICT N2015-2257 "Control, Electrónica e Instrumentación: Aplicaciones en Energías Alternativas y Bioingeniería", LEICI, Fac. Ingeniería, UNLP. 2016–2019; CONICET PIP 112-2015-0100496CO "Electrónica de potencia y sistemas de control avanzado aplicados a fuentes de energía no convencionales", LEICI, Fac. Ingeniería, UNLP. 2016–2019 and UNLP Proyect 11/I217 "Electrónica de potencia y sistemas de control avanzado aplicados a fuentes de energía no convencionales", LEICI, Fac. Ingeniería, UNLP. 2016–2019.

Acknowledgments: This research was supported by the Universidad Nacional de La Plata (UNLP); the Fac. Ingeniería, UNLP; the Consejo Nacional de Investigaciones Científicas y Técnicas (CONICET) and the Agencia Nacional de Promoción Científica y Tecnológica (ANPCyT) from Argentina.

Conflicts of Interest: The authors declare no conflict of interest. The funders had no role in the design of the study; in the collection, analyses, or interpretation of data; in the writing of the manuscript, or in the decision to publish the results.

Appendix A. Second Order Sliding Mode Power Control

As indicated in Section 3, a SOSM controller is used in this paper, although First Order Sliding Mode (FOSM) controllers could also be considered. In particular, SOSM algorithms have been widely accepted to control power system, due to their robust features while providing chattering amelioration [26]. It is not the objective of this paper to describe the controller design in depth, but an outline is provided in this Appendix.

A well-known dynamical averaged model of the power module based on boost converter with second-order low-pass filter (see Figure 10) can be described by:

$$\dot{x} = f(x) + h(x)u \quad \text{with} \quad x = \begin{bmatrix} i_s & v & i_c \end{bmatrix}' \tag{A1}$$

where

$$f(x) = \begin{bmatrix} -\dfrac{1}{L_f}v - \dfrac{R_s i_s}{L_f} + \dfrac{V_{oc}}{L_f} \\ \dfrac{1}{C_f}i_s - \dfrac{1}{C_f}i_c \\ \dfrac{1}{L_b}v - \dfrac{R_b}{L_b}i_c \end{bmatrix}, \quad h(x) = \begin{bmatrix} 0 \\ 0 \\ -\dfrac{V_{dc}}{L_b} \end{bmatrix}. \tag{A2}$$

The power module controller is designed to regulate the delivered electrical power P_f to the DC bus.

To developed the SM controller, firstly, the control objective must be defined through the sliding variable σ as:

$$\sigma(x) = vi_c - P_{fref}. \tag{A3}$$

Then, the SOSM control objective will be accomplished when system trajectories reach and remain on the second-order sliding surface $\sigma = \dot{\sigma} = 0$.

Please note that from (A3), the sliding variable σ has relative degree one with respect to the duty cycle u. Then, a Super-Twisting Algorithm (STA) is employed, because of its direct application to relative degree one systems.

The SOSM-STA controller structure is defined as:

$$u = -\alpha \cdot |\sigma(x)|^{1/2} \cdot \text{sign}(\sigma(x)) + \omega \tag{A4}$$
$$\dot{\omega} = -\beta \cdot \text{sign}(\sigma(x)) \tag{A5}$$

where α and β are the STA gains. They are computed from

$$\ddot{\sigma}(x,u) = \varphi(x,u) + \gamma(x) \cdot u \tag{A6}$$

by bounding functions $\varphi(x,u)$ and $\gamma(x)$ (see [26,27] for more details). Then, once the SOSM controlled system reaches $\sigma = 0$, the resulting ZD will be the one described in Section 2.1.2.

References

1. Dragičević, T.; Lu, X.; Vasquez, J.C.; Guerrero, J.M. DC Microgrids—Part II: A Review of Power Architectures, Applications, and Standardization Issues. *IEEE Trans. Power Electron.* **2016**, *31*, 3528–3549, doi:10.1109/TPEL.2015.2464277. [CrossRef]

2. Waqar, A.; Shahbaz Tanveer, M.; Ahmad, J.; Aamir, M.; Yaqoob, M.; Anwar, F. Multi-Objective Analysis of a CHP Plant Integrated Microgrid in Pakistan. *Energies* **2017**, *10*, 1625, doi:10.3390/en10101625. [CrossRef]

3. Gabriel Rullo, P.; Costa-Castelló, R.; Roda, V.; Feroldi, D. Energy Management Strategy for a Bioethanol Isolated Hybrid System: Simulations and Experiments. *Energies* **2018**, *11*, 1362, doi:10.3390/en11061362. [CrossRef]

4. Werth, A.; Kitamura, N.; Matsumoto, I.; Tanaka, K. Evaluation of centralized and distributed microgrid topologies and comparison to Open Energy Systems (OES). In Proceedings of the 2015 IEEE 15th International Conference on Environment and Electrical Engineering (EEEIC), Rome, Italy, 10–13 June 2015; pp. 492–497, doi:10.1109/EEEIC.2015.7165211. [CrossRef]

5. Li, S.; Yu, X.; Fridman, L.; Man, Z.; Wang, X. *Advances in Variable Structure Systems and Sliding Mode Control—Theory and Applications*; Springer: New York, NY, USA, 2018; Volume 115.

6. Fridman, L.; Barbot, J.P.; Plestan, F. *Recent Trends in Sliding Mode Control*; IET: London, UK, 2016.

7. Bandyopadhyay, B.; Sivaramakrishnan, J.; Spurgeon, S. *Advances in Sliding Mode Control. Concept, Theory and Implementation*; Springer: New York, NY, USA, 2013; Volume 440.

8. Fridman, L.; Moreno, J.; Iriarte, R. *Sliding Modes after the first Decade of the 21st Century*; Springer: New York, NY, USA, 2011.

9. Bartolini, G.; Fridman, L.; Pisano, A.; Usai, E. *Modern Sliding Mode Control Theory. New Perspectives and Applications*; Springer: New York, NY, USA, 2008; Volume 375.

10. Martínez-Treviño, B.; Jammes, R.; Aroudi, A.; Martinez-Salamero, L. Sliding-mode control of a boost converter supplying a constant power load. *IFAC-PapersOnLine* **2017**, *50*, 7807–7812, doi:10.1016/j.ifacol.2017.08.1055. [CrossRef]

11. Yasin, A.R.; Ashraf, M.; Bhatti, A.I. Fixed Frequency Sliding Mode Control of Power Converters for Improved Dynamic Response in DC Micro-Grids. *Energies* **2018**, *11*, 2799, doi:10.3390/en11102799. [CrossRef]

12. Ramos-Paja, C.A.; Bastidas-Rodríguez, J.D.; González, D.; Acevedo, S.; Peláez-Restrepo, J. Design and Control of a Buck–Boost Charger-Discharger for DC-Bus Regulation in Microgrids. *Energies* **2017**, *10*, 1847, doi:10.3390/en10111847. [CrossRef]

13. Serna-Garcés, S.I.; Gonzalez Montoya, D.; Ramos-Paja, C.A. Sliding-Mode Control of a Charger/Discharger DC/DC Converter for DC-Bus Regulation in Renewable Power Systems. *Energies* **2016**, *9*, 245, doi:10.3390/en9040245. [CrossRef]

14. Su, X.; Han, M.; Guerrero, J.M.; Sun, H. Microgrid Stability Controller Based on Adaptive Robust Total SMC. *Energies* **2015**, *8*, 1784–1801, doi:10.3390/en8031784. [CrossRef]

15. Singh, S.; Gautam, A.; Fulwani, D. Constant power loads and their effects in DC distributed power systems: A review. *Renew. Sustain. Energy Rev.* **2017**, *72*, 407–421, doi:10.1016/j.rser.2017.01.027. [CrossRef]

16. AL-Nussairi, M.K.; Bayindir, R.; Padmanaban, S.; Mihet-Popa, L.; Siano, P. Constant Power Loads (CPL) with Microgrids: Problem Definition, Stability Analysis and Compensation Techniques. *Energies* **2017**, *10*, 1656, doi:10.3390/en10101656. [CrossRef]

17. Huangfu, Y.; Pang, S.; Nahid-Mobarakeh, B.; Guo, L.; Rathore, A.K.; Gao, F. Stability Analysis and Active Stabilization of On-board DC Power Converter System with Input Filter. *IEEE Trans. Ind. Electron.* **2018**, *65*, 790–799, doi:10.1109/TIE.2017.2703663. [CrossRef]

18. Mishra, R.; Hussain, M.N.; Agarwal, V. A Sliding Mode Control based stabilization solution for multiple Constant Power Loads with identical input filters interfaced with the DC bus of a 'More Electric' Aircraft. In Proceedings of the 2016 IEEE International Conference on Power Electronics, Drives and Energy Systems (PEDES), Trivandrum, India, 14–17 December 2016; pp. 1–6, doi:10.1109/PEDES.2016.7914300. [CrossRef]

19. Li, A.; Zhang, D. Necessary and sufficient stability criterion and new forbidden region for load impedance specification. *Chin. J. Electron.* **2014**, *23*, 628–634,

20. Wu, M.; Lu, D.D.C. Active stabilization methods of electric power systems with constant power loads: A review. *J. Mod. Power Syst. Clean Energy* **2014**, *2*, 233–243. doi:10.1007/s40565-014-0066-y. [CrossRef]

21. Riccobono, A.; Santi, E. Comprehensive Review of Stability Criteria for DC Power Distribution Systems. *IEEE Trans. Ind. Appl.* **2014**, *50*, 3525–3535, doi:10.1109/TIA.2014.2309800. [CrossRef]

22. Riccobono, A.; Santi, E. A novel Passivity-Based Stability Criterion (PBSC) for switching converter DC distribution systems. In Proceedings of the 2012 Twenty-Seventh Annual IEEE Applied Power Electronics Conference and Exposition (APEC), Orlando, FL, USA, 5–9 February 2012; pp. 2560–2567, doi:10.1109/APEC.2012.6166184. [CrossRef]

23. Sanchez, S.; Molinas, M. Assessment of a stability analysis tool for constant power loads in DC-grids. In Proceedings of the 2012 15th International Power Electronics and Motion Control Conference (EPE/PEMC), Novi Sad, Serbia, 4–6 September 2012; pp. DS3b.2-1–DS3b.2-5, doi:10.1109/EPEPEMC.2012.6397317. [CrossRef]
24. Sanchez, S.; Ortega, R.; Griño, R.; Bergna, G.; Molinas, M. Conditions for Existence of Equilibria of Systems With Constant Power Loads. *IEEE Trans. Circuits Syst. I Regul. Pap.* **2014**, *61*, 2204–2211, doi:10.1109/TCSI.2013.2295953. [CrossRef]
25. Miyagi, H.; Munda, J.L.; Miyagi, N. Study on Lyapunov Functions for Lienard-type Nonlinear Systems. *IEEJ Trans. Electron. Inf. Syst.* **2001**, *121*, 748–755.
26. Levant, A. Principles of 2-sliding mode design. *Automatica* **2007**, *43*, 576–586. [CrossRef]
27. Shtessel, Y.; Edwards, C.; Fridman, L.; Levant, A. *Sliding Mode Control and Observation*; Springer: New York, NY, USA, 2014.

Article

Multi-Objective Optimisation-Based Tuning of Two Second-Order Sliding-Mode Controller Variants for DFIGs Connected to Non-Ideal Grid Voltage

Ana Susperregui [1,*], Juan Manuel Herrero [2], Miren Itsaso Martinez [1], Gerardo Tapia-Otaegui [1] and Xavier Blasco [2]

[1] Department of Automatic Control and Systems Engineering, Faculty of Engineering, Gipuzkoa, University of the Basque Country UPV/EHU, Plaza de Europa 1, 20018 Donostia, Spain; mirenitsaso.martinez@ehu.eus (M.I.M.); gerardo.tapia@ehu.eus (G.T.-O.)

[2] Instituto Universitario de Automática e Informática Industrial (ai2), Universitat Politècnica de València, Camí de Vera s/n, 46022 València, Spain; juaherdu@isa.upv.es (J.M.H.); xblasco@isa.upv.es (X.B.)

* Correspondence: ana.susperregui@ehu.eus

Received: 15 August 2019; Accepted: 28 September 2019; Published: 5 October 2019

Abstract: In this paper, a posteriori multi-objective optimisation (MOO) is applied to tune the parameters of a second-order sliding-mode control (2-SMC) scheme commanding the grid-side converter (GSC) of a doubly-fed induction generator (DFIG) subject to unbalanced and harmonically distorted grid voltage. Two variants (i.e., design concepts) of the same 2-SMC algorithm are assessed, which only differ in the format of their switching functions and which contain six and four parameters to be adjusted, respectively. A single set of parameters which stays valid for nine different operating regimes of the DFIG is also sought. As two objectives, related to control performances of grid active and reactive powers, are established for each operating regime, the optimisation process considers 18 objectives simultaneously. A six-parameter set derived in a previous work without applying MOO is taken as reference solution. MOO results reveal that both the six- and four-parameter versions can be tuned to overcome said reference solution in each and every objective, as well as showing that performances comparable to those of the six-parameter variant can be achieved by adopting the four-parameter one. Overall, the experimental results confirm the latter and prove that the performance of the reference parameter set can be significantly improved by using either of the six- or four-parameter versions.

Keywords: decision making; design concept; doubly-fed induction generator; grid-side converter; harmonic distortion; multi-objective optimisation; second-order sliding-mode control; tuning; unbalanced voltage; wind power generation

---

1. Introduction

As wind energy becomes a prevailing source of power generation, grid codes for interconnection of wind energy conversion systems (WECS), in order to ensure the reliable and safe operation of the electricity grid, have become more and more demanding. As a result, wind turbine technology must be developed accordingly.

The doubly-fed induction generator (DFIG) (refer to Figure 1) and the so-called full-scale converter wind generator are the dominating technologies in the present wind industry [1]. Both wind turbine configurations contain a power converter stage, which is usually comprised of two identical (three-phase, two-level) voltage source converters (VSCs). Thereby, the control system associated to the grid-side power converter (GSC) plays a critical role in the accomplishment of different grid codes, such as the capability to tolerate voltage and frequency deviations, control of active and reactive powers, fault ride-through (FRT) operation, and power quality-related requirements, such as low total harmonic distortion (THD) of the current fed into the grid.

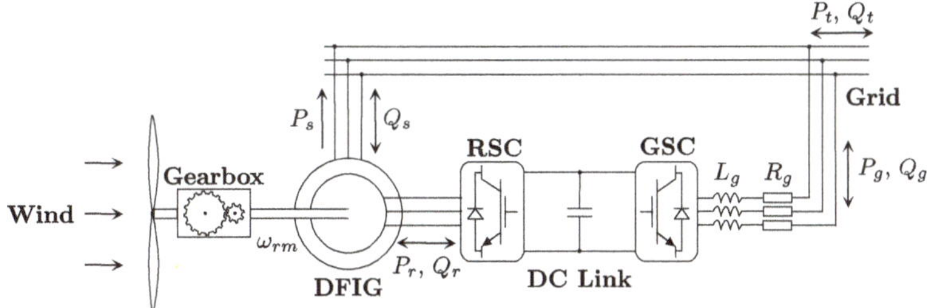

Figure 1. Structure of a doubly-fed induction generator (DFIG)-based wind turbine.

At present, to satisfy such demands, control systems of grid-connected VSCs must have the ability to control not only the fundamental component of their current positive sequence, but also other current components, such as the negative sequence, harmonics of any order, and subharmonics, that may arise due to grid disturbances.

In this context, proportional-integral (PI)- and PI+resonant (PI+R)-based control algorithms were, at first, predominant in the literature [2–4]. However, the main drawback of those kinds of solutions consists in the lack of versatility against uncertainties in the type of grid voltage disturbance. That is, said solutions require particularising at the beginning of the design phase, which are the specific disturbed grid voltage scenarios they are intended to cope with. Therefore, if a particular type of disturbance arises which was not contemplated in advance, it is more than likely that the control algorithm does not have enough bandwidth to perform well.

Hence, a less grid voltage-dependent solution, which is capable of dealing with diverse non-ideal grid voltage profiles, is desirable. In this sense, the high-performance dynamic response and robustness naturally conferred by the different variants of sliding-mode control (SMC)-based algorithms make them excellent candidates. The constant switching frequency imposed on the commanded power converter, as well as the ability to mitigate the chattering phenomenon, are probably the two main strengths of second-order SMC (2-SMC) algorithms and, therefore, they have become a reasonable choice for addressing the design of the GSC controller.

The following handicaps, however, arise with 2-SMC:

- It is complex to predict the expressions for the switching functions that lead to the best system performance; that is, to the best possible control of the active and reactive powers.
- They have a considerable number of parameters to be adjusted, whose tuning is not yet as intuitive as, for example, that of proportional-integral-derivative (PID)-type controllers.
- Simulation results obtained by running an empirically tuned controller have shown that, for each specific operation mode of the GSC (i.e., amount of active and reactive powers, wind turbine speed, degree and type of grid voltage disturbance, transient and steady-state of said grid voltage perturbation, and so on), there exists a different set of controller parameters giving rise to better performance, in terms of active and reactive power control.
- Tuning of a specific parameter may lead to improved behaviour of a given controlled variable (e.g., active power), while negatively affecting others (e.g., reactive power).

Thus, far from trial-and-error tuning methods, a more scientific adjustment procedure for 2-SMC-based algorithms needs to be approached, such that a unique set of controller parameters remains valid for a good number of representative GSC operating regimes. Certainly, this requirement can be met by posing a multi-objective optimisation problem (MOOP).

In this sense, there are few works published, at present, in the literature (which have been oriented towards very disparate applications) focused on optimally tuning a SMC-based algorithm under a multi-objective (MO) approach [5–10]. However, the MOOPs tackled by those papers considered between two and (at most) five objectives to be minimised, which may not cover all the possible operating regimes of the system under study. Moreover, the SMC variant adopted by practically all papers in the literature was the first-order SMC (1-SMC) in its different versions (i.e., combining every possibility: With/without equivalent control term and with/without boundary layer), whereas there has been a lack of solutions focused on the 2-SMC. In addition, most, though not all, have validated their results by simulation, while only a few proved that results derived from experimental tests were consistent with those obtained through simulation [8,9].

As a consequence, throughout this paper, a tuning analysis based on multi-objective optimisation (MOO) is tackled for a 2-SMC algorithm. The parameter tuning derived in [11] for the same system without applying any MOO approach, as well as the results to which such tuning leads, are adopted as baseline.

In particular, two versions (i.e., design concepts) of the same 2-SMC-based algorithm are compared under a MO approach: The first one containing six parameters to be tuned, including integral terms in its switching functions; whereas these integral terms have been removed from the second one, which contains just four parameters to adjust. To set the MOOP, two measures of the control performance, the integral of the absolute value of the error (IAE) for the active power and the standard deviation (SD) for the reactive power, in nine different operating regimes of the DFIG are taken into account. Therefore, 18 objectives are simultaneously considered.

An a posteriori MOO approach [12] is employed. First, both the Pareto front and set are obtained in the MOO stage and, second, the final solution is chosen in the decision-making stage. Under this approach, it is not necessary to aggregate objectives and, as a result, the designer avoids weighting them a priori. Furthermore, obtaining the Pareto front can help the designer to grasp the trade-off among objectives, as well as to select the final solution in a more informed way.

The MOO stage is solved by making use of the ev-MOGA algorithm [13], which is a multi-objective evolutionary algorithm (EA) capable of handling complex optimisation problems with non-convex and disjoint Pareto fronts. Thanks to the population nature of EAs, ev-MOGA obtains the Pareto front in a single run, as well as the majority of EAs [14].

Dealing with MOOPs with high number of objectives (18, in this particular case) makes the Pareto front analysis more difficult. In order to assist the designer in this task, the interactive tool of level diagrams (LDs) [15,16] is employed. LDs are a powerful graphical tool, allowing comparison of design concepts—for this paper, the two 2-SMC-based algorithms with four and six parameters, respectively—in a synchronised m-dimensional objective space. They have been successfully applied in a number of MOOPs, helping to analyse Pareto fronts in a more understandable way, such as multi-loop PI controller design [17], non-linear model identification [18], or for the tuning of biological synthetic devices [19].

The posed results of the MOOP corroborate that it is possible to tune the aforementioned 2-SMC algorithms for both of the proposed design concepts, such that they improve upon the performance of the reference 2-SMC scheme proposed in [11], in each and every one of the 18 objectives proposed. In addition, it is observed that the four-parameter variant of the 2-SMC algorithm exhibits similar behaviour to that of six-parameter version in practically all the objectives, hence leading us to conclude that the four-parameter version may be more suitable than its six-parameter counterpart, due to its greater simplicity.

With the aim of experimentally verifying these conclusions on a physical prototype, two specific controllers (one for each design concept) are selected, which present similar performances in simulation. These controllers, as well as the reference 2-SMC one, are tested 30 times each in the physical prototype. A statistical analysis of the obtained results is carried out, which confirms the conclusions derived from the simulation.

The rest of the paper is structured as follows: Section 2 is devoted to presenting the two variants of the 2-SMC algorithm adopted to command the GSCs of DFIGs, as well as the MOO tools to be used in order to tune their respective parameters. In Section 3, the framework designed to tackle the MOO-based tuning of said parameters is described in depth. Both simulation and experimental results derived from such MOO-assisted parameter tuning are provided and interpreted in Section 4. Finally, Section 5 draws the conclusions.

2. Theoretical Considerations

2.1. DFIG-Based Wind Turbine

Figure 1 shows the general structure of a DFIG-based wind turbine. Like any other wind turbine topology able to operate at variable speed, in addition to the electric generator, it is equipped with a power converter stage which, when adequately commanded, enables full control of the active and reactive powers interchanged with the electricity grid. Thereby, the stator of the generator is directly connected to the grid, whereas its rotor is linked to the power converter stage. Essentially, the latter comprises two identical three-phase, two-level VSCs—named the rotor-side converter (RSC) and the GSC—linked to each other by means of a DC bus. Likewise, the GSC is connected to the electricity grid through an L-type filter.

Although each power converter possesses its own control algorithm, certain co-ordination between them is required to satisfy the specific control targets, related to the overall wind turbine performance, that arise during electricity network disturbances.

Even if the present study is solely focused on the GSC control algorithm, the control goals of both converters are detailed next, aiming at providing a clear insight into the task of controller parameter tuning that is to be faced.

2.1.1. RSC and GSC Control Targets

The RSC control system is in charge of governing the active and reactive powers interchanged between the stator of the generator and the grid (P_s and Q_s, respectively). According to the maximum power point tracking (MPPT) curve [20], the higher the speed of rotation of the wind turbine, the higher the average value of the stator active power set-point, $P^*_{s\,av}$, should be.

During grid disturbances (e.g., imbalances, harmonics, or both) though, in order to prevent harmful fluctuations in the electromagnetic torque of the generator, it is necessary to add an oscillating active power component to the aforementioned set-point average value. Accordingly, the reference value of the stator active power can be expressed as the sum of two terms; that is, $P^*_s = P^*_{s\,av} + P^*_{s\,osc}$.

In contrast, the stator reactive power set-point, Q^*_s, does not fluctuate and, unless the system operator asks for a different value, it is kept near to zero most of the time. This guarantees a power factor close to unity.

With regard to the GSC control system, it is designed to command the instantaneous active and reactive powers flowing between the GSC and the grid (P_g and Q_g, respectively). In particular, the functional diagram displayed in Figure 2 corresponds to the GSC control algorithm adopted in this work, where "CLARKE" and "CLARKE^{-1}" stand for the Clarke's and inverse Clarke's transforms, respectively [21]. This algorithm must be implemented from the outer to the inner layer of the diagram; labelled, respectively, as "1st Step" and "3rd Step" at their bottom left-hand corners. In coherence with the latter, it is assumed that any variable present in a given layer of the diagram is also available to the layers inside.

As in many other works [11], the active power set-point, P^*_g, is established by an integral-proportional (I-P) controller aimed at keeping the DC-link voltage steady at its rated value. Again, the reactive power set-point, Q^*_g, is usually fixed to zero under non-faulty conditions.

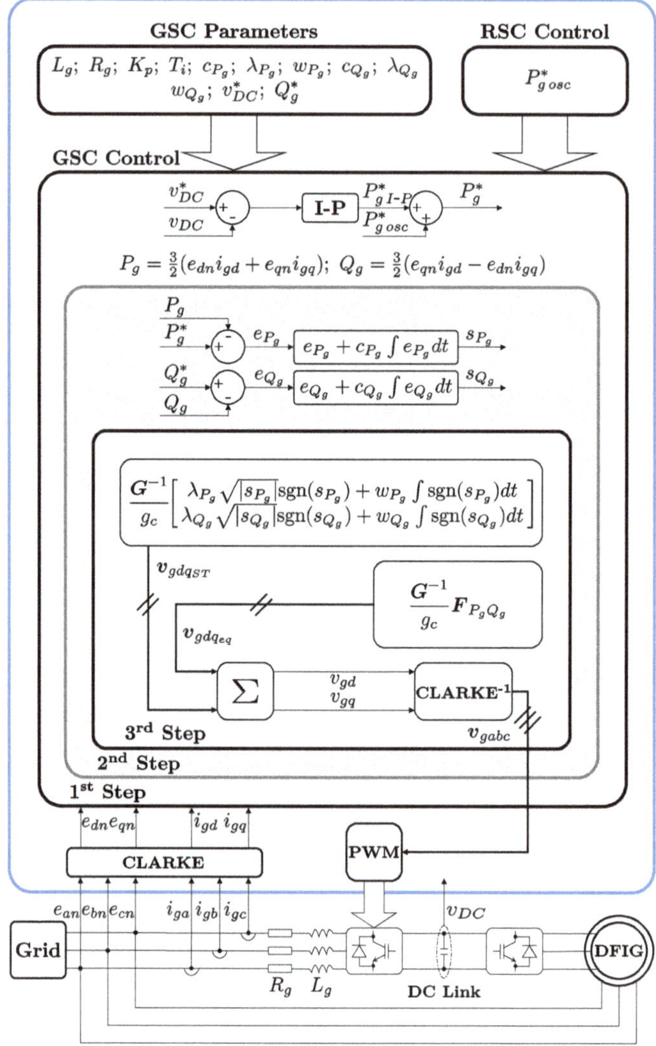

Figure 2. Functional diagram of the control scheme adopted for a DFIG grid-side converter (GSC).

Pushed by increasingly demanding grid codes, during grid voltages subject to imbalances or harmonic pollution, the GSC control system accomplishes additional control targets, the following two being the most common, as well as incompatible with each other [11,22,23]:

1. To add on an oscillating active power term, $P_{g\,osc}$, that compensates for the above-mentioned oscillatory component of the stator active power, $P_{s\,osc}$, at the point where the DFIG is connected to the grid. As a result, a non-fluctuating total active power, $P_t = P_s + P_g$, is achieved by the whole wind turbine.

2. To compensate the stator current imbalance and/or harmonic distortion, if any, thus balancing the overall current injected by the wind turbine into the grid and/or decreasing its THD as far as possible, respectively.

The first strategy is precisely the one adopted throughout this paper. As a result, not only the total active and reactive powers, P_t and Q_t, remain free of fluctuations, but also DC-link voltage oscillations are avoided (which is not possible with the second strategy). In return, in comparison with the approach numbered above as 2, the THD of the overall current injected into the grid turns out to be higher.

Thereby, for the selected strategy, the reference value of the grid-side active power is computed as follows [11]:

$$P_g^* = P_{g\,I\text{-}P}^* + P_{g\,osc}^*, \tag{1}$$

where $P_{g\,osc}^*$ depends on variables related to the electric generator, and may be estimated as

$$P_{g\,osc}^* = T_e\omega_{rm} - P_s, \tag{2}$$

with T_e and ω_{rm} being the generator electromagnetic torque and rotational speed, respectively.

The output of each power converter's control system is the three-phase voltage, to be applied by said converter at its AC side. Thus, fixing the appropriate three-phase AC voltage, the aforementioned active and reactive powers can be governed. However, as is usual in three-phase AC power systems, both the RSC and GSC control algorithms are designed, as well as run, in the so-called vector space.

Thus, it is important to clarify that, in the case at hand, the control signals generated by the 2-SMC algorithm under study correspond to the stationary-frame *d-q* components of the GSC output voltage.

2.1.2. GSC and Grid Filter Modelling

According to Figure 3, adopting the rectifier convention and expressing all variables in the stationary reference frame, the grid-side active and reactive powers can be derived as follows:

$$P_g = \frac{3}{2}\left(e_{dn}i_{gd} + e_{qn}i_{gq}\right) \tag{3}$$

$$Q_g = \frac{3}{2}\left(e_{qn}i_{gd} - e_{dn}i_{gq}\right), \tag{4}$$

with e_{dn}, e_{qn}, and i_{gd}, i_{gq} being, respectively, the direct- and quadrature-axis components of the grid voltage and current. The dynamics of the latter have been provided, in [11], as

$$\dot{i}_{gd} = \frac{1}{L_g}\left(e_{dn} - v_{gd} - R_g i_{gd}\right) \tag{5}$$

$$\dot{i}_{gq} = \frac{1}{L_g}\left(e_{qn} - v_{gq} - R_g i_{gq}\right), \tag{6}$$

where v_{gd} and v_{gq} denote the outputs of the GSC control algorithm, while R_g and L_g represent the equivalent resistance and inductance of the grid filter, respectively. Given that such filter is assumed to be of type L, R_g is typically close to zero.

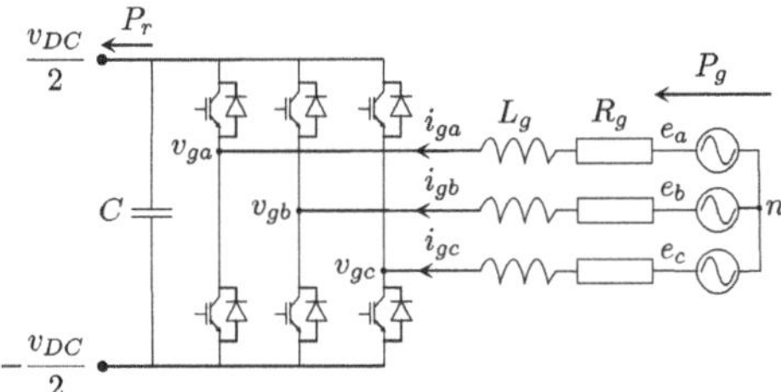

Figure 3. Scheme of the GSC and L-type grid filter.

2.2. 2-SMC Scheme Adopted for the GSC

2.2.1. Switching Functions Selected

Considering that P_g and Q_g are the variables to be controlled, the following two switching functions are defined:

$$s_{P_g} = \overbrace{P_g^* - P_g}^{e_{P_g}} + c_{P_g} \int_0^t e_{P_g}(\tau)d\tau \tag{7}$$

$$s_{Q_g} = \underbrace{Q_g^* - Q_g}_{e_{Q_g}} + c_{Q_g} \int_0^t e_{Q_g}(\tau)d\tau, \tag{8}$$

where the integral terms are aimed at steering possible steady-state errors to zero [24]. Regarding the weighting constants c_{P_g} and c_{Q_g}, which need to be tuned, two alternatives will be explored in this paper; namely:

1. To assume they both can take any strictly positive value. Specifically, MOO is applied in this work in order to select c_{P_g} and c_{Q_g} from within a wide range of possible values.
2. To force them to zero, hence simplifying both switching functions and, in turn, the global control scheme for the GSC.

2.2.2. Control Laws

Taking the time derivatives of Equations (7) and (8), and making use of Equations (3)–(6), the following dynamics arise for the switching functions s_{P_g} and s_{Q_g}:

$$\begin{bmatrix} \dot{s}_{P_g} \\ \dot{s}_{Q_g} \end{bmatrix} = \underbrace{\begin{bmatrix} F_{P_g} \\ F_{Q_g} \end{bmatrix}}_{F_{P_g Q_g}} - \frac{3}{2L_g} \underbrace{\begin{bmatrix} -e_{dn} & -e_{qn} \\ -e_{qn} & e_{dn} \end{bmatrix}}_{G} \begin{bmatrix} v_{gd} \\ v_{gq} \end{bmatrix}, \tag{9}$$

where

$$F_{P_g} = \dot{P}_g^* - \frac{3}{2}\left(\dot{e}_{dn}i_{gd} + \dot{e}_{qn}i_{gq}\right) - \frac{3}{2L_g}\left(e_{dn}^2 + e_{qn}^2\right) + \frac{R_g}{L_g}P_g + c_{P_g}e_{P_g} \tag{10}$$

$$F_{Q_g} = \dot{Q}_g^* - \frac{3}{2}\left(\dot{e}_{qn}i_{gd} - \dot{e}_{dn}i_{gq}\right) + \frac{R_g}{L_g}Q_g + c_{Q_g}e_{Q_g}. \tag{11}$$

As proposed in [11], the control signals v_{gd} and v_{gq} are computed as a summation of two terms; namely:

- The $v_{gdq_{ST}}$ "super-twisting" (ST) control term, intended to attain high-performance closed-loop dynamics, ability for disturbance rejection, and robustness in the face of uncertainties, both structured and unstructured.
- The $v_{gdq_{eq}}$ equivalent control term, incorporated with the main purpose of reducing the control effort to be made by the ST algorithm.

The preceding approach may be mathematically expressed as

$$\underbrace{\begin{bmatrix} v_{gd} \\ v_{gq} \end{bmatrix}}_{v_{gdq}} = \underbrace{\begin{bmatrix} v_{gd_{eq}} \\ v_{gq_{eq}} \end{bmatrix}}_{v_{gdq_{eq}}} + \underbrace{\begin{bmatrix} v_{gd_{ST}} \\ v_{gq_{ST}} \end{bmatrix}}_{v_{gdq_{ST}}}. \tag{12}$$

After forcing $\dot{s}_{P_g} = \dot{s}_{Q_g} = 0$ in Equation (9), the equivalent control term is derived by simply solving for v_{gd} and v_{gq} in said expression, which gives rise to

$$v_{gdq_{eq}} = \frac{2}{3}L_g G^{-1} F_{P_g Q_g} = \underbrace{\frac{2L_g}{3\left(e_{dn}^2 + e_{qn}^2\right)}}_{|e_n|^2} \begin{bmatrix} -e_{dn} & -e_{qn} \\ -e_{qn} & e_{dn} \end{bmatrix} F_{P_g Q_g}, \tag{13}$$

where the matrix G is invertible, except for the case in which $|e_n| = 0$, corresponding to a null grid voltage. Assuming that the sliding regime is reached (i.e., $s_{P_g} = s_{Q_g} = 0$), $v_{gdq_{eq}}$ would allow for preserving it in the absence of disturbances, as well as under both parametric and modelling uncertainties.

However, depending on the specific shapes of both P_g^* and Q_g^*, their respective $\dot{P}_g^*$ and $\dot{Q}_g^*$ time derivatives, present in $F_{P_g Q_g}$ by virtue of Equations (10) and (11), are likely to bring noise, and even derivative kicks, into the $v_{gdq_{eq}}$ equivalent control term. Therefore, in order to elude such a jeopardy, $\dot{P}_g^* = \dot{Q}_g^* = 0$ is considered in Equation (13) [22].

In any case, the inaccuracies made due to that design simplification, as well as the high parameter dependency evidenced by the equivalent control in Equation (13), do not compromise the robustness of the global control algorithm in Equation (12), as said robustness relies on the ST control term that follows:

$$v_{gdq_{ST}} = \frac{2}{3}L_g G^{-1} v_{P_g Q_g ST} = \frac{2L_g}{3|e_n|^2} \begin{bmatrix} -e_{dn} & -e_{qn} \\ -e_{qn} & e_{dn} \end{bmatrix} v_{P_g Q_g ST}, \tag{14}$$

with

$$v_{P_g Q_g ST} = \begin{bmatrix} v_{P_g ST} \\ v_{Q_g ST} \end{bmatrix} = \begin{bmatrix} \lambda_{P_g} \sqrt{\left|s_{P_g}\right|}\,\mathrm{sgn}\left(s_{P_g}\right) + w_{P_g}\int_0^t \mathrm{sgn}\left(s_{P_g}(\tau)\right)d\tau \\ \lambda_{Q_g} \sqrt{\left|s_{Q_g}\right|}\,\mathrm{sgn}\left(s_{Q_g}\right) + w_{Q_g}\int_0^t \mathrm{sgn}\left(s_{Q_g}(\tau)\right)d\tau \end{bmatrix}. \tag{15}$$

The terms of the form $\lambda_x \sqrt{|s_x|}\,\mathrm{sgn}\left(s_x\right)$, where $x = P_g$ or Q_g, are responsible for ensuring the achievement of the sliding regime in finite time.

It should be noted that there are six parameters to be tuned; namely: c_{P_g}, λ_{P_g}, w_{P_g}, c_{Q_g}, λ_{Q_g}, and w_{Q_g}. Nonetheless, as already stated at the end of Section 2.2.1, the option of forcing $c_{P_g} = c_{Q_g} = 0$ will also be explored, which leads to a simplified version of the GSC control algorithm with just four parameters: λ_{P_g}, w_{P_g}, λ_{Q_g}, and w_{Q_g}.

2.3. Multi-Objective Optimisation

A MOOP with m objectives to minimise can be stated as follows [25]:

$$\min_x f(x) \tag{16}$$

subject to

$$K(x) \leq 0 , \quad L(x) = 0 \tag{17}$$

$$\underline{x_i} \leq x_i \leq \overline{x_i} , \quad i = [1, 2 \ldots n], \tag{18}$$

where $x = [x_1, x_2 \ldots x_n] \in D$ is the decision vector, with $\dim(x) = n$; $f(x) = [f_1(x), f_2(x) \ldots f_m(x)]$ is the objective vector; $K(x)$ and $L(x)$ are the inequality and equality constraint vectors, respectively; and $\underline{x_i}$ and $\overline{x_i}$ are the lower and upper bounds in the D decision space, respectively.

As the objectives of a MOOP are usually in opposition, there is typically no single solution that minimises all the objectives. Instead, there will exist a set of Pareto optimal solutions (i.e., non-dominated solutions).

Definition 1. *(Pareto optimality [25]): An objective vector $f(x^2)$ is Pareto optimal if there is no other objective vector $f(x^1)$ such that $f_i(x^1) \leq f_i(x^2)$ for all $i \in [1, 2 \ldots m]$ and $f_j(x^1) < f_j(x^2)$, for at least one j, $j \in [1, 2 \ldots m]$.*

Definition 2. *(Dominance [26]): An objective vector $f(x^1)$ is dominated by another objective vector $f(x^2)$ iff $f_i(x^2) \leq f_i(x^1)$ for all $i \in [1, 2 \ldots m]$ and $f_j(x^2) < f_j(x^1)$, for at least one j, $j \in [1, 2 \ldots m]$. This is denoted as $f(x^2) \preceq f(x^1)$.*

Therefore, the set of solutions (the Pareto set) is defined as follows:

Definition 3. *(Pareto set, X_p): The Pareto set is the set of all solutions in D that are not dominated by any other solution in D:*

$$X_p := \{x \in D \mid \nexists x' \in D : f(x') \preceq f(x)\}.$$

Each solution in the Pareto set defines an objective vector in the Pareto front.

Definition 4. *(Pareto front, $f(X_p)$): Given a set of Pareto optimal solutions X_p, the Pareto front is defined as*

$$f(X_p) := \{f(x) \mid x \in X_p\}.$$

Usually, X_p contains an infinite number of solutions and, for this reason, it is not possible to completely obtain it. The way to proceed is to obtain a discrete set $X_p^* \subset X_p$, in such a way that X_p^* characterises X_p. Note that the set X_p^* is not unique. In this work, the ev-MOGA algorithm (Available at https://es.mathworks.com/matlabcentral/fileexchange/31080-ev-moga-multiobjective-evolutionary-algorithm) [13] will be used to obtain the Pareto front approximations. Figure 4 shows an example of characterisation of a bi-objective Pareto front and its corresponding Pareto set.

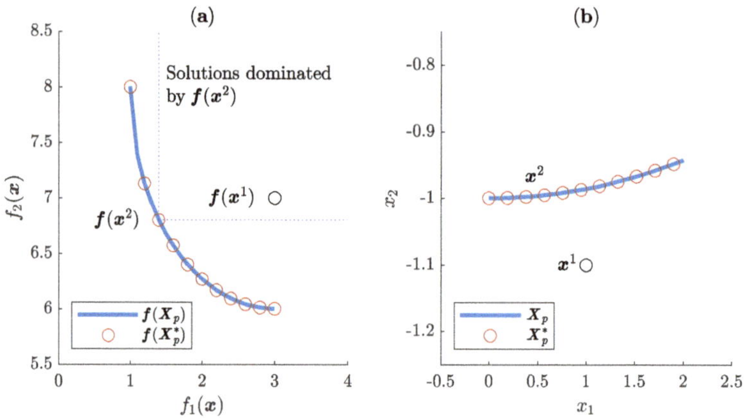

Figure 4. (a) Pareto front $f(X_p)$ for a bi-objective multi-objective optimisation problem (MOOP); and (b) the Pareto set X_p in the decision space. $f(X_p^*)$ and X_p^* represent a possible characterisation of $f(X_p)$ and X_p, respectively.

2.4. Comparison of Design Concepts Under MOO Approach

It is very common that several design alternatives (i.e., design concepts), C, are proposed, in order to solve a specific problem. Each design concept might, for example, represent a different control structure. Comparing the different concepts in a multi-objective scenario allows for differentiating the strengths and weaknesses of each of them, in relation to the chosen objectives [25,27]. To do so, a MOOP is set for each design concept, Cj, such that all MOOPs share the same objectives, f, but each of them has its own decision vector, x^{Cj}, related to the parameterisation of its corresponding design concept. Therefore, if s design concepts need to be compared, the MOOPs can be stated as

$$\min_{x^{Cj}} f(x^{Cj}) \tag{19}$$

subject to

$$K^{Cj}(x^{Cj}) \leq 0 \; , \quad L^{Cj}(x^{Cj}) = 0 \tag{20}$$

$$\underline{x_i^{Cj}} \leq x_i^{Cj} \leq \overline{x_i^{Cj}} \; , \quad i = [1, 2 \dots n^{Cj}], \tag{21}$$

with $j \in [1, 2 \dots s]$. For each design concept, $x^{Cj} = [x_1^{Cj}, x_2^{Cj} \dots x_{n^{Cj}}^{Cj}]$ is the decision vector; $K^{Cj}(x^{Cj})$ and $L^{Cj}(x^{Cj})$ are the inequality and equality constraint vectors, respectively; and $\underline{x_i^{Cj}}$ and $\overline{x_i^{Cj}}$ are the lower and upper bounds delimiting the searching space, respectively. In contrast, the objective vector $f(x^{Cj}) = [f_1(x^{Cj}), f_2(x^{Cj}) \dots f_m(x^{Cj})]$ is common to the s MOOPs.

After optimising each multi-objective problem, a discrete Pareto set, X_p^{*Cj}, and its corresponding Pareto front, $f(X_p^{*Cj})$, are obtained for each design concept. Thanks to the fact that all of the MOOPs share the same objectives, a comparison in the m-objective space can be made. This idea is illustrated in Figure 5, where the Pareto fronts of three design concepts are depicted in a bi-objective optimisation problem. By analysing the figure, it is possible to notice the following:

- Design concept 3 is dominated by design concepts 1 and 2. Therefore, the latter two will be preferred.
- Depending on designer preferences, design concept 1 or 2 may be preferred.
- Zone C (values of $f_2(x) < 6.06$) is only reachable by design concept 2. Consequently, if the designer demands such a trade-off, design concept 2 would be the right one.
- In Zone B ($f_1(x) \in [1.7, 2.5]$), design concept 2 dominates design concept 1. As a result, design concept 2 will be preferred over design concept 1.
- The opposite to what occurs in Zone B is observable in Zone A ($f_1(x) < 1.7$). Design concept 2 is dominated by design concept 1 and, thus, the latter will be preferred.

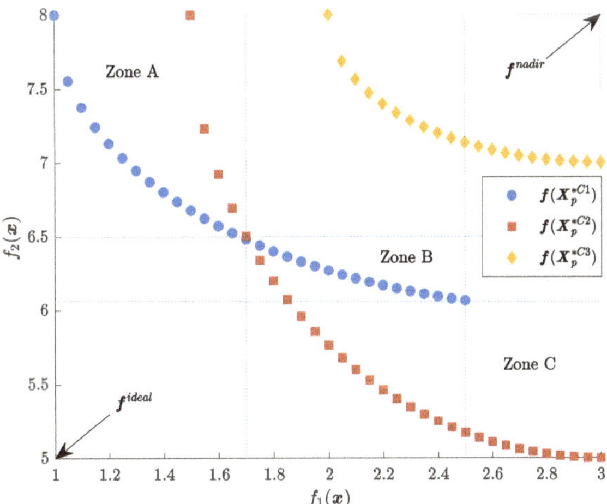

Figure 5. Three design concepts in a bi-objective optimisation problem. Example of comparison in the objective space.

2.5. LDs for Design Concept Comparison

In order to efficiently compare design concepts in an m-dimensional objective space, an adequate visualisation method is required. Among the several methods provided in the literature [28], the interactive tool referred to as LDs [15,16,29] is employed in this work.

The LD tool (Available at https://es.mathworks.com/matlabcentral/fileexchange/62224-interactive-tool-for-decision-making-in-multiobjective-optimization-with-level-diagrams) transforms the m-dimensional objective space and the n-dimensional decision space into $m+n$ two-dimensional separate (but synchronised) graphs. For that purpose, first, each point of the Pareto fronts $f(x^{Cj})$ is normalised with respect to the ideal f^{ideal} and nadir f^{nadir} points (see Figure 5), as given below:

$$\hat{f}_i(x^{Cj}) = \frac{f_i(x^{Cj}) - f_i^{ideal}}{f_i^{nadir} - f_i^{ideal}}, \quad i \in [1, 2 \ldots m]. \tag{22}$$

Second, the p-norm $\|\hat{f}(x^{Cj})\|_p$ is applied to each normalised point. Typical norms are: (1) Taxicab norm—also called Manhattan norm—, $p = 1$; (2) Euclidean norm, $p = 2$; and (3) infinity norm—also known as maximum norm—, $p = \infty$.

After that, the LD tool provides a two-dimensional graph for each objective and decision variable. On the abscissa axis of each graph, the values for each objective or decision variable are represented, while the ordinate axes of all graphs display the p-norm previously calculated for each solution. The latter allows graphics to stay synchronised, by means of their ordinate axes (meaning that each given solution of a design concept presents identical ordinate value in every graph) and, therefore, helps to compare solutions according to the selected norm.

Adopting the Euclidean norm, Figure 6 shows the LD corresponding to the same three design concepts presented in Figure 5. Given that, similarly to Figure 5, the search space is not contemplated, only two graphs associated to the objective space are provided, which corresponds to the bi-objective problem considered. The A, B, and C zones have been marked, in order to demonstrate their correspondence with the same zones displayed in Figure 5. It can be noticed that the solutions of design concept 2 are the closest to f^{ideal}, as they present lower values of $\|\hat{f}\|_2$.

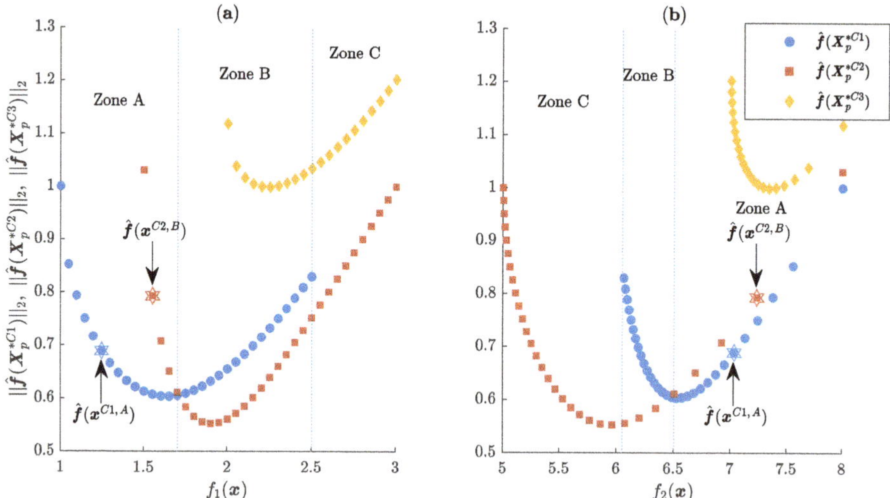

Figure 6. Comparison of design concepts 1, 2, and 3 by employing level diagrams (LDs) with the Euclidean norm. (**a**) LD graph for objective f_1; and (**b**) LD graph for objective f_2.

Two solutions, $x^{C1,A}$ from design concept 1 and $x^{C2,B}$ from design concept 2, have been highlighted. Although they both present low values of f_1 and high values of f_2, it is clearly observable that $x^{C1,A}$ dominates $x^{C2,B}$. It should be noted that, when more than three objectives are considered, it becomes difficult to appreciate such relations using classical visualisation tools.

3. Framework for MOO Tuning of the GSC Control Scheme

3.1. Simulation Test Designed

The proposed MOO-based tuning methodology, requiring a considerable amount of simulations to run, was applied on the 7 kW DFIG prototype employed in [11] for experimentation.

To that end, each new set of values to be tested for the GSC controller parameters (i.e., c_{P_g}, λ_{P_g}, w_{P_g}, c_{Q_g}, λ_{Q_g}, and w_{Q_g}) was evaluated on a simulation model reproducing the grid-connected GSC and the DC bus of the 7 kW DFIG prototype, as well as the DC bus v_{DC} voltage I-P regulator. Its parameters are provided in Table 1, where the equivalent R_g resistance of the L-type grid filter was assumed to be negligible.

Table 1. Parameters of the 7 kW DFIG grid filter, DC bus, and v_{DC} I-P regulator.

Parameter	Value
R_g	0 mΩ
L_g	2 mH
C	9.4 mF
v_{DC}	125 V
K_p	45.4333 W/V
T_i	103.4483 ms

Considering the high amount of simulation tests to run, it is essential to keep in mind that significantly higher simulation times are required if commutation of the GSC transistors is to be reproduced by the model. Consequently, the PWM–GSC set displayed in Figure 2 is treated as if its operation was ideal, by assuming that the three-phase v_{gabc} voltage applied by the GSC to the grid filter coincides exactly with that computed by its control scheme. In this way, the simulation times were drastically reduced while preserving impartiality of the comparisons, as the described simplification affected equally any parameter set to be evaluated.

It is intended that a unique set of controller parameters remains valid for a good number of representative DFIG operating regimes. For that purpose, the simulation test based on which the tuning process is tackled pushes the DFIG to transit, one after another, through the nine different stages collected and described in Table 2. The specific values assigned to the different time instants displayed in Table 2 are provided in Table 3.

In order to run simulations under realistic conditions of harmonic pollution, the three-phase e_{an}, e_{bn}, and e_{cn} grid voltage profile adopted for simulation was registered in the laboratory housing the 7 kW DFIG prototype. A detail suggesting the level of harmonic distortion present in said grid voltage profile is provided in Figure 7a. Furthermore, in accordance with Tables 2 and 3, such a grid voltage profile also presents a two-phase E-type imbalance of approximately 15% between time instants $t_6 = 6$ and $t_8 = 13$ s, as evidenced by Figure 7b,c.

Table 2. Stages of the designed test.

Time Range	Stage	Description
t_0–t_1	1	With the DFIG stator disconnected from the grid, transient of charge of the DC bus until its rated voltage is reached.
t_1–t_2	2	With its stator disconnected from the grid, initial positioning of the DFIG rotor.
t_2–t_3	3	With the DFIG stator disconnected from the grid, synchronisation of the voltage induced in the terminals of said open stator with the grid voltage.
t_3–t_4	4	Smooth connection (with no power or zero power exchange) of the DFIG stator to the grid at time t_3, and maintenance of said zero power for the entire t_3–t_4 interval.
t_4–t_5	5	Starting from zero power at time t_4, the power generated by the DFIG ramps up to its optimum value, which is reached at time t_5.
t_5–t_6	6	Generation of the optimum power corresponding to the DFIG rotor speed at which the test is carried out.
t_6–t_7	7	A two-phase E-type imbalance affects the grid voltage between time instants t_6 and t_8. This t_6–t_7 time interval corresponds to the transient following the start of said imbalance.
t_7–t_8	8	Steady state resulting from the two-phase E-type imbalance.
t_8–t_9	9	Transient following the conclusion of the two-phase E-type imbalance.

Table 3. Values for the time instants delimiting the stages of the designed test.

Time Instant	t_0	t_1	t_2	t_3	t_4	t_5	t_6	t_7	t_8	t_9
Value (s)	0	0.5	2	3	3.5	3.7033	6	11.3	13	13.5

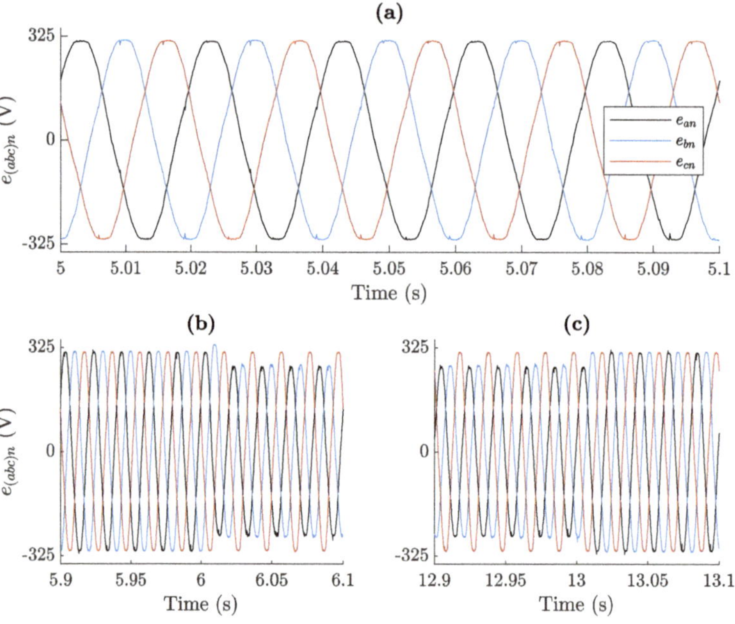

Figure 7. Grid voltage profile: (**a**) Harmonic distortion in the absence of imbalance; (**b**) zoom at the start of the imbalance; and (**c**) zoom at the end of the imbalance.

Concerning the effect of the RSC, it was incorporated into the so-far described simulation model by means of a disturbance representing the rotor active power, P_r, as shown in Figure 3. In particular, Figure 8a displays the specific P_r profile under which every set of GSC controller parameters considered was tested. Consequently, fairness of comparisons is preserved, as all possible sets of GSC controller parameters were evaluated under identical conditions.

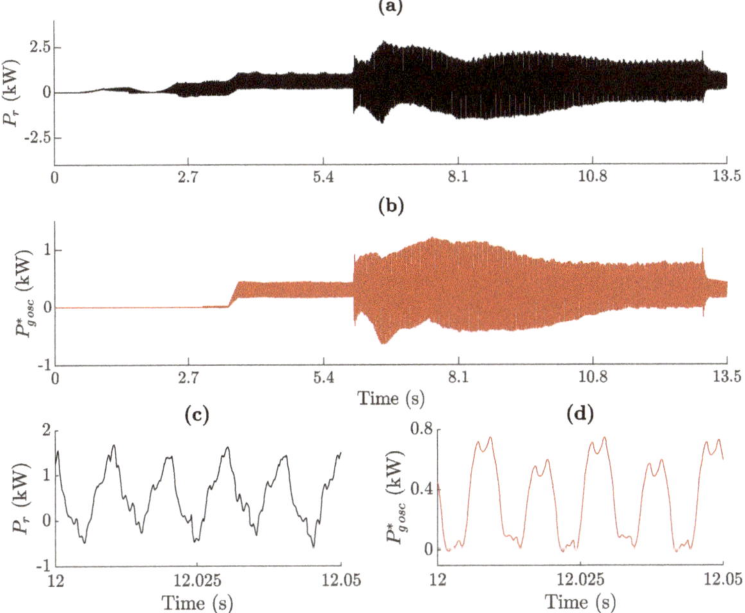

Figure 8. Profiles for P_r and $P^*_{g\,osc}$ throughout the test: (**a**) P_r; (**b**) $P^*_{g\,osc}$; (**c**) detail of P_r at the steady state of the imbalance; and (**d**) detail of $P^*_{g\,osc}$ at the steady state of the imbalance.

As far as grid power reference values are concerned, Q^*_g was set to zero, while P^*_g was derived by adding the feedforward $P^*_{g\,osc}$ term displayed in Figure 8b to the control signal generated by the DC bus voltage I-P regulator, as dictated by Equations (1) and (2), and as represented in Figure 2. An 100 Hz oscillation in both P_r and $P^*_{g\,osc}$ indicative of the presence of a negative sequence and, in turn, of an imbalance in the grid voltage, is made visible in the detail of Figure 8c,d.

In order to derive the P_r and $P^*_{g\,osc}$ profiles in Figure 8a,b, the test described in Table 2 was first run on a complete simulation model, reproducing not only the global 7 kW DFIG prototype considered in [11], but also its RSC and GSC control schemes, tuned as explicitly stated in Table 2 of said paper. It should be pointed out that the angular speed of the DFIG was kept constant, at 1320 rpm, during the entire test. This way, it was sought that the disturbance due to the wind speed variability affected the nine operating regimes equally, as the value of ω_{rm} is a direct consequence of the wind speed.

3.2. Indices Selected

Bearing in mind the two control targets specified for the GSC in Section 2.1.1, as well as the nine different DFIG operating regimes tackled, the following four considerations were made in order to define the performance indices on which the MOOP to be solved was based:

1. As evidenced by Equations (1) and (2), if the grid voltage is harmonically distorted or/and unbalanced, a strongly fluctuating P^*_g reference must be closely tracked by the active power P_g. Accordingly, the performance index referred to as IAE seems suited for determining the quality of tracking achieved.

2. Reactive power Q_g has to be regulated around 0. Consequently, deviations of Q_g from 0 and, as a result, the level of chatter in Q_g may be somehow quantified by means of a SD index.

3. The two indices suggested above are to be computed for each of the nine stages of the simulation test described in Table 2, thus giving rise to $2 \times 9 = 18$ indices in total. Given that the nine DFIG operating regimes considered are significantly different from each other, computing a single IAE and a single SD for the entire test leads to a loss of valuable information and skews the results [30].

4. As the IAE index is cumulative, it is highly dependent on the time interval over which it is calculated. For that reason, the IAE index computed for each of the nine test stages is divided by its corresponding time interval, so that the resulting nine "IAE per unit of time" indices are equitably comparable with each other.

As a consequence, the 18 performance indices considered were as follows:

$$f_{P_{g_i}} = \frac{IAE_{P_{g_i}}}{t_i - t_{i-1}} = \frac{\int_{t_{i-1}}^{t_i} \left| P_g^* - P_g \right| dt}{t_i - t_{i-1}}; \quad i \in [1, 2, 3 \ldots 9] \tag{23}$$

$$f_{Q_{g_i}} = SD_{Q_{g_i}}; \quad i \in [1, 2, 3 \ldots 9], \tag{24}$$

where the i subscript accompanying a given performance index indicates the index to correspond to the ith stage of the test.

3.3. Statement of the MOOP

In brief, the objective consists of minimising the 18 indices established in Equations (23) and (24) by properly tuning the parameters of the 2-SMC scheme, commanding both P_g and Q_g. As indicated at the end of Section 2.2.2, two alternative 2-SMC structures were actually considered; that is,

- **Design concept 1:** All the six controller parameters (explicitly listed at the end of Section 2.2.2) are assumed to be strictly positive (non-zero). Hence, the vector of controller parameters to be adjusted is given by

$$x^{C1_{6p}} = \begin{bmatrix} c_{P_g} & \lambda_{P_g} & w_{P_g} & c_{Q_g} & \lambda_{Q_g} & w_{Q_g} \end{bmatrix}. \tag{25}$$

- **Design concept 2:** The parameters c_{P_g} and c_{Q_g} are set to zero in Equation (25), thus removing the integral terms from the switching functions in Equations (7) and (8). As a result, only four parameters need to be tuned in this particular case, therefore yielding

$$x^{C2_{4p}} = \begin{bmatrix} \lambda_{P_g} & w_{P_g} & \lambda_{Q_g} & w_{Q_g} \end{bmatrix}. \tag{26}$$

On the other hand, the parameter set

$$x^{C1_{6p}, ref} = \begin{bmatrix} 96.6667 & 33.6256 \times 10^3 & 23.3611 \times 10^6 & 96.6667 & 10.6333 \times 10^3 & 2.3361 \times 10^6 \end{bmatrix}, \tag{27}$$

derived in [11] for the GSC 2-SMC scheme, was adopted as the baseline solution. In particular, only those parameter sets improving each and every one of the 18 indices resulting from application of the baseline solution in Equation (27) will be considered. The values for the indices corresponding to the baseline solution are reflected in Table 4.

Table 4. Values of the 18 indices produced by $x^{C1_{6p}, ref}$.

i	1	2	3	4	5	6	7	8	9
$f_{P_{g_i}}(x^{C1_{6p}, ref})$	83.803	10.144	10.458	10.838	11.312	13.204	25.749	22.653	13.995
$f_{Q_{g_i}}(x^{C1_{6p}, ref})$	69.702	3.383	3.576	3.64	5.814	7.368	12.389	11.853	7.852

Consequently, the two MOOPs to be solved are formally stated as follows:

- **MOOP for design concept 1:**

$$X_p^{*C16p} = \min_{x^{C16p}} f = \min_{x^{C16p}} \left[\left[f_{P_{g_1}},\ f_{P_{g_2}},\ f_{P_{g_3}} \cdots f_{P_{g_9}} \right],\ \left[f_{Q_{g_1}},\ f_{Q_{g_2}},\ f_{Q_{g_3}} \cdots f_{Q_{g_9}} \right] \right] \tag{28}$$

subject to constraints

$$f_{P_{g_i}}(x^{C16p}) \le f_{P_{g_i}}(x^{C16p,\,ref}); \quad i \in [1,\, 2,\, 3 \ldots 9] \tag{29}$$

$$f_{Q_{g_i}}(x^{C16p}) \le f_{Q_{g_i}}(x^{C16p,\,ref}); \quad i \in [1,\, 2,\, 3 \ldots 9] \tag{30}$$

$$\underline{x^{C16p}} \le x^{C16p} \le \overline{x^{C16p}}, \tag{31}$$

with

$$\underline{x^{C16p}} = \begin{bmatrix} 0 & 10^3 & 10^3 & 0 & 10^3 & 10^3 \end{bmatrix} \tag{32}$$

$$\overline{x^{C16p}} = \begin{bmatrix} 200 & 5 \times 10^4 & 3 \times 10^7 & 200 & 5 \times 10^4 & 3 \times 10^7 \end{bmatrix}. \tag{33}$$

- **MOOP for design concept 2:**

$$X_p^{*C24p} = \min_{x^{C24p}} f = \min_{x^{C24p}} \left[\left[f_{P_{g_1}},\ f_{P_{g_2}},\ f_{P_{g_3}} \cdots f_{P_{g_9}} \right],\ \left[f_{Q_{g_1}},\ f_{Q_{g_2}},\ f_{Q_{g_3}} \cdots f_{Q_{g_9}} \right] \right] \tag{34}$$

subject to constraints

$$f_{P_{g_i}}(x^{C24p}) \le f_{P_{g_i}}(x^{C16p,\,ref}); \quad i \in [1,\, 2,\, 3 \ldots 9] \tag{35}$$

$$f_{Q_{g_i}}(x^{C24p}) \le f_{Q_{g_i}}(x^{C16p,\,ref}); \quad i \in [1,\, 2,\, 3 \ldots 9] \tag{36}$$

$$\underline{x^{C24p}} \le x^{C24p} \le \overline{x^{C24p}}, \tag{37}$$

with

$$\underline{x^{C24p}} = \begin{bmatrix} 10^3 & 10^3 & 10^3 & 10^3 \end{bmatrix} \tag{38}$$

$$\overline{x^{C24p}} = \begin{bmatrix} 5 \times 10^4 & 3 \times 10^7 & 5 \times 10^4 & 3 \times 10^7 \end{bmatrix}. \tag{39}$$

4. Results and Evaluation

In order to perform the two MOOs defined in Equations (28) and (34), ev-MOGA was applied with the following configuration:

- $Nind_P = 1000,$
- $Nind_G = 8,$
- $Iterations = 5000,$ and
- $N_{box} = 15.$

For the definition of the remaining parameters, the default values suggested by [31] were adopted.

4.1. MOO Results and Analysis

As a result of the optimisation process, a Pareto front approximation with 13,649 solutions was obtained for design concept 1 ($f(X_p^{*C16p})$) and another one containing 6494 solutions for design concept 2 ($f(X_p^{*C24p})$), hence proving that it is possible to find 2-SMC controllers, for both the six- and four-parameter cases, that outperform the reference 2-SMC controller in each and every objective. Both Pareto fronts are simultaneously displayed in Figure 9 by means of the LD tool with ∞-norm, while their corresponding Pareto sets are provided in Figures 10 and 11 for design concepts 1 (X_p^{*C16p}) and 2 (X_p^{*C24p}), respectively. In addition, Tables 5 and 6 reflect the respective minimum values reached by $f(X_p^{*C16p})$ and $f(X_p^{*C24p})$ for each of the 18 performance indices.

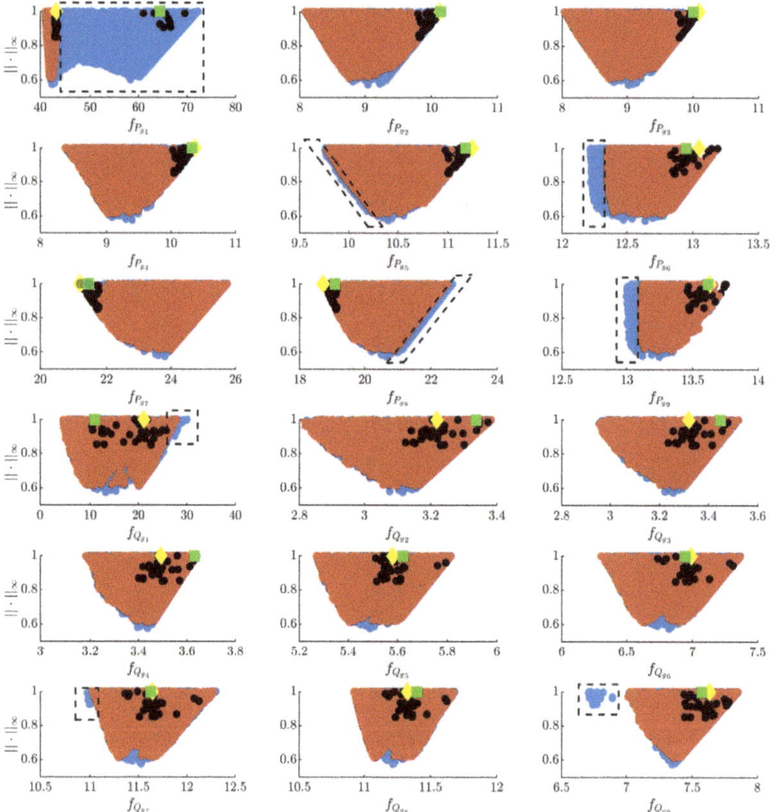

Figure 9. Comparison of Pareto fronts by means of LDs with the ∞-norm. Blue and red dots correspond to design concepts 1 ($f(X_p^{*C16p})$) and 2 ($f(X_p^{*C24p})$), respectively. Dashed lines delimit regions where differences between the two design concepts become more evident. Black dots denote solutions selected to illustrate the trade-off existing among the values of the objectives. The green square and yellow diamond mark, respectively, the preferred six-parameter ($f(x^{C16p,A})$) and four-parameter ($f(x^{C24p,B})$) solutions.

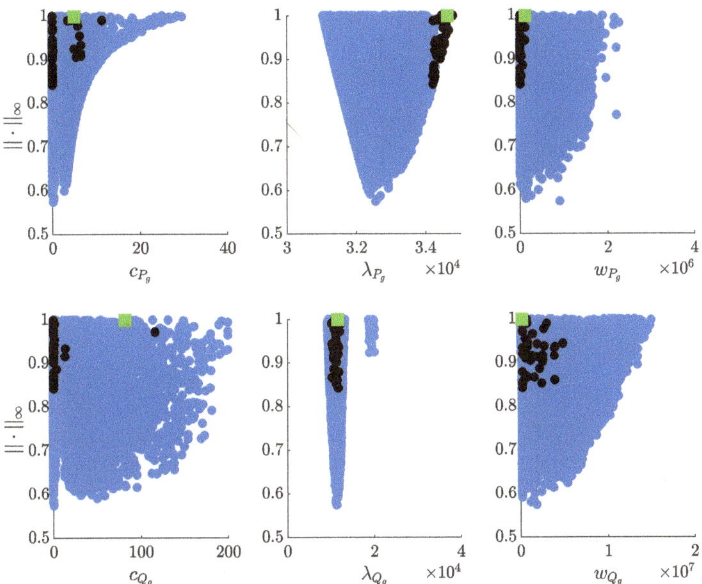

Figure 10. Pareto set of design concept 1 (X_p^{*C16p}) resulting from application of LD with ∞-norm. Black dots correspond to solutions selected to illustrate the trade-off among objectives. The green square marks the preferred six-parameter solution, $x^{C16p,\,A}$.

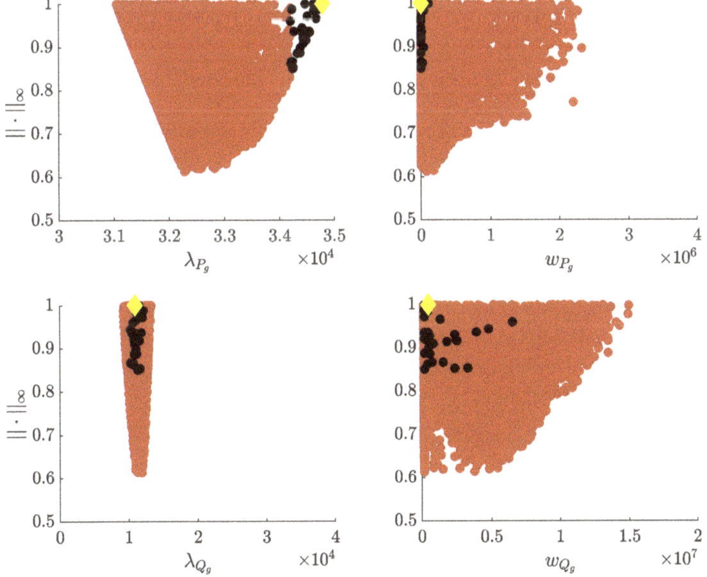

Figure 11. Pareto set of design concept 2 (X_p^{*C24p}) resulting from application of LD with ∞-norm. Black dots correspond to solutions selected to illustrate the trade-off among objectives. The yellow diamond marks the preferred four-parameter solution, $x^{C24p,\,B}$.

Table 5. Minimum values of the 18 performance indices achievable by the six-parameter controllers of design concept 1.

i	1	2	3	4	5	6	7	8	9
$f_{P_{g_i}}$	41.0553	8.0831	8.0372	8.3810	9.7569	12.2202	21.2045	18.7141	12.9973
$f_{Q_{g_i}}$	4.6689	2.8165	2.9554	3.1834	5.2693	6.3064	10.9742	10.9207	6.7105

Table 6. Minimum values of the 18 performance indices achievable by the four-parameter controllers of design concept 2. The five performance indices for which concept 2 does not reach the minimum values attainable by concept 1 are highlighted in bold.

i	1	2	3	4	5	6	7	8	9
$f_{P_{g_i}}$	41.0553	8.0831	8.0372	8.3810	**9.7749**	**12.3491**	21.2045	18.7141	**13.1103**
$f_{Q_{g_i}}$	4.6689	2.8165	2.9554	3.1834	5.2693	6.3064	**11.0244**	10.9207	**7.0129**

A thoughtful analysis of Figures 9–11, as well as of Tables 5 and 6, leads to the following conclusions:

- Figure 9 shows that the Pareto fronts corresponding to design concepts 1 ($f(X_p^{*C1_{6p}})$) and 2 ($f(X_p^{*C2_{4p}})$) practically overlap, their main differences being enclosed by dashed lines. It can be observed that there exist solutions of design concept 1 presenting a slight improvement, with respect to those of design concept 2, for the objectives $f_{P_{g_5}}$, $f_{P_{g_6}}$, $f_{P_{g_9}}$, $f_{Q_{g_7}}$, and $f_{Q_{g_9}}$, in accordance with that suggested by Tables 5 and 6. However, it was found that, in return, such solutions lose performance in the objectives $f_{P_{g_1}}$, $f_{P_{g_8}}$, and $f_{Q_{g_1}}$.
- The minimum values of the ∞-norm for design concepts 1 and 2 are, respectively, 0.575 and 0.613 (with a less than 4% difference), which means that the normalised distance to the ideal point, f^{ideal}, is practically the same.
- Aiming at illustrating the trade-off existing among objectives in more detail for both design concepts, the points of both Pareto fronts yielding lower values in $f_{P_{g_7}}$ were selected (see the black dots in Figure 9). Thanks to the synchronisation between objectives carried out by the LD tool, it becomes evident that the objectives $f_{P_{g_2}}$, $f_{P_{g_3}}$, $f_{P_{g_4}}$, $f_{P_{g_5}}$, $f_{P_{g_6}}$, and $f_{P_{g_9}}$ are in opposition to both $f_{P_{g_7}}$ and $f_{P_{g_8}}$, while no clear opposition is observable between objectives $f_{Q_{g_i}}$ and $f_{P_{g_i}}$.
- As the above-mentioned synchronisation also applies to the decision variables, the controller parameters marked with black dots in Figures 10 and 11 are precisely those leading to the solutions represented by black dots in Figure 9. In particular, analysis of the black dots in Figure 10 reveals that they are grouped around two different values of the parameter c_{P_g} (i.e., $c_{P_g} \simeq 0$ and $c_{P_g} \simeq 5$), whereas the great majority lead to a $c_{Q_g} \simeq 0$. The latter confirms that controllers with $c_{P_g} = c_{Q_g} = 0$, corresponding to the four-parameter 2-SMC variant, presented similar features to those of the six-parameter one.

Considering all four aspects above, it can be concluded that, although design concept 1 was slightly better than design concept 2, the greater simplicity of the four-parameter 2-SMC variant, compared to that with six parameters, may encourage the designer to eventually opt for the former.

4.2. Selection of 2-SMC Parameter Sets

In general terms, examination of the LDs displayed in Figure 9 reveals that, excluding the indices corresponding to the first stage of the test ($f_{P_{g_1}}$ and $f_{Q_{g_1}}$), the most unfavourable were those resulting from the seventh and eighth stages. Nevertheless, it should be considered that, while the latter two stages were intrinsic to common operation under non-ideal grid voltage, the former corresponded to a short-duration sporadic operating regime.

Accordingly, it is intended that the parameter sets selected for experimental evaluation correspond to solutions yielding outstanding values for $f_{P_{g_7}}$ and $f_{P_{g_8}}$, as those indices were related to the most demanding, though usual, operating conditions. Under this premise, two parameter sets giving rise to extremely similar $f_{P_{87}}$ and $f_{P_{88}}$ indices were chosen: One from design concept 1, referred to as $x^{C16p,A}$ henceforward, and the other from design concept 2, designated as $x^{C24p,B}$.

In particular, the parameter sets $x^{C16p,A}$ and $x^{C24p,B}$ are those leading, respectively, to the performance indices highlighted using green squares and yellow diamonds in the LDs of Figure 9. The exact values for the parameters of said two sets, displayed in Figures 10 and 11 following the same format, are those given as follows:

$$x^{C16p,A} = \begin{bmatrix} 5 & 34.64 \times 10^3 & 121.67 \times 10^3 & 82.3 & 11.606 \times 10^3 & 133.095 \times 10^3 \end{bmatrix} \tag{40}$$

$$x^{C24p,B} = \begin{bmatrix} 34.79 \times 10^3 & 10^3 & 10.91 \times 10^3 & 477.335 \times 10^3 \end{bmatrix}. \tag{41}$$

The precise values of the objectives that result from adopting parameter sets $x^{C16p,A}$ and $x^{C24p,B}$ are those provided in Tables 7 and 8, respectively.

Table 7. Values of the 18 indices produced by $x^{C16p,A}$.

i	1	2	3	4	5	6	7	8	9
$f_{P_{g_i}}(x^{C16p,A})$	64.432	10.141	10.005	10.328	11.180	12.950	21.450	19.076	13.616
$f_{Q_{g_i}}(x^{C16p,A})$	10.938	3.337	3.447	3.631	5.623	6.946	11.627	11.397	7.573

Table 8. Values of the 18 indices produced by $x^{C24p,B}$.

i	1	2	3	4	5	6	7	8	9
$f_{P_{g_i}}(x^{C24p,B})$	43.288	10.129	10.098	10.373	11.251	13.050	21.204	18.714	13.633
$f_{Q_{g_i}}(x^{C24p,B})$	21.180	3.218	3.319	3.493	5.581	6.993	11.645	11.327	7.634

4.3. Experimental Evaluation

4.3.1. Description of the Experimental Rig

As already pointed out at the beginning of Section 3.1, the whole tuning study presented in Section 4.1 was based on a simulation model of the 7 kW DFIG prototype adopted in [11] for experimentation. A diagram displaying how the main components of that prototype are connected to each other is depicted in Figure 12a, while the physical aspect of those main components is observable in Figure 12b,c.

As sketched in Figure 12a and evidenced by Figure 12b, a 15 kW armature-controlled DC motor, commanded by a commercial adjustable speed drive, is in charge of driving the 7 kW DFIG at the desired rotational speed. On the other hand, the low-cost equipment shown in Figure 12c was employed, so as to emulate two-phase voltage imbalances in a controlled manner.

In order to implement and run both the RSC and GSC control algorithms, rapid control prototyping was carried out by means of the Opal-RT OP5600 platform. As in the simulation test designed in Section 3.1, the adopted RSC control scheme (tuning included) was precisely that proposed in [11]. In good logic, the algorithm detailed in the functional diagram of Figure 2 was responsible for GSC control. In particular, the values for the K_p and T_i parameters of the DC bus voltage I-P controller were those provided in Table 1, while the x parameter set was modified according to the solution to be evaluated.

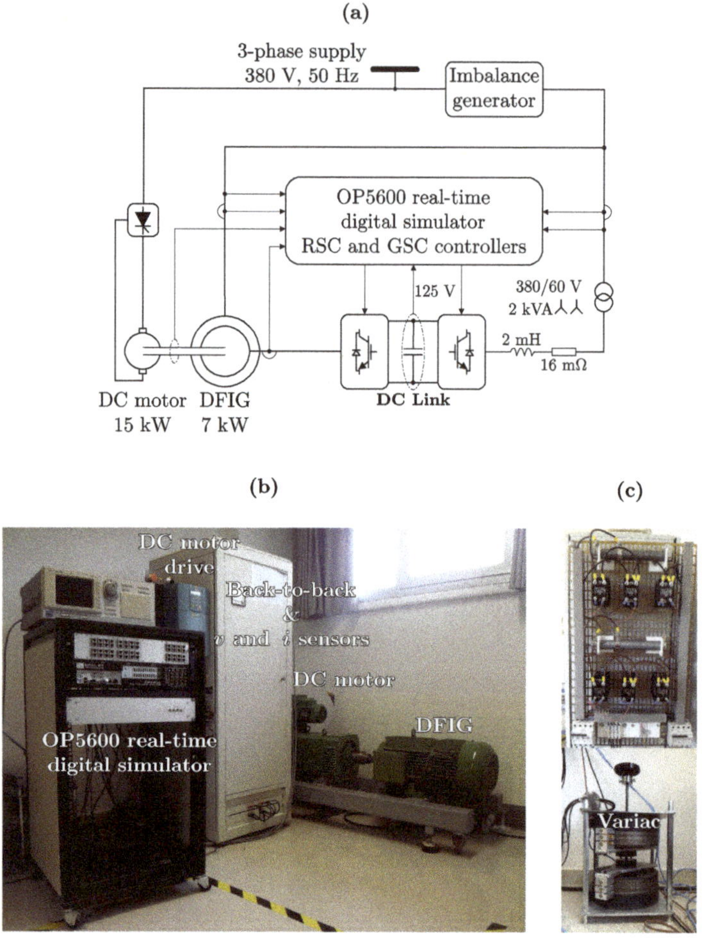

Figure 12. Experimental rig: (**a**) Connection diagram; (**b**) snapshot of the test bench containing the 7 kW DFIG prototype; and (**c**) controlled two-phase imbalance generator.

A 10 kHz switching frequency was adopted for both the GSC and the RSC, while their respective control algorithms were run at a 20 kHz sample rate.

4.3.2. Experimental Results

To conclude, the performances which the two parameter sets selected in Section 4.2 led to were experimentally evaluated and compared to each other, as well as to that resulting from applying the baseline solution. For that purpose, the simulation test described throughout Section 3.1 was reproduced experimentally, in the most faithful way possible. Nonetheless, specific features related to the generation of grid voltage imbalances and harmonic distortion needed to be accounted for, as well.

On one hand, it is well-known that the severity of the transients immediately following both the start and the conclusion of a given imbalance is highly dependent on the angles shown by the grid voltage space-vector at the initial and final instants of said imbalance [32]. Consequently, to prevent this factor from distorting the experimental results, the instants at which the imbalance begins and ends were controlled so that they always took place at the same angles of the grid voltage space-vector.

On the other hand, given that the available imbalance generator did not provide any control over the harmonic content of the grid voltage during the experimental tests, said grid voltage exhibited exactly the same harmonics naturally present in the grid of the laboratory that houses the DFIG prototype. This obviously implies that it was not possible to reproduce a grid voltage profile with identical harmonic content for any two different tests.

In order to minimise, as far as possible, the dispersion that differences in the harmonic content might cause in the performance indices, each of the three parameter sets under consideration did not undergo a unique experimental test, but a considerable number of them: 30 tests, specifically. Moreover, it was sought to perform the tests under grid conditions as similar as possible for each of the parameter sets under study. Accordingly, the trials for those three parameter sets were alternated with each other, repeating the $x^{C16p,ref}$, $x^{C16p,A}$, $x^{C24p,B}$ pattern 30 times; hence, completing 90 tests in total.

The results of those 90 tests are compiled in Figure 13, where each subfigure corresponds to one of the 18 performance indices. Three blue boxes are displayed in each subfigure, one for each parameter set assessed. Hence, for any given index, 30 data points lie behind each of such three boxes. The horizontal red line inside a certain box represents the median of those 30 data points, while its lower and upper edges delimit the 25th and 75th percentiles, respectively. Moreover, excluding outliers (shown as individual red crosses), the vertical black dashed lines outside the boxes extend up and down to the most extreme data points.

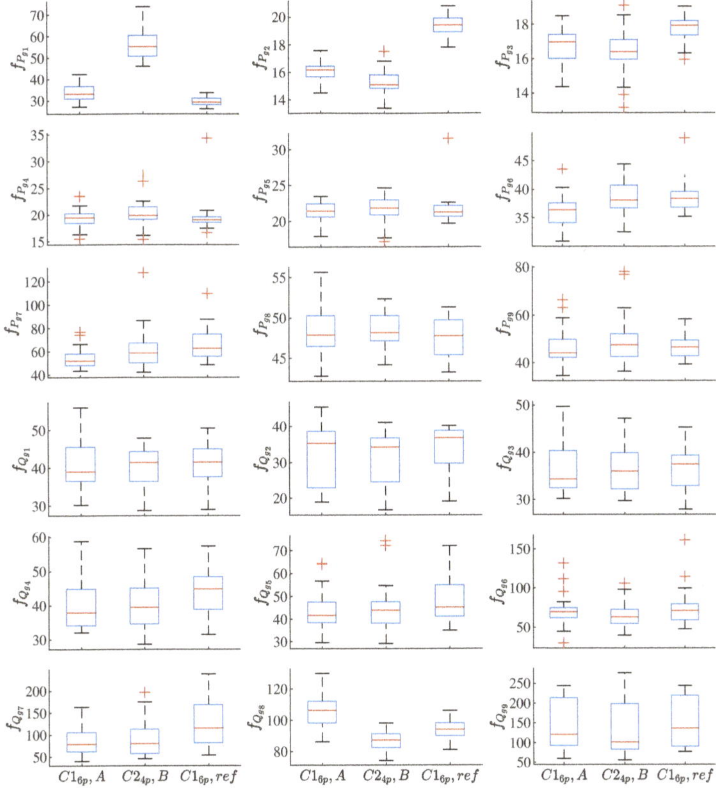

Figure 13. Comparison, for each performance index, of the boxplots representing the data points collected experimentally for the three parameter sets assessed.

As expected, the numerical values for the 18 indices differed from those obtained by running the simulation model, mainly because (as was already pointed out at the beginning of Section 3.1) the

Energies **2019**, *12*, 3782

latter treated the PWM–GSC set as ideal and did not reproduce the commutation of the GSC transistors. Furthermore, other non-idealities characteristic of experimentation, such as measurement errors and noise, also contributed towards increasing that discrepancy. In any case, it can be concluded that, generally speaking, the experimental results followed the same trends of the simulated ones for the different indices, except for those corresponding to the first stage.

Beyond confirming that the indices which the three parameter sets selected lead to were, in general, comparable to each other, a systematic analysis is required to discern, for each index, if any set of parameters performed significantly better than the other two. With that purpose, a multiple comparison test and a two-way analysis of variance (ANOVA2) were carried out for the three sets of 30 data points available, for each of the 18 performance indices.

The results of those studies are summarised in Table 9. Each of its three columns compares a different pair of the three parameter sets selected, while each of its 18 rows specifies the performance index for which the comparison was made. Black, blue, and red cells identify when significantly better indices were obtained by adopting the parameter sets $x^{C16p,ref}$, $x^{C16p,A}$, and $x^{C24p,B}$, respectively. In contrast, white cells indicate that the resulting indices were not significantly different from each other.

Table 9. Comparison, in pairs, of the parameter sets $x^{C16p,ref}$, $x^{C16p,A}$, and $x^{C24p,B}$ for each performance index. Black, blue, and red cells highlight those indices for which the solutions $x^{C16p,ref}$, $x^{C16p,A}$, and $x^{C24p,B}$ are significantly better, respectively.

	$x^{C16p,A}$ & $x^{C16p,ref}$	$x^{C24p,B}$ & $x^{C16p,ref}$	$x^{C16p,A}$ & $x^{C24p,B}$
$f_{P_{g1}}$	black	black	blue
$f_{P_{g2}}$	blue	red	red
$f_{P_{g3}}$	blue	red	–
$f_{P_{g4}}$	–	–	–
$f_{P_{g5}}$	–	–	–
$f_{P_{g6}}$	blue	–	blue
$f_{P_{g7}}$	blue	–	blue
$f_{P_{g8}}$	–	–	–
$f_{P_{g9}}$	–	–	–
$f_{Q_{g1}}$	–	–	–
$f_{Q_{g2}}$	–	–	–
$f_{Q_{g3}}$	–	–	–
$f_{Q_{g4}}$	blue	red	–
$f_{Q_{g5}}$	–	–	–
$f_{Q_{g6}}$	–	–	–
$f_{Q_{g7}}$	blue	red	–
$f_{Q_{g8}}$	black	red	red
$f_{Q_{g9}}$	–	–	–

Consequently, the first column of Table 9 reveals that, compared to the baseline solution, the parameter set $x^{C16p,A}$ led to poorer $f_{P_{g1}}$ and $f_{Q_{g8}}$ performance indices, but to significantly better $f_{P_{g2}}$, $f_{P_{g3}}$, $f_{P_{g6}}$, $f_{P_{g7}}$, $f_{Q_{g4}}$, and $f_{Q_{g7}}$ indices. It can, therefore, be concluded that the parameter set $x^{C16p,A}$ was overall better than the baseline one. Identical reasoning applied to the second column yields that the parameter set $x^{C24p,B}$ was also globally better than the baseline solution. Similarly, the last column demonstrates that the performance of solution $x^{C24p,B}$ was overall comparable to that of the parameter set $x^{C16p,A}$.

As a whole, it can thus be considered that both $x^{C1_{6p},A}$ and $x^{C2_{4p},B}$ parameter sets were better than the baseline solution, as well as comparable to each other, according to what the simulation results predicted.

5. Conclusions

With the aim of tuning the parameters of a 2-SMC scheme commanding the GSC of a DFIG, an a posteriori MOO approach has been presented and successfully applied in this paper, both in simulation and experimentally. Two variants (i.e., design concepts) of the same 2-SMC algorithm, which only differed in the switching functions adopted, were tuned and their respective performances were compared to each other. The first algorithm contained six parameters to be tuned, while the second, whose switching functions were simplified versions of those defined for the first one, contained just four. The grid voltage was assumed to be continuously harmonically polluted, as well as subject to imbalances. In this context, the tuning process was carried out in such a way that a single set of controller parameters was valid for nine possible operating regimes of the DFIG, three of which were directly related to the appearance of imbalances in the grid voltage.

In particular, two performance indices, f_{P_g} and f_{Q_g}, were defined for each of those nine operating regimes, which, respectively, quantify to what extent the reference values set for the grid active and reactive powers were complied with. As a result, the MOOP, on which the tuning is based, was set out by considering 18 indices in total. Driven by the high number of indices to be accounted for, the interactive tool of LDs was employed during the decision-making stage, with the purpose of facilitating analysis of the Pareto fronts (trade-off among objectives) and assisting selection of the preferred parameter sets.

The optimisation process gave rise to a Pareto front for each of the two design concepts considered. Analysis of those two Pareto fronts led to the following conclusions:

- Taking a set of experimentally validated parameters as starting point, multiple solutions to the MOO-based tuning problem were found, through simulation, by demanding that each and every one of the 18 performance indices they lead to were better than those obtained when applying the baseline parameter set.
- As expected, trade-offs among some of the $f_{P_{g_i}}$ performance indices, with $i = 1, 2, 3 \ldots 9$, became evident. In contrast, the compromise between indices $f_{P_{g_i}}$ and $f_{Q_{g_i}}$ was found to be not as marked as intuitively thought beforehand.
- Although a number of solutions for the six-parameter 2-SMC algorithm behaved slightly better than those corresponding to the four-parameter variant for five of the performance indices, they also gave poorer values for another three. In summary, the six-parameter variant of the 2-SMC algorithm does not dominate that of four-parameter variant.

Considering the designer preferences, two sets of parameters (one from each design concept) were selected and compared experimentally to each other, as well as to the baseline parameter set. To that end, aiming at reducing the impact that the variability of the harmonic distortion present in the grid voltage can have on the performance indices, each of those three parameter sets underwent the same test 30 times.

A statistical analysis of the results derived from the total of 90 experimental tests carried out allows us to draw the following main conclusions:

- In good logic, it has been corroborated that the two solutions selected globally improve the performance of the parameter set adopted as a baseline solution.
- Performances comparable to those resulting from application of the six-parameter 2-SMC algorithm are achievable by using its simplified four-parameter version.

Author Contributions: Conceptualisation, G.T.-O. and X.B.; Data curation, A.S. and J.M.H.; Formal analysis, J.M.H. and X.B.; Funding acquisition, J.M.H., G.T.-O., and X.B.; Investigation, A.S., J.M.H., and M.I.M.; Methodology, A.S. and M.I.M.; Software, A.S., J.M.H., M.I.M., and G.T.-O.; Supervision, G.T.-O. and X.B.; Validation, A.S. and M.I.M.; Visualisation, A.S. and J.M.H.; Writing—original draft, J.M.H., M.I.M., and G.T.-O.; Writing—review & editing, A.S., M.I.M., G.T.-O., and X.B.

Funding: This research was co-funded by the Spanish Ministry of Economy and Competitiveness—project codes DPI2015-64985-R and RTI2018-096904-B-I00—and FEDER Funds, EU. The authors from the University of the Basque Country UPV/EHU are with the "Intelligent Systems and Energy (SI+E)" research group, funded by UPV/EHU—research grant GIU16/54—and the Basque Government—research grant IT1256-19.

Acknowledgments: The authors wish to express their gratitude to Imanol Bardají for his always valuable advice when dealing with the experimental rig.

Conflicts of Interest: The authors declare no conflict of interest. The founding sponsors had no role in the design of the study; in the collection, analyses, or interpretation of data; in the writing of the manuscript, and in the decision to publish the results.

References

1. Yaramasu, V.; Wu, B.; Sen, P.C.; Kouro, S.; Narimani, M. High-power wind energy conversion systems: state-of-the-art and emerging technologies. *Proc. IEEE* **2015**, *103*, 740–788. [CrossRef]
2. Xu, L. Coordinated control of DFIG's rotor and grid side converters during network unbalance. *IEEE Trans. Power Electron.* **2008**, *23*, 1041–1049.
3. Hu, J.; He, Y.; Xu, L.; Williams, B.W. Improved control of DFIG systems during network unbalance using PI-R current regulators. *IEEE Trans. Ind. Electron.* **2009**, *56*, 439–451. [CrossRef]
4. Xu, H.; Hu, J.; He, Y. Operation of wind-turbine-driven DFIG systems under distorted grid voltage conditions: analysis and experimental validations. *IEEE Trans. Power Electron.* **2012**, *27*, 2354–2366. [CrossRef]
5. Sharifi, M.; Shahriari, B.; Bagheri, A. Optimization of sliding mode control for a vehicle suspension system via multi-objective genetic algorithm with uncertainty. *J. Basic Appl. Sci. Res.* **2012**, *2*, 6724–6729.
6. Mahmoodabadi, M.J.; Arabani Mostaghim, S.; Bagheri, A.; Nariman-zadeh, N. Pareto optimal design of the decoupled sliding mode controller for an inverted pendulum system and its stability simulation via Java programming. *Math. Comput. Modell.* **2013**, *57*, 1070–1082. [CrossRef]
7. Alitavoli, M.; Taherkhorsandi, M.; Mahmoodabadi, M.J.; Bagheri, A.; Miripour-Fard, A. Pareto design of sliding-mode tracking control of a biped robot with aid of an innovative particle swarm optimization. In Proceedings of the International Symposium on Innovations in Intelligent Systems and Applications, Trabzon, Turkey, 2–4 July 2012; pp. 1–5.
8. Elgammal, A.A.A.; El-naggar, M.F. MOPSO-based optimal control of shunt active power filter using a variable structure fuzzy logic sliding mode controller for hybrid (FC-PV-Wind-Battery) energy utilisation scheme. *IET Renew. Power Gener.* **2017**, *11*, 1148–1156. [CrossRef]
9. Qin, Z.C.; Xiong, F.R.; Ding, Q.; Hernández, C.; Fernandez, J.; Schütze, O.; Sun, J.Q. Multi-objective optimal design of sliding mode control with parallel simple cell mapping method. *J. Vib. Control* **2017**, *23*, 46–54. [CrossRef]
10. Trebi-Ollennu, A.; White, B.A. Multiobjective fuzzy genetic algorithm optimisation approach to nonlinear control system design. *IEE Proc.-Control Theory Appl.* **1997**, *144*, 137–142. [CrossRef]
11. Martínez, M.I.; Susperregui, A.; Tapia, G. Second-order sliding-mode-based global control scheme for wind turbine-driven DFIGs subject to unbalanced and distorted grid voltage. *IET Electr. Power Appl.* **2017**, *11*, 1013–1022. [CrossRef]
12. Mattson, C.A.; Messac, A. Pareto frontier based concept selection under uncertainty, with visualization. *Optim. Eng.* **2005**, *6*, 85–115. [CrossRef]
13. Herrero, J.M.; Reynoso-Meza, G.; Martínez, M.; Blasco, X.; Sanchis, J. A smart-distributed Pareto front using the ev-MOGA evolutionary algorithm. *Int. J. Artif. Intell. Tools* **2014**, *23*, 1450002. [CrossRef]
14. Coello, C.A.; Lamont, G.L.; van Veldhuizen, D.A. *Evolutionary Algorithms for Solving Multi-Objective Problems*, 2nd ed.; Genetic and Evolutionary Computation; Springer: Berlin/Heidelberg, Germany, 2007.
15. Blasco, X.; Herrero, J.M.; Sanchis, J.; Martínez, M. A new graphical visualization of *n*-dimensional Pareto front for decision-making in multiobjective optimization. *Inf. Sci.* **2008**, *178*, 3908–3924. [CrossRef]

16. Blasco, X.; Herrero, J.M.; Reynoso-Meza, G.; Martínez Iranzo, M.A. Interactive tool for analyzing multiobjective optimization results with level diagrams. In Proceedings of the Genetic and Evolutionary Computation Conference Companion, Berlin, Germany, 15–19 July 2017; pp. 1689–1696.

17. Huilcapi, V.; Blasco, X.; Herrero, J.M.; Reynoso-Meza, G. A loop pairing method for multivariable control systems under a multi-objective optimization approach. *IEEE Access* **2019**, *7*, 81994–82014. [CrossRef]

18. Huilcapi, V.; Lima, B.; Blasco, X.; Herrero, J.M. Optimización multiobjetivo en modelado y control de un péndulo invertido rotatorio. *Rev. Iberoam. Autom. Inf. Ind.* **2018**, *15*, 363–373. [CrossRef]

19. Boada, Y.; Reynoso-Meza, G.; Picó, J.; Vignoni, A. Multi-objective optimization framework to obtain model-based guidelines for tuning biological synthetic devices: an adaptive network case. *BMC Syst. Biol.* **2016**, *10*, 27. [CrossRef] [PubMed]

20. Beltran, B.; Benbouzid, M.E.H.; Ahmed-Ali, T. Second-order sliding mode control of a doubly fed induction generator driven wind turbine. *IEEE Trans. Energy Convers.* **2012**, *27*, 261–269. [CrossRef]

21. Vas, P. *Sensorless Vector and Direct Torque Control*, 1st ed.; Oxford University Press: New York, NY, USA, 1998.

22. Martinez, M.I.; Susperregui, A.; Tapia, G.; Xu, L. Sliding-mode control of a wind turbine-driven double-fed induction generator under non-ideal grid voltages. *IET Renew. Power Gener.* **2013**, *7*, 370–379. [CrossRef]

23. Rodriguez, P.; Timbus, A.V.; Teodorescu, R.; Liserre, M.; Blaabjerg, F. Flexible active power control of distributed power generation systems during grid faults. *IEEE Trans. Ind. Electron.* **2007**, *54*, 2583–2592. [CrossRef]

24. Utkin, V.; Guldner, J.; Shi, J. *Sliding Mode Control in Electro-Mechanical Systems*, 2nd ed.; CRC Press: Boca Raton, FL, USA, 2009.

25. Miettinen, K.M. *Nonlinear Multiobjective Optimization*, 1st ed.; Kluwer Academic Publishers: Boston, MA, USA, 1999.

26. Coello, C.A.; Lamont, G.B. *Applications of Multi-Objective Evolutionary Algorithms*, 1st ed.; World Scientific: Singapore, 2004.

27. Reynoso-Meza, G.; Blasco, X.; Sanchis, J.; Herrero, J.M. Comparison of design concepts in multi-criteria decision-making using level diagrams. *Inf. Sci.* **2013**, *221*, 124–141. [CrossRef]

28. Tušar, T.; Filipič, B. Visualization of Pareto front approximations in evolutionary multiobjective optimization: A critical review and the prosection method. *IEEE Trans. Evol. Comput.* **2015**, *19*, 225–245. [CrossRef]

29. Reynoso, G.; Blasco, X.; Sanchis, J.; Herrero, J.M. *Controller Tuning with Evolutionary Multiobjective Optimization: A Holistic Multiobjective Optimization Design Procedure*, 1st ed.; Springer International Publishing AG: Gewerbestrasse, Switzerland, 2017.

30. Pajares, A.; Blasco, X.; Herrero, J.M.; Reynoso-Meza, G. A new point of view in multivariable controller tuning under multiobjetive optimization by considering nearly optimal solutions. *IEEE Access* **2019**, *7*, 66435–66452. [CrossRef]

31. Herrero, J.M. Identificación Robusta de Sistemas no Lineales mediante Algoritmos Evolutivos. Ph.D. Thesis, Universitat Politècnica de València, València, Spain, October 2006.

32. López, J.; Gubía, E.; Sanchis, P.; Roboam, X.; Marroyo, L. Wind turbines based on doubly-fed induction generator under asymmetrical voltage dips. *IEEE Trans. Energy Convers.* **2008**, *23*, 321–330. [CrossRef]

Article

Adaptive-Gain Second-Order Sliding Mode Direct Power Control for Wind-Turbine-Driven DFIG under Balanced and Unbalanced Grid Voltage

Yaozhen Han * and Ronglin Ma *

School of Information Science and Electrical Engineering, Shandong Jiaotong University, Jinan 250357, China
* Correspondence: hyz125@163.com (Y.H.); maronglin@sdjtu.edu.cn (R.M.)

Received: 26 August 2019; Accepted: 10 October 2019; Published: 14 October 2019

Abstract: In a wind turbine system, a doubly-fed induction generator (DFIG), with nonlinear and high-dimensional dynamics, is generally subjected to unbalanced grid voltage and unknown uncertainty. This paper proposes a novel adaptive-gain second-order sliding mode direct power control (AGSOSM-DPC) strategy for a wind-turbine-driven DFIG, valid for both balanced and unbalanced grid voltage. The AGSOSM-DPC control scheme is presented in detail to restrain rotor voltage chattering and deal with the scenario of unknown uncertainty upper bound. Stator current harmonics and electromagnetic torque ripples can be simultaneously restrained without phase-locked loop (PLL) and phase sequence decomposition using new active power expression. Adaptive control gains are deduced based on the Lyapunov stability method. Comparative simulations under three DPC schemes are executed on a 2-MW DFIG under both balanced and unbalanced grid voltage. The proposed strategy achieved active and reactive power regulation under a two-phase stationary reference frame for both balanced and unbalanced grid voltage. An uncertainty upper bound is not needed in advance, and the sliding mode control chattering is greatly restrained. The simulation results verify the effectiveness, robustness, and superiority of the AGSOSM-DPC strategy.

Keywords: DFIG; adaptive-gain second-order sliding mode; direct power control; balanced and unbalanced grid voltage

---

1. Introduction

Over the past decade, renewable energy generation has continued to grow rapidly due to widely known problems such as environmental pollution and resource shortage [1]. Wind power generation accounted for 21% of renewable generating capacity until the end of 2018 [2]. The doubly-fed induction generator (DFIG) has become the most widely used electric generator in wind turbine systems, owing to its inherent advantages including high system efficiency, low converter rating, four-quadrant active and reactive power capability, variable wind speed operation, and controllable power factor [3,4].

DFIG control is one of the most difficult issues in a wind turbine system because DFIG dynamics is intrinsically nonlinear and high-dimensional; system model parameters are uncertain; and the encountered wind speed is random [5,6]. Currently, the main control techniques for DFIG in industrial application are vector control (VC) and direct power control (DPC) [7]. Although the VC method has outstanding steady-state performance, its dynamic performance is rather disillusionary due to the hysteresis of proportional integral (PI) control [8]. Many studies improved the traditional VC [9–12], but some inherent problems still exist, such as that the control algorithm is complex, and synchronous coordinate transform and phase-locked loop (PLL) are still prerequisites [13,14].

DPC is a useful alternative to improve dynamic performance. Current rotor control loops are not required for a DPC scheme, and a switching table is directly used to select a suitable switching vector. Hence, the control structure is easy to implement, and excellent dynamic performance is achieved [15]. Many modified achievements have been published based on this traditional DPC idea [16–18]. However, the switching frequency is unfixed, unacceptable power ripples still exist, and system robustness should be improved [19].

Among all the nonlinear control approaches, sliding mode control, which is used in wind turbine systems, is a robust control method capable of providing finite-time convergence, disturbance suppression, fast response, and simple implementation [20,21]. Some studies have evaluated the direct power sliding mode control for DFIG [22–27]. Although good results were achieved, some drawbacks still exist, and positive and negative sequence decomposition is needed [23]; all these studies focused on the conventional first-order sliding mode (FOSM), with unsatisfactory control switching and variable switching frequency. These intrinsic drawbacks caused by the FOSM may produce torque ripple, harmonic current, overheating of the windings, etc. [28].

The well-known super-twisting second order sliding mode (SOSM)method hides a discontinuous item under the integral and generates continuous control action to propel the sliding mode vector and its derivative to the origin in finite time, and then the control chattering can be greatly attenuated [29]. Some literature discussed the applications for DFIG [30–32], which need synchronous rotating frame transformation and were mainly concentrated on SOSMVC control.

In recent years, some scholars set about studying the second-order sliding mode(SOSM) DPC under stationary reference frames. Susperregui et al. [33] proposed a fixed-frequency PWM-based rotor converter control and a reactive power control under balanced grid voltage. In the study following [33], Reference [34] achieved power regulation and grid synchronization based on a SOSM control scheme. Yet, it was also verified under a balanced grid voltage. Martinez presented a SOSM global control scheme for DFIG suffering from unbalanced and distorted grid voltage [35]. Both rotor side and grid side power converters were controlled via a super-twisting algorithm. However, the mechanism of dealing with unbalanced grid voltage condition was not mentioned and the upper bound of uncertainty, which cannot be easily estimated in many practical cases, was hypothetically known. Reference [36] presented a super-twisting DPC scheme for adjusting active and reactive power in detail. Yet, the strategy only focused on a balanced voltage scenario. Unbalanced grid voltage, natural flux, and uncertainty upper bound conditions were not considered. As the continuity of the study presented in Reference [36], the controller in Reference [37] was designed with adaptive control gain to handle the unknown uncertainty upper bound. Yet, the unbalanced grid voltage condition, which may cause a severe power harmonic, was still undiscussed.

As mentioned above, some key issues should be considered simultaneously in a DPC scheme, including the following: (a) The scheme should deal with the uncertainty upper bound scenario. The overestimation of uncertainty can produce redundant control gain. (b) Switching frequency is fixed and control chattering can be significantly restrained. (c) The scheme needs a simple control structure, needless of phase sequence decomposition, coordinate transformation, and PLL. (d) The scheme should be valid for both balanced and unbalanced grid voltage, and stator current harmonic and electromagnetic torque ripple can be restrained simultaneously. (e) The scheme needs to filter the roughly static stator flux component. Hence, this paper proposes an adaptive-gain SOSM (AGSOSM-DPC) scheme for DFIG subjected to balanced and unbalanced grid voltage. First, a new active power expression is applied based on a detailed phase sequence analysis. Then, an adaptive control gain law of SOSM DPC is designed via the Lyapunov stability method to solve the unknown upper bound of uncertainty. A band-pass filter was also used to calculate stator flux and avoid stator natural flux. The main contributions of the paper include the following: (1) Under a two-phase stationary reference frame, a novel AGSOSM-DPC strategy for DFIG is proposed that can solve the abovementioned key issues simultaneously; (2) using adaptive control gain, the upper bound of uncertainty is not necessarily known in advance; and (3) rotor voltage control chattering is highly suppressed via a super-twisting algorithm.

The rest of the paper is organized as follows: An elaborated model analysis is presented as Section 2. Section 3 presents the controller design and stability analysis. Section 4 shows the comparative simulation results obtained on a 2-MW DFIG. Finally, Section 5 provides the conclusions.

2. Model Analysis

The diagrammatic drawings of DFIG-based wind turbine system and DFIG equivalent circuit under a two-phase stationary reference frame are shown in Figure 1. To facilitate the analysis and design, the DFIG equivalent model under two-phase stationary frame can be presented as follows [22]:

$$\begin{cases} \boldsymbol{\psi}_{s\alpha\beta} = L_s \boldsymbol{I}_{s\alpha\beta} + L_m \boldsymbol{I}_{r\alpha\beta} \\ \boldsymbol{\psi}_{r\alpha\beta} = L_r \boldsymbol{I}_{r\alpha\beta} + L_m \boldsymbol{I}_{s\alpha\beta} \\ \boldsymbol{U}_{s\alpha\beta} = R_s \boldsymbol{I}_{s\alpha\beta} + \dfrac{d\boldsymbol{\psi}_{s\alpha\beta}}{dt} \\ \boldsymbol{V}_{r\alpha\beta} = R_r \boldsymbol{I}_{r\alpha\beta} + \dfrac{d\boldsymbol{\psi}_{r\alpha\beta}}{dt} - j\omega_r \boldsymbol{\psi}_{r\alpha\beta} \\ T_e = \frac{3}{2} p \mathrm{Im}\left\{ \boldsymbol{\psi}^*_{s\alpha\beta} \boldsymbol{I}_{s\alpha\beta} \right\} \\ S_s = P_s + jQ_s = \frac{3}{2} \boldsymbol{I}^*_{s\alpha\beta} \boldsymbol{U}_{s\alpha\beta} \end{cases} \tag{1}$$

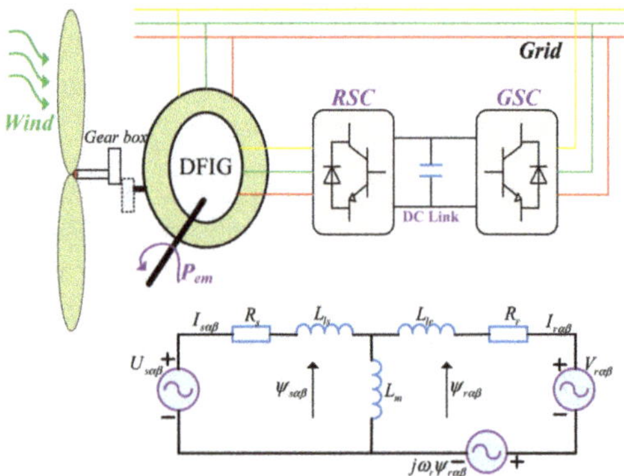

Figure 1. Doubly-fed induction generator (DFIG)-based wind power system schematic diagram and DFIG equivalent circuit under a two-phase stationary coordinate frame.

Although phase sequence decomposition is not needed in the proposed control scheme, both positive and negative sequence components of stator voltage and current are listed to analyze the effect of DPC when DFIG is subjected to unbalanced grid voltage.

$$\begin{cases} \boldsymbol{U}_{s\alpha\beta} = u_{s\alpha} + j u_{s\beta} = \boldsymbol{U}^+_{s\alpha\beta} + \boldsymbol{U}^-_{s\alpha\beta} \\ \boldsymbol{U}^+_{s\alpha\beta} = u^+_{s\alpha} + j u^+_{s\beta},\, \boldsymbol{U}^-_{s\alpha\beta} = u^-_{s\alpha} + j u^-_{s\beta} \\ u^+_{s\alpha} = \left|\boldsymbol{U}^+_{s\alpha\beta}\right|\cos(\omega_s t + \theta^+_u),\, u^+_{s\beta} = \left|\boldsymbol{U}^+_{s\alpha\beta}\right|\sin(\omega_s t + \theta^+_u) \\ u^-_{s\alpha} = \left|\boldsymbol{U}^-_{s\alpha\beta}\right|\cos(\omega_s t + \theta^-_u),\, u^-_{s\beta} = \left|\boldsymbol{U}^-_{s\alpha\beta}\right|\sin(\omega_s t + \theta^-_u) \end{cases} \tag{2}$$

$$\begin{cases} \boldsymbol{I}_{s\alpha\beta} = i_{s\alpha} + j i_{s\alpha} = \boldsymbol{I}^+_{s\alpha\beta} + \boldsymbol{I}^-_{s\alpha\beta} \\ \boldsymbol{I}^+_{s\alpha\beta} = i^+_{s\alpha} + j i^+_{s\beta},\, \boldsymbol{I}^-_{s\alpha\beta} = i^-_{s\alpha} + j i^-_{s\beta} \\ i^+_{s\alpha} = \left|\boldsymbol{I}^+_{s\alpha\beta}\right|\cos(\omega_s t + \theta^+_i),\, i^+_{s\beta} = \left|\boldsymbol{I}^+_{s\alpha\beta}\right|\sin(\omega_s t + \theta^+_i) \\ i^-_{s\alpha} = \left|\boldsymbol{I}^-_{s\alpha\beta}\right|\cos(\omega_s t + \theta^-_i),\, i^-_{s\beta} = \left|\boldsymbol{I}^-_{s\alpha\beta}\right|\sin(\omega_s t + \theta^-_i) \end{cases} \tag{3}$$

where $\left|U_{s\alpha\beta}^{+}\right|$, $\left|U_{s\alpha\beta}^{-}\right|$, $\left|I_{s\alpha\beta}^{+}\right|$, and $\left|I_{s\alpha\beta}^{-}\right|$ are the amplitudes of positive and negative sequence components of stator voltage and stator current. The values θ_{u}^{+}, θ_{u}^{-}, θ_{i}^{+}, and θ_{i}^{-} are the corresponding initial phase angles.

Substituting Formulas (2) and (3) into Formula (1), the stator instantaneous power can be deduced as follows:

$$\begin{cases} P_s = P_{s0} + P_{s1} + P_{s2} \\ Q_s = Q_{s0} + Q_{s1} + Q_{s2} \end{cases} \tag{4}$$

where P_{s0} and Q_{s0} are the average values of active and reactive power; P_{s1}, P_{s2}, Q_{s1}, and Q_{s2} are the oscillating components at twice the grid frequency of active and reactive power, respectively. According to Formulas (2) and (3), these oscillating components can be represented as follows:

$$\begin{cases} P_{s1} = \frac{3}{2}(u_{s\alpha}^{+}i_{s\alpha}^{-} + u_{s\beta}^{+}i_{s\beta}^{-}) = \frac{3}{2}\left|U_{s\alpha\beta}^{+}\right|\left|I_{s\alpha\beta}^{-}\right|\cos(2\omega_s t + \theta_{u+} - \theta_{i-}) \\ P_{s2} = \frac{3}{2}(u_{s\alpha}^{-}i_{s\alpha}^{+} + u_{s\beta}^{-}i_{s\beta}^{+}) = \frac{3}{2}\left|U_{s\alpha\beta}^{-}\right|\left|I_{s\alpha\beta}^{+}\right|\cos(2\omega_s t + \theta_{i+} - \theta_{u-}) \end{cases} \tag{5}$$

$$\begin{cases} Q_{s1} = \frac{3}{2}(u_{s\beta}^{+}i_{s\alpha}^{-} - u_{s\alpha}^{+}i_{s\beta}^{-}) = \frac{3}{2}\left|U_{s\alpha\beta}^{+}\right|\left|I_{s\alpha\beta}^{-}\right|\sin(2\omega_s t + \theta_{u+} - \theta_{i-}) \\ Q_{s2} = \frac{3}{2}(u_{s\beta}^{-}i_{s\alpha}^{+} - u_{s\alpha}^{-}i_{s\beta}^{+}) = \frac{3}{2}\left|U_{s\alpha\beta}^{-}\right|\left|I_{s\alpha\beta}^{+}\right|\sin(2\omega_s t + \theta_{i+} - \theta_{u-}) \end{cases} \tag{6}$$

This clearly indicates that P_{s1}, Q_{s1} and P_{s2}, Q_{s2}, caused by positive sequence voltage and negative sequence current and negative sequence voltage and positive sequence current, respectively, are the oscillating parts with twice the grid frequency. A third harmonic current is generated and causes severe harmonic distortion if active and reactive power are both simultaneously maintained as a constant under the unbalanced grid voltage condition.

Considering Formulas (2) and (3), the electromagnetic torque can be expressed as follows:

$$T_e = T_{e0} + T_{e1} + T_{e2}, \tag{7}$$

where T_{e0} is the average value of electromagnetic torque; T_{e1} and T_{e2} are the oscillating parts with twice the grid frequency. To neglect the effect of stator resistance, T_{e1} and T_{e2} can be denoted as follows:

$$\begin{cases} T_{e1} = \frac{3}{2\omega_1}(u_{s\alpha}^{+}i_{s\alpha}^{-} + u_{s\beta}^{+}i_{s\beta}^{-}) = \frac{3}{2\omega_1}\left|U_{s\alpha\beta}^{+}\right|\left|I_{s\alpha\beta}^{-}\right|\cos(2\omega_s t + \theta_{u+} - \theta_{i-}) \\ T_{e2} = \frac{3}{2\omega_1}(u_{s\alpha}^{-}i_{s\alpha}^{+} + u_{s\beta}^{-}i_{s\beta}^{+}) = \frac{3}{2\omega_1}\left|U_{s\alpha\beta}^{-}\right|\left|I_{s\alpha\beta}^{+}\right|\cos(2\omega_s t + \theta_{i+} - \theta_{u-}) \end{cases} \tag{8}$$

In Formula (8), it is clearly indicated that T_{e1} and T_{e2} are caused by positive sequence voltage and negative sequence current, and negative sequence voltage and positive sequence current, respectively, and are the oscillating parts with twice the grid frequency, which may generate bearing chattering and influence the service life.

To achieve active and reactive power tracking under balanced grid voltage, and also to suppress stator current harmonics and electromagnetic torque ripples under unbalanced grid voltage, a new active power expression is used to track instead of the traditional active expression [18,25,38]. The new active power can be expressed as follows:

$$P_{sn} = -\frac{3}{2}\mathrm{Im}(\hat{I}_{s\alpha\beta}\widehat{U}_{s\alpha\beta}) \tag{9}$$

In Formula (9), $\widehat{U}_{s\alpha\beta}$ is the value which lags $U_{s\alpha\beta}$ by 90 electrical degrees.

The delayed value of stator voltages can be represented as follows:

$$\begin{cases} \widehat{u}_{s\alpha}^{+} = \left|U_{s\alpha\beta}^{+}\right|\cos(\omega_s t + \theta_u^{+} - \pi/2) = \left|U_{s\alpha\beta}^{+}\right|\sin(\omega_s t + \theta_u^{+}) = u_{s\beta}^{+} \\ \widehat{u}_{s\beta}^{+} = \left|U_{s\alpha\beta}^{+}\right|\sin(\omega_s t + \theta_u^{+} - \pi/2) = -\left|U_{s\alpha\beta}^{+}\right|\cos(\omega_s t + \theta_u^{+}) = -u_{s\alpha}^{+} \\ \widehat{u}_{s\alpha}^{-} = \left|U_{s\alpha\beta}^{-}\right|\cos(-\omega_s t + \theta_u^{-} + \pi/2) = -\left|U_{s\alpha\beta}^{-}\right|\sin(-\omega_s t + \theta_u^{-}) = -u_{s\beta}^{-} \\ \widehat{u}_{s\beta}^{-} = \left|U_{s\alpha\beta}^{-}\right|\sin(-\omega_s t + \theta_u^{-} + \pi/2) = \left|U_{s\alpha\beta}^{-}\right|\cos(-\omega_s t + \theta_u^{-}) = u_{s\alpha}^{-} \end{cases} \tag{10}$$

According to Formulas (1), (9), and (10), the representations of new active power and traditional active power are consistent when only positive sequence voltage exists. This indicates that the power tracking strategy can still work well via the new active power under balanced grid voltage.

P_{sn} can be then further denoted as follows:

$$P_{sn} = P_{sn0} + P_{sn1} + P_{sn2} \tag{11}$$

The equation $P_{sn0} = P_{s0}$ can be easily satisfied, and P_{sn1} and P_{sn2} are represented as follows:

$$\begin{cases} P_{sn1} = \frac{3}{2}\left(\widehat{u}_{s\alpha}^{+}i_{s\beta}^{-} - \widehat{u}_{s\beta}^{+}i_{s\alpha}^{-}\right) = \frac{3}{2}\left(u_{s\beta}^{+}i_{s\beta}^{-} + u_{s\alpha}^{+}i_{s\alpha}^{-}\right) = \frac{3}{2}\left|U_s^{+}\right|\left|I_s^{-}\right|\cos(2\omega_s t + \theta_u^{+} - \theta_i^{-}) \\ P_{sn2} = \frac{3}{2}\left(\widehat{u}_{s\alpha}^{-}i_{s\beta}^{+} - \widehat{u}_{s\beta}^{-}i_{s\alpha}^{+}\right) = -\frac{3}{2}\left(u_{s\beta}^{-}i_{s\beta}^{+} + u_{s\alpha}^{-}i_{s\alpha}^{+}\right) = -\frac{3}{2}\left|U_s^{-}\right|\left|I_s^{+}\right|\cos(2\omega_s t + \theta_i^{+} - \theta_u^{-}) \end{cases} \tag{12}$$

According to Formulas (8) and (12):

$$\begin{cases} T_{e1} = \frac{P_{sn1}}{\omega_s} \\ T_{e2} = \frac{P_{sn2}}{\omega_s} \end{cases} \tag{13}$$

Because θ_u^{+}, θ_u^{-}, θ_i^{+}, and θ_i^{-} are the initial phase angles of positive and negative parts of voltage and current, then:

$$\theta_i^{+} - \theta_u^{+} = -(\theta_i^{-} - \theta_u^{-}) \tag{14}$$

Then:

$$\sin(\theta_u^{+} + \theta_u^{-} - \theta_i^{+} - \theta_i^{-}) = \sin(2\omega_s t + \theta_u^{+} - \theta_i^{-})\cos(2\omega_s t + \theta_i^{+} - \theta_u^{-}) \\ - \cos(2\omega_s t + \theta_u^{+} - \theta_i^{-})\sin(2\omega_s t + \theta_i^{+} - \theta_u^{-}) \quad = 0 \tag{15}$$

Therefore:

$$\frac{\cos(2\omega_s t + \theta_i^{+} - \theta_u^{-})}{\sin(2\omega_s t + \theta_i^{+} - \theta_u^{-})} = \frac{\cos(2\omega_s t + \theta_u^{+} - \theta_i^{-})}{\sin(2\omega_s t + \theta_u^{+} - \theta_i^{-})} \tag{16}$$

$P_{sn1} + P_{sn2} = 0$ and $Q_{s1} + Q_{s2} = 0$ can be satisfied simultaneously according to Formulas (6), (12), and (16). Thus, if the reference active power is placed as a constant and to track P_{sn}, first, $P_{sn1} + P_{sn2} = 0$ is satisfied, and then $Q_{s1} + Q_{s2} = 0$ is established. Thus, a harmonic attenuation for reactive power can be achieved. Moreover, in terms of Formula (13), $T_e = T_{e0}$ is satisfied and electromagnetic torque ripples are restrained.

According to Formula (9), the derivatives of P_s and Q_s can be represented as follows:

$$\begin{cases} \frac{dP_s}{dt} = -\frac{3}{2}\mathrm{Im}(\frac{d\widehat{i}_{s\alpha\beta}}{dt}\widehat{U}_{s\alpha\beta} + \widehat{I}_{s\alpha\beta}\frac{d\widehat{U}_{s\alpha\beta}}{dt}) \\ \frac{dQ_s}{dt} = \frac{3}{2}\mathrm{Im}(\frac{d\widehat{i}_{s\alpha\beta}}{dt}U_{s\alpha\beta} + \widehat{I}_{s\alpha\beta}\frac{dU_{s\alpha\beta}}{dt}) \end{cases} \tag{17}$$

Considering Formula (1) and neglecting the effect of stator resistance and rotor resistance, the derivative of stator current can be deduced as follows:

$$\frac{dI_{s\alpha\beta}}{dt} = -\frac{1}{\rho L_m}\left[V_{r\alpha\beta} - \frac{L_r}{L_m}U_{s\alpha\beta} + j\omega_r\left(-\rho L_m I_{s\alpha\beta} + \frac{L_r}{L_m}\psi_{s\alpha\beta}\right)\right], \tag{18}$$

where $\rho = (L_r L_s / L_m^2 - 1)$.

The derivative of stator voltage under stationary reference frame can be denoted as follows:

$$\frac{dU_{s\alpha\beta}}{dt} = -j\omega_s\left|U_{s\alpha\beta}^+\right|e^{j(\omega_s t + \theta_u^+)} - j\omega_s\left|U_{s\alpha\beta}^-\right|e^{j(-\omega_1 t + \theta_u^-)} = j\omega_s U_{s\alpha\beta}^+ - j\omega_1 U_{s\alpha\beta}^- = -\omega_s \widehat{U}_{s\alpha\beta} \tag{19}$$

$$\frac{d\widehat{U}_{s\alpha\beta}}{dt} = (-j)j\omega_s U_{s\alpha\beta}^+ + j(-j\omega_1 U_{s\alpha\beta}^-) = \omega_1 U_{s\alpha\beta}^+ + \omega_1 U_{s\alpha\beta}^- = \omega_1 U_{s\alpha\beta} \tag{20}$$

By substituting Formulas (18)–(20) into Formula (17), and converting to the matrix form:

$$\begin{aligned}
\frac{d}{dt}\begin{bmatrix} P_{sn} \\ Q_s \end{bmatrix} =& -\frac{3}{2\rho L_m}\begin{bmatrix} -\widehat{u}_{s\beta} & \widehat{u}_{s\alpha} \\ u_{s\beta} & -u_{s\alpha} \end{bmatrix}\begin{bmatrix} V_{r\alpha} \\ V_{r\beta} \end{bmatrix} + \frac{3L_r}{2\rho L_m^2}\begin{bmatrix} u_{s\beta}\widehat{u}_{s\alpha} - u_{s\alpha}\widehat{u}_{s\beta} \\ 0 \end{bmatrix} \\
& -\frac{3L_r\omega_r}{2\rho L_m^2}\begin{bmatrix} \psi_{s\beta}\widehat{u}_{s\alpha} - \psi_{s\alpha}\widehat{u}_{s\beta} \\ \psi_{s\alpha}u_{s\beta} - \psi_{s\beta}u_{s\alpha} \end{bmatrix} + \begin{bmatrix} -\omega_1 Q_s - 3/2\omega_r(\widehat{u}_{s\alpha}i_{s\beta} - \widehat{u}_{s\beta}i_{s\alpha}) \\ \omega_1 P_{sn} + 3/2\omega_r(u_{s\alpha}i_{s\alpha} + u_{s\beta}i_{s\beta}) \end{bmatrix}
\end{aligned} \tag{21}$$

3. Second-Order Sliding Mode Direct Power Controller Design

Control tasks for DPC are to track the new active power, P_{sn}, and reactive power, Q_s. The tracking errors are as follows:

$$\begin{cases} e_P = P_{snref} - P_{sn} \\ e_Q = Q_{sref} - Q_s \end{cases} \tag{22}$$

For the sake of reducing steady state error and maintaining good dynamic performance, the integral form sliding mode surface can be adopted. Therefore, sliding mode surfaces are designed as follows:

$$\begin{cases} \sigma_P = e_P + k_P \int_0^t e_P(\tau)d\tau \\ \sigma_Q = e_Q + k_Q \int_0^t e_Q(\tau)d\tau \end{cases}, \tag{23}$$

where $\sigma = \begin{bmatrix} \sigma_P & \sigma_Q \end{bmatrix}^T$, $k_P k_Q$ are respectively integral gains of new active power and reactive power. The first-order time derivative of the sliding mode function is calculated as follows:

$$\dot{\sigma} = F + GV_{r\alpha\beta} \tag{24}$$

$$F = -\frac{3L_r}{2\rho L_m^2}\begin{bmatrix} u_{s\beta}\widehat{u}_{s\alpha} - u_{s\alpha}\widehat{u}_{s\beta} \\ 0 \end{bmatrix} + \frac{3L_r\omega_r}{2\rho L_m^2}\begin{bmatrix} \psi_{s\beta}\widehat{u}_{s\alpha} - \psi_{s\alpha}\widehat{u}_{s\beta} \\ \psi_{s\alpha}u_{s\beta} - \psi_{s\beta}u_{s\alpha} \end{bmatrix} - \begin{bmatrix} -\omega_1 Q_s - 3/2\omega_r(\widehat{u}_{s\alpha}i_{s\beta} - \widehat{u}_{s\beta}i_{s\alpha}) \\ \omega_1 P_{sn} + 3/2\omega_r(u_{s\alpha}i_{s\alpha} + u_{s\beta}i_{s\beta}) \end{bmatrix} + \begin{bmatrix} K_p e_p \\ K_q e_q \end{bmatrix} + \begin{bmatrix} \dot{P}_{snref} \\ \dot{Q}_{sref} \end{bmatrix}$$

$$G = \frac{3}{2\rho L_m}\begin{bmatrix} -\widehat{u}_{s\beta} & \widehat{u}_{s\alpha} \\ u_{s\beta} & -u_{s\alpha} \end{bmatrix}, \quad V_{r\alpha\beta} = [V_{r\alpha} \ V_{r\beta}]^T.$$

As most known parts of F and G in Formula (24) are regarded as uncertainties, control chattering can be serious if the constant speed FOSM or the super-twisting SOSM are directly applied to Formula (24). Therefore, the controller is constructed as two parts. Formula (24) is firstly represented as known and unknown parts, as follows:

$$\begin{aligned}
\dot{\sigma} &= \overline{F} + \Delta_F + (\overline{G} + \Delta_G)V_{r\alpha\beta} \\
&= \overline{F} + \overline{G}u_{r\alpha\beta} + \Delta_F + \Delta_G V_{r\alpha\beta} \\
&= \overline{F} + \overline{G}V_{r\alpha\beta} + \Delta
\end{aligned} \tag{25}$$

where $\overline{F}, \overline{G}$ are the known part and $\Delta = [\Delta_1 \ \Delta_2]^T$ contain uncertainty parameters, measuring errors, unmodeled dynamics, and so on. The value Δ is related to physical parameters; thus, $|\dot{\Delta}_1| \leq L_{\Delta 1}$, $|\dot{\Delta}_2| \leq L_{\Delta 2}$ are undoubtedly satisfied. $L_{\Delta 1}$ and $L_{\Delta 2}$ are constants.

Feedback control is designed as follows:

$$V_{ra\beta} = \begin{bmatrix} V_{ra} \\ V_{r\beta} \end{bmatrix} = \overline{G}^{-1}\left(-F + \begin{bmatrix} u_{ra} \\ u_{r\beta} \end{bmatrix}\right) \tag{26}$$

where u_{ra} and $u_{r\beta}$ are auxiliary control. Then:

$$\dot{\sigma} = \begin{bmatrix} \dot{\sigma}_P \\ \dot{\sigma}_Q \end{bmatrix} = \overline{F} + \overline{G}V_{ra\beta} + \Delta = \begin{bmatrix} u_{ra} \\ u_{r\beta} \end{bmatrix} + \Delta, \tag{27}$$

where u_{ra} and $u_{r\beta}$ are designed based on a super-twisting algorithm [28], as follows:

$$\begin{cases} u_{ra} = -\lambda_{ra}|\sigma_P|_{1/2}sign(\sigma_P) + u_{rav} \\ \dot{u}_{rav} = -\gamma_{ra}sign(\sigma_P) \end{cases}, \tag{28}$$

$$\begin{cases} u_{r\beta} = -\lambda_{r\beta}|\sigma_Q|_{1/2}sign(\sigma_Q) + u_{r\beta v} \\ \dot{u}_{r\beta v} = -\gamma_{r\beta}sign(\sigma_Q) \end{cases}, \tag{29}$$

where λ_{ra}, γ_{ra}, $\lambda_{r\beta}$, and $\gamma_{r\beta}$ are control parameters of the super-twisting SOSM.

Finite time stability can be achieved as long as the SOSM, with respect to σ, can be established and maintained in finite time. The control parameters λ_{ra}, γ_{ra}, $\lambda_{r\beta}$, and $\gamma_{r\beta}$ can be chosen according to Reference [39], in which the parameters are required as follows:

$$\begin{cases} \gamma_{ra} > L_{\Delta 1}, \quad \lambda_{ra} > \sqrt{\gamma_{ra} + L_{\Delta 1}} \\ \gamma_{r\beta} > L_{\Delta 2}, \quad \lambda_{r\beta} > \sqrt{\gamma_{r\beta} + L_{\Delta 2}} \end{cases} \tag{30}$$

The values $L_{\Delta 1}$ and $L_{\Delta 2}$ should be calculated and analyzed according to actual operating environment in the wind turbine system. Yet, the accurate values of $L_{\Delta 1}$ and $L_{\Delta 2}$ are usually difficult to acquire. Thus, it has practical meaning to design the adaptive control parameters $\lambda_{ra}, \gamma_{ra}, \lambda_{r\beta}$, and $\gamma_{r\beta}$.

The next procedure is to construct an adaptive law for $\lambda_{ra}, \gamma_{ra}, \lambda_{r\beta}$, and $\gamma_{r\beta}$, establish SOSM with respect to σ_P, σ_Q in finite time, and track P_{sn}, Q_s. The design procedure of $\lambda_{r\beta}\gamma_{r\beta}$ are similar to $\lambda_{ra}\gamma_{ra}$.

Combining Formulas (27) and (28) and introducing state variable $\sigma_{Pv} = \Delta_1 - \gamma_{ra}\int_0^t sign(\sigma_P)d\tau$, then:

$$\begin{cases} \dot{\sigma}_P = -\lambda_{ra}|\sigma_P|_{1/2}sign(\sigma_P) + \sigma_{Pv} \\ \dot{\sigma}_{Pv} = -\gamma_{ra}sign(\sigma_P) + \dot{\Delta}_1 \end{cases} \tag{31}$$

To choose vector $\xi^T = [sign(\sigma_P)|\sigma_P|_{1/2} \quad \sigma_{Pv}]$, an inequation $\frac{d|x|}{dt} = \dot{x}sign(x)$ is adopted, then:

$$\begin{cases} \dot{\xi}_1 = \frac{1}{2|\sigma_P|^{1/2}}(-\lambda_{ra}|\sigma_P|_{1/2}sign(\sigma_P) + \sigma_{Pv}) \\ \dot{\xi}_2 = -\gamma_{ra}sign(\sigma_P) + \dot{\Delta}_1 \end{cases} \tag{32}$$

To define $\widetilde{\dot{\Delta}}_1 = |\sigma_P|_{1/2}\dot{\Delta}_1$, $A = \begin{bmatrix} -\frac{\lambda_{ra}}{2} & \frac{1}{2} \\ -\gamma_{ra} & 0 \end{bmatrix}$, $B = \begin{bmatrix} 0 \\ 1 \end{bmatrix}$, and $C = \begin{bmatrix} 1 & 0 \end{bmatrix}$, according to Formula (32):

$$\dot{\xi} = \frac{1}{|\sigma_P|^{1/2}}(A\xi + B\widetilde{\dot{\Delta}}_1) \tag{33}$$

Considering the Lyapunov function:

$$V(\xi, \lambda_{ra}, \gamma_{ra}) = V_0(\xi) + \frac{1}{2a_1}(\lambda_{ra} - \lambda_{ra}^*)^2 + \frac{1}{2a_2}(\gamma_{ra} - \gamma_{ra}^*)^2 \tag{34}$$

where λ_{ra}^* and γ_{ra}^* are positive constants; $V_0(\xi) = \xi^T P \xi$; $P = \begin{bmatrix} \frac{m_P^2 + 4\mu_P}{2} & -\frac{m_P}{2} \\ -\frac{m_P}{2} & 1 \end{bmatrix}$; a_1, a_2, and μ_P are positive constants, and m_P is an arbitrary constant.

Notice that P is a positive definite symmetric matrix, so then the derivative of $V_0(\xi)$ is as follows:

$$\dot{V}_0(\xi) = 2\dot{\xi}^T P \xi = \frac{1}{|\sigma_P|^{1/2}}(2\xi^T A^T + 2\widetilde{\Delta}_1 B^T)P\xi \leq \frac{1}{|\sigma_P|^{1/2}}(2\xi^T A^T P \xi + 2\widetilde{\Delta}_1 B^T P \xi + L_{\Delta 1}^2 \left| \sigma_P \left| -\widetilde{\Delta}_1^2 \right. \right.)$$

$$= \frac{1}{|\sigma_P|^{1/2}}(2\xi^T A^T P \xi + 2\widetilde{\Delta}_1 B^T P \xi + L_{\Delta 1}^2 \xi^T C^T C \xi - \widetilde{\Delta}_1^2) \tag{35}$$

$$\leq \frac{1}{|\sigma_P|^{1/2}}(2\xi^T A^T P \xi + L_{\Delta 1}^2 \xi^T C^T C \xi + \xi^T P B B^T P \xi)$$

$$= \frac{1}{|\sigma_P|^{1/2}}\xi^T(A^T P + P A^T + L_{\Delta 1}^2 C^T C + P B B^T P)\xi$$

To define $Q = -(A^T P + P A^T + L_{\Delta 1}^2 C^T C + P B B^T P)$, Formula (35) can be written as follows:

$$V_0(\xi) \leq -\frac{1}{|\sigma_P|^{1/2}}\xi^T Q \xi \tag{36}$$

A, B, C, P are substituted in Q, then:

$$Q = \begin{bmatrix} 2\lambda_{ra}\mu_P + \frac{\lambda_{ra}m_P^2}{2} - \gamma_{ra}m_P - L_{\Delta 1}^2 - \frac{m_P^2}{4} & -\mu_P - \frac{m_P^2}{4} - \frac{\lambda_{ra}m_P}{4} + \gamma_{ra} + \frac{m_P}{2} \\ -\mu_P - \frac{m_P^2}{4} - \frac{\lambda_{ra}m_P}{4} + \gamma_{ra} + \frac{m_P}{2} & \frac{m_P}{2} - 1 \end{bmatrix} \tag{37}$$

In order to guarantee positive definiteness, define:

$$\gamma_{ra} = \mu_P + \frac{m_P^2}{4} + \frac{\lambda_{ra}m_P}{4} \tag{38}$$

Formula (37) is substituted into Formula (38), then:

$$Q - \frac{m_P I}{4} = \begin{bmatrix} 2\lambda_{ra}\mu_P + \frac{\lambda_{ra}m_P^2}{2} - \mu_P m_P - \frac{m_P^3}{4} - L_{\Delta 1}^2 - \frac{m_P^2}{4} - \frac{m_P}{4} & \frac{m_P}{2} \\ \frac{m_P}{2} & \frac{m_P}{4} - 1 \end{bmatrix} \tag{39}$$

According to properties of the *Schur complement*, the conditions to guarantee a positive definiteness of Q and a minimum eigenvalue $\lambda_{\min}(Q) > \frac{m_P}{4}$ are as follows:

$$\begin{cases} \lambda_{ra} > \dfrac{\frac{m_P^2}{4} + (\mu_P m_P + \frac{m_P^3}{4} + L_{\Delta 1}^2 + \frac{m_P^2}{4} + \frac{m_P}{4})(\frac{m_P}{4} - 1)}{(2\mu_P + \frac{m_P^2}{4})(\frac{m_P}{4} - 1)} \\ m_P > 4 \end{cases} \tag{40}$$

According to Formula (36):

$$V_0(\xi) \leq -\frac{1}{|\sigma_P|^{1/2}}\xi^T Q \xi \leq -\frac{m_P}{4|\xi_1|}\xi^T \xi = -\frac{m_P}{4|\xi_1|}\left\|\xi\right\|_2 = -\frac{m_P \|\xi\|}{4|\xi_1|}\|\xi\| \tag{41}$$

According to $\|\xi\|_2^2 = \xi_1^2 + \xi_2^2 = |\sigma_P| + \xi_2^2$, then:

$$\|\xi\|_2 \geq |\xi_1| \tag{42}$$

Then, Formula (41) can be represented as follows:

$$V_0(\xi) \le -\frac{m_P}{4}\|\xi\|_2 \tag{43}$$

According to positive definite quadratic form $V_0(\xi) = \xi^T P \xi$:

$$\lambda_{\min}(P)\|\xi\|_2^2 \le V_0(\xi) = \xi^T P \xi \le \lambda_{\max}(P)\|\xi\|_2^2 \tag{44}$$

In view of Formula (44):

$$\left(\frac{V_0(\xi)}{\lambda_{\max}(P)}\right)^{1/2} \le \|\xi\|_2 \tag{45}$$

Considering Formulas (43) and (45):

$$V_0(\xi) \le -r V_0^{1/2}(\xi), \tag{46}$$

where $r = \frac{m_P}{4\lambda_{\max}^{1/2}(P)}$, then:

$$
\begin{aligned}
\dot{V}(\xi, \lambda_{ra}, \gamma_{ra}) &= -r V_0^{1/2}(\xi) + \tfrac{1}{a_1}(\lambda_{ra} - \lambda_{ra}^*)\dot{\lambda}_{ra} + \tfrac{1}{a_2}(\gamma_{ra} - \gamma_{ra}^*)\dot{\gamma}_{ra}\\
&= -r V_0^{1/2}(\xi) - \tfrac{\beta_{P1}}{\sqrt{2a_1}}\left|\lambda_{ra} - \lambda_{ra}^*\right| - \tfrac{\beta_{P2}}{\sqrt{2a_2}}\left|\gamma_{ra} - \gamma_{ra}^*\right| + \tfrac{1}{a_1}(\lambda_{ra} - \lambda_{ra}^*)\dot{\lambda}_{ra} + \tfrac{1}{a_2}(\gamma_{ra} - \gamma_{ra}^*)\dot{\gamma}_{ra}\\
&\quad + \tfrac{\beta_{P1}}{\sqrt{2a_1}}\left|\lambda_{ra} - \lambda_{ra}^*\right| + \tfrac{\beta_{P2}}{\sqrt{2a_2}}\left|\gamma_{ra} - \gamma_{ra}^*\right|\\
&\le -\min(r, a_1, a_2)\left((V_0(\xi) + \tfrac{1}{2a_1}(\lambda_{ra} - \lambda_{ra}^*)^2 + \tfrac{1}{2a_2}(\gamma_{ra} - \gamma_{ra}^*)^2\right)^{1/2} + \tfrac{1}{a_1}(\lambda_{ra} - \lambda_{ra}^*)\dot{\lambda}_{ra}\\
&\quad + \tfrac{1}{a_2}(\gamma_{ra} - \gamma_{ra}^*)\dot{\gamma}_{ra} + \tfrac{\beta_{P1}}{\sqrt{2a_1}}\left|\lambda_{ra} - \lambda_{ra}^*\right| + \tfrac{\beta_{P2}}{\sqrt{2a_2}}\left|\gamma_{ra} - \gamma_{ra}^*\right|
\end{aligned}
\tag{47}
$$

where β_{P1} and β_{P2} are positive constants.

The values λ_{ra} and γ_{ra} are both bounded. Therefore, Formula (47) is written as follows:

$$\dot{V}(\xi, \lambda_{ra}, \gamma_{ra}) \le -\min(r, a_1, a_2)V^{1/2} + \zeta, \tag{48}$$

where $\zeta = -\left|\lambda_{ra} - \lambda_{ra}^*\right|\left(\tfrac{1}{a_1}\dot{\lambda}_{ra} - \tfrac{\beta_{P1}}{\sqrt{2a_1}}\right) - \left|\gamma_{ra} - \gamma_{ra}^*\right|\left(\tfrac{1}{a_2}\dot{\gamma}_{ra} - \tfrac{\beta_{P2}}{\sqrt{2a_2}}\right)$.

In order to achieve finite time convergence, to make $\zeta = 0$, then the adaptive law for λ_{ra} and γ_{ra} are as follows:

$$
\begin{cases}
\dot{\lambda}_{ra} = \beta_{P1}\sqrt{\tfrac{a_1}{2}}\\
\dot{\gamma}_{ra} = \beta_{P2}\sqrt{\tfrac{a_2}{2}}
\end{cases}
\tag{49}
$$

For the sake of the uniformity of Formulas (49) and (38), choose:

$$m_P = \frac{4\beta_{P2}}{\beta_{P1}}\sqrt{\frac{a_2}{a_1}} \tag{50}$$

Formula (48) is rewritten as follows:

$$\dot{V}(\xi, \lambda_{ra}, \gamma_{ra}) \le -\min(r, a_1, a_2)V^{1/2} \tag{51}$$

Thus, $V(\xi, \lambda_{ra}, \gamma_{ra})$ can converge to zero in finite time, and it can be observed that $V_0(\xi)$ can also converge to zero in finite time. Hence, when control gains λ_{ra} and γ_{ra} are chosen as follows:

$$
\begin{cases}
\dot{\lambda}_{ra} = \begin{cases} \beta_{P1}\sqrt{\tfrac{a_1}{2}} & \lambda_{ra} \ne 0\\ 0 & \lambda_{ra} = 0 \end{cases}\\
\gamma_{ra} = \mu_P + \tfrac{m_P^2}{4} + \tfrac{\lambda_{ra}m_P}{4}
\end{cases}
,
\tag{52}
$$

the values σ_P and $\dot{\sigma}_P$ can converge to zero in finite time and SOSM with respect to σ_P can be established. Then, active power tracking is achieved.

Similarly, when control gains $\lambda_{r\beta}$ and $\gamma_{r\beta}$ are designed as follows:

$$
\begin{cases}
\dot{\lambda}_{r\beta} = \begin{cases} \beta_{Q1}\sqrt{\frac{a_3}{2}} & \lambda_{r\beta} \neq 0 \\ 0 & \lambda_{r\beta} = 0 \end{cases} \\
\gamma_{r\beta} = \mu_Q + \frac{m_Q^2}{4} + \frac{\lambda_{r\beta}m_Q}{4}
\end{cases}
\tag{53}
$$

SOSM with respect to σ_Q can be established in finite time and the reactive power tracking objective can be achieved.

4. Simulation Results

4.1. Control System Overview

The system control diagram can be described as in Figure 2 according to the aforementioned design procedure.

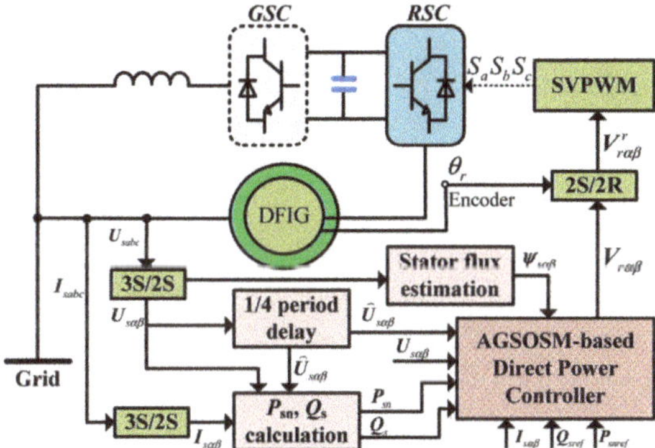

Figure 2. System control diagram.

Firstly, the measured three-phase stator voltage, U_{sabc}, and current, I_{sabc}, are converted to $U_{s\alpha\beta}$ and $I_{s\alpha\beta}$ under two phase stationary frame, and $\hat{U}_{s\alpha\beta}$ is calculated via $U_{s\alpha\beta}$. Secondly, $\psi_{s\alpha\beta}$ is estimated according to stator voltage, and the active power, P_{sn}, and reactive power, Q_s, are calculated by Formulas (1) and (9). Furthermore, the reference values of power, stator voltage, stator flux linkage, and stator current are applied as inputs of the AGSOSM direct power controller. Then, $V_{r\alpha\beta}$ can be obtained via related controller Formulas (26), (28), (29), (52), and (53). Then, $V_{r\alpha\beta}$ is converted to $V^r_{r\alpha\beta}$ under the rotor reference frame. Finally, space vector pulse width modulation (SVPWM) is controlled by $V^r_{r\alpha\beta}$ to generate S_a, S_b, and S_c. It is observed that the control system is rather simple, and phase sequence, PLL, and the upper bounds of the uncertainties are not needed.

It should be noted that, though stator flux linkage is sometimes calculated via $\psi_{s\alpha\beta} = \int_0^t \left[U_{s\alpha\beta} - R_s I_{s\alpha\beta} \right]$, an integral operator is usually substituted by band-pass filter to avoid the drift phenomenon. A band-pass filter can help filter static state stator flux, which is a component that appears on account of instantaneous voltage dips, since the stator flux cannot change suddenly. The natural flux gives rise to a large voltage in the rotor and may induce an out of control state for the electric generator. Thus, stator flux linkage is expressed as follows:

$$
\psi_{s\alpha\beta}(p) = \frac{p}{(p + \omega_c)^2} U_{s\alpha\beta}(p)
\tag{54}
$$

where p is a Laplace operator and ω_c is the cut-off frequency of the filter.

4.2. Simulation Experiment

Simulations were carried out under the MATLAB/Simulink platform for a 2MW DFIG with the characteristics shown in Table 1. In order to verify the performance of the proposed AGSOSM-DPC strategy, comparative simulations based on FOSM-DPC [22] and FOSM-EDPC [25] were also executed. The dc-link voltage was maintained as 1200 V via a method mentioned in Reference [22], which is not included here. The sampling frequency was set as 4 kHz for both the control strategies. The controller parameters were $k_P = 3500$, $k_Q = 3500$, $\beta_{P1} = 5.7$, $a_1 = 3.5$, $\mu_P = 6.5$, $m_P = 2.1$, $\beta_{Q1} = 4.5$, $a_3 = 2.2$, $\mu_Q = 6.2$, and $m_Q = 3.5$.

Table 1. DFIG parameters.

Rated Power (MW)	2
Line-to-line voltage (rms)(V)	690
Stator frequency (Hz)	50
Stator-to-rotor ratio	3
Rs (ohm)	0.001518
Rr (ohm)	0.002087
Ls (mH)	0.059906
Lr (mH)	0.08206
Lm (mH)	2.4
Pole pairs	2
Lumped inertia constant (kg·m^2)	17.23

Figures 3–5 show the variations of active power, reactive power, electromagnetic torque, stator current, and rotor current when active power reference changed from 1MW to 2MW, and reactive power changed from 1 MVar to 0 MVar under balanced grid voltage for the three control strategies. The response curves demonstrate excellent steady state and dynamic characteristics under all the three control strategies. Table 2 shows a quantitative comparison of the transitory response and power ripples of active power and reactive power, and total harmonic distortion (THD) of stator current and rotor current. It is evident that better dynamic performance is achieved under the proposed control strategy.

Table 2. Quantitative comparison under the three control strategies.

Control Strategy	Transitory r Response (ms)		Power Ripple (%)		THD (%)	
	P	*Q*	*P*	*Q*	*Is*	*Ir*
AGSOSM-DPC	1.3	1.6	12.7	17.4	1.9	2.7
FOSM-EDPC	1.3	1.7	15.3	19.7	2.3	3.2
FOSM-DPC	1.5	2.1	19.5	21.2	5.2	6.1

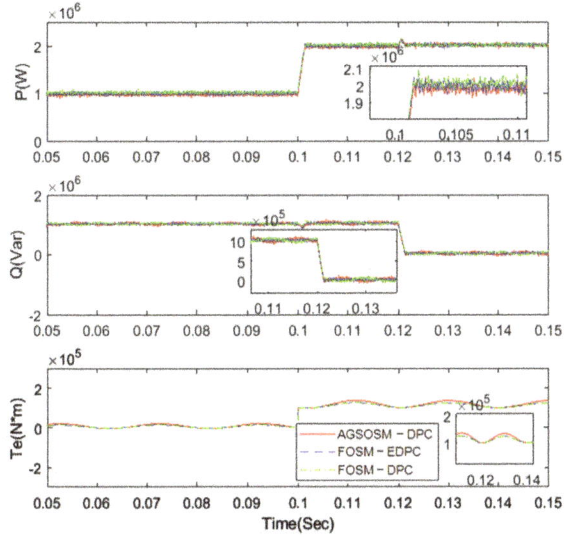

Figure 3. Active power, reactive power, and electromagnetic torque under balanced grid voltage.

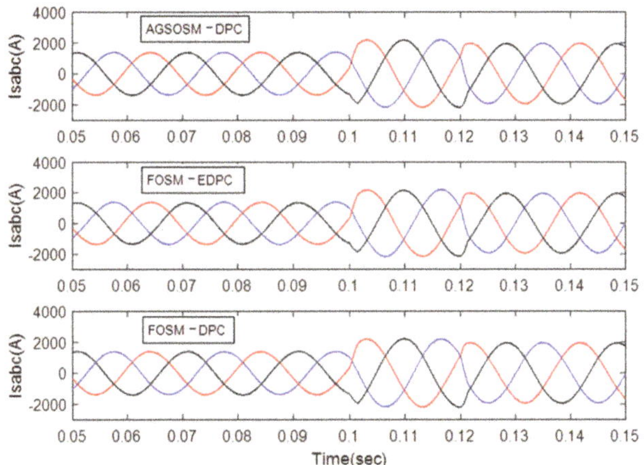

Figure 4. Stator current under balanced grid voltage.

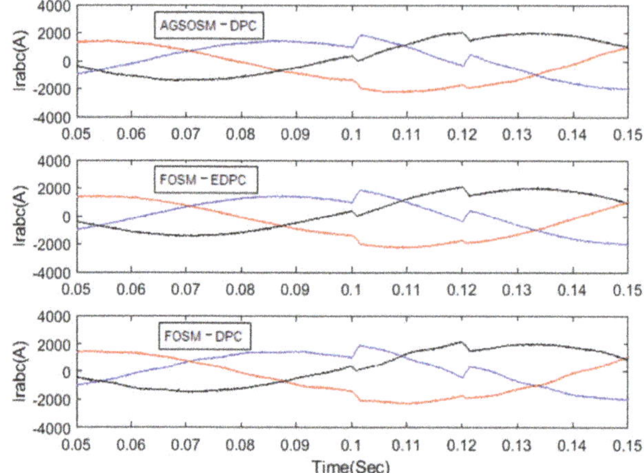

Figure 5. Rotor current under balanced grid voltage.

Figure 6 shows control voltages under the three control strategies. It can be observed that control action is continuous because the SOSM method is adopted in the proposed AGSOSM control strategy. The control chattering is smaller, which means a longer service life. Figure 7 displays control parameters for the proposed AGSOSM control law. The control parameters can be adaptively adjusted according to system variation.

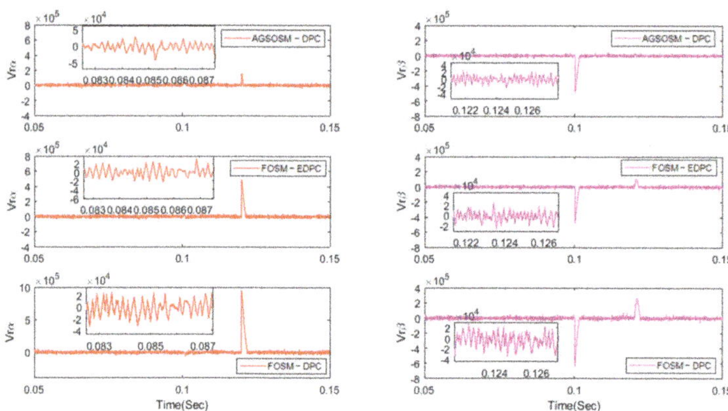

Figure 6. Control voltage under two phase stationary frame under balanced grid voltage.

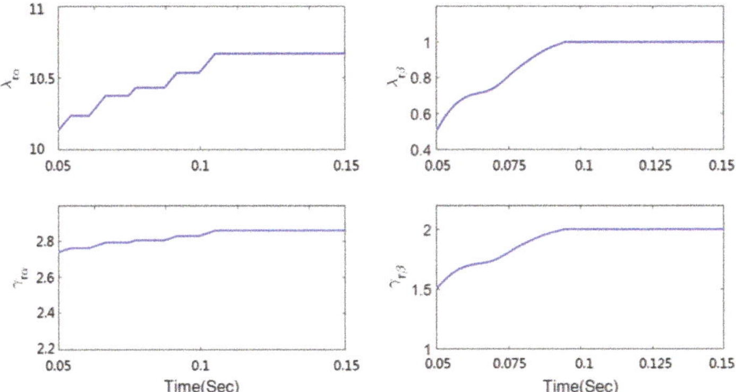

Figure 7. Control parameters for the proposed AGSOSM control law.

Steady state responses under unbalanced grid voltage for the three control strategies are shown in Figures 8–10. Active and reactive power are respectively set as 2MW and 0.5MVAr. Under the FOSM-DPC scheme, though active power and reactive power are maintained as the reference values, electromagnetic torque ripples are bigger and the stator current contains more harmonic components than the other two schemes. For the FOSM-EDPC and AGSOSM-DPC schemes, active power contains more ripples because the new active power is selected as the control target to obtain sinusoidal stator currents. As is shown in Figure 11, control chattering is smaller under the proposed AGSOSM-DPC scheme than that under the other two control schemes.

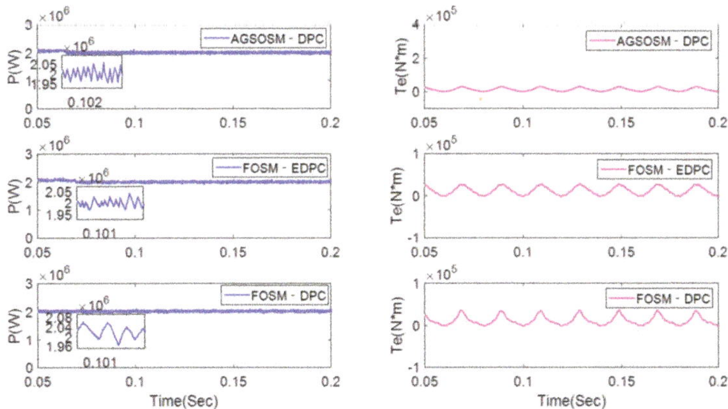

Figure 8. Active power and electromagnetic torque under unbalanced grid voltage.

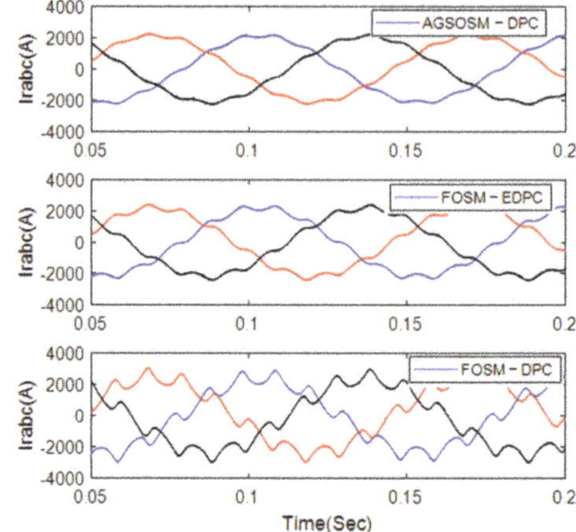

Figure 9. Rotor current under unbalanced grid voltage.

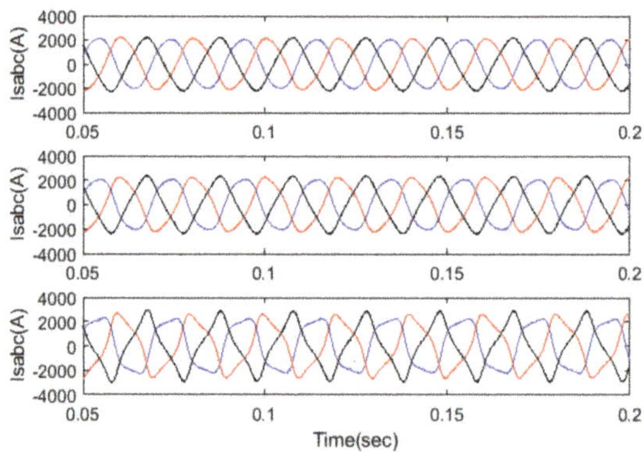

Figure 10. Stator current under unbalanced grid voltage.

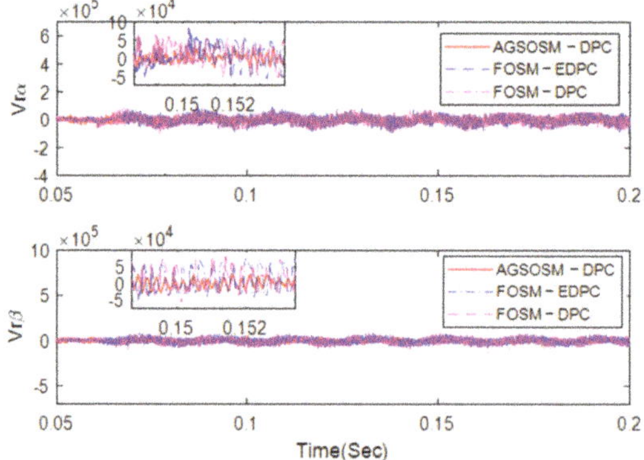

Figure 11. Control voltage under the proposed AGSOSM-DPC strategy.

To expediently verify the robustness of the proposed AGSOSM-DPC scheme, the uncertainty of mutual inductance, which is often influenced by stator and rotor cores, is taken into account. In addition to this, the variations of stator resistance and rotor resistance should also be specially considered. The mutual inductance, stator resistance, and rotor resistance are reduced and increased to 50% L_m, 50% Rs, 50% R_r and 120% L_m, 120% Rs, and 120% R_r, respectively. The stator current, active power, and electromagnetic torque are shown in Figure 12, which displays that the relative responses are almost the same as that of Figures 8 and 10. This means the system is robust under the proposed AGSOSM-DPC.

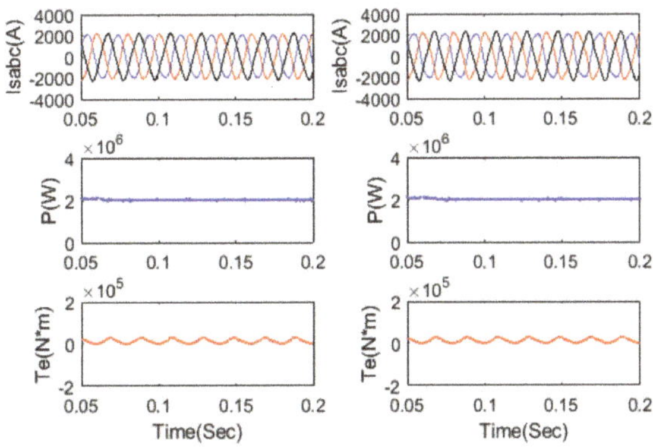

Figure 12. Responses with 50% L_m, 50% Rs, 50% R_r and 120% L_m, 120% Rs, and 120% R_r.

5. Conclusions

This study proposes a novel DPC scheme for a wind turbine driven DFIG based on an AGSOSM super-twisting algorithm. First, SOSM direct power controllers were designed based on the detailed analysis for a DFIG model under a two-phase stationary reference frame. Then, adaptive control gains were constructed considering the unknown upper bound of uncertainty. The simulation results indicate that the proposed scheme is valid for balanced and unbalanced grid voltage. Remarkable steady-state performance and dynamic performance can be achieved under balanced grid voltage, and control chattering is significantly reduced. Under unbalanced grid voltage, electromagnetic torque ripples are restrained, and stator currents are sinusoidal. They can be simultaneously achieved without PLL and phase sequence decomposition. The more important contributions are that severe control chattering is significantly reduced, and the upper bound of uncertainty is not necessary during the operational process.

Author Contributions: Conceptualization, Y.H.; Funding acquisition, Y.H.; Methodology, Y.H.; Software, R.M.; Validation, Y.H. and R.M.; Writing—original draft, Y.H.; Writing—review & editing, R.M.

Funding: This work was funded by National Natural Science Foundation of China under Grant 61803230, 61773015 and 61803154; A Project of Shandong Province Higher Educational Science and Technology Program under Grant J18KA330;University outstanding youth innovation team development plan of Shandong Province under Grant 2019KJN023; Key research and development plan project of Shandong Province under Grant 2018GGX105003.

Conflicts of Interest: The authors declare no conflict of interest.

Nomenclature

U_s, V_r	Stator, rotor voltage vectors.
I_s, I_r	Stator, rotor current vectors.
Q_s, Q_r	Stator output active and reactive powers.
ψ_s, ψ_r	Stator, rotor flux linkage vectors.
R_s, R_r	Stator, rotor resistances
L_m, L_s, L_r	Mutual inductance, Stator, rotor self-inductances.
α, β	Stator α, β axis

References

1. Djilali, L.; Sanchez, E.N.; Belkheiri, M. Real-time neural sliding mode field oriented control for a DFIG-based wind turbine under balanced and unbalanced grid conditions. *IET Renew. Power Gener.* **2019**, *13*, 618–632. [CrossRef]
2. Available online: https://www.irena.org/ (accessed on 11 August 2019).
3. Liu, Y.; Wang, Z.; Xiong, L.; Wang, J.; Jiang, X.; Bai, G.; Li, R.; Liu, S. DFIG wind turbine sliding mode control with exponential reaching law under variable wind speed. *Int. J. Electr. Power Energy Syst.* **2018**, *96*, 253–260. [CrossRef]
4. Xiong, L.; Li, P.; Wu, F.; Ma, M.; Khan, M.W.; Wang, J. A coordinated high-order sliding mode control of DFIG wind turbine for power optimization and grid synchronization. *Int. J. Electr. Power Energy Syst.* **2019**, *105*, 679–689. [CrossRef]
5. Merabet, A.; Eshaft, H.; Tanvir, A.A. Power-current controller based sliding mode control for DFIG-wind energy conversion system. *IET Renew. Power Gener.* **2018**, *12*, 1155–1163. [CrossRef]
6. Yuan, T.; Wang, J.; Guan, Y.; Liu, Z.; Song, X.; Che, Y.; Cao, W. Virtual Inertia Adaptive Control of a Doubly Fed Induction Generator (DFIG) Wind Power System with Hydrogen Energy Storage. *Energies* **2018**, *11*, 904. [CrossRef]
7. Villanueva, I.; Rosales, A.; Ponce, P.; Molina, A. Grid-voltage-oriented sliding mode control for DFIG under balanced and unbalanced grid Faults. *IEEE Trans. Sustain. Energy* **2018**, *9*, 1090–1098. [CrossRef]
8. Nian, H.; Cheng, P.; Zhu, Z.Q. Coordinated direct power control of DFIG system without phase-locked loop under unbalanced grid voltage conditions. *IEEE Trans. Power Electron.* **2015**, *31*, 2905–2918. [CrossRef]
9. Cheng, P.; Nian, H.; Wu, C.; Zhu, Z.Q. Direct stator current vector control strategy of DFIG without phase-locked loop during network unbalance. *IEEE Trans. Power Electron.* **2016**, *32*, 284–297. [CrossRef]
10. Ayyarao, T.S.L.V. Modified vector controlled DFIG wind energy system based on barrier function adaptive sliding mode control. *Prot. Control Mod. Power Syst.* **2019**, *4*, 4. [CrossRef]
11. Mehdipour, C.; Hajizadeh, A.; Mehdipour, I. Dynamic modeling and control of DFIG-based wind turbines under balanced network conditions. *Int. J. Electr. Power Energy Syst.* **2016**, *83*, 560–569. [CrossRef]
12. Bektache, A.; Boukhezzar, B. Nonlinear predictive control of a DFIG-based wind turbine for power capture optimization. *Int. J. Electr. Power Energy Syst.* **2018**, *101*, 92–102. [CrossRef]
13. Alsmadi, Y.M.; Xu, L.; Blaabjerg, F.; Ortega, A.J.; Abdelaziz, A.Y.; Wang, A.; Albataineh, Z. Detailed investigation and performance improvement of the dynamic behavior of grid-connected DFIG-based wind turbines under LVRT conditions. *IEEE Trans. Ind. Appl.* **2018**, *54*, 4795–4812. [CrossRef]
14. Kashkooli, M.R.A.; Madani, S.M.; Lipo, T.A. Improved Direct Torque Control for a DFIG under Symmetrical Voltage Dip with Transient Flux Damping. *IEEE Trans. Ind. Electron.* **2019**. [CrossRef]
15. Xu, L.; Cartwright, P. Direct active and reactive power control of DFIG for wind energy generation. *IEEE Trans. Energy Conver.* **2006**, *21*, 750–758. [CrossRef]
16. Zandzadeh, M.J.; Vahedi, A. Modeling and improvement of direct power control of DFIG under unbalanced grid voltage condition. *Int. J. Electr. Power Energy Syst.* **2014**, *59*, 58–65. [CrossRef]
17. Debouza, M.; Al-Durra, A.; Errouissi, R.; Muyeen, S.M. Direct power control for grid-connected doubly fed induction generator using disturbance observer based control. *Renew. Energy* **2018**, *125*, 365–372. [CrossRef]
18. Zhang, Y.; Jiao, J.; Xu, D. Direct Power Control of Doubly Fed Induction Generator Using Extended Power Theory Under Unbalanced Network. *IEEE Trans. Power Electron.* **2019**. [CrossRef]
19. Benamor, A.; Benchouia, M.T.; Srairi, K.; Benbouzid, M.E. A novel rooted tree optimization apply in the high order sliding mode control using super-twisting algorithm based on DTC scheme for DFIG. *Int. J. Electr. Power Energy Syst.* **2019**, *108*, 293–302. [CrossRef]
20. Han, Y.; Ma, R.; Cui, J. Adaptive higher-order sliding mode control for islanding and grid-connected operation of a microgrid. *Energies* **2018**, *11*, 1459. [CrossRef]
21. Mystkowski, A. Lyapunov sliding-mode observers with application for active magnetic bearing operated with zero-bias flux. *J. Dyn. Sys. Meas. Control* **2019**, *141*, 041006. [CrossRef]
22. Hu, J.; Nian, H.; Hu, B.; He, Y.; Zhu, Z.Q. Direct active and reactive power regulation of DFIG using sliding-mode control approach. *IEEE Trans. Energy Convers.* **2010**, *25*, 1028–1039. [CrossRef]

23. Shang, L.; Hu, J. Sliding-mode-based direct power control of grid-connected wind-turbine-driven doubly fed induction generators under unbalanced grid voltage conditions. *IEEE Trans. Energy Convers.* **2012**, *27*, 362–373. [CrossRef]

24. Li, S.; Wang, H.; Tian, Y.; Aitouch, A.; Klein, J. Direct power control of DFIG wind turbine systems based on an intelligent proportional-integral sliding mode control. *ISA Trans.* **2016**, *64*, 431–439. [CrossRef] [PubMed]

25. Sun, D.; Wang, X.; Nian, H.; Zhu, Z.Q. A sliding-mode direct power control strategy for DFIG under both balanced and unbalanced grid conditions using extended active power. *IEEE Trans. Power Electron.* **2017**, *33*, 1313–1322. [CrossRef]

26. Martinez, M.I.; Susperregui, A.; Tapia, G.; Xu, L. Sliding-mode control of a wind turbine-driven double-fed induction generator under non-ideal grid voltages. *IET Renew. Power Gener.* **2013**, *7*, 370–379. [CrossRef]

27. Martinez, M.I.; Tapia, G.; Susperregui, A.; Camblong, H. Sliding-mode control for DFIG rotor-and grid-side converters under unbalanced and harmonically distorted grid voltage. *IEEE Trans. Energy Convers.* **2012**, *27*, 328–339. [CrossRef]

28. Liu, X.; Han, Y.; Wang, C. Second-order sliding mode control for power optimisation of DFIG-based variable speed wind turbine. *IET Renew. Power Gener.* **2016**, *11*, 408–418. [CrossRef]

29. Behera, A.K.; Chalanga, A.; Bandyopadhyay, B. A new geometric proof of super-twisting control with actuator saturation. *Automatica* **2018**, *87*, 437–441. [CrossRef]

30. Beltran, B.; Benbouzid, M.E.H.; Ahmed-Ali, T. Second-order sliding mode control of a doubly fed induction generator driven wind turbine. *IEEE Trans. Energy Convers.* **2012**, *27*, 261–269. [CrossRef]

31. Benbouzid, M.; Beltran, B.; Amirat, Y.; Yao, G.; Han, J.; Mangel, H. Second-order sliding mode control for DFIG-based wind turbines fault ride-through capability enhancement. *ISA Trans.* **2014**, *53*, 827–833. [CrossRef]

32. Evangelista, C.; Puleston, P.; Valenciaga, F.; Fridman, L.M. Lyapunov-designed super-twisting sliding mode control for wind energy conversion optimization. *IEEE Trans. Ind. Electron.* **2012**, *60*, 538–545. [CrossRef]

33. Susperregui, A.; Martinez, M.I.; Zubia, I.; Tapia, G. Design and tuning of fixed-switching-frequency second-order sliding-mode controller for doubly fed induction generator power control. *IET Electr. Power Appl.* **2012**, *6*, 696–706. [CrossRef]

34. Susperregui, A.; Martinez, M.I.; Tapia, G.; Vechiu, I. Second-order sliding-mode controller design and tuning for grid synchronisation and power control of a wind turbine-driven doubly fed induction generator. *IET Renew. Power Gener.* **2013**, *7*, 540–551. [CrossRef]

35. Martinez, M.I.; Susperregui, A.; Tapia, G. Second-order sliding-mode-based global control scheme for wind turbine-driven DFIGs subject to unbalanced and distorted grid voltage. *IET Electr. Power Appl.* **2017**, *11*, 1013–1022. [CrossRef]

36. Shah, A.P.; Mehta, A.J. Direct power control of DFIG using super-twisting algorithm based on second-order sliding mode control. In Proceedings of the 2016 14th International Workshop on Variable Structure Systems (VSS), Nanjing, China, 1–4 June 2016; pp. 136–141. [CrossRef]

37. Shah, A.P.; Mehta, A.J. Direct power control of grid-connected DFIG using variable gain super-twisting sliding mode controller for wind energy optimization. In Proceedings of the IECON 2017-43rd Annual Conference of the IEEE Industrial Electronics Society, Beijing, China, 29 October–1 November 2017; pp. 2448–2454. [CrossRef]

38. Zhang, Y.; Qu, C. Direct power control of a pulse width modulation rectifier using space vector modulation under unbalanced grid voltages. *IEEE Trans. Power Electron.* **2014**, *30*, 5892–5901. [CrossRef]

39. Seeber, R.; Horn, M. Stability proof for a well-established super-twisting parameter setting. *Automatica* **2017**, *84*, 241–243. [CrossRef]

Article

A Dual-Stator Winding Induction Generator Based Wind-Turbine Controlled via Super-Twisting Sliding Mode

Juan I. Talpone [1,2], Paul F. Puleston [2], Marcelo G. Cendoya [2,*] and José. A. Barrado-Rodrigo [3]

[1]   CIDEI, Instituto Tecnológico de Buenos Aires (ITBA), C1106ACD Buenos Aires, Argentina;
     jtalpone@itba.edu.ar
[2]   Instituto LEICI, Universidad Nacional de La Plata (UNLP)/CONICET, 1900 La Plata, Argentina;
     puleston@ing.unlp.edu.ar
[3]   GAEI, DEEEA-ETSE, Universitat Rovira i Virgili (URV), 43007 Tarragona, Spain; joseantonio.barrado@urv.cat
*   Correspondence: cendoya@ing.unlp.edu.ar; Tel.: +54-022-1425-9306

Received: 28 September 2019; Accepted: 3 November 2019; Published: 25 November 2019

Abstract: The dual-stator winding induction generator (DWIG) is a promising electrical machine for wind energy conversion systems, especially in the low/mid power range. Based on previous successful results utilising feed forward control, in this article, a super-twisting (ST) sliding mode improved control set-up is developed to maximise power extraction during low wind regimes. To accomplish this objective, via constant volts/hertz implementation, a ST controller was designed to command the DWIG control winding, such that the tip-speed ratio is robustly maintained at its optimal value. The proposed super-twisting control set-up was experimentally assessed to analyse its performance and to verify its efficiency in an actual generation test bench. The results showed a fast convergence to maximum power operation, avoiding chattering and offsets due to model uncertainties.

Keywords: wind energy; control; dual-stator winding induction generator; second order sliding mode

---

1. Introduction

Variable speed wind turbines have high efficiency in a wide range of wind speeds. This kind of wind energy conversion system (WECS) can use different types of electric generators and control techniques. In variable speed operation, when wind turbines are connected to an electrical network, it is necessary to include some frequency conversion stages [1,2]. In practice, there are two widespread power topologies: (a) a power converter connected between the stator winding of the generator and the grid, and (b) a power converter connected between the wound rotor of the generator and the electrical network.

The wind power systems of the first group usually use brushless machines, such as the squirrel cage induction generator (SCIG) or the permanent magnet synchronous generator (PMSG); and they require a converter of the same power as the generator (full power converter). While for the second group, usually a doubly fed induction generator (DFIG) is used. In this asynchronous machine, the stator is connected directly to the grid, and the rotor is fed through a bidirectional converter. This system has the advantage that it requires a converter which must deal with only a fraction of the total generator power (fractional power converter). On the downside, the wound rotor presents maintenance problems with the rings and brushes.

The dual-stator winding induction generator (DWIG) with brushless rotor seems to be a good option that combines the advantages of the two groups of generators mentioned above, since it is a very robust and reliable electrical machine, in which one of the stator windings is fed via a fractional controlled power converter, while the other winding can be directly connected to the grid. There are some options of induction generator with two sets of stator windings and brushless rotor that can be used in variable speed wind turbine systems [3]. In all these options, the two sets of stator windings are electrically isolated, in some cases with different pole numbers and various rotor configurations. Thus, in practice, the DWIG types can be

Case (I) a dual-stator induction machine having the two sets of three-phase windings, with the same pole number, but with a spatial shift of 30 electrical degrees. The rotor is a standard squirrel cage.

Case (II) a dual-stator induction machine having the two sets of stator windings with dissimilar pole numbers and the rotor is a nested-loop arrangement. One of the stator windings is directly connected to the grid (called power winding-PW), and the other stator winding (called control winding-CW) is connected to the grid via a fractionally rated frequency converter.

Case (III) a dual-stator induction machine having the two sets of three-phase windings with different numbers of poles in a 1:3 ratio. This configuration is usually chosen from the viewpoint of better magnetic utilization and to eliminate magnetic coupling between windings [4]. In this case, the rotor is a standard squirrel cage rotor.

Regarding the control strategies of variable speed wind energy conversion systems based on a dual-stator winding induction generator (WECS-DWIG), in [5], various approaches for DWIG with similar pole number are compared. The analysed control strategies are instantaneous slip frequency control (ISFC), field oriented control (FOC), voltage oriented control (VOC) and direct power control (DPC); whereas, in [6], a first order sliding mode (FOSM) controller for this type of induction generator is described.

There are some works about DWIG with dissimilar pole numbers, and a nested loop rotor, applied in variable speed wind turbines. Usually in this wind generator, the PW is connected directly to the grid, and CW is supplied via a bidirectional power converter. This topology is called a brushless doubly fed induction generator (BDFIG). In [7], a direct torque control (DTC) strategy for this BDFIG system is shown. Likewise, Ref. [8] developed a field oriented control (FOC), and [9] developed FOSM control strategies.

Additionally, for the third case of DWIG (dissimilar pole numbers and squirrel cage rotor), a high-performance control of a DC generating system was proposed by [10]. That paper shows two topological structures, using series (or parallel) connected AC-DC pulse width modulation rectifiers between each stator winding and the DC bus. A wind turbine system with a stator winding of DWIG connected directly to the grid was presented in [11]. Up to a certain value of wind speed, the induction generator works only with CW and its power converter. Under these conditions, a feed forward scalar control is applied to the generator. For higher wind speed, the PW is connected directly to the grid. In this zone of operation, with both stator windings working together, the wind turbine turns at quasi-constant speed. As a complementary work, Ref. [12] studied the capability of this DWIG to grid disturbances.

Encouraged by the good results that the authors have obtained in [11] using a feed forward action to control the WECS-DWIG, as a next step in that research, this paper proposes an improved control set-up based on a sliding mode (SM) control strategy [13]. Specifically, a controller that combines a feed forward action with a feedback second order sliding mode (SOSM) super-twisting (ST) algorithm [14,15]. This technique has been chosen because SOSM based controllers have shown numerous advantages to control nonlinear systems under heavy disturbances [16–22], in particular WECS [23–28]. Some of these proven advantages are: robustness to several bounded parameter variations, uncertainties and external disturbances; reduction of mechanical stresses and chattering, thanks to applying the discontinue control action at the output second-derivative level; and control laws of relatively low computational cost.

This paper is organized as follows. In Section 2, the DWIG based WECS under study is described, its operation zones are explained and a dynamic model of the system is presented. In Section 3, the ST control set-up to robustly achieve MPPT is designed. In Section 4, experimental results are shown and analysed. Finally, in Section 5, the work's conclusions and future research lines are discussed.

2. DWIG Based Wind Energy Conversion System

2.1. System Description

The wind energy conversion topology under study is shown in Figure 1. A three-bladed horizontal axis wind turbine drives the rotor of the DWIG by means of a multiplier gearbox (GB), so that the rotational speed of the generator remains in a useful operating range. As mentioned, the DWIG has a squirrel cage rotor and two stator windings of different pole numbers, the power winding PW and the control winding CW. The PW can be connected directly to the network via a power contactor, commanded by an upper level supervisory system depending on the operation zone. There is a capacitor bank in parallel with this winding to improve the power factor. As for the CW, it is indirectly linked to the grid by means of two three-phase inverters arranged in back to back connection sharing a common DC link. With this electronic conversion chain, the CW can be fed with a frequency and voltage different from that of the grid. The supply voltage of the CW is varied, through sinusoidal PWM modulation, accompanying the variation of the frequency following a constant V/f ratio. In this way, the air gap rotating magnetic field produced by the CW is maintained at its rated value throughout the operating range of the system. The inverter that connects the CW to the grid has a control loop associated with it, whose objective is to keep the DC bus voltage constant.

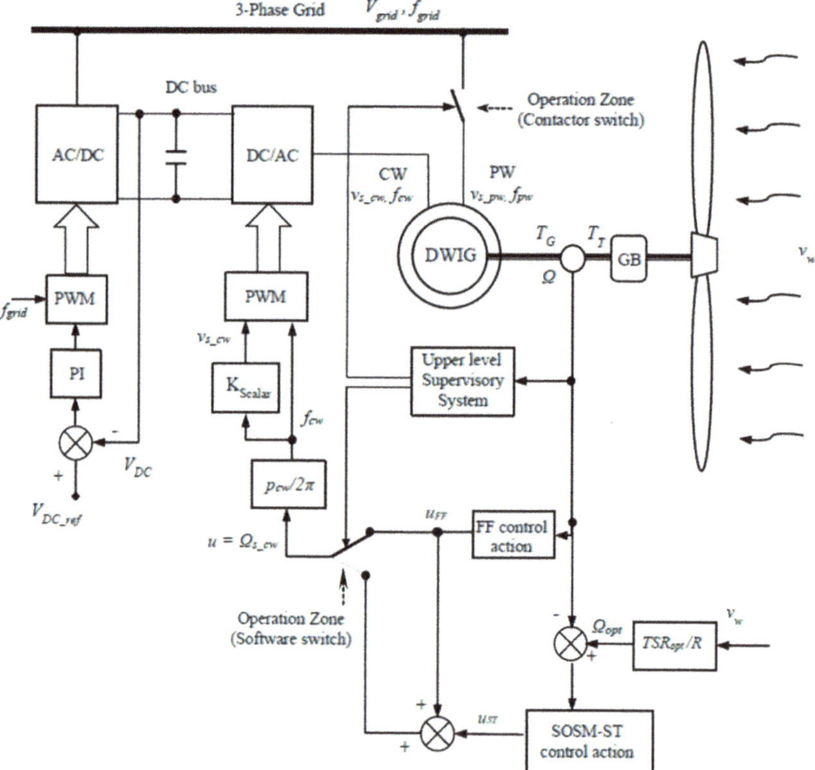

Figure 1. Structure of the wind energy conversion system based on a dual-stator winding induction generator (DWIG).

2.2. Wind Turbine Model

The wind turbine extracts a fraction of the wind power, depending on its aerodynamic efficiency given by the power coefficient C_p. Then, the turbine power can be expressed as ([27])

$$P_T(\Omega, v_w) = C_p(TSR)P_{wind}(v_w) = C_p(TSR)\frac{1}{2}\rho\pi R^2 v_w^3 \tag{1}$$

where Ω is the mechanical rotational speed, v_w is the wind speed, P_{wind} is the kinetic power of the wind, ρ is the air density and R is the blades' length. Coefficient $Cp(TSR)$ depends on the topology and dimensions of the blades and it is, in fact, a nonlinear function of the tip speed ratio:

$$TSR = \frac{R\Omega}{v_w} \tag{2}$$

The $Cp(TSR)$ of the horizontal axis wind turbine considered in this paper is depicted in Figure 2 (referred to the generator side). It has been assumed fixed blade pitch angle given that, in the operation zones under study, it is fixed at its optimal value of maximum power extraction (operation when the DWIG rated power is reached is beyond the scope of this work. In that situation, the supervisory system should turn into variable pitch operation to restrain the power extraction and to protect the generator).

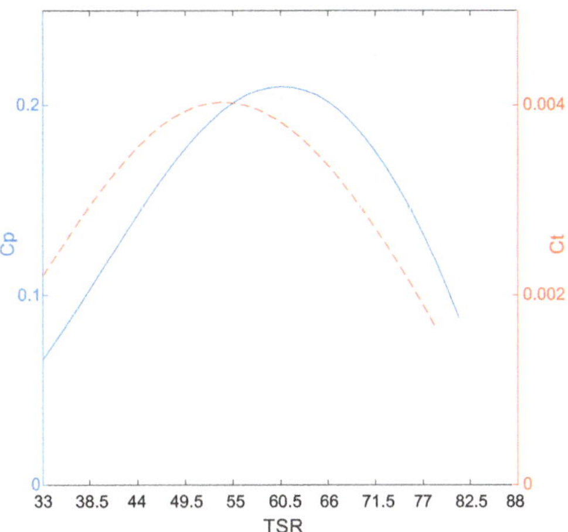

Figure 2. Wind turbine power (blue) and torque coefficients (dashed-red) versus tip speed ratio (both referred to the generator side).

In Figure 2 it can be appreciated that $Cp(TSR)$ presents a unique maximum at $TSR_{opt} = 60.5$. Therefore, the objective of maximum power point tracking or maximum wind power extraction would be accomplished by tracking a variable optimum speed reference Ω_{opt}, designed to maintain $TRS = TSR_{opt}$. From Equation (2):

$$\Omega_{opt} = \frac{TSR_{opt}}{R} v_w \tag{3}$$

As for the turbine torque, it is given by

$$T_T(\Omega, v_w) = \frac{P_T(\Omega, v_w)}{\Omega} = C_T(TSR)\frac{1}{2}\rho\pi R^3 v_w^2 \tag{4}$$

where $C_T(TSR) = \frac{C_P(TSR)}{TSR}$ is the turbine torque coefficient (see Figure 2). Turbine torque curves as a function of the rotation speed (referred to the generator side), for different wind speeds, can be appreciated in Figure 3 in the blue-dashed line.

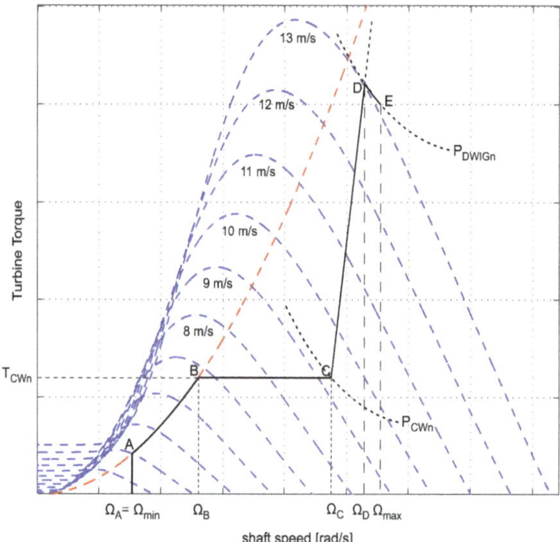

Figure 3. Operating zones of the wind energy conversion system based on a DWIG.

Then, from Equation (4) and considering TSR_{opt} , the expression of the optimal torque can be obtained:

$$T_{opt}(\Omega) = \frac{C_T\left(TSR_{opt}\right)\frac{1}{2}\rho\pi R^5}{TSR_{opt}{}^2}\Omega^2 = K_{opt}\Omega^2 \tag{5}$$

that is, the expression of the T_T corresponding to the maximum power point tracking (MPPT) as a function of the rotational speed Ω (depicted in the red-dashed line in Figure 3).

To conclude this subsection, it is worth mentioning that, in this paper, the high-speed side of the gear box has been chosen to work with, so the turbine variables in the article are referring to the generator side through the transmission ratio, GB.

2.3. WECS-DWIG System Operation Zones

In the complete range of operation, the DWIG based WECS's functioning modes are associated to four different zones which can be illustrated in the shaft speed–torque (see black line in Figure 3). From the measurement of Ω, an upper level supervisory system must identify the current zone and, consequently, set the pertinent operation mode.

This paper focuses on improving the conversion efficiency in the zone corresponding to low wind regimes, to ensure the best use of the scarce resource. However, to frame the design in a comprehensive context, in this subsection a succinct outline of all four operations zones is provided (a detailed description and analysis can be found in [11]).

- *Operation in Zone AB*. During low wind speed periods the objective is to maximize the energy extraction from the wind. Thus, a maximum power point tracking (MPPT) strategy is established for the CW whereas, in this operation zone, the PW remains disconnected from the grid (points of maximum wind power generation are depicted in red-dashed line in Figure 3). This zone starts at point A (at minimum speed Ω_{min}) and ends at point B (corresponding speed Ω_B), where the rated current of the control winding is attained and its nominal torque is reached.

- *Operation in Zone BC*: From point B on, if the wind speed increases, the DWIG's control system abandons the MPPT and changes the control objective to CW current regulation $i_{s\_cw} = i_{s\_cw\_RATED}$ (equivalently, constant torque), to prevent generator damages. When the shaft speed reaches Ω_C the control winding is operating at its rated power P_{CWn} (point C).

- *Operation in Zone CD*: corresponds to operation during high wind speed regimes. From point C, if the wind speed keep increasing, the CW would not be able to process the full amount of the wind power, thus, the supervisory system maintains current regulation for the CW, but also connects the PW directly to the grid to extract the wind energy surplus. When the shaft speed reaches Ω_D, the DWIG is working at point D, delivering the generator rated power P_{DWIGn}. This power is the maximum that can be obtained with both windings, CW and PW, in operation.

- *Operation beyond point D*: an appropriate mechanism to limit DWIG power and a maximum speed Ω_{max} (for instance, a variable pitch control system) is needed to avoid WECS damage or even destruction.

2.4. Wind Energy Conversion System Dynamics

- *DWIG Electrical Dynamics*

The DWIG has a squirrel cage rotor, the latter designed to be magnetically coupled to both the control and power windings of the stator, respectively. Regarding the stator windings, the number of CW pole pairs is three times the number of PW pole pairs, to eliminate the magnetic coupling between both windings, separating their influence on the generator torque [4]. In the experimental equipment under study, the pair of poles are $p_{cw} = 3$ and $p_{pw} = 1$, respectively (see windings details in Figure 4).

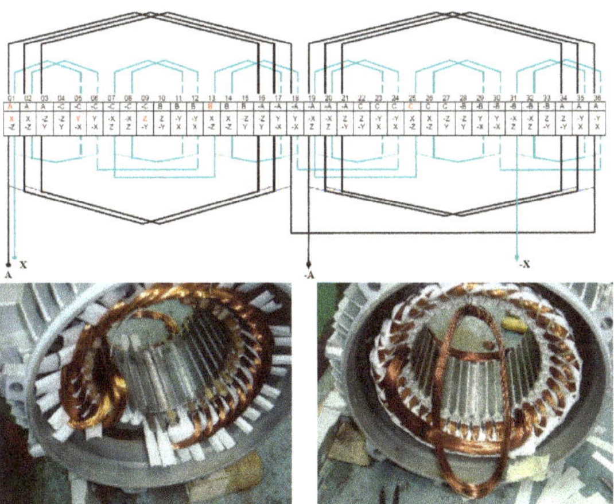

Figure 4. Schematic and images of the DWIG's control and power stator windings. Control winding (CW; 1/3 of maximum power): XYZ. 6 poles, 36 coils of 28 turns. Power winding (PW; 2/3 of maximum power): ABC. 2 poles, 18 coils of 36 turns.

The aforementioned decoupling effect makes the DWIG work as two separate induction machines that share the same mechanical shaft. Consequently, the dynamic electrical model of the DWIG in the time domain, considering a *dq* synchronous reference frame (after applying the Park transformation to the model expressed in phase variables), can be written as ([11])

$$\begin{cases} v_{qds\_pw} = R_{s\_pw}i_{qds\_pw} + p_{pw}\Omega_{s\_pw}\lambda_{dqs\_pw} + \dfrac{d\lambda_{qds\_pw}}{dt} \\ v_{qdr\_pw} = 0 = R_{r\_pw}i_{qdr\_pw} + p_{pw}\left(\Omega_{s\_pw} - \Omega\right)\lambda_{dqr\_pw} + \dfrac{d\lambda_{qdr\_pw}}{dt} \end{cases} \tag{6}$$

$$\begin{cases} v_{qds\_cw} = R_{s\_cw}i_{qds\_cw} + p_{cw}\Omega_{s\_cw}\lambda_{dqs\_cw} + \dfrac{d\lambda_{qds\_cw}}{dt} \\ v_{qdr\_cw} = 0 = R_{r\_cw}i_{qdr\_cw} + p_{cw}(\Omega_{s\_cw} - \Omega)\lambda_{dqr\_cw} + \dfrac{d\lambda_{qdr\_cw}}{dt} \end{cases} \tag{7}$$

In Equations (6) and (7) R are winding resistances, where subscripts s and r refer to the stator and rotor side, while pw and cw refer to the PW and the CW. p are the pole pairs and λ is the magnetic flux. Ω is the shaft speed and Ω_s is the mechanical synchronous speed, defined for each winding as

$$\Omega_{s\_pw} = \frac{2\pi f_{pw}}{p_{pw}}, \quad \Omega_{s\_cw} = \frac{2\pi f_{cw}}{p_{cw}} \tag{8}$$

where f_{pw} and f_{cw} are the electrical frequency of the power supplies that feed each stator winding.

The flux linkage equations are

$$\begin{cases} \lambda_{qds\_n} = \left(L_{ls\_n} + \frac{3}{2}L_{ms\_n}\right)i_{qds\_n} + \frac{3}{2}L_{ms\_n}i_{qdr\_n} \\ \lambda_{qdr\_n} = \left(L_{lr\_n} + \frac{3}{2}L_{mr\_n}\right)i_{qdr\_n} + \frac{3}{2}L_{ms\_n}i_{qds\_n} \end{cases} \tag{9}$$

In Equation (9), L_l are leakage inductances and L_m mutual inductances, where the sub-index "n" stands for "pw" or "cw" depending on the stator winding.

The total electromagnetic torque of the DWIG, is given by

$$\begin{aligned} T_G = \; & T_{Gpw} + T_{Gcw} \\ = \; & \tfrac{3}{2}p_{pw}L_{ms\_pw}\left(i_{qs\_pw}i_{dr\_pw} - i_{ds\_pw}i_{qr\_pw}\right) + \tfrac{3}{2}p_{cw}L_{ms\_cw}\left(i_{qs\_cw}i_{dr\_cw} - i_{ds\_cw}i_{qr\_cw}\right) \end{aligned} \tag{10}$$

- *WECS Mechanical Dynamics*

The mechanical dynamics of the WECS-DWIG is determined by Newton's law:

$$J\frac{d\Omega}{dt} = T_T - T_G - T_r \tag{11}$$

where T_r is the friction torque and J is the combined inertia of the whole rotating parts.

3. WECS-DWIG Super-Twisting Based Proposed Control

As it was stated, the super-twisting based control set-up designed in this paper focuses on optimizing the WECS operation during low wind speeds regimes, when maximum power extraction is required (Zone AB in Figure 3). In this zone, only the DWIG's control winding is functioning. Recalling (5), to fulfil the control objective of MPPT, the system must operate at $T_{opt}(\Omega)$, which can be attained by tracking the optimal speed reference, given by $\Omega_{opt} = \frac{TSR_{opt}}{R}v_w$.

The control past point B is treated in detail in [11] and it is beyond the scope of this paper. In particular, the control winding in Zone BC and Zone CD is controlled for power limitation (consequently, constant current control), which is successfully implemented through a simple feed forward (FF) action. Whereas, the power winding is only operative in Zone CD, where it is directly connected to the grid [11]. Above the rated wind speed (i.e., when the maximum power of the DWIG is reached), an external power limiting mechanism must exist (possibly involving active or passive modification of the aerodynamic characteristics of the blades).

3.1. WECS Model for the Control Design

A reduced order model can be used for the design of the proposed controller. In WECS, the electrical dynamics are considerably faster than the mechanical ones, then a practical assumption for the design of the controller is to neglect the electrical dynamics. Under this consideration, the torque of the DWIG in zones AB and BC becomes ([29])

$$T_G = T_{Gcw} = 3\frac{p_{cw}R_{r\_cw}}{s_{cw}(2\pi f_{cw})}\frac{v_{s\_cw}^2}{\left[\left(R_{s\_cw} + \frac{R_{r\_cw}}{s_{cw}}\right)^2 + (2\pi f_{cw})^2(L_{ls\_cw} + L_{lr\_cw})^2\right]} \tag{12}$$

where s_{cw} is the CW slip relative to Ω_{s_cw}, defined as

$$s_{cw} = \frac{\Omega - \Omega_{s\_cw}}{\Omega_{s\_cw}} \tag{13}$$

If $s_{cw} << 1$, then Equation (12) can be approximated by the linear expression [29]:

$$T_G = 3\frac{p_{cw}^2}{R_{r\_cw}}\left(\frac{v_{s\_cw}}{2\pi f_{cw}}\right)^2 (\Omega - \Omega_{s\_cw}) \tag{14}$$

Moreover, the DWIG is commanded using a scalar technique, i.e., the CW is fed with sinusoidal voltages whose frequency f_{cw} is varied to maintain a constant $\frac{v_{s\_cw}}{f_{cw}}$ ratio, then Equation (14) can be expressed as

$$T_G = K_T(\Omega - \Omega_{s\_cw}) \tag{15}$$

The numerical value of the torque constant K_T in Equation (15) can be experimentally obtained or computed from the DWIG parameters using Equation (14).

Finally, substituting Equation (15) into the equation of the dominant dynamics Equation (11) and neglecting the friction, it yields the following reduced order model for the control design:

$$\dot{x} = f(x, v_w) + gu \tag{16}$$

$$f(x, v_w) = \left(\frac{T_T(x, v_w)}{J} - \frac{K_T}{J}x\right) \text{ and } g = \frac{K_T}{J}$$

with the state variable $x = \Omega$ and the control input $u = \Omega_{s\_cw}$. Note that it would have been possible to include in Equation (16) a term to model the nominal torque friction, leaving a reduced uncertainty for T_r. However, it was preferred to consider the latter completely unknown for the design. In this way, the experimental results obtained with the *ST* control set-up will better prove the effectiveness and robustness of the proposed controller in real operation, even facing such assumed unknown friction together with the other existing uncertainties/disturbances.

3.2. Super-Twisting Based Control Set-Up Design

The proposed *ST* control set-up comprises two control terms:

$$u = u_{FF} + u_{ST} \tag{17}$$

The first one is a feed forward (FF) control action, in charge of steering the system to the neighborhood of the desired $T_{opt}(\Omega)$ zone. The second one is a SOSM ST control action, responsible for accurately performing MPPT during scarce wind regimes (i.e., Zone AB), even in the presence of disturbances and uncertainties with respect to the nominal WECS-DWIG model.

3.2.1. Feed-Forward Control Term

Firstly, for the design of the feed forward control action, the steady state torque balance of the system is obtained from Equation (11), i.e., $T_T = T_G + T_r$. Then, assuming that the optimal torque is attained, $T_T = T_{opt}(\Omega)$, under ideal operation (undisturbed system and neglecting friction) the torque balance gives

$$T_{opt}(\Omega) = K_{opt}\Omega^2 = T_G = K_T(\Omega - \Omega_{s\_cw}) \tag{18}$$

and the expression of the proposed Feed Forward action results in

$$\Omega_{s\_cw}\big|_{T_T = T_{opt}} = u_{FF} = \Omega - \frac{K_{opt}}{K_T}\Omega^2 \tag{19}$$

Note that, even though u_{FF} cannot accurately deal with the uncertainties of the real WECS, this feed forward approach has proven to be successful to lead the system to the vicinity of T_{opt} [11].

3.2.2. SOSM Super-Twisting Control Term

During low wind regimes, the resource is scarce, so it is of paramount importance to extract as much energy from the wind as possible. As previously said, the proposed feed forward action is an effective technique to guide the system to the proximity of T_{opt}, although to attain a highly precise MPPT in a real generation system, the incorporation of a robust feedback control action is essential.

To this end, a SOSM technique has been selected to implement such robust feedback control term for the dual-stator winding induction generator based WECS under consideration. SOSM has demonstrated to be a suitable design tool applicable to several WECS topologies based on conventional induction or synchronous generators.

To track the optimum shaft speed in Equation (3), the following smooth sliding variable σ can be defined:

$$\sigma = \Omega - \Omega_{opt} = \Omega - \frac{TSR_{opt}}{R}v_w \tag{20}$$

which is of relative degree (RD) 1 with respect to the input u in Equation (16).

Then, a super-twisting algorithm is selected to synthetize the SOSM control term u_{ST}. In addition to be suitable for σ of RD 1, this robust control SOSM algorithm also avoids direct discontinuous control action, reducing mechanical stresses and diminishing chattering. Plus, it does not require measurement of the sliding variable time derivative $\dot{\sigma}$. The selected ST control term is given by ([30])

$$u_{ST} = -\beta|\sigma|^{\frac{1}{2}}sign(\sigma) - \alpha\int_0^t sign(\sigma(\tau))d\tau \tag{21}$$

where β and α are the control gains.

With an appropriate tuning of those gains, the ST guarantees the zeroing of σ and $\dot{\sigma}$ in finite time, which implies robust MPPT operating at $\Omega = \Omega_{opt}(t)$ for variable wind speeds.

For the design of such gains, it is required to compute and obtain bounds for the second time-derivative of the sliding variable. The first step is to substitute Equation (16) into Equation (20) and to differentiate σ twice:

$$\ddot{\sigma} = \left[\dot{f}(\Omega, v_w) + g\dot{u}\right] - \ddot{\Omega}_{opt} = \left[\dot{f}(\Omega, v_w) - \ddot{\Omega}_{opt}\right] + \frac{K_T}{J}\dot{u} \tag{22}$$

Using Equation (17), it can be written in affine form with respect to $\dot{u}_{ST}$ as

$$\ddot{\sigma} = \left[\dot{f}(\Omega, v_w) - \ddot{\Omega}_{opt} + \frac{K_T}{J}\dot{u}_{FF}\right] + \frac{K_T}{J}\dot{u}_{ST} = \varphi + \gamma\dot{u}_{ST} \tag{23}$$

where functions φ and γ are bounded by positive constants Φ, Γ_m, and Γ_M as

$$-\Phi \leq \varphi \leq \Phi \text{ and } \Gamma_m \leq \gamma \leq \Gamma_M \tag{24}$$

including, in the bounding process, the uncertainties and disturbances to be rejected.

Then, the *ST* control gains must be tuned to fulfil the following conditions:

$$\alpha > \frac{\Phi}{\Gamma_m} \text{ and } \beta^2 \geq \frac{4\Phi\,\Gamma_M(\alpha + \Phi)}{\Gamma_m^2\,\Gamma_m(\alpha - \Phi)} \tag{25}$$

To obtain the bounds for the WECS-DWIG under study, an essentially practical procedure has been followed:

- Firstly, primary bounds were obtained through systematic simulation tests, complemented with a detailed analysis of the actual system topology and limitations. In this framework, simulations were run to thoroughly assess the behavior of functions $\varphi = f(\Omega, v_w) - \ddot{\Omega}_{opt} + \frac{K_T}{J}\dot{u}_{FF}$, and $\gamma = \frac{K_T}{J}$ under the effect of several wind profiles, disturbances, and model uncertainties, covering the operation range of Zone AB.
- To conclude the design, the main phase, i.e., the experimental tuning phase, was performed. The ST-SOSM controller was implemented in the testing workbench. Based on the previous bounds, different sets of preliminary control gains β and α fulfilling Equation (25) were programmed. Then, progressive refinement of the control gains was undertaken, conducting iterative laboratory tests, and gains β and α were chosen, prioritising the chattering reduction in the definitive selection. The resulting gains for the *ST* implementation are

$$\alpha = 0.038 \text{ and } \beta = 0.12$$

It is worth noting that these two gains are computed off-line during the tuning procedure, so on-line operation of the *ST* control term is rather simple.

4. Experimental Results

This section presents the experimental results of the proposed ST control set-up for the DWIG based WECS. The features of the controller aiming to MPPT are examined through two sets of experiments. Case Study 1 shows the performance of the controlled system under a realistic variable wind speed profile, while in Case Study 2, a stepped wind profile is considered, to better analyse and compare the transient response of the controlled system.

Before presenting the results, a brief description of the equipment used in the tests, whose picture is shown in Figure 5, is given in the next few paragraphs.

Figure 5. Experimental set-up.

The experimental set-up is made up of two fundamental subsystems: the DWIG generation system module and the wind turbine emulator module.

- *DWIG Generation System Module*

The generation system module consists of a DWIG prototype, designed and constructed in the URV laboratory, with a power electronic conversion chain and its associated control system, to feed the stator CW with variable voltage and frequency, keeping a constant V/f ratio.

The DWIG prototype's most significant technical data—rated power: 5.5 kW (CW: 1.8 kW and PW: 3.7 kW), rated synchronous speed: 314 rad/s (3000 rpm), rated torque: 18 Nm, rated voltages for both PW and CW: 400 V_{RMS} (wye connection), rated frequency: 50 Hz, rated current: CW = 2.6 A/phase and PW = 5.3 A/phase. The electrical parameters of the windings—PW: $R_{s\_pw} = 2.9\ \Omega$, $R_{r\_pw} = 1.2\ \Omega$, $L_{ls\_pw} = L_{lr\_pw} = 9.8$ mH, $L_{m\_pw} = 470$ mH; CW: $R_{s\_cw} = 5.5\ \Omega$, $R_{r\_cw} = 2.4\ \Omega$, $L_{ls\_cw} = L_{lr\_cw} = 4.6$ mH, $L_{m\_cw} = 175$ mH. Then, the numerical value of the torque constant K_T in Equation (15) is 1.105 [Nm/(rad/s)] and was experimentally obtained and verified from the DWIG parameters using Equation (14).

The CW is fed by a three-phase inverter made with a Semikron Semistack SKS-35F power module that comprises a bridge rectifier and a 1200 V, 35 A three-legged IGBT CC-CA converter that supports a 15 kHz maximum switching frequency. The three-phase inverter output voltage can be varied by means of a symmetric regular sampled sinusoidal PWM method, implemented in a Cortex® M4 120 MHz Texas Instruments Tiva C Series TM4C1294NCPDT Microcontroller, with a carrier frequency of 5 kHz and with overmodulation capability.

A Hewlett-Packard programmable DC electronic power supply (1000 V_{DCmax}, 15 kW, with bidirectional current capability), working as a constant voltage source, is connected to the inverter DC-link. The power supply behaves as a dummy load, absorbing the active power delivered by the CW. The DC-link is regulated at 600 V_{DC}, which is an adequate value so that the three-phase inverter can deliver the CW rated voltage.

- *Wind Turbine Emulator Module*

The wind turbine emulator is based on a Texas Instruments DSP F28335-TMS320 which computes the turbine torque from a programmed turbine characteristic, the measurement of the DWIG shaft speed and the desired wind speed. Such variable wind profile can be either downloaded to memory or entered via keypad. This computed torque acts as reference for the driving electric machine, a SIEMENS three-phase induction motor (model: 1LA71632AA) of 11 kW, 400 V, 50 Hz and 2 poles (3000 rpm synchronous speed), fed by a SIEMENS Micromaster MM440 frequency converter.

For these tests, a 5.6 kW three-blade horizontal axis wind turbine coupled to the DWIG trough a gearbox with a 1:11 speed ratio is emulated, with $R = 2.5$ m and $TSR_{opt} = 60.5$ (generator side).

4.1. Case Study 1: Variable Wind Speed Profile

The first series of tests was performed using a realistic wind speed profile, which is shown in Figure 6. Note that its maximum wind speed was below 7.5 m/s to ensure operation under a scarce wind regime, so the WECS-DWIG is functioning in conditions that maximum wind power extraction is of paramount importance (i.e., Zone AB with MPPT objective).

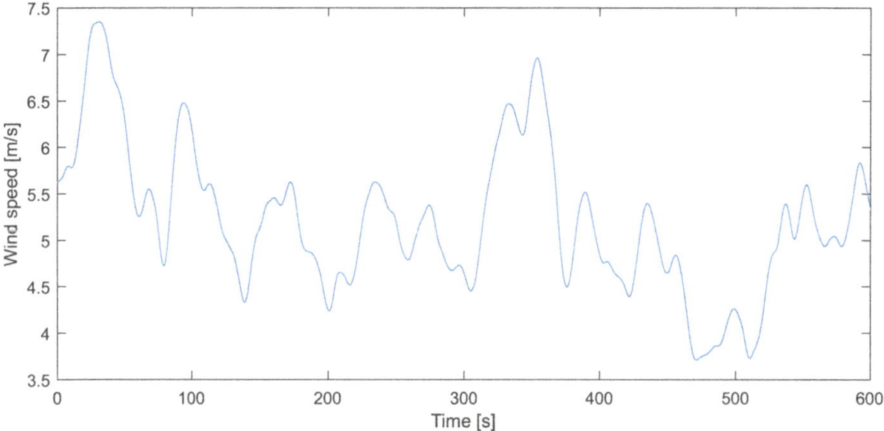

Figure 6. Variable wind speed profile.

Figure 7 shows the corresponding optimum speed Ω_{opt} in orange dashed line. That is, the temporal evolution of the reference speed for maximum power extraction in accordance with Equation (18). In addition, in blue, the real DWIG shaft speed when only the FF controller is employed is depicted [11]. It can be appreciated that the feed forward approach provides good tracking of the optimum speed. However, due to disturbances and uncertainties with respect to the real system, it is evident that a certain persistent error exists, being of particular interest to avoid it when the wind resource is scarce (in several periods, for instance around 475 s, the speed error reaches values of approximately 20%).

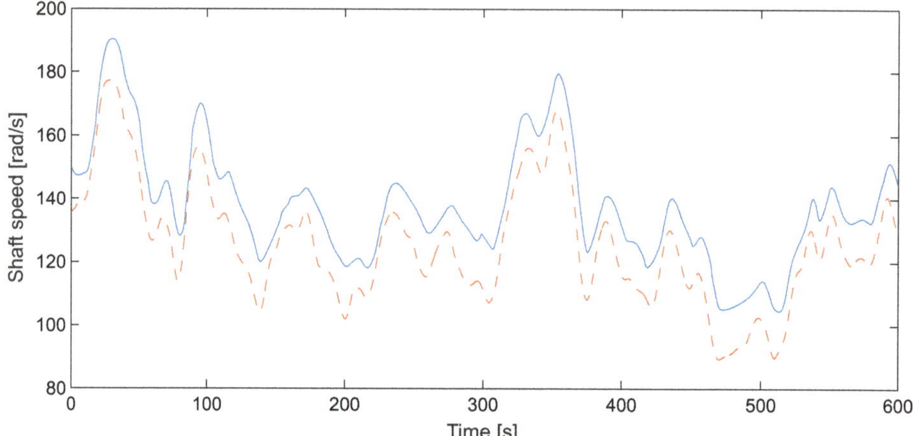

Figure 7. Optimal rotational speed reference for the variable wind profile (orange dashed line) and experimental speed tracking using only the feed forward (FF) controller (blue line).

As a counterpart, Figure 8 displays the experimental results obtained with the ST control set-up designed in Section 3.2. It can be observed that the behavior of the WECS-DWIG under the combine action of the proposed ST + FF strategy greatly improves. The shaft rotational, in practice, precisely overlaps the desired optimum speed Ω_{opt}, consequently excellent MPPT is attained. It is interesting to compare the system evolution in the DWIG speed-torque plane (Figure 9b), with the previous case, plotted in Figure 9a. It is notable how the SOSM based controller steers the system to travel almost over the optimum torque locus.

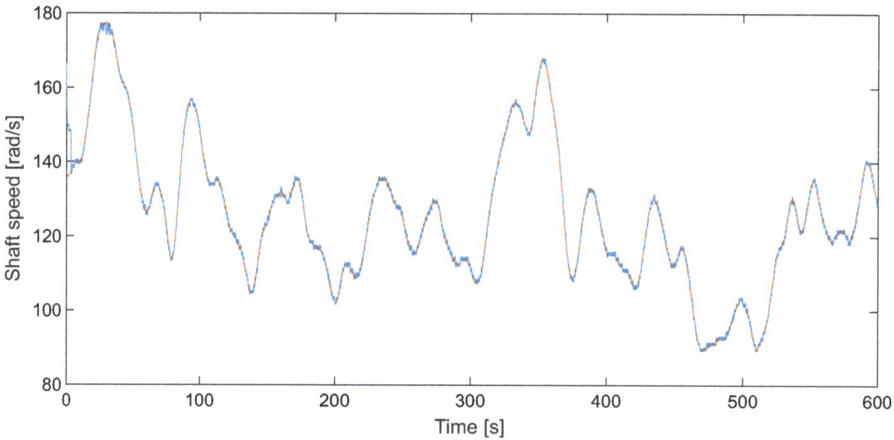

Figure 8. Optimal rotational speed reference for the random wind profile and experimental speed tracking using the proposed FF + ST (super-twisting) control set-up (overlapped).

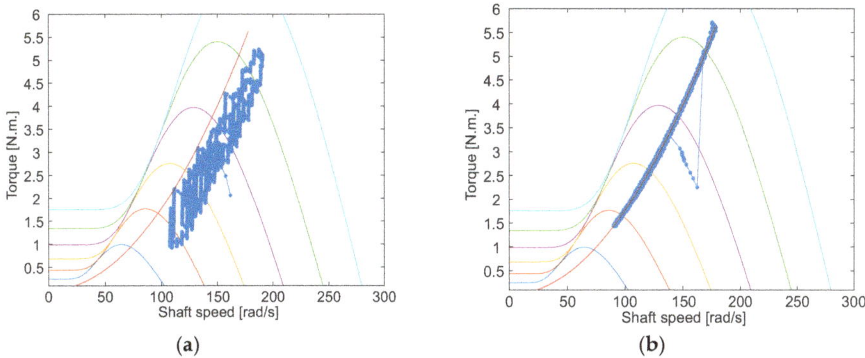

Figure 9. Evolution of the system operating points in the speed-torque plane (blue line) and optimal torque locus $T_{opt}(\Omega)$ (red parabola). (**a**) FF controller; (**b**) Proposed ST + FF control set-up.

Such reduction in the wind turbine conversion efficiency can be clearly visualised in the actual rotational speed-torque plane (see Figure 9a), where the operating points do not precisely evolve over the parabola of optimum torque $T_{opt}(\Omega)$, given by Equation (19). This, in turn, means that the power extraction is lower than the maximum available power in the wind.

Figure 10 shows the theoretical maximum power that the turbine can extract from the wind (red line), obtained from Equation (1) with $Cp(TSR_{opt})$, together with the actual power extracted by both controllers. It is clear that the proposed FF + ST controller practically allows full advantage of the wind resource (dashed blue) to be taken; whereas, using the FF controller (yellow line), some power available in the wind is wasted (e.g., about 23% around 510 s).

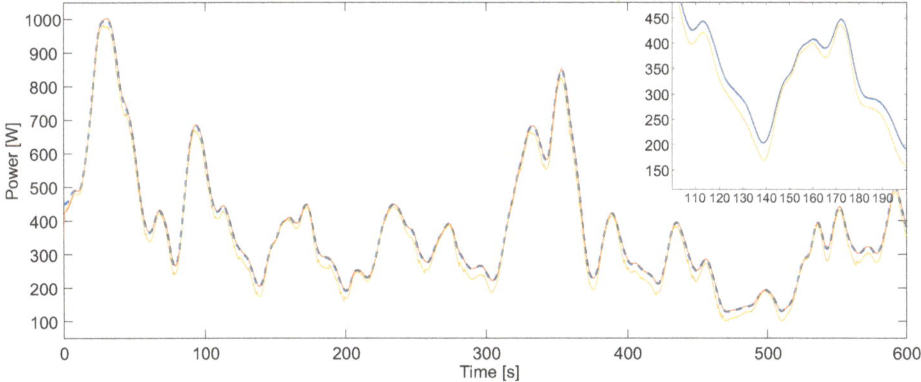

Figure 10. Maximum power the turbine can extract from the wind (red line). Actual extracted power using the FF + ST controller (dashed blue). Actual extracted power using the FF controller (yellow line).

Finally, the controllers control actions (namely, the mechanical synchronous speed of the CW, $\Omega_{s\_cw} = \frac{2\pi f_{cw}}{p_{cw}}$) are depicted in Figure 11a,b.

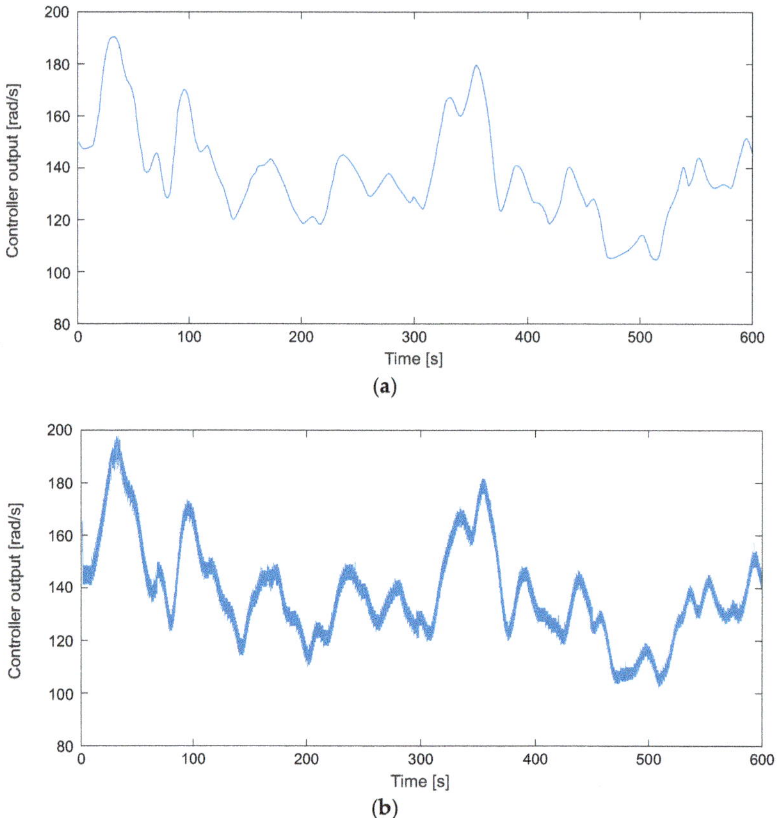

(a)

(b)

Figure 11. Control action $\Omega_{s\_cw} = \frac{2\pi f_{cw}}{p_{cw}}$ (a) FF controller and (b) FF + ST controller.

4.2. Case Study 2 Stepped Wind Profile

The second series of experimental tests is done using the stepped wind speed profile depicted in Figure 12. This is not a realistic profile, but it is of interest for the visualization of the transient and steady state responses of the controlled WECS-DWIG under study.

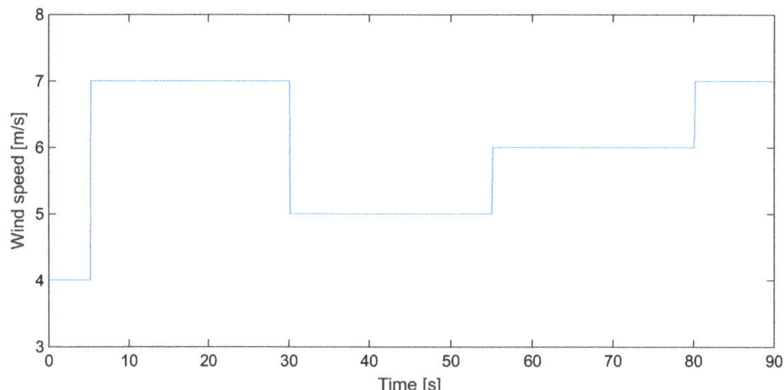

Figure 12. Stepped wind profile used in the tests.

The first experimental test corresponds to the feed forward controller on its own (blue line in Figure 13). The optimal shaft speed reference Ω_{opt} is plotted in orange dashed line. Similarly to Case Study 1, even though the FF controller proves its capability to steer the system to the neighborhood of reference Ω_{opt}, a remnant offset is clearly visible (approximately 15 rad/s in excess, corresponding to 17% at the lower speed), due to disturbances and model errors with respect to the real WECS-DWIG. Besides, in the zoomed view it can also be observed a noticeable settling-time (approximately 2 s) until steady state is reached.

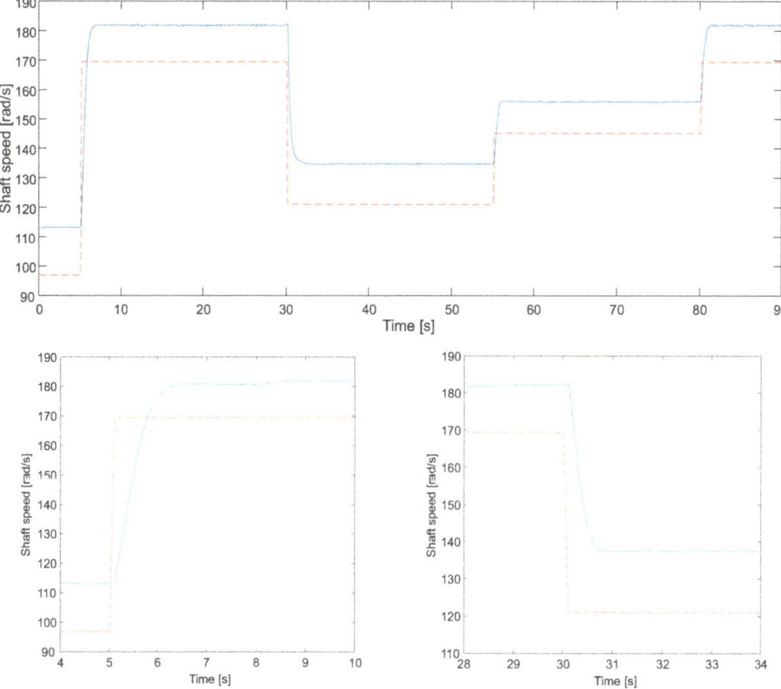

Figure 13. Optimal rotational speed reference for the variable wind profile (orange dashed line) and experimental speed tracking using only the FF controller (blue line).

To enrich the comparison analysis, the following experimental test presents the behavior of the system, adding a PI control term (PI + FF control in blue line in Figure 14). A two stages tuning procedure was used. Firstly, preliminary values for the PI's gains were computed using Ziegler–Nichols rules and, in a second tuning stage, those values were refined through simulation and experimental tests, obtaining the following gains: $k_p = 0.012$ and $k_i = 5$. It can be appreciated that the PI control action greatly improves the steady state offset error. However, the transient responses present overshoot, the first being the largest, of the order of 6%. In the zoomed view, it can also be observed a noticeable settling-time, of the order of 2 s.

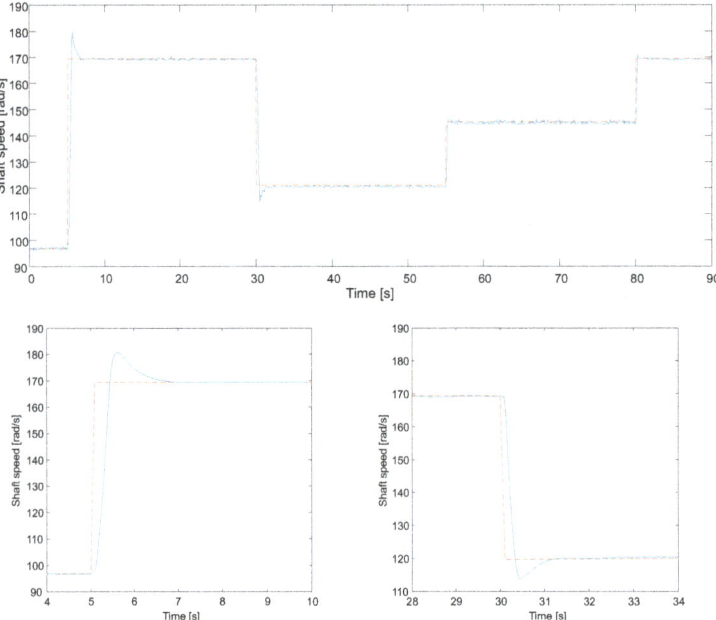

Figure 14. Optimal rotational speed reference for stepped wind profile (orange dashed line) and experimental speed tracking using the FF with a classic PI controller (blue line).

Finally, Figure 15 displays the experimental results obtained with the ST + FF control set-up. The proposed robust control strategy practically eliminates the steady state error, presents a negligible transient overshoot and very good settling-time (in less than 0.5 s the steady state is practically reached), achieving an excellent tracking of the optimal speed Ω_{opt}.

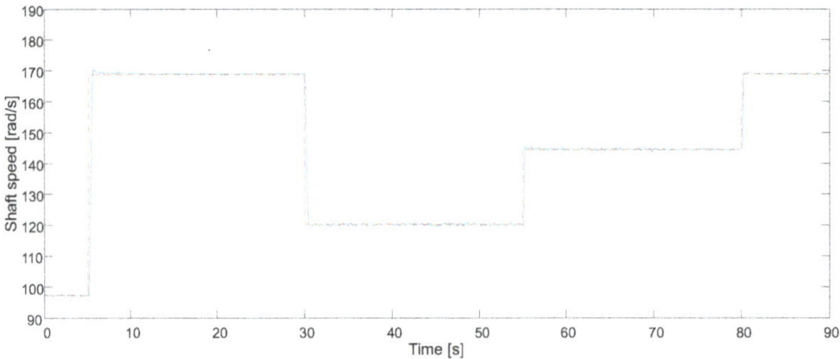

Figure 15. *Cont.*

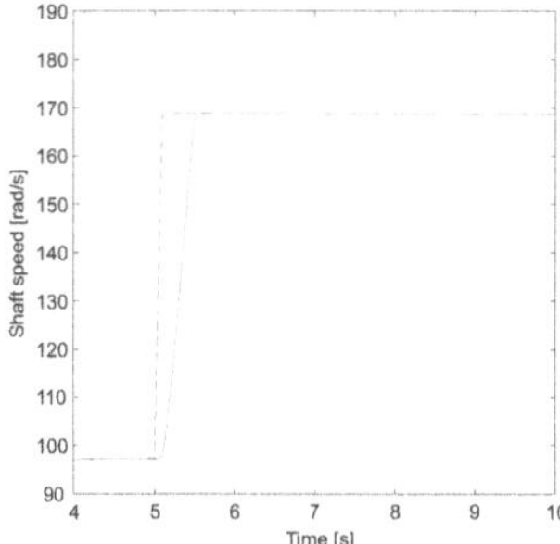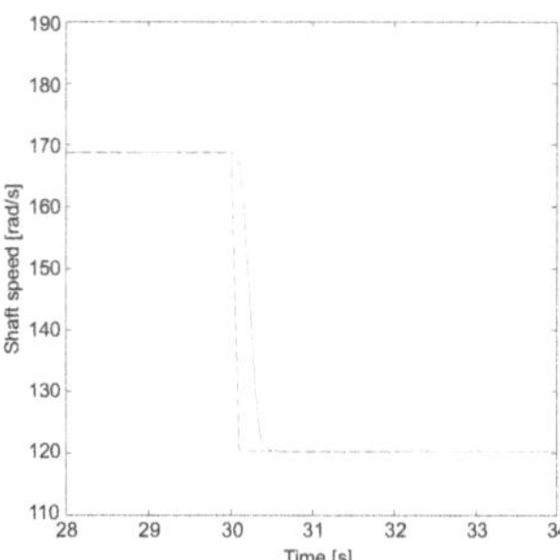

Figure 15. Optimal rotational speed reference for stepped wind profile and experimental speed tracking using the proposed ST + FF control set-up (overlapped).

5. Conclusions

Experimental results of the ST control set-up have been implemented and thoroughly assessed on a 5.5 kW WECS-DWIG laboratory test station, including comparison with a previous successful FF based control and with a classic PI controller. The proposed control strategy for this variable speed wind power generator proved to have excellent dynamic transient and steady-state performance.

The highly satisfactory results using the aforementioned approach confirmed the feasibility of the solution for implementation in real generation plants based on this type of generator. The principal advantages of the ST control set-up for a DWIG based wind turbine can be summarized as follows: guaranteed extended range of operation in spite of the nonlinear nature of the system; fast finite time convergence for MPPT; reduced mechanical stresses and chattering; robustness against real WECS parameter uncertainties/variations; on-line operation of the proposed controller not requiring of measured signals differentiation; and relatively simple controller structure, resulting in moderate real time computational burden.

As a future research line in this project, the encouraging results obtained with the proposed ST control set-up for the WECS-DWIG will lead to the development of a hybrid micro-grid (MG) topology. It will be designed to take advantage of the electrical features of the DWIG. The MG will present a DC bus, for generation and energy storage module interconnection, and an AC bus, for power exchange with a weak grid and the load. In addition to the main DWIG based wind power source, the MG will consider an ancillary power module (e.g., a photovoltaic module (PV)) and an energy storage module (ESM), with combined storage devices (high-power and high-energy density ones).

It is expected that this MG topology will unite the DWIG squirrel cage's robustness and its capacity to simultaneously generate to the DC and the AC buses, with the complementary capability of a storage module, resulting in a versatile MG.

Author Contributions: Conceptualization, P.F.P.; Formal analysis, J.I.T., P.F.P. and M.G.C.; Funding acquisition, J.A.B.-R. and P.F.P.; Investigation, J.I.T., P.F.P., M.G.C. and J.A.B.-R.; Methodology, P.F.P. and M.G.C.; Resources, J.A.B.-R.; Software, J.I.T.; Supervision, P.F.P.; Writing—original draft, M.G.C.; Writing—review & editing, J.I.T., P.F.P., M.G.C. and J.A.B.-R.

Funding: This research was funded by the Spanish Agencia Estatal de Investigación (AEI) and the Fondo Europeo de Desarrollo Regional (FEDER) under research projects DPI2015-67292-R (AEI/FEDER, UE) and the Argentinean projects ANPCyT PICT N2015-2257; UNLP 11/I217 and CONICET PIP 112-2015-0100496CO.

Acknowledgments: Many thanks to GAEI, DEEEA-ETSE, Universitat Rovira i Virgili (URV), Agencia Estatal de Investigación, from Spain, Fondo Europeo de Desarrollo Regional; and Universidad Nacional de La Plata (UNLP), Consejo Nacional de Investigaciones Científicas y Técnicas (CONICET), Agencia Nacional de Promoción Científica y Tecnológica (ANPCyT), Facultad de Ingeniería (FI-UNLP) and CIDEI, ITBA, from Argentina.

Conflicts of Interest: The authors declare no conflict of interest.

References

1. Bhutto, D.K.; Ansari, J.A.; Bukhari, S.H.; Chachar, F.A. Wind energy conversion systems (WECS) generators: A review. In Proceedings of the International Conference Computing, Mathematics and Engineering Technologies, Pakistan, Sukkur, Pakistan, 30–31 January 2019. [CrossRef]
2. Akinrinde, A.; Swanson, A.; Tiako, R. Dynamic Behavior of Wind Turbine Generator Configurations during Ferroresonant Conditions. *Energies* **2019**, *12*, 639. [CrossRef]
3. Basak, S.; Chakraborty, C. Dual stator winding induction machine: Problems, progress, and future scope. *IEEE Trans. Ind. Electron.* **2015**, *62*, 4641–4652. [CrossRef]
4. Muñoz, A.R.; Lipo, T.A. Dual stator winding induction machine drive. *IEEE Trans. Ind. Appl.* **2000**, *36*, 1369–1379. [CrossRef]
5. Bu, F.; Liu, H.; Huang, W.; Xu, H.; Hu, Y. Recent advances and developments in dual stator winding induction generator and system. *IEEE Trans. Energy Convers.* **2018**, *33*, 1431–1442. [CrossRef]
6. Amimeur, H.; Aouzellag, D.; Abdessemed, R.; Ghedamsi, K. Sliding mode control of a dual-stator induction generator for wind energy conversion systems. *Electr. Power Energy Syst.* **2012**, *42*, 60–70. [CrossRef]
7. Sarasola, I.; Poza, J.; Rodriguez, M.A.; Abad, G. Predictive direct torque control for brushless doubly fed machine with reduced torque ripple at constant switching frequency. In Proceedings of the IEEE International Symposium on Industrial Electronics (ISIE), Vigo, Spain, 4–7 June 2007; pp. 1074–1079.
8. Shao, S.; Abdi, E.; Barati, F.; McMahon, R. Stator-flux-oriented vector control for brushless doubly fed induction generator. *IEEE Trans. Ind. Electron.* **2009**, *56*, 4220–4228. [CrossRef]
9. Mahboub, M.A.; Drid, S.; Sid, M.A.; Cheikh, R. Sliding mode control of grid connected brushless doubly fed induction generator driven by wind turbine in variable speed. *Int. J. Syst. Assur. Eng. Manag.* **2017**, *8*, 788–798. [CrossRef]
10. Wu, Z.; Ojo, O.; Sastry, J. High-performance control of a dual stator winding DC power induction generator. *IEEE Trans. Ind. Appl.* **2007**, *43*, 582–592. [CrossRef]
11. Barrado-Rodrigo, J.A.; Talpone, J.I.; Martinez-Salamero, L. Variable-speed wind energy conversion system based on a dual stator-winding induction generator. *IET Renew. Power Gener.* **2017**, *11*, 73–80. [CrossRef]
12. Barrado-Rodrigo, J.A.; Valderrama-Blavi, H.; Ayad, M.; Talpone, J.I. Low-voltage ride through capability testing of a DWIG for wind power generation. In Proceedings of the IEEE International Conference on Compatibility, Power Electronics and Power Engineering, Cadiz, Spain, 4–6 April 2017. [CrossRef]
13. Utkin, V. Survey paper variable structure systems with sliding modes. *IEEE Trans. Autom. Control* **1977**, *22*, 212–222. [CrossRef]
14. Levant, A. Sliding order and sliding accuracy in sliding mode control. *Int. J. Control* **1993**, *58*, 1247–1263. [CrossRef]
15. Fridman, L.; Levant, A. Sliding mode control in engineering. In *Higher Order Sliding Modes*; Perruquetti, W., Barbot, J., Eds.; Marcel Dekker, Inc.: New York, NY, USA; Basel, Switzerland, 2002; Chapter 3; pp. 53–101.
16. Bartolini, G.; Pisano, A.; Punta, E.; Usai, E. A survey of applications of second-order sliding mode control to mechanical systems. *Int. J. Control* **2003**, *76*, 875–892. [CrossRef]
17. Fridman, L.; Moreno, J.; Iriarte, R. (Eds.) *Sliding Modes after the First Decade of the 21st Century*; Ser. Lecture Notes in Control and Information Sciences; Springer: Berlin/Heidelberg, Germany, 2012; Volume 412. [CrossRef]
18. Bandyopadhyay, B.; Janardhanan, S.; Spurgeon, S.K. (Eds.) *Advances in Sliding Mode Control. Concept, Theory and Implementation*; Ser. Lecture Notes in Control and Information Sciences; Springer: Heidelberg, Germany, 2013; Volume 440, p. 380. [CrossRef]
19. Fridman, L.; Barbot, J.-P.; Plestan, F. (Eds.) *Recent Trends in Sliding Mode Control*; IET: London, UK, 2016. [CrossRef]
20. Li, S.; Yu, X.; Fridman, L.; Man, Z.; Wang, X. (Eds.) *Advances in Variable Structure Systems and Sliding Mode Control Theory and Applications*; Ser. Studies in Systems, Decision and Control; Springer International Publishing: Cham, Switzerland, 2018; Volume 115. [CrossRef]
21. Shtessel, Y.; Edwards, C.; Fridman, L.; Levant, A. (Eds.) *Sliding Mode Control and Observation*; Springer: New York, NY, USA, 2013. [CrossRef]

22. Huangfu, Y.; Zhuo, S.; Rathore, A.K.; Breaz, E.; Nahid-Mobarakeh, B.; Gao, F. Super-twisting differentiator-based high order sliding mode voltage control design for DC-DC buck converters. *Energies* **2016**, *9*, 494. [CrossRef]

23. Beltran, B.; Benbouzid, M.; Ahmed-Ali, T. Second-order sliding mode control of a doubly fed induction generator driven wind turbine. *IEEE Trans. Energy Convers.* **2012**, *27*, 261–269. [CrossRef]

24. Martinez, M.I.; Susperregui, A.; Tapia, G. Second-order sliding-mode-based global control scheme for wind turbine-driven DFIGs subject to unbalanced and distorted grid voltage. *IET Electr. Power Appl.* **2017**, *11*, 1013–1022. [CrossRef]

25. Chavira, F.; Ortega-Cisneros, S.; Rivera, J. A Novel Sliding Mode Control Scheme for a PMSG-Based Variable Speed Wind Energy Conversion System. *Energies* **2017**, *10*, 1476. [CrossRef]

26. Benamor, A.; Benchouia, M.T.; Srairi, K.; Benbouzid, M. A novel rooted tree optimization apply in the high order sliding mode control using super-twisting algorithm based on DTC scheme for DFIG. *Int. J. Electr. Power Energy Syst.* **2019**, *108*, 293–302. [CrossRef]

27. Evangelista, C.A.; Pisano, A.; Puleston, P.; Usai, E. Receding-Horizon Adaptive Second-Order Sliding Mode Control for Doubly Fed Induction Generator Based Wind Turbines. *IEEE Trans. Control Syst. Technol.* **2017**, *25*, 73–84. [CrossRef]

28. Plestan, F.; Evangelista, C.; Puleston, P.; Guenoune, I. Control of a Twin Wind Turbines System without Wind Velocity Information. In Proceedings of the IEEE 15th International Workshop on Variable Structure Systems and Sliding Mode Control, VSS, Graz, Austria, 9–11 July 2018.

29. Bose, B.K. *Modern Power Electronics and AC Drives*; Prentice Hall: Upper Saddle River, NJ, USA, 2001.

30. Bartolini, G.; Ferrara, A.; Levant, A.; Usai, E. On second order sliding mode controllers. In *Variable Structure Systems, Sliding Mode and Nonlinear Control*; Young, K., Özgüner, Ü., Eds.; Lecture Notes in Control and Information Sciences; Springer: London, UK, 1999; Volume 247. [CrossRef]

Article

A Loss-Free Resistor-Based Versatile Ballast for Discharge Lamps

Hugo Valderrama-Blavi *, Antonio Leon-Masich, Carlos Olalla and Àngel Cid-Pastor

Department of Electrical, Electronic, and Automatic Control Engineering (DEEEA), Universitat Rovira i Virgili, 43007 Tarragona, Spain; antonio.leon-masich@eu.panasonic.com (A.L.-M.); carlos.olalla@urv.cat (C.O.); angel.cid@urv.cat (À.C.-P.)
* Correspondence: hugo.valderrama@urv.cat; Tel.: +34-977-558-523

Received: 4 March 2019; Accepted: 7 April 2019; Published: 11 April 2019

Abstract: This paper presents a versatile ballast for discharge lamps, whose operation is based on the notion of a loss-free resistor (LFR). The ballast consists of two stages: (1) a boost converter operating in continuous conduction mode (CCM) and exhibiting an LFR behavior imposed by sliding-mode control; and (2) a resonant inverter supplying the discharge lamp at high frequencies. Thanks to this mode of operation, the power transferred to the lamp is regulated by the LFR input resistance, allowing successful ignition, warm-up, nominal, and dimming operation of a range of discharge lamps, with no need for complex regulation schemes based on lamp models. The versatility of the ballast has been experimentally proven for both conventional and electrodeless discharge lamps. Tests include induction electrodeless fluorescent (IEFL), high-pressure sodium (HPS) vapor, and metal-halide lamps.

Keywords: Induction Electrodeless Fluorescent Lamps (IEFL); High-Intensity Discharge Lamps (HID); sliding-mode control; loss-free resistor (LFR)

---

1. Introduction

Lighting from electricity accounts for approximately 15–19% of global energy consumption and over five percent of worldwide greenhouse gas (GHG) emissions. It has been estimated that replacing all inefficient on-grid lighting would result in 939 TWh of electricity savings annually, which correspond to approximately 490 million tons of CO_2 [1,2]. For that reason, efficient lighting sources, such as light-emitting diodes (LEDs) and discharge lamps, are increasingly used as a simple and cheap procedure to reduce the contribution of electricity consumers to global warming [3–7]. In this sense, although LEDs are steadily increasing their rated power and luminous efficacy in terms of lumen/Watt, discharge lamps are still competitive in applications requiring high power and long lifespan. Concretely, for a given discharge lamp type, higher-power bulbs usually have better efficacy than lower-power ones [8]. In contrast, due to the efficiency droop, the situation is normally the opposite in LEDs, making them more suitable for low-power applications [9]. Regarding lifetime, although high-performance LEDs last about 50,000 h, and high-pressure sodium (HPS) or metal-halide (MH) lifespan is between 15,000 and 25,000 h, external coil induction electrodeless fluorescents (IEFLs) can last up to 100,000 h with a very little light output depreciation until the very end of its life.

Many types of discharge lamps are available in the market, differing in the type of gas, filling pressure, operating voltage, color rendering index, or efficiency. One of the main differences among them is the way the electrical energy is transferred to the bulb. Whereas conventional lamps use internal electrodes to transfer the energy to the bulb, the magnetic induction electrodeless lamps use internal or external coils to do so. This coil behaves as the primary winding of a transformer, whereas the discharge operates as the secondary one-turn coil.

Although single-stage solutions exist [10–13], ballasts for discharge lamps are typically composed of two stages [14]. The first stage is a buck, boost, or buck-boost stage performing power factor correction (PFC) and providing an approximately constant DC-link voltage. The second stage is a full or half-bridge LCC resonant converter driving the lamp.

In this context, power regulation is the subject of many papers because dimming operation can provide about a 50–60% energy saving [7,12,15]. The issue is still open to discussion because discharge lamp systems are expensive, and light-dimming normally requires a more complex circuitry than traditional systems without light regulation. In high-intensity discharge (HID) lamps, current mode control is the most extended technique [10], such that power control can be realized by analog multiplication of voltages and currents within the lamp. Also, other techniques have been used, such as controlling the resonant tank input current [11]. Regarding IEFLs, J. Tae-Kun et al. [16] present a comparison between two power control methods: the variable switching frequency control and the variable DC-link voltage control method, to conclude that the latter is more suitable for linear dimming operation of the lamp. Later, several authors presented power conversion topologies dedicated to regulating the light in the IEFL [17–20] with the aim to compare performance and efficiency in the converters. In all these cases dimming control is achieved by adjusting the DC-link voltage. The common characteristic in all the previous works is the necessity of exactly understanding the lamp behavior, and its electrical characteristics at different power ratings [21] to achieve power regulation. However, precise lamp modelling also implies that each ballast is designed for a given lamp type, which results in the opposite concept of a versatile design.

In this paper, a novel approach for power control in electronic ballasts is proposed. The approach is based on a conventional two-stage ballast: besides of performing PFC, the first stage has the characteristics of a loss-free resistor (LFR), whereas the second stage simply converts the DC input power into the required AC waveforms at the desired frequency. The LFR is a POPI device [22,23] whose output port is a power source which delivers to the load the power absorbed by the emulated input resistance, no matter what type of lamp is connected at the output. Consequently, the power injected in the lamp can be accurately regulated with a simple controller and with no stability issues, no matter what impedance the lamp connected at the output presents. In other words, the proposed ballast does not require lamp modelling. The second stage operates as an inverter, whose switching frequency can be selected depending on the requirements of the lamp. In this sense, the paper also presents a method to design a resonant tank that is compatible with several discharge lamp types. Thanks to these features, the only expected limits on the ballast versatility is the rating of the components used in the power stages. The proposed approach results in three advantageous characteristics. First, the proposed ballast is extremely versatile. Secondly, stability is ensured irrespective of the frequency response of the lamp. Finally, light-dimming control can be achieved easily by adjusting the LFR input power.

The proposed approach could be applied to any discharge lamp. In the paper, we have demonstrated the ballast with a prototype that can drive some of the most common types. Specifically, the paper shows results for an IEFL, a HPS vapor lamp, and two MH lamps.

The remaining part of this article is organized as follows: the proposed ballast and the lamp start-up procedure are described in Section 2. An appropriate sliding-mode control is applied to the boost converter to impose an LFR behavior in Section 3. The effect of different lamp models in the resonant inverter design and the ballast stability is analyzed in Section 4. Experimental results demonstrating the feasibility of the proposed ballast are shown in Section 5. The conclusions and future research are presented in Section 6.

2. Proposed LFR-Based Ballast

The operation of the proposed ballast in steady-state combines the features of its two stages, i.e., LFR and resonant inverter depicted in Figure 1. The LFR stage assures the lamp supply with both constant and adjustable power to regulate the luminosity and protect the lamp. As an advantage, the LFR behavior precludes the instabilities caused by the negative incremental impedance of many discharge lamps.

The resonant inverter stage is required to supply the lamps in AC at high frequency (90–250 kHz). AC supply is needed to avoid electrode wear in voltage driven discharge lamps, and to make the IEFL coils behave like a transformer.

Figure 2 depicts the schematic of the proposed ballast. The LFR is implemented by an appropriate sliding-mode control in a boost converter supplied by low-voltage DC source (such as a car battery) ranging from 12 V to 15 V. The figure shows a DC input, but the proposed approach can also be applied in the case that V_g is the AC input from a rectifier. Also in the figure, a boost converter is considered, but other topologies (such as the buck-boost, flyback or forward) could be employed if the ballast is supplied from the grid.

The regulation of the lamp requires very different modes of operation: start-up, warm-up, nominal output, and dimming. Thanks to the LFR behavior, the ballast operates as a power source regardless of the mode, and control is straightforward.

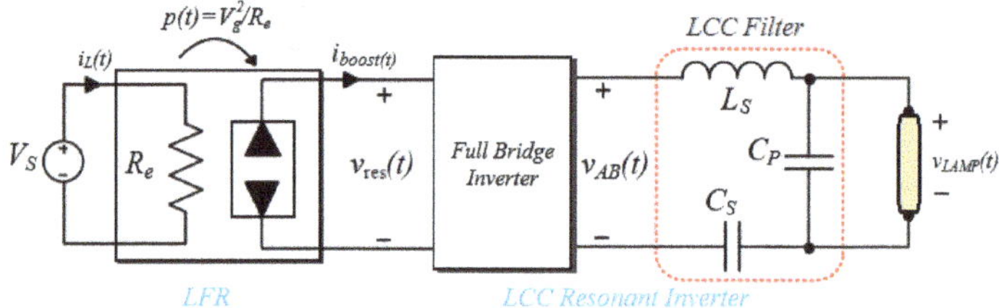

Figure 1. Block diagram of the LFR-based electronic ballast.

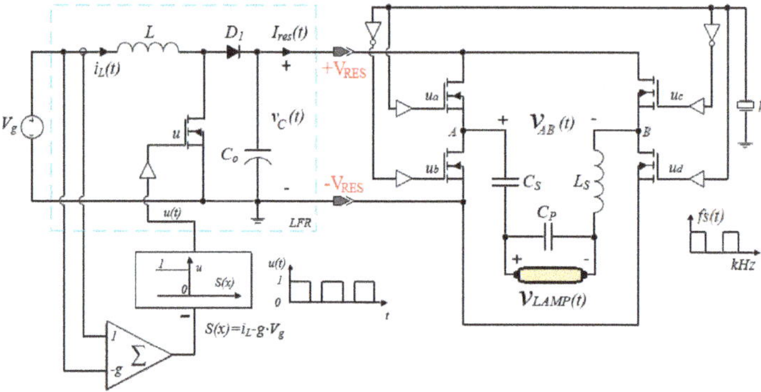

Figure 2. Two-stage ballast block diagram. Sliding-mode controlled boost converter in cascade with LsCsCp full bridge resonant inverter.

2.1. Lamp Start-Up

In electroded lamps, many factors influence the peak voltage required to start the lamp. Among them the distance between electrodes, the gas type and pressure, and in the case of MH lamps, the existence or nonexistence of a third electrode to reduce the strike voltage. When an electroded lamp is OFF, its impedance is an open circuit, and the ballast is unloaded. In pulse-start lamps, a high peak voltage is required, ranging from 1.5 kV up to 30 kV (hot re-strike). In probe-start lamps, due to the third electrode, start voltages are in the range of 500 V or less.

External coil IEFLs are slightly different. When a IEFL lamp is OFF, the lamp input impedance is purely reactive, and equal to the external coil inductance. Thus, the lamp behaves like an unloaded transformer.

In the proposed ballast, the first and the second stages contribute to reach the required voltage to start the lamps, directly inducing the required electric field inside the bulb with the electrodes, or through Faraday's Law in the external coil IEFL. A single resonant tank is used for all lamps, and the inverter frequency is adjusted to operate the tank a a frequency that provides the required voltage gain from the second stage, which in our case is 224 kHz. As the boost converter is unloaded, its output voltage increases continuously until the lamp starts-up, or the boost output reaches the maximum voltage allowed.

2.2. Lamp Warm-Up

Immediately after the start-up, the lamp voltage drops, and warm-up begins. During this process light intensity increases and, in some lamps, color changes are produced. The voltage increases slowly while the current decreases at similar rate, until the steady-state is reached at the warm-up end. It is important to remark that, although the lamp impedance changes during warm-up, the supplied power is controlled by the LFR stage, so damaging the lamp is precluded. In high-pressure devices like metal-halide (MH) and sodium vapor (HPS) lamps, warm-up can last up to 10 min. In contrast, the steady-state luminous flux of IEFLs is reached in few seconds.

In this period, the lamp impedance is no longer an open circuit, and the gain peak of the resonant inverter decreases. However, the impedances of electroded lamps are quite lower than those of IEFL lamps for the same power, and the resonant inverter gain reduction is less severe with IEFL lamps. In the proposed ballast, this gain reduction is automatically compensated by an increment of the LFR stage gain, but as more severe be the resonant gain reduction, more important will be the increase of the LFR gain, and higher will be the component stress in the boost stage, affecting the global ballast efficiency. To avoid a significant gain increase of the LFR stage, the inverter frequency is reduced from 224 kHz to 90 kHz, when the ballast is driving an electroded lamp. This frequency reduction assures at least, a minimum gain of 0 dB for the resonant inverter, thus limiting the gain increase of the LFR stage. The inverter frequency change is not applied with IEFL lamps because they cannot operate below 150 kHz, and the resonant gain reduction is quite less severe.

2.3. Nominal Operating Point and Dimming

After warm-up, the lamp impedance is approximately constant, lamp voltage and current reach their steady state values, and the ballast continues to deliver constant power. If dimming is required, the LFR stage can reduce the power transferred to the lamp, adjusting the LFR input conductance, and dimming takes place independently of the lamp impedance.

3. Sliding-Mode Dynamics of the First Power Stage

This section shows how the LFR can be realized with a sliding-mode controller. A small-signal model is derived, and its stability is proven.

The LFR behavior of the first stage in Figure 2 is achieved by a sliding-mode control loop that makes the boost converter input current proportional to the input voltage. Specifically, the switching surface in (1) forces the input current to track a slow reference $k(t)$ proportional to the ballast input voltage v_g. Therefore, the lamp active power P_{ac} is indirectly regulated by controlling the converter input power P_{in} through the adjustment of either the LFR input resistance R_e or the LFR input conductance g,

$$s(x) = i_L(t) - k(t), \quad \text{where } k(t) = g \cdot v_g(t) \tag{1}$$

$$g = \frac{1}{R_e} \tag{2}$$

$$p_{in} = g \cdot v_g^2 \approx p_{ac} \tag{3}$$

Assuming that the boost converter operates in continuous conduction mode (CCM), only two topological changes occur in a period, as shown in Figure 3. Each converter topology can be represented by means of two linear vector differential Equation (4), where the corresponding state vector is given in (5).

$$\dot{x}(t) = A_1 x(t) + B_1, \quad \text{for } u = 1 \ (\text{ON State})$$
$$\dot{x}(t) = A_2 x(t) + B_2, \quad \text{for } u = 0 \ (\text{OFF State}) \tag{4}$$

$$\dot{x}(t) = [i_L, v_{res}]^T \tag{5}$$

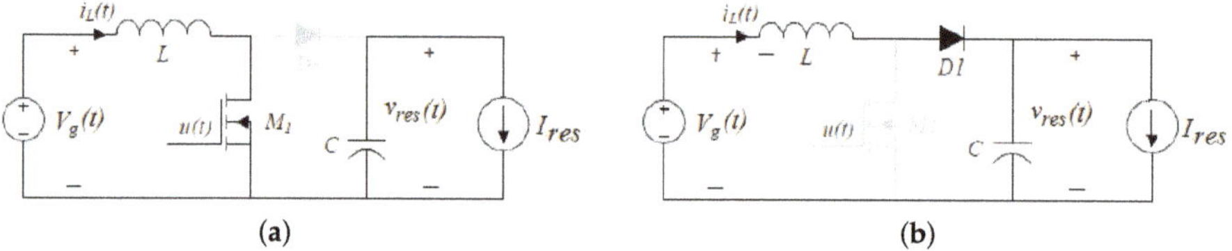

Figure 3. Boost converter states. (**a**) ON-state; and (**b**) OFF-state.

Matrices A_1, A_2, B_1 and B_2 are given in (6), where, $I_{res}(t)$ represents the current delivered by the boost-based LFR to the resonant inverter stage.

$$A_1 = \begin{bmatrix} 0 & 0 \\ 0 & 0 \end{bmatrix}, \quad A_2 = \begin{bmatrix} 0 & -\dfrac{1}{L} \\ \dfrac{1}{C} & 0 \end{bmatrix}, \quad B_1 = B_2 = \begin{bmatrix} \dfrac{v_g(t)}{L} \\ \dfrac{-i_{res}}{C} \end{bmatrix}. \tag{6}$$

The converter dynamics can be described using the bilinear Equation (7), where $A = A_2$, $B = A_1 - A_2$, $\delta = B_2$ and $\gamma = B_1 - B_2$.

$$\dot{x}(t) = \{A \cdot x(t) + \delta\} + \{B \cdot x(t) + \gamma\} \tag{7}$$

From (6) and (7) we obtain

$$\begin{cases} \dfrac{di_L(t)}{dt} = \dfrac{u - 1}{L} \cdot v_{res}(t) + \dfrac{v_g(t)}{L} \\ \dfrac{dv_{res}(t)}{dt} = \dfrac{1 - u}{C} \cdot i_L(t) - \dfrac{i_{res}(t)}{C} \end{cases} \tag{8}$$

If the invariance conditions [24] $s(x) = 0$ and $ds(x)/dt = 0$ are applied in (1), the equivalent control $u_{eq}(t)$ can be derived. The equivalent control $u_{eq}(t)$ is bounded by the maximum and minimum values of $u(t): 0 < u_{eq}(t) < 1$. In such case, a switching law of the type

$$\begin{cases} u(t) = 0 \quad \text{if} \quad s(x) > 0 \\ u(t) = 1 \quad \text{if} \quad s(x) < 0 \end{cases} \tag{9}$$

induces a sliding regime on the switching surface, ensuring the sliding-mode existence because

$$s(x)\dfrac{ds(x)}{dt} < 0. \tag{10}$$

The dynamics are then derived as follows. First, the equilibrium point is found assuming constant values of input voltage $v_g = V_g$ and LFR input conductance $g = G$ in the control loop. Second, the influence of the time-varying components of the mentioned input variables is analyzed as linearized low-frequency signals superposed on their corresponding values V_g and G. Considering constant values for V_g and G leads to the equivalent control u_{eq}, ideal sliding dynamics, and equilibrium point $X^* = [I_L, V_{res}]^T$ given in (11)–(13) respectively.

$$u_{eq}(t) = 1 - \frac{V_g}{v_{res}} \tag{11}$$

$$\begin{cases} i_L(t) = G \cdot V_g \\ \dfrac{dv_{res}(t)}{dt} = \dfrac{G \cdot V_g}{C}(1 - u_{eq}) - \dfrac{i_{res}(t)}{C} \end{cases} \tag{12}$$

$$X^* = \left\{ G \cdot V_g, \; \frac{V_g^2 \cdot G}{I_{res}} \right\}^T \tag{13}$$

Assuming now that $k(t)$ is time-varying, the corresponding equivalent control, and ideal dynamics are given by (14) and (15) respectively

$$u_{eq}(t) = 1 - \frac{v_g(t)}{v_{res}(t)} + L\frac{g(t)}{v_{res}(t)} \cdot \frac{dv_g(t)}{dt} + L\frac{v_g(t)}{v_{res}(t)} \cdot \frac{dg(t)}{dt} \tag{14}$$

$$\begin{cases} i_L(t) = k(t) = g(t) \cdot v_g(t) \\ \dfrac{dv_{res}(t)}{dt} = \dfrac{g(t)v_g^2(t)}{Cv_{res}(t)} - L\left(\dfrac{g^2(t)v_g(t)}{Cv_{res}(t)}\dfrac{dv_g(t)}{dt} + \dfrac{g(t)v_g^2(t)}{Cv_{res}(t)}\dfrac{dg(t)}{dt} \right) - \dfrac{i_{res}(t)}{C} \end{cases} \tag{15}$$

Variables can be expressed as (16), as a sum of its DC component and a small-signal term denoted with a hat sign:

$$\begin{aligned} g(t) &= G + \hat{g}(t), & v_g(t) &= V_g + \hat{v}_g(t), \\ i_L(t) &= I_L + \hat{i}_L(t), & v_{res}(t) &= V_{res} + \hat{v}_{res}(t), \end{aligned} \tag{16}$$

Linearizing the differential equation of the output capacitor (15) around the equilibrium point $X^* = [I_L, V_{res}]^T$ leads to the following expression

$$f(x) = \frac{d\hat{v}_{res}(t)}{dt} \approx a \cdot \hat{g}(t) + b \cdot \frac{d\hat{g}(t)}{dt} + c \cdot \hat{v}_g(t) + d \cdot \frac{d\hat{v}_g(t)}{dt} + e \cdot \hat{v}_{res}(t) + f \cdot \hat{i}_{res}(t) \tag{17}$$

where coefficients a, b, c, d, e, and f are as follows

$$\begin{aligned} a &= \left. \frac{\partial f(x)}{\partial g(t)} \right|_{x=X^*} = \frac{I_{res}}{C \cdot G}, & b &= \left. \frac{\partial f(x)}{\partial a(t)} \right|_{a(t)=\frac{dg(t)}{dt},x=X^*} = -\frac{L \cdot I_{res}}{C}, \\ c &= \left. \frac{\partial f(x)}{\partial v_g(t)} \right|_{x=X^*} = \frac{2I_{res}}{C \cdot V_g}, & d &= \left. \frac{\partial f(x)}{\partial b(t)} \right|_{b(t)=\frac{dv_g(t)}{dt},x=X^*} = -\frac{L \cdot I_{res} \cdot G}{C \cdot V_g}, \\ e &= \left. \frac{\partial f(x)}{\partial v_{res}(t)} \right|_{x=X^*} = -\frac{I_{res}^2}{C \cdot V_g^2 \cdot G}, & f &= \left. \frac{\partial f(x)}{\partial i_{res}(t)} \right|_{x=X^*} = -\frac{1}{C}. \end{aligned} \tag{18}$$

Figure 4 shows the block diagram of the boost converter-based LFR small-signal model. The transfer functions of the block diagram are as follows

$$H(s) = \frac{\hat{v}_{res}(s)}{\hat{g}(s)} = \frac{a + s \cdot b}{s - e} = -\frac{LI_{res}}{C}\frac{s - 1/(LG)}{s + I_{res}^2/(CV_g^2 G)} \tag{19}$$

$$A(s) = \frac{\hat{v}_{res}(s)}{\hat{v}_g(s)} = \frac{c + s \cdot d}{s - e} = -\frac{LI_{res}G}{CV_g}\frac{s - 2/(LG)}{s + I_{res}^2/(CV_g^2 G)} \tag{20}$$

$$Z_0(s) = \frac{\hat{v}_{res}(s)}{\hat{i}_o(s)} = \frac{\hat{v}_{res}(s)}{-\hat{i}_{res}(s)} = \frac{1/C}{s + I_{res}^2/(CV_g^2 G)} \tag{21}$$

Considering that $\frac{I_{res}^2}{V_g^2 G} = \frac{1}{Z_{DC}(P)}$, it can be seen that the control to output transfer function $H(s)$ given above is stable, because the following condition is always satisfied

$$\frac{I_{res}^2}{CV_g^2 G} = \frac{I_{res}^2}{C\,P_{in}} \approx \frac{1}{C\,Z_{DC}(P)} > 0 \tag{22}$$

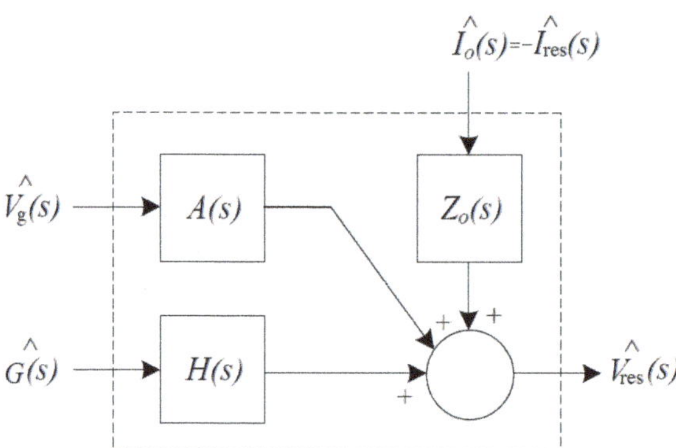

Figure 4. Boost converter, LFR-based small-signal model.

4. Design of the Resonant Tank and System Stability

In the literature there are some works analyzing the $L_s C_s C_p$ resonant filters of electronic ballasts supplying IEFL lamps with dimming operation [16]. Nonetheless, most of these papers do not consider the lamp model, except for [21] where a precise IEFL equivalent model was introduced.

In fact, the resonant converter cannot be analyzed in a classical way, i.e., when the load is a resistor, because its behavior depends on the nonlinear nature of the lamp. Consequently, the input-output voltage transfer function $H_T(P,s)$ of the converter will depend on the lamp impedance $Z_{lamp}(P,s)$, which, in turn, is also function of the power P handled by the lamp.

$$H_T(P,s) = \frac{\hat{v}_{lamp}(s)}{\hat{v}_{AB}(s)} = \frac{\dfrac{Z_{lamp}(P,s)}{C_p s}\left(Z_{lamp}(P,s) + \dfrac{1}{C_p s}\right)^{-1}}{\dfrac{Z_{lamp}(P,s)}{C_p s}\left(Z_{lamp}(P,s) + \dfrac{1}{C_p s}\right)^{-1} + \dfrac{1}{C_s s} + L_s s} \tag{23}$$

In the design of the proposed ballast, two lamp models are used: (a) a general first-order model with a right half-plane zero and a stable pole, and (b) the IEFL model introduced in [21]. Four lamps have been tested in this paper: a 150 W IEFL from Osram, a 150 W HPS vapor lamp from Sylvania, and two 250 W probe-start MH lamps, from Osram and I-Quarium, respectively. The latter three lamps have internal electrodes and can be described with the first-order model.

4.1. First-Order Lamp Model

All lamps with negative incremental resistance can be described with this model. The impedance parameters are the equivalent DC resistance R_L, a real half-plane pole p_L, a right-plane zero z_L, and a negative resistance λ_n

$$Z_{HPS-MH}(P,s) = R_L(P)\frac{s + z_L}{s + p_L}, \quad \lambda_n = R_L(P)\frac{z_L}{p_L} < 0 \tag{24}$$

The zero is located between 10 and 10^3 rad/s, and the pole between 100 and 10^4 rad/s. In reference [25] the parameters of an specific HPS lamp with $z_L = -3141.5$ rad/s and $p_L = 18.8$ rad/s are given, whereas in reference [26], similar parameters for a 150 W MH lamp with $z_L = -1850$ rad/s and $p_L = 10,690$ rad/s can be found. This illustrates the great variability in the location of p_L and z_L. The ballast operation with HPS and MH lamps is simulated using the values of [25,26] and the real resonant tank values ($L_s = 150$ µH, $C_s = 22$ nF, and $C_p = 3.3$ nF) of the ballast prototype. The corresponding results are given in Table 1, where $Z_{DC}(P) = \frac{V_{res}}{I_{res}}$.

Table 1. Simulated ballast behavior with a given HPS and MH lamp.

P	30 W	60 W	90 W	120 W	150 W
$f_s = 2\pi\omega_s$			90 kHz		
HPS		$z_L = -3141.5$ rad/s and $p_L = 18.8$ rad/s			
MH		$z_L = -1850$ rad/s and $p_L = 10,690$ rad/s			
R_L	225 Ω	135 Ω	100 Ω	79.5 Ω	65.4 Ω
V_{lamp}	82.2 V	90.4 V	95 V	97.8 V	99 V
$|H_T(j\omega_s)|$	1.13	1.05	0.96	0.88	0.81
$|V_{res}| = V_c$	80.7 V	95.5 V	110 V	123 V	135 V
I_{res}	0.37 A	0.673 A	0.82 A	0.98 A	1.11 A
$Z_{DC}(P)$	218 Ω	151 Ω	134 Ω	125 Ω	121 Ω

4.2. IEFL Lamp Model

The IEFL model proposed here is directly derived from the model presented in [21] but has been improved by considering the saturation of the IEFL transformer core. As a result, the parameter L_c (which was constant in [21]) is now power dependent. The remaining model parameters continue to be power dependent as they were in [21]. According to the circuit in Figure 5b, the parameters that can be seen from the transformer primary side are: the core losses R_C, the inductance L_c, the lamp resistance R_L which represents the power transformed into light, and the lamp capacitance C_L. Expression (25) models the IEFL lamp impedance. According to the procedure given in [21], the parameters of the model of the 150 W IEFL lamp used in the experiments were obtained. Then, these parameters were used to simulate the behavior of the proposed ballast using, as in the previous case, the resonant tank values ($L_s = 150$ µH, $C_s = 22$ nF, and $C_p = 3.3$ nF). The results of the simulation are given in Table 2.

$$Z_{IEFL}(P,s) = \frac{R_C \, R_L \, L_C \, s}{R_C \, R_L \, L_C \, C_L \, s^2 + L_C(R_C + R_L)s + R_L \, R_C} \tag{25}$$

Table 2. Simulated ballast behavior for the IEFL lamp used in the experiments.

P	30 W	60 W	90 W	120 W	150 W
$f_s = 2\pi\omega_s$			220–230 kHz		
R_C	29 kΩ	34 kΩ	33 kΩ	32 kΩ	30 kΩ
R_L	5 kΩ	1.47 kΩ	570 Ω	410 Ω	425 Ω
L_C	66 µH	59 µH	56 µH	48 µH	40 µH
C_L	380 pF	340 pF	240 pF	100 pF	110 pF
V_{lamp}	357 V	291 V	224 V	221 V	250 V
$|H_T(j\omega_s)|$	8.59	6.45	2.87	2.08	2.12
V_{res}	46.1 V	50.1 V	86.7 V	118 V	131 V
I_{res}	0.65 A	1.2 A	1.03 A	1.02 A	1.14 A
$Z_{DC}(P)$	71 Ω	42 Ω	84 Ω	120 Ω	115 Ω

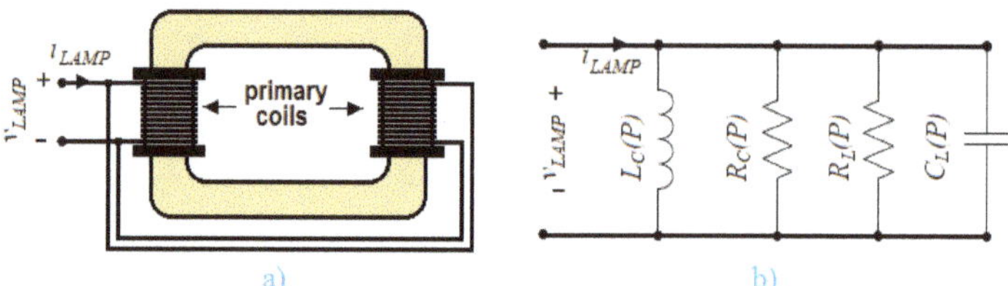

Figure 5. IEFL: (**a**) lamp construction, (**b**) electrical model (from [21]).

4.3. Design of the Resonant Tank

In the proposed ballast, the LFR behavior controls the power delivered to the lamp and makes it possible to provide light-dimming, if necessary. On the other hand, the resonant inverter stage is required to supply the lamps in AC, but also to increase the ballast voltage gain during the lamp start-up by means of the tank resonant gain peak.

The resonant tank voltage gain $G(P,\omega)$, can be found from the inverter transfer function in (23) by $G(P,\omega) = |H_T(P,s)|_{s=j\omega}$. Nevertheless, to simplify tank design, a reduced gain function $G_s(P,\omega)$ is used. This function can be easily derived from (23) considering the lamp as a power dependent resistor $R_L(P)$. By defining the following normalized parameters,

$$Q_L(P) = \omega_0 \frac{C_P}{1+A} R_L(P), \quad \omega_0^2 = \frac{1}{L_s(C_p||C_s)}, \quad A = \frac{C_p}{C_s}$$
$$Q_S(P) = \frac{Z_S}{R_L(P)}, \qquad Z_S = \frac{1}{\omega_s C_s}, \qquad \omega_s^2 = \frac{1}{L_s C_s} \tag{26}$$

two different expressions are obtained for the resonant tank voltage gain $G_s(P)$.

$$G_S(P,\omega) = \frac{1}{\sqrt{(1+A)^2 \left(1 - \left(\frac{\omega}{\omega_0}\right)^2\right)^2 + \frac{1}{Q_L^2(P)}\left(\frac{\omega}{\omega_0} - \frac{\omega_0 A}{\omega(1+A)}\right)^2}} \tag{27}$$

$$G_S(P,\omega) = \frac{1}{\sqrt{\left(1 - \left(\left(\frac{\omega}{\omega_s}\right)^2 - 1\right)A\right)^2 + Q_S^2(P)\left(\frac{\omega}{\omega_s} - \frac{\omega_s}{\omega}\right)^2}} \tag{28}$$

where A is the parallel to series capacitance ratio, ω_0 is the undamped resonant frequency, $Q_L(P)$ is the loaded quality factor, ω_s is the series resonant frequency, Z_S is the series characteristic impedance, and $Q_S(P)$ is the series quality factor.

Figure 6 depicts the frequency response of $G_S(P,\omega)$ for different values of lamp power and resistance. There are two different resonant frequencies, i.e., the series resonant frequency ω_s, around 90 kHz, and the main resonant frequency ω_0, slightly below 250 kHz. The values of the tank components, namely $L_s = 150\ \mu H$, $C_s = 22$ nF, and $C_p = 3.3$ nF are obtained by solving a set of equations as explained below.

The resistive part (active power) of any lamp impedance $R_L(P)$ is extremely variable with the lamp power, aging, and type of lamp. In fact, in common lamp types, $10\ \Omega < R_L(P) < 500\ \Omega$. As the lamp power regulation is ensured by the LFR stage, the inverter stage is only used to provide the appropriate high-frequency AC signal to the lamp, once the lamp is started.

If the lamps are supplied at the series resonant frequency ω_s, the series capacitor C_s and inductor L_s impedances are mutually cancelled, and the resonant inverter has little effect on the lamp regulation because the gain is $G_S(P, \omega_s) = 0$ dB at any load. This choice is the first design constraint, and can be posed as follows

$$Z_{L_sC_s} = \frac{L_sC_ss^2 + 1}{C_ss}, \quad Z_{L_sC_s}(j\omega_s) \approx 0 \rightarrow G_S(\omega_s) \approx 0 \text{ dB}, \tag{29}$$

$$\omega_s = \sqrt{\frac{1}{L_sC_s}} = 2\pi 87 \cdot 10^3 \text{ rad/s}. \tag{30}$$

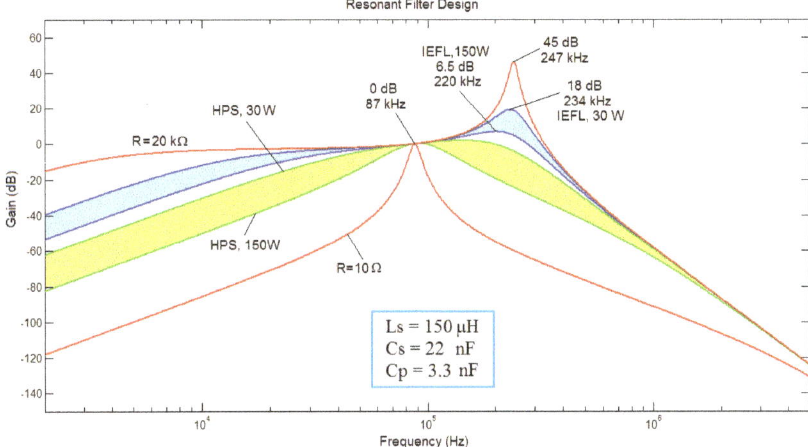

Figure 6. Resonant tank gain and design criteria.

A notable exception is the IEFL lamps. These lamps include a transformer coil to transfer the energy to the plasma, which is usually designed to work optimally in the range of 200–250 kHz. Besides, they also exhibit a higher input impedance ($500 \ \Omega < R_L(P) < 10$ kΩ) than other lamp types, working at higher voltages. As a result, to avoid a boost stage extreme voltage gain, part of the required gain can be supplied by the resonant tank through the second resonance. The frequency of this second resonance is noted ω_o, and the voltage gain at this frequency $G_S(\omega_o)$ corresponds to the second and third tank design constraints, shown next:

$$\omega_o = \sqrt{\frac{C_s + C_p}{L_sC_sC_p}} \approx 2\,\pi\,245 \text{ kHz}, \tag{31}$$

$$\left.\begin{aligned}
G_S(\omega_o) &= \omega_oC_pR_L(P)|_{R_L=10 \text{ k}\Omega} \approx 50.77 \rightarrow 34 \text{ dB},\\
V^{max}_{lamp}|_{strike} &= V^{max}_{LFR}G_s(\omega_o) \approx 230x50.77 > 10 \text{ kV}.
\end{aligned}\right\} \tag{32}$$

The voltage given by the third constraint up to 10 kV should be sufficiently high to start-up not only probe-start lamps as proposed in Section 2, but also in most of pulse-start lamps. In fact, higher voltage peaks can be obtained by increasing C_P. Nevertheless, this increases the slope (as shown below in (33)) of $G_S(\omega)$ around ω_S, making the operation at 90 kHz with gain of 0 dB, in a wide range of loads, more difficult. This mode corresponds to the flat region in Figure 6. There is a trade-off in the value of C_p, so that increasing C_p produces higher resonant peak gains $|G_s(\omega_o)|$, but deteriorates the flat region around ω_s for any given load, where the gain $|G_s(\omega_s)|$ is approximately 0 dB.

To maximize the bandwidth of the flat region, $\partial G_s(\omega)|_{\omega=\omega_s}$ must be minimized. In the following expression, the effect of increasing $\frac{C_p}{C_s} = A$ can be clearly identified.

$$
\begin{cases}
|G_s(\omega_s)| \approx 1 \ (0 \ \text{dB}). \\
\left. \dfrac{\partial G_s(\omega)}{\partial \omega} \right|_{\omega=\omega_s} = \dfrac{2A}{\omega_s} = \dfrac{2C_p}{C_s \omega_s} \\
|G_s(\omega_s)| \approx Q_L(P)(1+A)
\end{cases}
\tag{33}
$$

4.4. Ballast Stability Analysis

The stability of the cascade connection of the two converters will be ensured if the output impedance modulus of the first stage is smaller than the input impedance modulus of the second stage for all operating frequencies [27]. Figure 7 depicts the Thévenin's model of the cascade connection of the LFR stage and the resonant inverter. $Z_o(P,s)$ is the output impedance of the LFR and $Z_i(P,s)$ is the input impedance of the inverter. From Figure 7, the following expression holds

$$
\frac{\hat{v}_{inr}(s)}{\hat{v}_{res}(s)} = \frac{Z_i(P,s)}{Z_i(P,s) + Z_o(P,s)} = \frac{1}{1 + \frac{Z_o(P,s)}{Z_i(P,s)}}
\tag{34}
$$

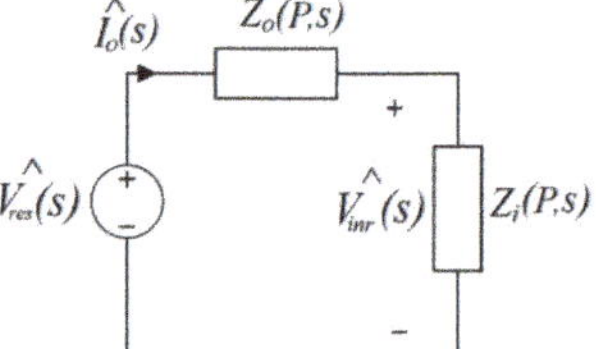

Figure 7. Thévenin's model of the cascade connection of both ballast stages, the boost-LFR, and the LCC resonant inverter.

Therefore, the stability condition can be expressed as follows

$$
\begin{cases}
\forall \omega, \quad \left| \dfrac{Z_o(P,j\omega)}{Z_i(P,j\omega)} \right| < 1 \\
\text{if } \exists \, \omega_p \quad \text{where} \quad \left| \dfrac{Z_o(P,j\omega)}{Z_i(P,j\omega)} \right| = 1 \rightarrow \arg\left(\dfrac{Z_o(P,j\omega)}{Z_i(P,j\omega)} \right) \neq 180 \ \text{deg.}
\end{cases}
\tag{35}
$$

Moreover, from (27) and (28) it is derived

$$
\hat{v}_{res}(s) = A(s)\hat{v}_g(s) + B(s)\hat{g}(s)
\tag{36}
$$

$$
Z_o(P,s) = \frac{\hat{v}_C(s)}{\hat{i}_o(s)} = \frac{\hat{v}_{res}(s)}{-\hat{i}_{res}(s)} = \frac{\frac{1}{C}}{s + \frac{I_{res}^2}{CV_g^2 G}} = \frac{\frac{1}{C}}{s + \frac{1}{Z_{DC}(P)C}}
\tag{37}
$$

Besides, the resonant inverter input impedance $Z_i(P,s)$ is given by

$$
Z_i(P,s) = \frac{Z_{lamp}(P,s)\left(L_s C_s s^2 + (C_p + C_s)s\right) + L_s C_s s^2 + 1}{C_s s \left(Z_{lamp}(P,s)C_p s + 1\right)}
\tag{38}
$$

where $Z_{lamp}(P,s)$ must be replaced by expression (24) in case of using a HPS or MH lamp, or by (25) in case of an IEFL lamp.

It is worth remarking that the validity of the small-signal analysis depends on the linearity of the dynamics of the system. In this sense, when a lamp is operated at high frequency its dynamic behavior is much more linear than when operated at small frequency, thus extending the validity of the small-signal models. As an example, the voltage and current waveforms in a discharge lamp are much less distorted, and more sinusoidal, when they are supplied at 50 kHz than when are driven at 50 Hz [15].

Also, it is worth mentioning that the boost output impedance has low-pass filter characteristics, which ensures the system rejection to high-frequency perturbations, i.e., the higher is the frequency of a possible perturbation, less is the boost converter affected. Moreover, the resonant tank is used to transfer the power to the lamp at high frequency, between ω_s and ω_o, just within that frequency band where the boost output impedance is much lower than at zero frequency.

The Bode diagrams depicted in Figures 8a and 9a have been plotted using the data of Tables 1 and 2 respectively. Two cases can be distinguished in Figure 8a depending on the position of the pole p_L and zero z_L of the lamp. As it can be seen in the Bode diagram of Figure 8b, and according to (35), the stability of the ballast is not affected by changes in z_L and p_L over a wide range. As an example, according to Table 1, the ratio between the pole positions of the MH and HPS cases is $p_L(MH)/p_L(HPS) \approx 570$ and the ratio between zeros is $z_L(MH)/z_L(HPS) \approx 0.588$. The values of Table 2 and Figure 9b correspond to an ENDURA IEFL lamp used in the experiments. The overall result, shown in Figures 8b and 9b proves that $|Z_o(\omega)/Zi(\omega)|$ is well below 0 dB, which ensures the stability of the cascade connection.

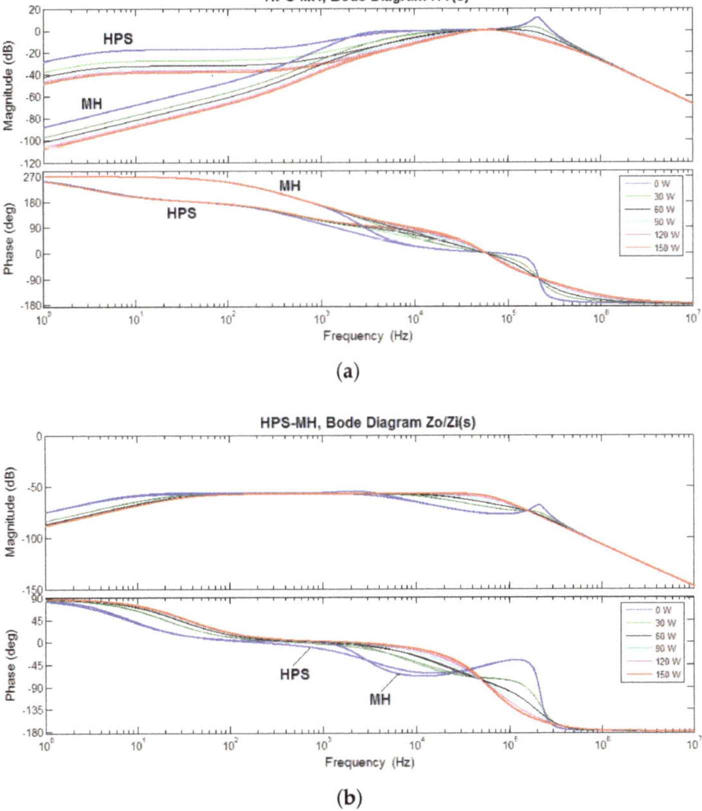

Figure 8. Frequency response of $H_T(P,s)$ and $Z_o(s)/Z_i(s)$ for the HPS and MH cases. (a) Bode Diagram of $H_T(P,s) = \hat{v}_{lamp}(s)/\hat{V}_{AB}(s)$; (b) Bode Diagram of $Z_o(s)/Z_i(s)$.

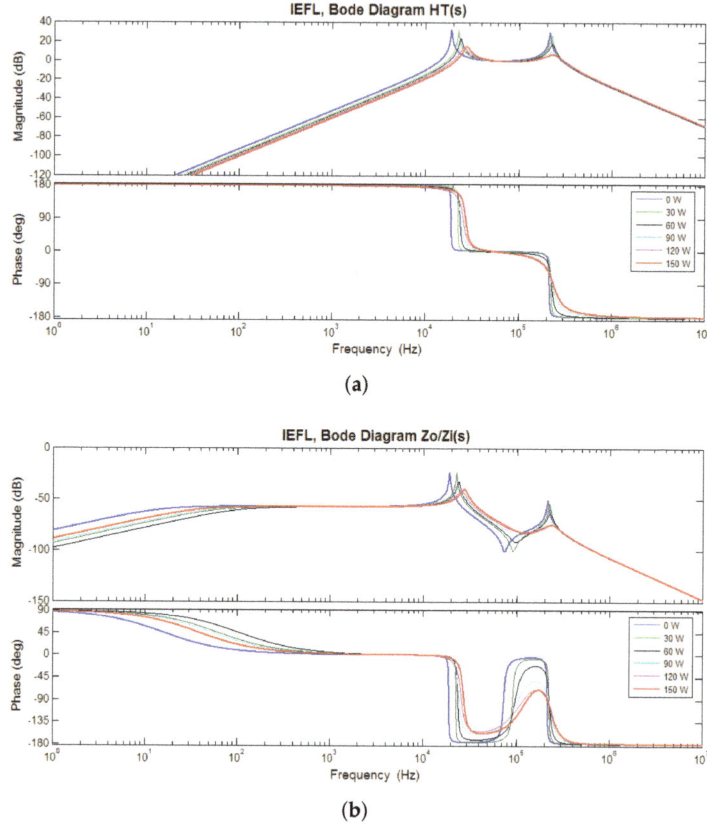

Figure 9. Frequency response of $H_T(P,s)$ and $Z_o(s)/Z_i(s)$ for the IEFL case. (**a**) Bode Diagram of $H_T(P,s) = \hat{v}_{lamp}(s)/\hat{V}_{AB}(s)$; (**b**) Bode Diagram of $Z_o(s)/Z_i(s)$.

5. Ballast Realization and Experimental Results

This section shows the experimental realization of the proposed universal ballast and demonstrates its performance for a set of representative lamps. The details of the different parts of the ballast are described first. Then, the steady-state and transient waveforms of the ballast in different operating conditions, including dimming, are shown.

5.1. Realization of the First Stage

As shown in Figure 2, the ballast consists of two power stages. The first stage includes two parts: the boost converter and its controller. The boost converter, which is shown in Figure 10, behaves like an LFR thanks to the controller depicted in Figure 11. The converter main switch is based on the complementary action of two silicon carbide devices, the MOSFET Q1 (IRFP4768), and the diode D1 (SDP20S120D). The driver is implemented with two bipolar transistors, a npn transistor (ZTX 653) with a Vce of 100 V and a collector current of 2 A, and its complementary pair, the pnp transistor (ZTX 753). The MOSFET gate resistance is an E24 resistor of 2.2 Ω, and the parallel protection network consists of an E24 10 kΩ resistor and a 16 V 500 mW zener diode, model C16PH. The input series inductor of 20 μH has been made with 50 parallel strands of 0.07 mm^2 to reduce the skin-effect and 17 turns around the Kool μ 77109-A7 core. The output capacitor involves two polypropylene capacitors of

20 µF each one (EPCOS B32926E3206M) with a breakdown voltage of 500 V connected in parallel with four ceramic capacitors of 1 µF and the same breakdown voltage.

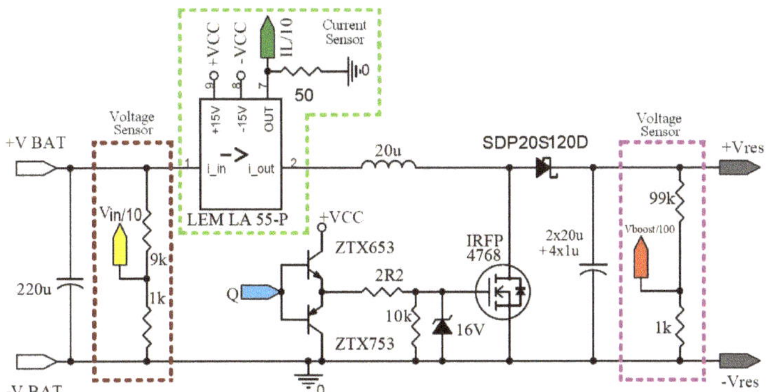

Figure 10. Detail of the circuit realization of the first stage (boost converter) of the ballast.

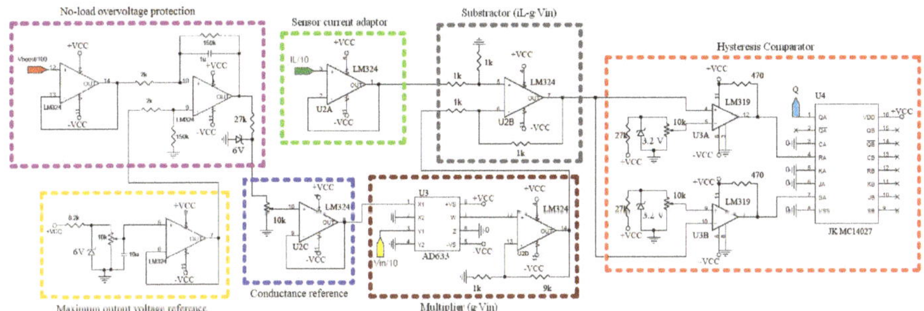

Figure 11. Schematic of the controller of the first stage.

The switch-driver input signal Q is activated by the hysteretic comparator as illustrated in Figure 11. In the control circuit, the input and output voltages are sensed by two voltage dividers, and the input current sensor is a Hall effect transductor model (LEM LA 55-P) with a bandwidth of 200 kHz. Signals Vin/10, Vboost/100, and IL/10 are proportional to the input voltage V_g, the output voltage V_{res}, and the inductor current i_L respectively.

The controller, which can be seen in Figure 11, includes a maximum voltage reference block with overvoltage protection that sets the maximum boost voltage to V_{res} = 230 V. This reference has been included to preclude the voltage increase in case of failure in the lamp strike, because in this case, the boost converter would remain unloaded. If the lamp start-up is successful, the boost voltage will never reach 230 V, and the integrator of the protection block will be saturated at 15 V supplying energy to the conductance block. On the contrary, if the lamp start-up fails when the boost voltage reaches 230 V, the input voltage of the conductance reference block will be set to zero, and the converter input current will become zero, avoiding output voltage overvoltage.

The controller also includes a potentiometer to adjust G, that is, the power delivered to the lamp. The remaining blocks of the controller include, an AD633 multiplier, a subtraction circuit to implement the sliding surface and the hysteretic comparator. This last part employs two LM319 comparators with a response time of 80 ns, and a JKMC14027 flip flop with a bandwidth of 13 MHz.

5.2. Realization of the Second Stage

The second stage, which is shown in Figure 12, consists of a full bridge and a $L_sC_sC_p$ resonant tank. The full bridge employs four MOSFETs (IRFP4768), featuring E24 gate resistances of 5 Ω, and a parallel protection network with a 1 kΩ resistance and the 16 V zener diode. Regarding the resonant tank, the L_s series inductor has been built in-house, using a 14-strands 0.07 mm^2 litz wire. The winding has 18 turns around a Molypermalloy 55868-A2 core. The series capacitor C_s comprises two polypropylene capacitors of 12 nF and 10 nF (Vyshay) in parallel and with a breakdown voltage of 2 kV. Finally, the parallel capacitor C_p consists of two 4 kV 6.8 nF polypropylene capacitors connected in series.

The switches are activated by two IR2110 drivers, whose inputs are given by a logic network, shown in Figure 13, which assures a minimum blanking-time between the high-side and low-side MOSFETs. Depending on the lamp and the mode of operation, the blanking delay network is activated at a given constant frequency by a voltage-controlled oscillator. As an example, conventional lamps (HPS or MH) can be started at the main resonant frequency $\omega_0 = 245$ kHz, but are operated in steady-state at the series resonant frequency $\omega_s = 90$ kHz. The proposed oscillator is based on the ICL8038 analog function generator integrated circuit. By adding the npn transistor BC547, the 1N4148 diode, three resistors, and an LDR (light dependent resistance) the frequency of the oscillator is reduced automatically to 90 kHz during the first seconds of the lamp warm-up, before the lamp begins to change the color. Since the transformer coil of the IEFL lamp is designed to operate around 240 kHz, the LDR used to reduce the inverter switching frequency is not used in this case.

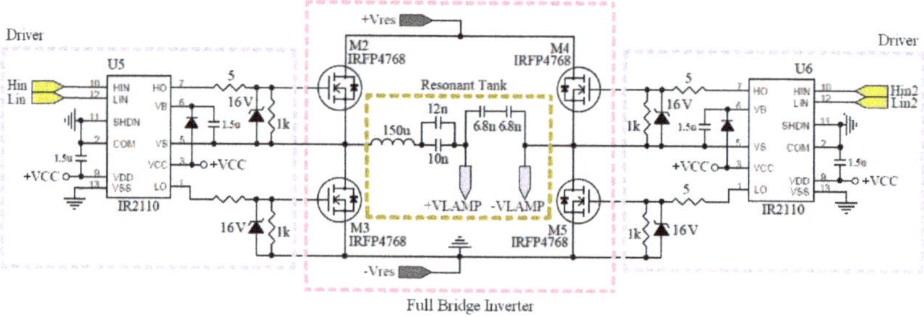

Figure 12. Detail of the circuit realization of the second stage (inverter) of the ballast.

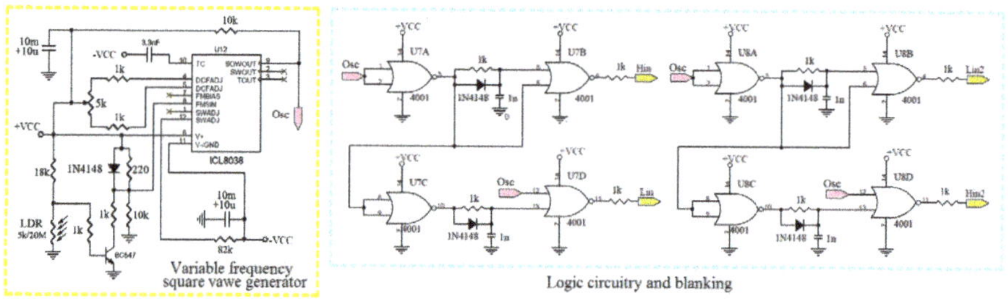

Figure 13. Schematic of the controller of the second stage.

5.3. Experimental Results

Figure 14a shows the power stage of the ballast, whereas the four different discharge lamps used to test the ballast can be seen in Figure 14b. These lamps are: two 250 W metal-halide lamps (the HQI-T PRO with greenish light, and the I-Quarium DE-MH with bluish light), the 150 W HPS-T sodium vapor lamp with orange light, and the 150 W Endura IEFL lamp.

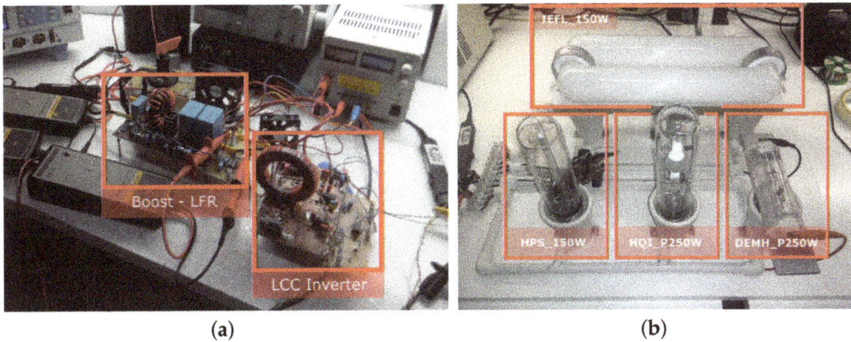

(**a**)   (**b**)

Figure 14. Pictures of the experimental setup. (**a**) Converters of the ballast; (**b**) Lamp test-bed.

Figure 15 depicts the waveforms of voltages and currents at the input of the resonant converter and at the lamp, for four different lamp types. Three of the operational modes of the lamps are shown: strike, warm-up, and steady-state operation. The LFR controller ensures that the power delivered is 150 W for all lamps. This means that the IEFL and HPS lamps are tested at full power, whereas both 250 W MH lamps are tested at 60% of their rated power. As expected, the steady-state voltage of the IEFL lamp is higher than the voltages of other non-coil driven lamps. Also, it can be seen that HPS, HQI-T and DE-MH lamps suffer a sudden voltage and current transient when the inverter switching frequency changes from 220 kHz to 90 kHz.

To analyze the performance of the proposed ballast with different lamps, the ballast has been tested in steady-state at different power levels, which correspond to dimming operation: 50 W, 70 W, 100 W and 150 W of input power. To simplify the analysis, the input voltage is constant and equal to 12 V, and input power is adjusted by means of the LFR controller. Figure 16 shows: (a) the efficiency of the first boost-LFR stage, (b) the efficiency of the LCC inverter, (c) the overall ballast efficiency, (d) the boost-LFR switching frequency, (e) the boost-LFR output voltage, and (f) the voltage of the lamp. The overall efficiency of the lamp is well above 90% for most of the cases and power outputs, except for the IEFL case at full power. One of the reasons behind this is the voltage variation range of the IEFL lamp that under dimming operation is wider than in other lamps. The power efficiency of the proposed approach at similar power levels can be compared with single-stage ballasts ranging from 83% [12] to 93% [13] and conventional two-stage ballasts whose efficiencies are typically around 85% to 90% [14].

Details of the waveforms of the ballast under dimming operation are shown in Figure 17. Figure 17a,b show the IEFL and the HPS lamp operating at 75 W and 150 W (50% and 100% of their rated power). As expected, the resonant inverter works at 220 kHz with the IEFL lamp, whereas the switching frequency is 90 kHz when testing the HPS lamp. The figure also includes: (c) the 250 W HQI-T lamp shortly after the strike, at the warm-up beginning, (d) the same lamp operating at 150 W (60% of rated power), and (e) the aquarium 250 W DE-MH lamp operating at 150 W.

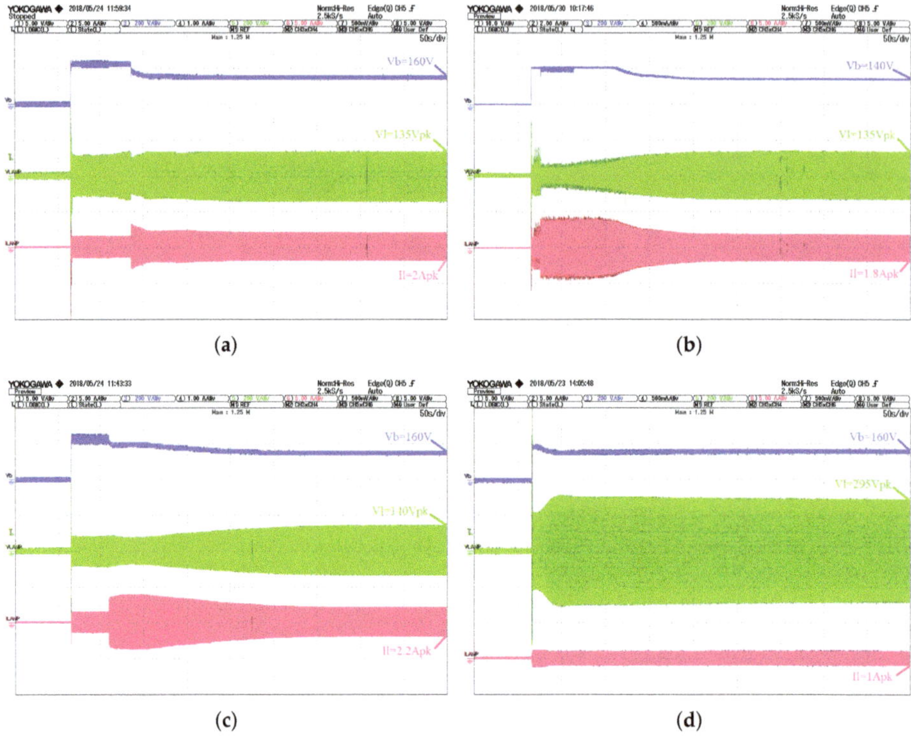

Figure 15. Waveforms of V_C, V_{lamp} and I_{res} showing strike, warm-up, and steady operation at 150 W with the four different lamps: (**a**) 250 W HQI-T PRO (greenish light MH); (**b**) 250 W I-Quarium DE-MH (bluish light MH); (**c**) 150 W HPS-T Basic Plus (orange light); and (**d**) 150 W IEFL Endura Luminarc (white light).

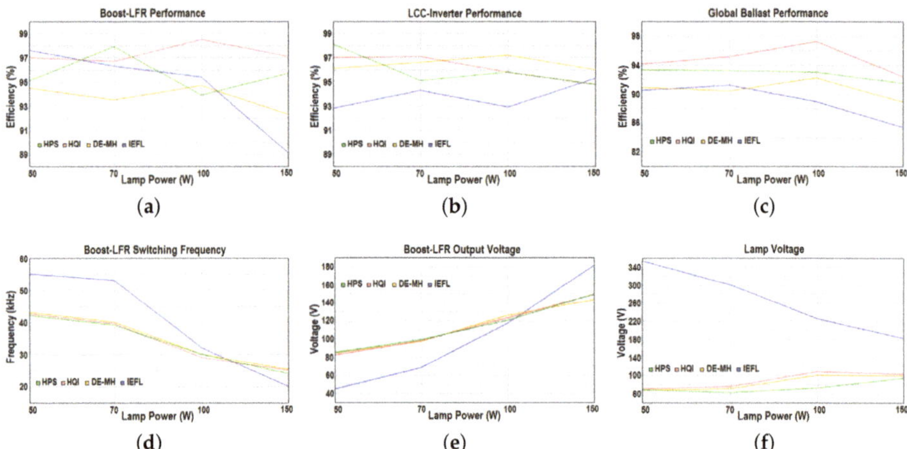

Figure 16. Figures of merit of the proposed universal ballast. (**a**) First Stage (Boost-LFR) Efficiency; (**b**) Second Stage (Inverter) Efficiency; (**c**) Overall Efficiency; (**d**) Switching Frequency of the First Stage; (**e**) Output Voltage of the First Stage; (**f**) Output Voltage of the Second Stage.

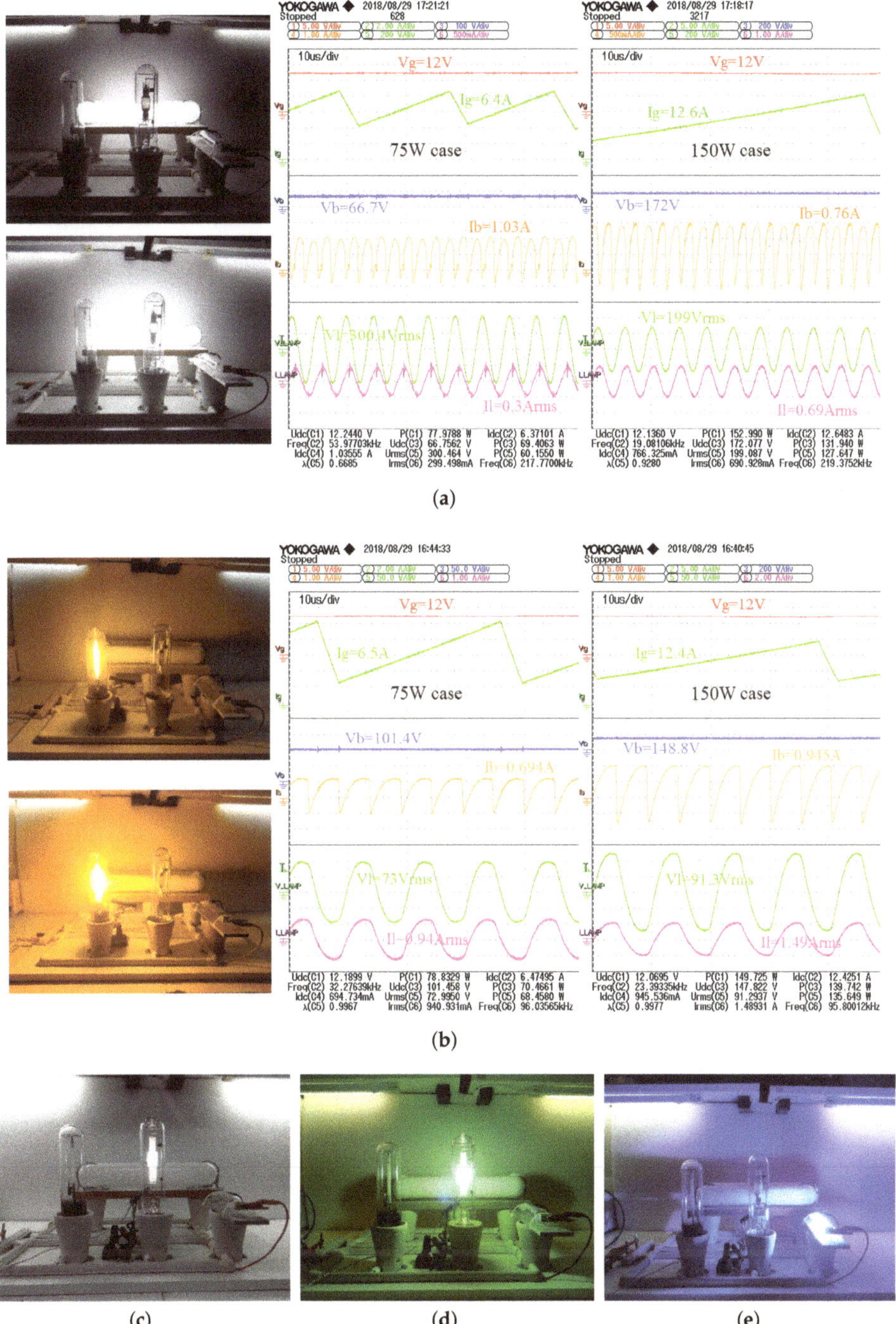

Figure 17. Detail of dimming operation of the ballast with different lamps. (**a**) 150 W IEFL under dimming operation: 75 W (half) and 150 W (full power); (**b**) 150 W HPS under dimming operation: 75 W (half) and 150 W (full power); (**c**) 250 W HQI-T after strike; (**d**) 250 W HQI-T at 60% of the rated power; (**e**) 250 W DE-MH at 60% of the rated power.

6. Conclusions

A two-stage electronic ballast to drive discharge lamps based on the LFR concept has been presented in this paper. The ballast behaves like an LFR, so it delivers the desired power to the lamp irrespective of the lamp impedance. This feature ensures the lamp operation at nominal power during all the lamp lifetime and allows an easy light-dimming. Design key issues have been discussed, and an experimental prototype has been realized. The tests with diverse HID lamps have proven the feasibility and versatility of the proposed approach, showing the design robustness, and a good agreement with the theoretical predictions. The ballast has been tested with IEFL, HPS and MH lamps. These experiments include start-up, warm-up transients, and steady-state operation. The ballast versatility has been shown also at full power and under dimming operation. The ballast global efficiency is in the range of 86–96% , and can be improved by means of using silicon carbide (SiC) or Gallium Nitride (GaN) devices.

Author Contributions: Conceptualization, H.V.-B.; Investigation, H.V.-B. and A.L.-M.; Validation, A.L.-M.; Visualization, C.O. and À.C.-P.; Writing—original draft, H.V.-B. and A.L.-M.; Writing—review & editing, C.O. and À.C.-P.

Funding: The research leading to these results has received funding from the Spanish AEI agency under grants DPI2015-67292-R and DPI2017-84572-C2-1-R.

Acknowledgments: The authors would like to acknowledge the contribution of Xavier Genaro-Muñoz for his support in the experimental validation of the ballast.

Conflicts of Interest: The authors declare no conflict of interest. The founding sponsors had no role in the design of the study; in the collection, analyses, or interpretation of data; in the writing of the manuscript, and in the decision to publish the results.

References

1. UN Environment—Global Environmental Facility. Accelerating the Global Adoption of Energy-Efficient Lighting. Available online: http://wedocs.unep.org/bitstream/handle/20.500.11822/20406/Energy_efficient_lighting.pdf (accessed on 27 March 2019).

2. Steiner, A. The Transition to Energy Efficient Lighting is One of the Most Straightforward and Cost-Effective Approaches to Significantly Reduce the Threat of Global Climate Change. Available online: https://www.unenvironment.org/news-and-stories/story/brighten-making-switch-efficient-lighting (accessed on 27 March 2019).

3. Philips QL Induction Lighting Systems: Information for Original Equipment Manufacturers. Available online: https://www.stefanslichtparade.de/files/ql_oem_guide.pdf (accessed on 20 January 2019).

4. Piper, J.E. *Operations and Maintenance Manual for Energy Management*; Sharpe Professional; Routledge: New York, NY, USA, 2009.

5. Osram Endura Datasheet. The High Performance Electrodelesss Fluorescent Lamp. Available online: https://www.osram.com/media/resource/HIRES/333886/554635/ENDURA-QUICKTRONIC-System-QT-ENDURA.pdf (accessed on 21 January 2019).

6. Godyak, V.A. Bright idea, radio-frequency light sources. *IEEE Ind. Appl. Mag.* **2002**, *8*, 42–49. [CrossRef]

7. Rubinstein, F.; Siminovitch, M.; Verderber, R. Fifty percent energy savings with automatic lighting controls. *IEEE Trans. Ind. Appl.* **1993**, *29*, 768–773. [CrossRef]

8. Waymouth, J.F. *Electric Discharge Lamps*; MIT Press: Cambridge, MA, USA, 1971.

9. Stevenson, R. The LED's Dark Secret: Solid-state lighting won't supplant the lightbulb until it can overcome the mysterious malady known as droop. *IEEE Spectrum* 2009. Available online: https://spectrum.ieee.org/semiconductors/optoelectronics/the-leds-dark-secret (accessed on 9 April 2019).

10. Alonso, J.M.; Rico, M.; Blanco, C.; Lopez, E. A novel low-loss clamped-mode LCC resonant inverter for HID lamp supply. In Proceedings of the PESC'95-Power Electronics Specialist Conference, Atlanta, GA, USA, 18–22 June 1995; Volume 2, pp. 736–742.

11. Cardesin, J.; Alonso, J.M.; Lopez-Corominas, E.; Calleja, A.J.; Ribas, J.; Rico-Secades, M.; Garcia, J. Small-signal analysis of a low-cost power control for LCC series-parallel inverters with resonant current mode control for HID lamps. *IEEE Trans. Power Electron.* **2005**, *20*, 1205–1212. [CrossRef]

12. Marchesan, T.B.; Dalla-Costa, M.A.; Alonso, J.M.; Prado, R.N.D. Integrated Zeta-Flyback Electronic Ballast to Supply High-Intensity Discharge Lamps. *IEEE Trans. Ind. Electron.* **2007**, *54*, 2918–2921. [CrossRef]
13. Giezendanner, F.; Biela, J.; Kolar, J.W. Optimization and Performance Evaluation of an AC-Chopper Ballast for HPS Lamps. *IEEE Trans. Ind. Electron.* **2014**, *61*, 2236–2243. [CrossRef]
14. Azcondo, F.J.; Diaz, F.J.; Brañas, C.; Casanueva, R. Microcontroller Power Mode Stabilized Power Factor Correction Stage for High Intensity Discharge Lamp Electronic Ballast. *IEEE Trans. Power Electron.* **2007**, *22*, 845–853. [CrossRef]
15. Brañas, C.; Azcondo, F.J.; Bracho, S. Electronic ballast for HPS lamps with dimming control by variation of the switching frequency. Soft start-up method for HPS and fluorescent lamps. In Proceedings of the 24th Annual Conference of the IEEE Industrial Electronics Society (IECON), Aachen, Germany, 31 August–4 September 1998; pp. 953–958.
16. Jang, T.; Kim, H.; Kim, H. Dimming Control Characteristics of Electrodeless Fluorescent Lamps. *IEEE Trans. Ind. Electron.* **2009**, *56*, 93–100. [CrossRef]
17. Fraytag, J.; Schlittler, M.E.; Costa, M.A.D.; Seidel, A.R.; Alonso, J.M.; Prado, R.N.D.; Silva, M.F.D. A Comparative Performance Investigation of Single-Stage Dimmable Electronic Ballasts for Electrodeless Fluorescent Lamp Applications. *IEEE Trans. Power Electron.* **2015**, *30*, 2239–2252. [CrossRef]
18. Silva, M.F.D.; Fraytag, J.; Marchesan, R.; Rosa, V.L.; Costa, M.A.D.; Alonso, J.M.; Prado, R.N.D. A dimmable Cuk half-bridge single-stage converter applied to electrodeless fluorescent lamps. In Proceedings of the 2012 15th International Power Electronics and Motion Control Conference (EPE/PEMC), Novi Sad, Serbia, 4–6 September 2012; pp. DS1b.6-1–DS1b.6-5.
19. Silva, M.F.D.; Fraytag, J.; Schlittler, M.E.; Marchesan, T.B.; Costa, M.A.D.; Alonso, J.M.; Prado, R.N.D. Analysis and Design of a Single-Stage High-Power-Factor Dimmable Electronic Ballast for Electrodeless Fluorescent Lamp. *IEEE Trans. Ind. Electron.* **2013**, *60*, 3081–3091. [CrossRef]
20. Schlittler, M.E.; Fraytag, J.; Seidel, A.R.; Alonso, J.M.; Prado, R.N.D.; Chagas, N.B.; Silva, M.F.D. Comparison between integrated and non-integrated SEPIC half-bridge electronic ballasts for electrodeless fluorescent lamp applications. In Proceedings of the 2013 Brazilian Power Electronics Conference, Gramado, Brazil, 27–31 October 2013; pp. 1201–1206.
21. Silva, M.F.D.; Chagas, N.B.; Schlittler, M.E.; Fraytag, J.; Marchesan, T.B.; Bisogno, F.E.; Alonso, J.M.; Prado, R.N.D. Electric Equivalent Model for Induction Electrodeless Fluorescent Lamps. *IEEE Trans. Power Electron.* **2013**, *28*, 3603–3613. [CrossRef]
22. Singer, S. Realization of loss-free resistive elements. *IEEE Trans. Circuits Syst.* **1990**, *37*, 54–60. [CrossRef]
23. Singer, S.; Erickson, R.W. Canonical modeling of power processing circuits based on the POPI concept. *IEEE Trans. Power Electron.* **1992**, *7*, 37–43. [CrossRef]
24. Haroun, R.; Cid-Pastor, A.; Aroudi, A.E.; Martinez-Salamero, L. Synthesis of Canonical Elements for Power Processing in DC Distribution Systems Using Cascaded Converters and Sliding-Mode Control. *IEEE Trans. Power Electron.* **2014**, *29*, 1366–1381. [CrossRef]
25. Kirsten, A.L.; Hansen, J.; Luz, P.C.V.D.; Rech, C.; Prado, R.N.D.; Costa, M.A.D. Modeling and control strategy of HPS electronic ballast considering the dynamic model of the lamp. In Proceedings of the XI Brazilian Power Electronics Conference, Praiamar, Brazil, 11–15 September 2011; pp. 1–6.
26. Costa, M.A.D.; Kirsten, A.L.; Alonso, J.M.; Garcia, J.; Gacio, D. Analysis, Design, and Experimentation of a Closed-Loop Metal Halide Lamp Electronic Ballast. *IEEE Trans. Ind. Appl.* **2006**, *48*, 28–36. [CrossRef]
27. Leon-Masich, A.; Valderrama-Blavi, H.; Bosque-Moncusi, J.M.; Maixe-Altes, J.; Martinez-Salamero, L. Sliding-Mode-Control-Based Boost Converter for High-Voltage Low-Power Applications. *IEEE Trans. Ind. Electron.* **2014**, *62*, 229–237. [CrossRef]

Article

Analysis of Sliding-Mode Controlled Impedance Matching Circuits for Inductive Harvesting Devices

Juan A. Garriga-Castillo [1], Hugo Valderrama-Blavi [2,*], José A. Barrado-Rodrigo [2] and Àngel Cid-Pastor [2]

[1] Department of Informatics and Industrial Engineering, Universitat de Lleida, c/Jaume II, 69.25001 Lleida, Spain; garriga@diei.udl.es
[2] Laboratory GAEI, ETSE-Universitat Rovira i Virgili, Av. Països Catalans 26, 43007 Tarragona, Spain; joseantonio.barrado@urv.cat (J.A.B.-R.); angel.cid@urv.cat (À.C.-P.)
* Correspondence: hugo.valderrama@urv.cat; Tel.: +34-977-558-523

Received: 3 September 2019; Accepted: 6 October 2019; Published: 12 October 2019

Abstract: A sea-wave energy harvesting, articulated device is presented in this work. This hand-made, wooden device is made combining the coil windings of an array of three single transducers. Taking advantage of the sea waves sway, a linear oscillating motion is produced in each transducer generating an electric pulse. Magnetic fundamentals are used to deduce the electrical model of a single transducer, a solenoid-magnet device, and after the model of the whole harvesting array. The energy obtained is stored in a battery and is used to supply a stand-alone system pay-load, for instance a telecom relay or weather station. To maximize the harvested energy, an impedance matching circuit between the generator array and the system battery is required. Two dc-to-dc converters, a buck-boost hybrid cell and a Sepic converter are proposed as impedance adaptors. To achieve this purpose, sliding mode control laws are introduced to impose a loss free resistor behavior to the converters. Although some converters operating at discontinuous conduction mode, like the buck-boost converter, can exhibit also this loss free resistor behavior, they usually require a small input voltage variation range. By means of sliding mode control the loss free resistor behavior can be assured for any range of input voltage variation. After the theoretical analysis, several simulation and experimental results to compare both converters performance are given.

Keywords: harvesting; inductive transducer; sliding mode control; loss free resistor; dc-to-dc converter

---

1. Introduction

Modern technology consumes large amounts of electrical energy. This energy is usually generated in power plants where different energy sources are converted into electrical energy. Each power plant type and energy source have their own advantages, drawbacks, and conversion efficiencies, but in addition, many of them raise environmental concerns due to the excess of pollutants produced in the conversion procedure.

Although the energy conversion efficiency of power plants can be improved, the pollution and residues generated cannot always be reduced. In this context, renewable energy sources, as hydro power, wind farms, PV (photovoltaic) plants, and the oceans energy, must be profitable to favor a more sustainable development that is respectful with the natural environment.

Depending on the physical characteristic considered, the oceans and sea waters offer different ways to collect the energy of water in movement, namely: Tides, sea-waves, and marine currents. Each water displacement type requires different technologies and transducers for collecting its energy [1].

Wave energy can be a promising energy source. Indeed, the areas of the world having larger wave energy resources are those subjected to regular wind fluxes. There are different harvesting methods.

Linear-oscillating magnetic transducers can be used for direct wave energy conversion to electricity. These transducers are usually made with a coil and a permanent magnet [2].

Sea waves are generated when wind passes over the water surface. As the sea waves propagate slower than the wind speed causing them, the energy is transferred from the wind to the waves. Wind friction on the water surface, and the difference of air pressure between the two wave sides, makes stress on the water, causing the waves growth [3]. Then, generated waves propagate on the sea surface, and transport their energy to the shore with the group velocity c_{wg} [4].

The oscillatory motion is higher at the sea surface, and decreases exponentially with depth, making the waves more independent of sea floor contour conditions as the sea depth increases. For this reason, floating harvesters are more competitive than bottom standing ones. Besides, the available energy is higher in near-shore and off-shore locations than in on-shore placements. Figure 1 depicts possible placements for floating harvesting devices.

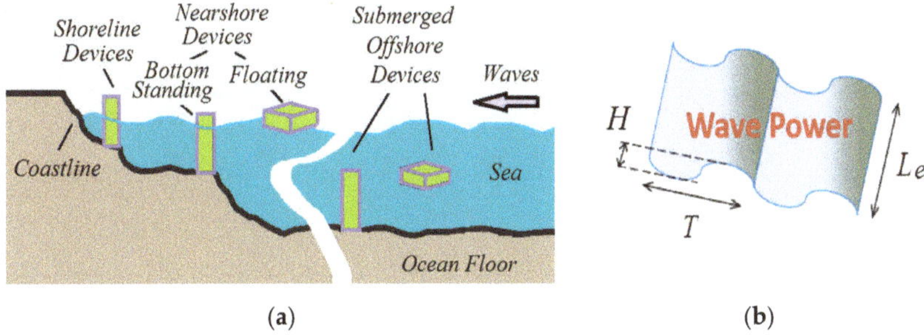

Figure 1. Wave-energy harvesting placements, and wave parameters: (**a**) possible harvesting locations and (**b**) marine wave scheme.

The wave height H is given by the wind speed, the sea depth, the fetch angle, and the seafloor topography. The average energy density $d_s[E_w]$ per unit area (1) is the sum of the kinetic and potential energy of the wave [3], where ρ is the density of sea water, H is the wave height, and g is the acceleration of gravitational force. According to the equipartition theorem, both energy types contribute equally to the wave energy. Parameter k_w accounts for the wave periodicity level. Thus, $k_w = 1$ for random waves, and $k_w = 2$ for periodic ones.

$$E_W = E_K + E_P \rightarrow d_S[E_W] = \frac{k_w}{16}\rho g H^2 \ [\text{Joule}/\text{m}^2] \tag{1}$$

The energy flux, per unit-width, through a vertical plane that is perpendicular to the wave propagation direction is the average power density (2) per unit-width $d_l[P_w]$. Then, the average power available $P_w(L_e)$ in a wave-train with T period and L_e width, can be accounted for with (3). As an example, in a wave train of $H = 1$ m, with a period of $T = 10$ s, the power density is around 5 kW/m.

$$d_l[P_w] = c_{wg} \cdot d_s[E_w] = \frac{k_w c_{wg}}{16}\rho g H^2 \ [W/m] \tag{2}$$

$$P_w(L_e) = \frac{k_w \rho g^2}{64\pi} T L_e H^2 \ [W], \quad where \ c_{wg} \approx \frac{g}{4\pi}T \tag{3}$$

Notice that although wave parameters $\{T, L_e, H\}$ determine the available wave energy, the real electric power extracted from the sea will depend additionally on the transducer kind, rating, and conversion efficiency, and finally on the efficiency of the power processing circuits used.

Most systems converting wave energy into longitudinal or angular mechanical energy use the upward and downward movement of the waves at a fixed sea point. Thus, in [4–6] electricity is produced by an angular movement using a pulley and a rod to convert a longitudinal movement into an angular one. In this sense, different floating buoys with its respective efficiencies are given in [7].

A difficulty with extracting energy from inductive magnetic energy harvesters is that they normally produce a low AC voltage magnitude at a very low frequency (<10 Hz), and boost transformers cannot be directly used because of their large size. To step-up the voltage many applications propose using a step-up converter after the rectifying stage. Thus, to maximize the harvested energy, the rectifier losses must be reduced. Some literature works propose using voltage multipliers [8], others propose reducing the rectifier voltage drop using mosfet active rectifiers [9,10] as depicted in Figure 2a. The use of two equal generators in counter-phase connected to a three-wire, two-diode full-wave rectifier, has also been proposed in several works, for example in [11,12], and depicted in Figure 2b.

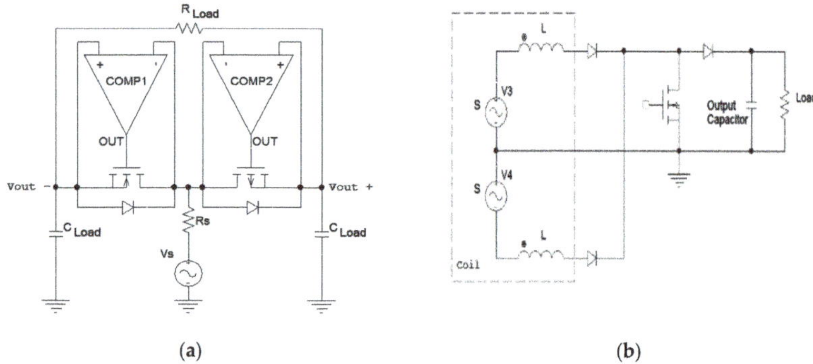

Figure 2. Rectifying circuits for inductive harvesting devices: (**a**) active rectifier [9–11] and (**b**) two-generators full-wave rectifier [11–13].

The harvesting device presented here, shown in Figure 3, is made with three inductive transducers with a high output voltage peak. As this device delivers a high voltage output, in clear contrast with previous works, no special rectifier circuits are required. Nevertheless, to reduce the rectifier losses, Schottky diodes have been considered for the full bridge rectifier input stage. Indeed, instead of maximizing the device output voltage combining the three transducer coils in series, a different interconnection has been preferred to reduce the device output impedance, as explained in Section 3.

Impedance matching has been proposed in different works to maximize the energy transfer between the harvesting generator and the load. References [13–18] propose the use of a buck-boost power converter operating in discontinuous conduction mode (DCM) as an impedance adaptor, some of them without voltage sensors [13–15] because input and output adaptor voltages are quite constant; or including voltage feedback to compensate the generator voltage variations [15–17]. Nevertheless, none of them use a harvesting generator with a similar degree of generator voltage variation as proposed in this work. We propose the use of a sliding-mode control to force a switching converter to behave like a loss free resistor (*LFR*) [19] for the impedance adaptor. The inherent robustness to parametric variations of sliding control, can assure a good impedance matching for a wide range of input voltage variation voltage as here.

After this introduction, this work continues as follows. In Section 2, the operation of the harvesting device presented in this work is analyzed, modelled, and verified experimentally. The principle of using an impedance matching circuit to maximize the collected energy is described in Section 3. Next, two sliding mode control laws are proposed to force two different switching converters to behave like a *LFR*-based impedance adaptors. The first converter, the hybrid buck-boost (HBB) is analyzed in Section 4, and next, the Sepic converter is studied in Section 5. The realization of both converters is described in Section 6. Section 7 is dedicated to the experimental results, and finally in Section 8, some conclusions and future research lines, are given.

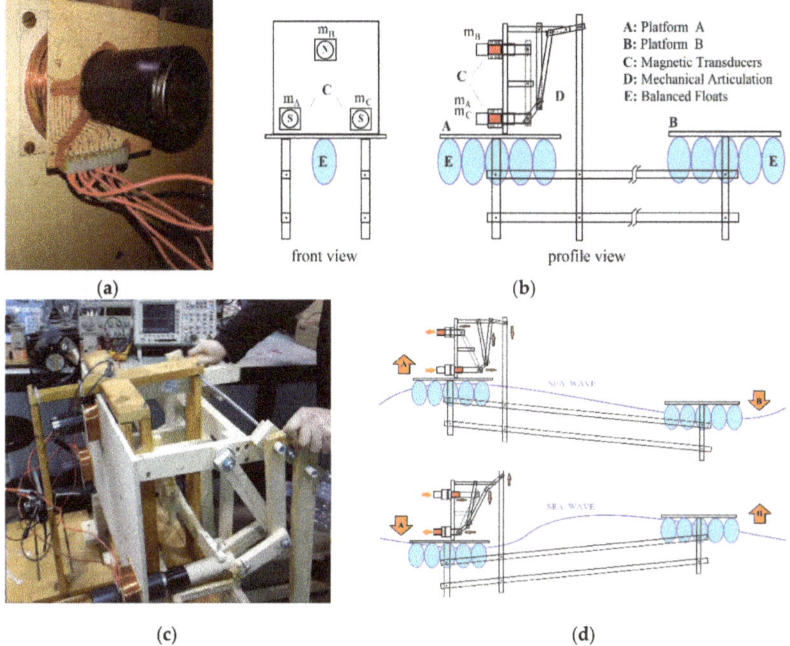

Figure 3. Prototype of a wooden articulated harvesting device made with three transducers: (**a**) detail of a single transducer, (**b**) device scheme: Front and profile views, (**c**) device photograph, and (**d**) operation principle.

2. Transducer and Harvesting Device Description

In this work we propose to use this wave energy for small-power, stand-alone applications. The extraction of small amounts of energy from each wave, conveniently stored in a battery, is a solution to supply intermittent operation small equipment, where the peak power consumption is occasional and brief: Weather stations, telecom relays, and similar equipment.

Figure 3c shows the wooden hand-made, articulated harvesting device, developed in our laboratory. This device includes three magnet-coil transducers (mA, mB, and mC). Figure 3a is depicts one of these transducers. To understand the operation of the real harvesting device appearing in Figure 3c, Figure 3b depicts in a simplified way, the front and profile views of the whole generator, and Figure 3d shows schematically the interaction between the sea waves and the developed generator. To track appropriately the train waves, the balloon floats must have a given amount of water inside, as the floats must only compensate the device weight.

Most harvesting systems that convert wave energy into a longitudinal displacement to drive a coil-magnet based linear generator, use the upward and downward movement of the wave at a fixed point [11,19]. In contrast, the system presented in Figure 3, uses the differential movement between two points of the water surface.

By means of levers, the differential movement caused by a sea wave propagating over the sea is converted into a synchronized horizontal displacement of each magnet through the corresponding coil, inducing a given voltage pulse according to Faraday's law. As seen in Figure 3b,d, the transducer mB is moving in counterphase compared to mA and mC, that are moving in phase. This can be compensated electrically by modifying the coils connections, or placing two magnets facing a given magnetic pole, and the remaining magnet, to the opposite one."

2.1. Harvesting Transducer Operation Principle

The transducer shown in photograph Figure 3a, a permanent magnet linear generator [20], is depicted schematically in Figure 4. According to its operation principle, a permanent magnet is moving inside and outside of a transducer coil winding. This causes a magnetic flux variation through it, that considering the Faraday's Law, creates a voltage difference across its terminals.

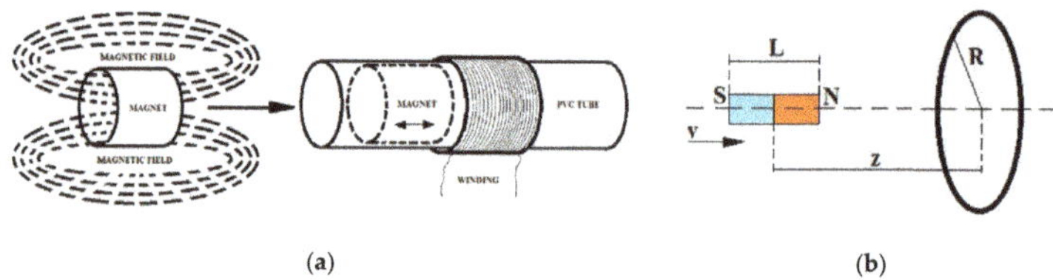

(a)   (b)

Figure 4. Transducer principle operation and description: (**a**) N-turns coil with a moving magnet inside and (**b**) equivalent scheme of the magnet-coil with a single-turn coil carrying N-times the real coil current.

The N-turns coil in Figure 4a, has been idealized in the equivalent diagram of Figure 4b, where all the turns are collapsed in a single turn, of zero thickness and radius R. To produce electricity a magnet of length L, crosses the coil at a variable speed $v(t)$. The distance from the magnet center to the coil center is the variable z. The magnet is modeled as two fictitious magnetic charges of equal value and opposite sign, realizing a magnetic dipole, with both charges separated by the distance L, the magnet length. This approach shown in Figure 5, is sufficient for our purposes.

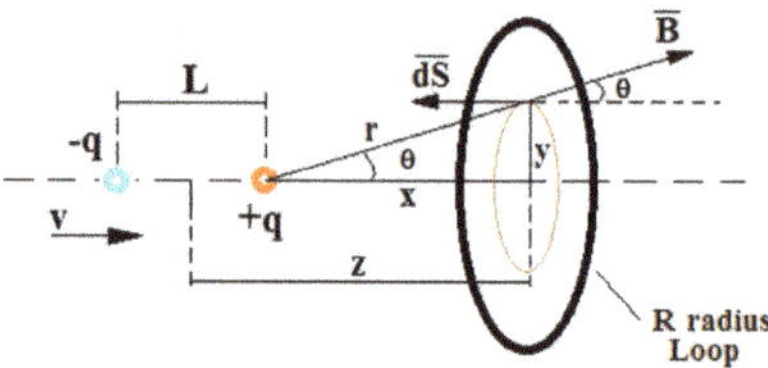

Figure 5. Transducer scheme using the magnetic dipole.

Considering the magnet length L, and its dipolar moment m, the equivalent magnetic charges are $q = \pm m/L$, and the magnetic flux density $B_q(z)$ in the vicinity of a magnetic charge/pole is

$$\vec{B}_q = \frac{\mu_0 q}{4\pi\, r^2}\hat{r} \;\Rightarrow\; B_q(z) = \frac{\mu_0 m}{4\pi\, L r^2} \tag{4}$$

The total magnetic field $B(z)$ created by both model charges at a given point of the permanent magnet axis placed at a distance z from its center (5), is the sum of contributions of both charges.

$$B(z) = B_{+q} + B_{-q} = \frac{\mu_0 m}{4\pi\, L}\left[\left(z - \frac{L}{2}\right)^{-2} - \left(z + \frac{L}{2}\right)^{-2}\right] \tag{5}$$

The electromagnetic force (EMF) generated by the transducer can be calculated from the total magnetic flux φ created by the dipole charges $\pm q$ through the equivalent collapsed loop.

$$\Phi(z) = \int_S \vec{B}\cdot d\vec{S} = -\int_S B\cdot dS\,\cos(\theta) = \Phi_{+q}(z) + \Phi_{-q}(z) \tag{6}$$

$$\Phi_{+q}(z) = -\mathrm{sgn}\left(z - \frac{L}{2}\right)\frac{\mu_0 q N}{2}\left\{1 - \left|z - \frac{L}{2}\right|\cdot\left[\left(z - \frac{L}{2}\right)^2 + R^2\right]^{-1/2}\right\} \tag{7}$$

$$\Phi_{-q}(z) = \text{sgn}\left(z + \frac{L}{2}\right)\frac{\mu_0 qN}{2}\left\{1 - \left|z + \frac{L}{2}\right| \cdot \left[\left(z + \frac{L}{2}\right)^2 + R^2\right]^{-1/2}\right\} \tag{8}$$

This flux φ is variable because the magnet position z in respect to the coil is changing because the magnet is moving with a certain speed $v(t)$, along the axis spire. Applying Faraday's law to the whole flux produced (6), the EMF induced $\varepsilon(t)$ in the spire is given by (9), where the charges values $\pm q$ have been changed by their equivalent value in terms of the magnetic dipolar moment ($\pm m/L$). When the transducer is placed in a vertical position, the velocity $v(t)$ and the position $z(t)$ are the result of gravitational acceleration, and the potential $\varepsilon(t)$ becomes (10)

$$\varepsilon(t) = -\frac{d\Phi}{dt} = \frac{v(t)N\mu_0 mR^2}{2L}v(t) \cdot \left(\left[\left(z(t) - \frac{L}{2}\right)^2 + R^2\right]^{\frac{-3}{2}} - \left[\left(z(t) + \frac{L}{2}\right)^2 + R^2\right]^{\frac{-3}{2}}\right) \tag{9}$$

$$\varepsilon(t) = \frac{gtN\mu_0 mR^2}{2L}\left(\left[\left(\frac{gt^2}{2} - \frac{L}{2}\right)^2 + R^2\right]^{\frac{-3}{2}} - \left[\left(\frac{gt^2}{2} + \frac{L}{2}\right)^2 + R^2\right]^{\frac{-3}{2}}\right) \tag{10}$$

The theoretical voltage waveform $\varepsilon(t)$ given by expression (10) is shown in Figure 6a, and the real transducer voltage at no-load conditions is given in Figure 6b. The magnet parameters are $R = 2.5$ cm, $L = 6$ cm, $m = 5.15 \cdot 10^{-4}$ Am^2, and the coil has $N = 7000$ turns of a copper wire with a diameter of $\varphi = 0.3$ mm. The measured inductance and resistance are respectively $L_T = 3$ H and $R_T = 424$ Ω.

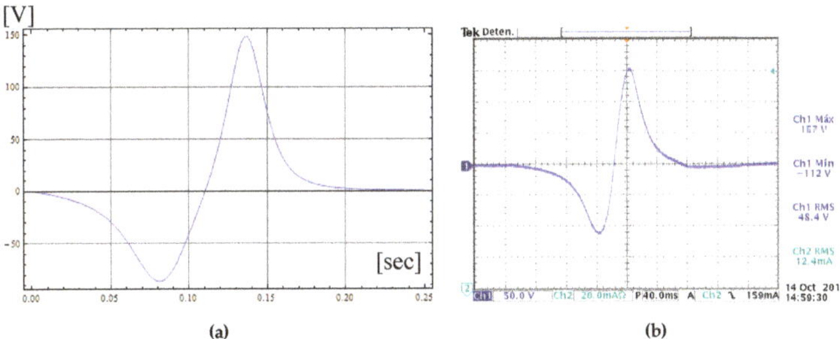

Figure 6. Pulse voltage generated by the inductive transducer: (**a**) theoretical case and (**b**) real case.

The experimental pulse waveform shown in Figure 6b corresponds to a single transducer with all the partial coil windings are connected in series, and therefore $N = 7000$ turns. These 7000 turns are organized in five partial windings, three of 1000 turns, and two of 2000 turns. These windings are organized as described in Table 1. Realize that the resistance and inductance of each partial winding (at equal number of turns) is smaller in the inner windings than in the outer ones.

Table 1. Winding parameters of the coil of a single n-transducer, n = {a, b, c}.

Winding	Winding Name	Winding Terminals	Turns	Resistance (Ω)	Inductance (mH)
1st	L1n	1n–2n	1000	43	50
2nd	L2n	3n–4n	2000	104	250
3rd	L3n	5n–6n	1000	61	85
4th	L4n	7n–8n	2000	138	390
5th	L5n	9n–10n	1000	78	120
full coil	L$_T$	1n–10n	7000	434	3000

2.2. Final Harvesting Device

The harvesting device has three interconnected single transducers. The connections realized between the different windings and coils are given in Figure 7. The final goal is to reduce the device output impedance, increasing also the energy yield.

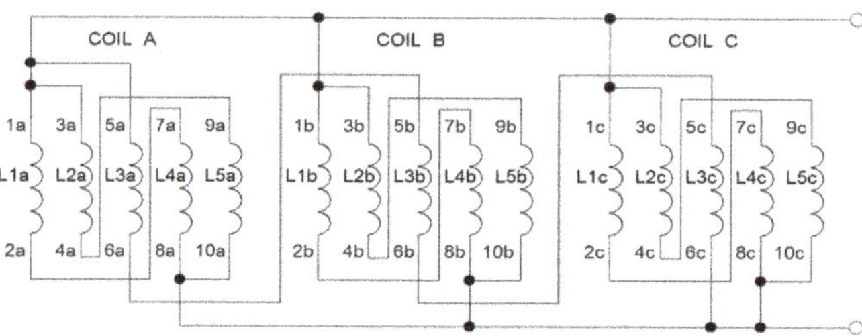

Figure 7. Final harvesting device, and scheme of the array.

According to the circuit of Figure 7, the specific details of the coil-interconnection parameters are given in Table 2.

Table 2. Final Harvesting Device Parameters.

From Coil	Windings Association	Resistance (Ω)	Inductance (mH)
A	$W_A = $ (L1a + L4a) $\|$ (L2a + L5a)	91.15	200
B	$W_B = $ (L1b + L4b) $\|$ (L2b + L5b)	91.15	200
C	$W_C = $ (L1c + L4c) $\|$ (L2c + L5c)	91.15	200
A, B, C	$W_D = $ L3a + L3b + L3c	183.3	255
device	$W_A \| W_B \| W_C \| W_D$	26.06	52.86

At no-load conditions, the voltage pulse produced by the proposed device is shown in Figure 8. The final parameters of this generator are: $R_{int} = 26\ \Omega$, $L_{int} = 52.8$ mH.

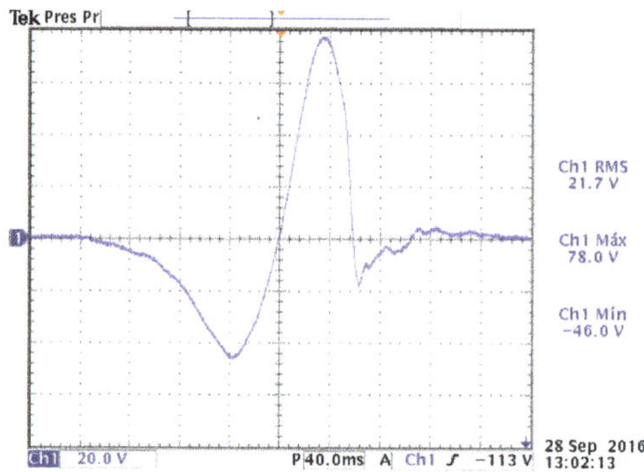

Figure 8. Experimental pulse voltage of the harvesting. Device/generator at no-load condition.

3. Impedance Matching with a Loss Free Resistor

The main purpose of the work presented here is to charge the pay-load system battery taking as much as energy as possible from the harvesting generator. If the battery is connected directly to the generator, using only a bridge rectifier, the transferred energy cannot be maximized, in part due to impedance mismatching between the battery and the generator, and in addition, because only the parts of the input voltage pulse waveform exceeding the battery voltage will be profited. Figure 9 depicts the proposed solution, a matching circuit between the generator and the battery.

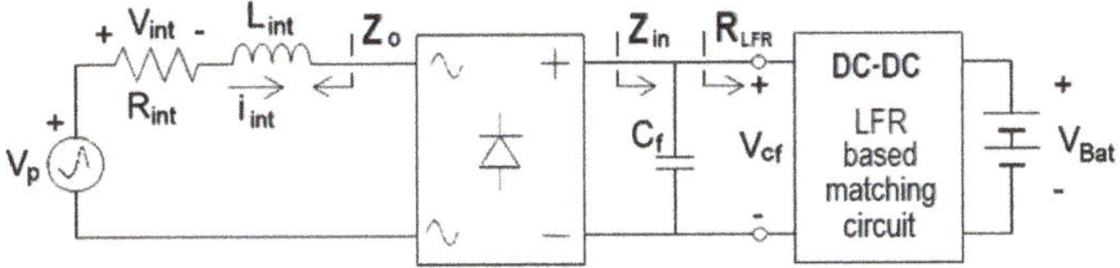

Figure 9. Proposed matching idea using the generator equivalent circuit.

Therefore, the impedance adaptation is the only solution to maximize the energy transferred at any working condition. However, even in this case, only 50% of the energy produced in the transducer will be transferred to the load. Indeed, if the conjugate input impedance of the power adaptor were equal to the generator output impedance (11), a perfect adaptation would occur.

For its simplicity, a DC-DC converter behaving as a loss free resistor [21] is proposed as impedance matching circuit. The *LFR* is a two-port circuit with a resistive input impedance R_{LFR} that can be adjusted. The output port, with a power source characteristic, delivers to the load, in this case the system battery, all the power absorbed by the *LFR* input resistance.

$$\left. \begin{array}{l} Z_o = R_{int} + j\omega L_{int} = 26 + j0.33 \ [\Omega] \\ Z^*_{in} = \left(R_{LFR} \parallel \dfrac{-j}{\omega\,C_f} \right)^* = \dfrac{jR_{LFR}/\omega\,C_f}{R_{LFR}+j/\omega\,C_f} = 25.96 + j0.045 \ [\Omega] \end{array} \right\} \ then \ Z_o = Z^*_{in} \Leftrightarrow R_{int} \approx R_{LFR} \qquad (11)$$

Expression (11) shows the generator output impedance Z_o, and Z^*_{in}, that is the conjugate of the matching circuit input impedance. The numerical values for Z_o and Z^*_{in} appearing in (11) have been calculated for $C_f = 10$ µF, and considering the worst case shown in Table 3. This means a time lapse between consecutive waves of one second ($\Delta T = 1$ s), although the real one is longer, between 10 and 30 seconds."

Table 3. Real and imaginary parts of Z_o and Z^*_{in} for $C_f = 10$ µF, and different lapse times ΔT between waves.

[Ω]	∞	30 s	10 s	3 s	1 s
$Re\,[Z_o]$	26	26	26	26	26
$Re\,[Z^*_{in}]$	R_{LFR}	R_{LFR}	R_{LFR}	R_{LFR}	R_{LFR}
$Im\,[Z_o]$	0	10^{-2}	$3.3 \cdot 10^{-2}$	10^{-1}	0.33
$Im\,[Z^*_{in}]$	0	$1.5 \cdot 10^{-3}$	$4.5 \cdot 10^{-3}$	$1.5 \cdot 10^{-2}$	$4.5 \cdot 10^{-2}$

The design of filter capacitor C_f implies three different concerns: (a) the low-pass filter effect must be small because its voltage $v_{cf}(t)$ must track appropriately the generator pulses $v_p(t)$, (b) the capacitor must compensate the inductive impedance of the generator, and (c) the capacitor must filter the high frequency switching noise. Finally, the selected value for the capacitor is $C_f = 10$ µF.

According to circuit of Figure 8, the instantaneous capacitor voltage $v_{cf}(t)$ is given in (12). Next, neglecting the rectifier bridge voltage drop $2V_d$, and assuming impedance matching ($R_{LFR} = 26\ \Omega$), the capacitor voltage $v_{cf}(t)$ should be (13) the half of the input pulse absolute value.

$$v_{Cf}(t) \approx \left| v_p(t) - R_{int}I_{int}(t) - 2V_d \right| \qquad (12)$$

$$v_{Cf}(t) \approx \left| v_{int}(t) \right| = \frac{\left| v_p(t) \right|}{2} \qquad (13)$$

The *LFR* behavior is required for impedance matching, but the appropriate converter to implement the *LFR* depends on its input and output voltage ranges.

According to the pulse waveform in Figure 7, the capacitor voltage $v_{Cf}(t)$ experiences a large variation following the voltage pulse v_p evolution. During 250 ms, the generator voltage varies from $-100\text{ V} < v_p(t) < 160\text{ V}$. Once rectified, if impedance matching occurs, the capacitor voltage changes from $0 < v_{Cf}(t) < V_{max} = 80\text{ V}$. Thus, using a common 12 V battery, the matching circuit input voltage $v_{Cf}(t)$ can be greater ($v_{Cf}(t) > V_{bat}$) or smaller ($v_{Cf}(t) < V_{bat}$) than the battery voltage. Consequently, any converter proposed as matching circuit must exhibit a buck-boost characteristic.

3.1. DCM Operated Buck-Boost Converter

Since the buck-boost operated at DCM [19] has a natural *LFR* behavior, that converter could be proposed as an impedance matching circuit. From the buck-boost input power in DCM, the expression of the converter input impedance R_{LFR} can be easily calculated (14)

$$P_{in} = \frac{v_{cf}^2 D^2 T_S}{2L} = \frac{v_{cf}^2}{R_{LFR}} \Rightarrow R_{LFR} = \frac{2L}{D^2(t)T_S(t)} \approx R_{int} = 26\,\Omega \tag{14}$$

where D is the converter duty ratio, and T_S is the switching period. Theoretically, any pair of constant D and T_S can be used to regulate the input impedance $R_{LFR} = 26\ \Omega$. However, to guarantee that the converter is working in the discontinuous mode, the converter duty ratio $D(t)$ must change continuously (15) to follow the $v_{Cf}(t)$. Therefore, to keep constant, at 26 Ω, the switching frequency $f_S(t) = 1/T_S(t)$ must also be adapted continuously.

$$\left.\begin{array}{c} \text{DCM} \\ < v_L >= 0 \end{array}\right\} \Rightarrow D(t) < \frac{V_{bat}}{V_{bat} + v_{Cf}(t)} \tag{15}$$

It can be easily concluded that designing a low consumption control circuit able to adjust continuously the values of duty cycle $D(t)$ and $T_S(t)$ to fulfill simultaneously nonlinear Equation (14) and in Equation (15) is very complicated. Besides, the buck-boost converter has output voltage sign inversion, input and output pulsating currents, and, therefore, other solutions must be explored.

3.2. Buck/Boost Hybrid Converter (HBB)

The buck/boost hybrid converter is shown in Figure 10. While $v_{Cf}(t)$ be smaller than the battery voltage V_{bat}, the circuit will operate in boost mode, but when $v_{Cf}(t)$ be greater than the battery one, it will work in buck mode. When the converter is in step-up mode, S_{Buck} is permanently at ON-state, whereas in the buck working mode S_{Boost} is permanently at OFF-state.

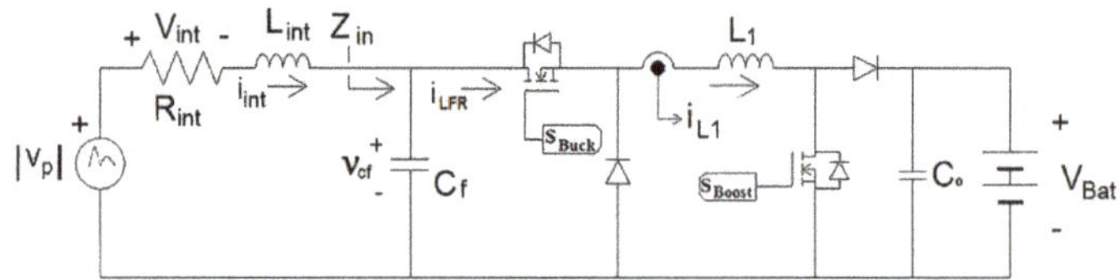

Figure 10. Buck/Boost Hybrid Converter (HBB).

This converter has a pulsating output current in boost mode and a pulsating input current in buck mode, so there is an electromagnetic interference (EMI) reduction.

To impose a *LFR* behavior to both working modes, assuring impedance matching, two sliding control laws must be used, one per each mode. Both control surfaces are analyzed in Section 4.

The mode selection is done with a comparator, but there is a small dead-zone around V_{bat} where the converter is not switching to reduce switching losses. This is an advantage compared to a classic step-up/down converter like the buck-boost, Ćuk, or Sepic, that are switching continuously regardless the value of $v_{Cf}(t)$ and V_{bat}. Conversely, in the dead zone, there is no control, and the impedance matching is lost.

3.3. Sepic Converter

Figure 11 depicts the Sepic converter. As in the HBB converter case, there is no output voltage sign change. As happens with the HBB converter in the boost mode, the input current has a triangular waveform whereas the output current is pulsating. In this case, as there is no dead-zone around V_{bat}, the impedance matching occurs for any $v_{cf}(t)$ value. Besides, as additional advantage, only one sliding-mode control law is required to impose the *LFR* behavior for all $v_{cf}(t)$ voltage range.

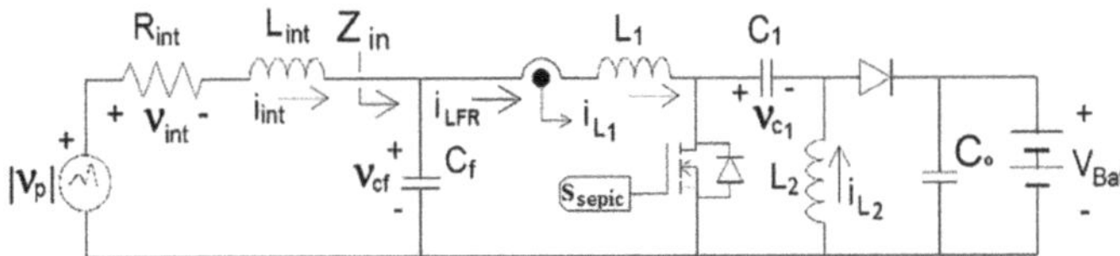

Figure 11. Sepic Converter.

4. HBB Converter Matching Circuit

The hybrid converter is a combination of two converters. Therefore, it has two operation modes, buck and boost. Two different sliding mode surfaces are required to force this circuit to behave like a loss free resistor. The controllable inductor is L_1. In the boost mode, the inductor current $i_{L1}(t)$ is the converter input current, so $i_{L1}(t) = i_{LFR}(t)$. Conversely, in the buck mode, $i_{L1}(t)$ is the converter output current, that is, the current charging the battery, and therefore $i_{LFR}(t) = i_{L1}(t) \cdot [V_{bat}(t)/v_{cf}(t)]$.

4.1. Boost Mode Operation (HBB)

When $v_{Cf}(t) < V_{bat}$, the switch S_{Buck} is always at ON state, and $i_{L1}(t) = i_{int}(t)$. The resulting circuit is an input filtered boost converter. The output capacitor C_o is not included because it is in parallel with the battery and has no dynamics. For simplicity, we use $v_p(t)$ in the analysis, instead of $|v_p(t)|$. Figure 12 depicts the two circuit topologies, ON and OFF.

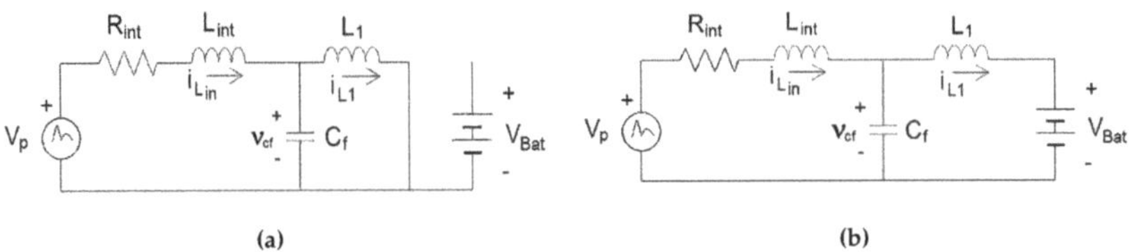

(a)   (b)

Figure 12. Boost mode, HBB converter topologies: (**a**) ON and (**b**) OFF.

Using the switching variable $u(t) = \{0, 1\}$ the equations of ON and OFF topologies can be compacted, and the converter dynamics at any instant is described with a sole set of Equations, (16)

$$x(t) = \begin{bmatrix} i_{int} \\ i_{L1} \\ v_{cf} \end{bmatrix} \Rightarrow \begin{cases} L_{int}\frac{di_{int}}{dt} = v_P - R_{int}i_{int} - v_{cf} \\ L_1\frac{di_{L1}}{dt} = v_{cf} - V_{Bat}(1-u) \\ C_f\frac{dv_{cf}}{dt} = i_{int} - i_{L1} \end{cases} \tag{16}$$

The sliding surface (17) imposed to the inductor current $i_{L1}(t)$, guarantees the impedance matching. From the Equation set (16), taking into account the control surface $S(x)$ and its existence conditions, the equivalent control (18) is obtained.

$$S(x) = i_{L1} - \frac{v_{cf}}{R_{int}} = 0 \Rightarrow i_{LFR} = \frac{v_{cf}}{R_{int}} \quad S(x)\dot{S}(x) < 0 \tag{17}$$

$$u_{eq} = 1 - \frac{v_{cf}}{V_{Bat}} + \frac{L_1}{R_{int}V_{Bat}} \cdot \frac{dv_{cf}}{dt} \tag{18}$$

Forcing to zero the dynamics of Equation set (16), and considering the equivalent control (18), the system equilibrium point (19) and the related ideal sliding dynamics (20) can be obtained. The resulting dynamics is linear. At the equilibrium point, the capacitor voltage results to be the half part of the input voltage pulse ($V_{cf}^* = V_p^*/2$), as expected from an impedance matching system.

$$X^* = \left[I_{int}^*, V_{cf}^*, I_{L_1}^* \right]^T = \left[\frac{V_P^*}{2R_{int}}, \frac{V_P^*}{2}, \frac{V_P^*}{2R_{int}} \right]^T \tag{19}$$

$$\begin{cases} g_1(x) = \frac{di_{int}}{dt} = \frac{1}{L_{int}}v_P - \frac{R_{int}}{L_{int}}i_{int} - \frac{1}{L_{int}}v_{cf} \\ g_2(x) = \frac{dv_{cf}}{dt} = \frac{1}{C_f}i_{int} - \frac{v_{cf}}{R_{int}C_f} \end{cases} \tag{20}$$

The ideal sliding dynamics is linear, as can be seen in (20). Consequently, the Laplace transform can be directly applied in the ideal dynamics (20), and a small signal model is not required. After some manipulations the transfer function $V_{cf}(s)/V_P(s)$ (21) is obtained. This transfer function evidences the impedances matching at DC, because $V_{Cf}/V_P = 1/2$. The tracking of the input pulse $v_p(t)$ is controlled by a second order low-pass filter, with a characteristic polynomial $P(s)$ is unconditionally stable, because all its coefficients exist and are positive. Considering that $C_f = 10\ \mu$F, the system natural frequency (21) is $\omega_n \approx 1900$ rad/s, and the bandwidth is around BW$_{-3dB}$–310 Hz.

$$\frac{V_{cf}(s)}{V_P(s)} = \frac{\hat{V}_{cf}(s)}{\hat{V}_P(s)} = \frac{K}{P(s)} = \frac{\frac{1}{L_{in}C_f}}{s^2 + \left(\frac{R_{in}}{L_{in}} + \frac{1}{R_{in}C_f} \right)s + \frac{2}{L_{in}C_f}} \tag{21}$$

4.2. Buck Mode Operation (HBB)

When $v_{Cf}(t) < V_{bat}$, the switch S_{Boost} will be permanently at OFF state. The resulting circuit is a common buck converter with an input filter. The output capacitor C_o is not depicted because is in parallel with the battery. The topologies ON and OFF, are shown respectively in Figure 13a,b.

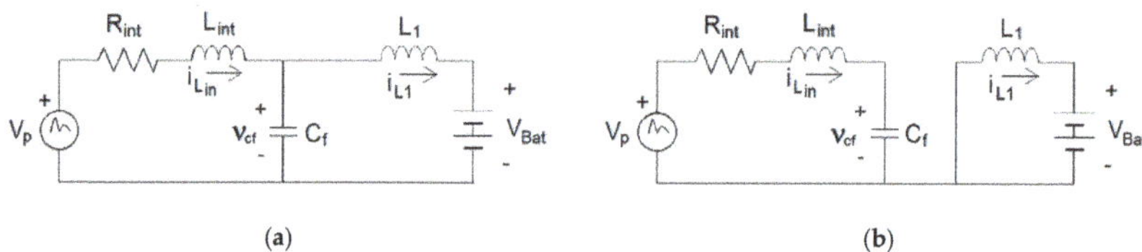

(a)   (b)

Figure 13. Buck mode, HBB converter topologies: (a) ON, (b) OFF.

Repeating the process described for the boost mode, a single set of differential Equation (22) describes, at any instant, the dynamics of the converter.

$$x(t) = \begin{bmatrix} i_{int} \\ i_{L1} \\ v_{cf} \end{bmatrix} \Rightarrow \begin{cases} L_{int}\frac{di_{int}}{dt} = v_P - R_{int}i_{int} - v_{cf} \\ L_1\frac{di_{L1}}{dt} = v_{cf}\,u - V_{Bat} \\ C_f\frac{dv_{cf}}{dt} = i_{int} - i_{L1}\,u \end{cases} \tag{22}$$

To assure impedance matching, the converter input current should be forced to be proportional to the input voltage $v_{Cf}(t)$. However, the only state variable leading to a stable control surface is the output current $i_{L1}(t)$. As a result, the proposed sliding surface (23) regulates the output current $i_{L1}(t)$ to a given value that, considering the converter voltage gain, imply the desired input current, that is $i_{LFR}(t)$. From the Equation set (22), considering the surface $S(x)$ and its existence conditions, the equivalent control (24) is obtained.

$$S(x) = i_{L1} - \frac{v_{cf}}{R_{int}} \cdot \frac{v_{cf}}{V_{Bat}} = 0 \Rightarrow i_{LFR} = \frac{v_{cf}}{R_{int}} \quad S(x)\dot{S}(x) < 0 \tag{23}$$

$$u_{eq} = 1 - \frac{v_{cf}}{V_{Bat}} + \frac{L_1}{R_{int}V_{Bat}} \cdot \frac{dv_{cf}}{dt} \tag{24}$$

Forcing to zero the system dynamics in (22), and considering the equivalent control (24), we obtain the equilibrium point (25) and the corresponding ideal sliding dynamics (26), that is non-linear. As happened in the boost mode, the equilibrium point capacitor voltage is the half part of the pulse voltage is ($V^*_{cf} = V^*_p/2$), as expected from impedance matching.

$$X^* = \left[I^*_{in}, V^*_{cf}, I^*_{L_1} \right]^T = \left[\frac{V_P^*}{2R_{int}}, \frac{V_P^*}{2}, \frac{V^{*2}_{cf}}{4R_{int}V_{Bat}} \right]^T \tag{25}$$

$$\begin{cases} g_1(x) = \frac{di_{int}}{dt} = \frac{1}{L_{int}}v_P - \frac{R_{int}}{L_{int}}i_{int} - \frac{1}{L_{int}}v_{cf} \\ g_2(x) = \frac{di_{L1}}{dt} = \frac{2v_{cf}}{R_{int}V_{Bat}} \cdot \frac{dv_{cf}}{dt} \\ g_3(x) = \frac{dv_{cf}}{dt} = \frac{1}{C_f}i_{int} - \frac{2}{C_f}\frac{L_1}{R_{int}V_{Bat}}\frac{dv_{cf}}{dt}i_{L1} - \frac{V_{Bat}}{v_{cf}}\frac{i_{L1}}{C_f} \end{cases} \tag{26}$$

To evaluate the local stability, the ideal dynamics (26) is linearized around the equilibrium point X^* using the typical small signal perturbation model (27), (28), whose coefficients are given in (29)

$$x(t) = X^* + \hat{x}(t) \tag{27}$$

$$\begin{cases} g_1(x) \approx a \cdot \hat{v}_P(t) + b \cdot \hat{i}_{int}(t) + c \cdot \hat{v}_{cf}(t) \\ g_2(x) \approx d \cdot \hat{v}_{cf}(t) + e \cdot \hat{w}(t) \; where \; w = \frac{dv_{cf}}{dt} \\ g_3(x) \approx f \cdot \hat{i}_{int}(t) + g \cdot \hat{i}_{L1}(t) + h \cdot \hat{v}_{cf}(t) + k \cdot \hat{w}(t) \end{cases} , \; W^* = 0 \tag{28}$$

$$\begin{array}{lll} a = \left.\frac{\partial g_1}{\partial v_P}\right|_{x^*} = \frac{1}{L_{in}} & b = \left.\frac{\partial g_1}{\partial i_{int}}\right|_{x^*} = -\frac{R_{int}}{L_{in}} & c = \left.\frac{\partial g_1}{\partial v_{cf}}\right|_{x^*} = -\frac{1}{L_{in}} \\ d = \left.\frac{\partial g_2}{\partial v_{cf}}\right|_{x^*} = 0 & e = \left.\frac{\partial g_2}{\partial w}\right|_{x^*} = \frac{V_P^*}{R_{int}V_{Bat}} & f = \left.\frac{\partial g_3}{\partial i_{int}}\right|_{x^*} = \frac{1}{C_f} \\ g = \left.\frac{\partial g_3}{\partial i_{L1}}\right|_{x^*} = \frac{-2V_{Bat}}{V_P^*C_f} & h = \left.\frac{\partial g_3}{\partial v_{cf}}\right|_{x^*} = \frac{1}{R_{int}C_f} & k = \left.\frac{\partial g_3}{\partial w}\right|_{x^*} = \frac{-L_1}{2C_f}\left(\frac{V_P^*}{R_{int}V_{Bat}}\right)^2 \end{array} \tag{29}$$

Applying the Laplace transform to the linearized model (28) with the coefficients of (29), the $\hat{V}_{cf}(s)/\hat{V}_p(s)$ transfer function (30) is obtained. This function evidences the impedance matching at

DC because $V_{Cf}/V_p = 1/2$. As in the previous case, the tracking of $v_p(t)$ is given by a low-pass filter of second order. The system stability is verified because all coefficients of $P(s)$ exist and are positive.

$$\frac{\hat{V}_{cf}(s)}{\hat{V}_P(s)} = \frac{\frac{1}{L_{int}C_f}}{s^2\left[1 + \frac{L_1}{2C_f}\left(\frac{V_P^*}{R_{int}V_{Bat}}\right)^2\right] + s\left\{\frac{1}{R_{int}C_f}\left[1 + \frac{L_1}{2L_{int}}\left(\frac{V_P^*}{V_{Bat}}\right)^2\right] + \frac{R_{int}}{L_{int}}\right\} + \frac{2}{L_{int}C_f}} \tag{30}$$

As expected from the nonlinear sliding dynamics (26), in the buck mode, the bandwidth B_{-3d}, and the natural frequency (31) are not constant, depending on the generator pulse V_P and the converter gain. Realize that in the boost mode (18) the natural frequency was constant.

$$\omega_n^2 = \frac{2}{L_{int}C_f} \cdot \left[1 + \frac{L_1}{2C_f}\left(\frac{V_P^*}{R_{int}V_{Bat}}\right)^2\right]^{-1} \tag{31}$$

5. Sepic Converter Matching Circuit

In the Sepic converter the controllable inductor is L_1. As in the HBB boost mode, the current through this inductor $i_{L1}(t)$ corresponds to the converter input current $i_{LFR}(t)$. This circumstance allows using the same control surface $S(x)$ to force the *LFR* behavior that was proposed for the HBB boost mode in Section 4. Figure 14 depicts the two converter topologies.

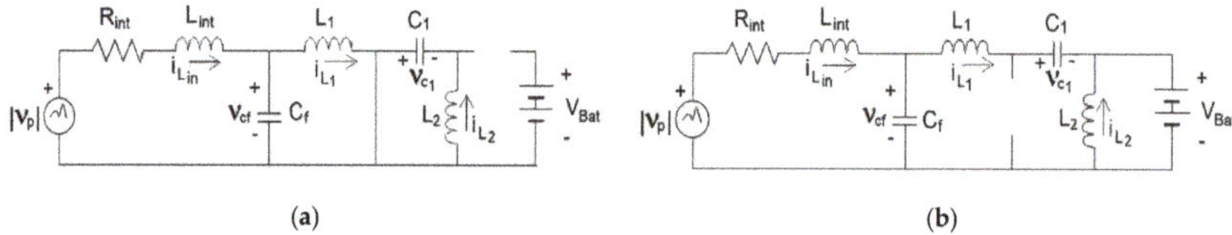

(a)   (b)

Figure 14. Sepic converter topologies: (a) ON and (b) OFF.

Again, in the previous cases by means of the switching variable $u(t) = \{0, 1\}$, the dynamics of the converter at any instant can be described with a single set of differential Equation (32).

$$x(t) = \begin{bmatrix} i_{int} \\ i_{L1} \\ i_{L2} \\ v_{c1} \\ v_{cf} \end{bmatrix} \Rightarrow \begin{cases} L_{int}\frac{di_{int}}{dt} = v_P - R_{int}i_{int} - v_{cf} \\ L_1\frac{di_{L1}}{dt} = v_{C_f} - (1-u)\left(v_{C_1} + V_{Bat}\right) \\ L_2\frac{di_{L2}}{dt} = -V_{Bat} + u\left(v_{C_1} + V_{Bat}\right) \\ C_f\frac{dv_{C_f}}{dt} = i_{int} - i_{L_1} \\ C_1\frac{dv_{C_1}}{dt} = i_{L_1} - u\left(i_{L_1} + i_{L_2}\right) \end{cases} \tag{32}$$

The sliding surface (33) imposed to the inductor current $i_{L1}(t)$ assures the impedance matching. The equivalent control (34) is deduced from the Equation set (32), using the surface $S(x)$ and its related existence conditions.

$$S(x) = i_{L1} - \frac{v_{C_f}}{R_{int}} = 0 \Rightarrow i_{LFR} = \frac{v_{cf}}{R_{int}} \quad S(x)\dot{S}(x) < 0 \tag{33}$$

$$u_{eq}(t) = 1 - \frac{v_{C_f}}{\left(v_{C_1} + V_{Bat}\right)} + \frac{L_1}{R_{int}\left(v_{C_1} + V_{Bat}\right)} \cdot \frac{dv_{C_f}}{dt} \tag{34}$$

Forcing to zero the dynamics shown in (32), and considering the equivalent control (34), the equilibrium point (35) and the corresponding ideal dynamics (36) are deduced, which result to be non-linear. As in boost mode, the capacitor voltage at the equilibrium point is again, the half part of the pulse voltage is ($V_{cf}{}^* = V_p{}^*/2$), corresponding to a situation of impedance matching.

$$X^* = \left[I_{int}^*, I_{L_1}^*, I_{L_2}^*, V_{C_f}^*, V_{C_1}^* \right]^T = \left[\frac{V_P^*}{2 \cdot R_{int}}, \frac{V_P^*}{2 \cdot R_{int}}, \frac{V_P^{*2}}{4 \cdot R_{int} V_{Bat}}, \frac{V_P^*}{2}, \frac{V_P^*}{2} \right]^T \tag{35}$$

$$\begin{cases} g_1(x) = \frac{di_{int}}{dt} = \frac{1}{L_{int}} \left[v_P - R_{int} i_{int} - v_{cf} \right] \\ g_2(x) = \frac{di_{L_1}}{dt} = \frac{1}{R_{int}} \frac{dv_{C_f}}{dt} \\ g_3(x) = \frac{di_{L_2}}{dt} = \frac{1}{L_2} \left[v_{C_1} - v_{C_f} + \frac{L_1}{R_{int}} \cdot \frac{dv_{C_f}}{dt} \right] \\ g_4(x) = \frac{dv_{C_f}}{dt} = \frac{1}{C_f} \left[i_{int} - i_{L_1} \right] \\ g_5(x) = \frac{dv_{C_1}}{dt} = \frac{1}{C_1} \left\{ \frac{i_{L_1} + i_{L_2}}{v_{C_1} + V_{Bat}} \cdot \left[v_{C_f} - \frac{L_1}{R_{int}} \cdot \frac{dv_{C_f}}{dt} \right] - i_{L_2} \right\} \end{cases} \tag{36}$$

To evaluate the local stability of the proposed surface, the ideal dynamics (32) should be linearized around the equilibrium point X^* using the well-known small signal perturbation model (37), (38). The coefficients of the small signal model are given in (39).

$$x(t) = X^* + \hat{x}(t) \tag{37}$$

$$\begin{cases} g_1(x) = \frac{di_{int}}{dt} = a \cdot \hat{v}_P(t) + b \cdot \hat{i}_{int}(t) + c \cdot \hat{v}_{cf}(t) \\ g_2(x) = \frac{di_{L_1}}{dt} = d \cdot \hat{w}(t) \ where, w(t) = \frac{dv_{C_f}}{dt} \\ g_3(x) = \frac{di_{L_2}}{dt} = e \cdot \hat{v}_{C_1}(t) + f \cdot \hat{w}(t) + h \cdot \hat{v}_{C_f}(t) \\ g_4(x) = \frac{dv_{C_f}}{dt} - j \cdot \hat{i}_{int}(t) + k \cdot \hat{i}_{L_1}(t) \\ g_5(x) = \frac{dv_{C_1}}{dt} \approx m \cdot \hat{i}_{L_2}(t) + n \cdot \hat{i}_{L_1}(t) + p \cdot \hat{w}(t) + q \cdot \hat{v}_{C_1}(t) + r \cdot \hat{v}_{C_f}(t) \end{cases} \tag{38}$$

$$\begin{cases} a = \left. \frac{\partial g_1}{\partial v_P} \right|_{X^*} = \frac{1}{L_{int}} & e = \left. \frac{\partial g_3}{\partial v_{C_1}} \right|_{X^*} = \frac{1}{L_2} & m = \left. \frac{\partial g_5}{\partial i_{L_2}} \right|_{X^*} = \frac{-2V_{Bat}}{C_1 (V_P^* + 2V_{Bat})} \\ b = \left. \frac{\partial g_1}{\partial i_{int}} \right|_{X^*} = \frac{-R_{int}}{L_{int}} & f = \left. \frac{\partial g_3}{\partial w} \right|_{X^*} = \frac{L_1}{L_2 R_{int}} & n = \left. \frac{\partial g_5}{\partial i_{L_1}} \right|_{X^*} = \frac{V_P^*}{C_1 (V_P^* + 2V_{Bat})} & k = \left. \frac{\partial g_4}{\partial i_{L_1}} \right|_{X^*} = \frac{-1}{C_f} \\ c = \left. \frac{\partial g_1}{\partial v_{C_f}} \right|_{X^*} = \frac{-1}{L_{int}} & h = \left. \frac{\partial g_3}{\partial v_{C_f}} \right|_{X^*} = \frac{-1}{L_2} & q = \left. \frac{\partial g_5}{\partial v_{C_1}} \right|_{X^*} = \frac{-V_P^{*2}}{2R_{int} C_1 V_{Bat} (2V_{Bat} + V_P^*)} & p = \left. \frac{\partial g_5}{\partial w} \right|_{X^*} = \frac{-V_P^*}{2C_1 R_{int}^2 V_{Bat}} \\ d = \left. \frac{\partial g_2}{\partial w} \right|_{X^*} = \frac{1}{R_{int}} & j = \left. \frac{\partial g_4}{\partial i_{int}} \right|_{X^*} = \frac{1}{C_f} & r = \left. \frac{\partial g_5}{\partial v_{C_f}} \right|_{X^*} = \frac{V_P^*}{2R_{int} C_1 V_{Bat}} \end{cases} \tag{39}$$

After linearizing (38) the ideal dynamics, the small signal $\hat{V}_{cf}(s)/\hat{V}_p(s)$ transfer function (40) is obtained, that evidences the perfect impedance matching is at DC, and good at low frequency, like the sea waves-trains. Realize in (40) that at DC $V_{Cf} = 1/2V_p$. When frequency increases the module of $\hat{V}_{cf}(s)/\hat{V}_p(s)$ decreases from its maximum value at DC (1/2) and the phase-shift between V_{Cf} and V_p increases. As in the HBB boost mode, the natural frequency is given by L_{int} and the capacitor C_f.

$$H_{C_f P}(s) = \frac{\hat{V}_{C_f}(s)}{\hat{V}_P(s)} = \frac{1/L_{int} C_f}{s^2 + s \left(\frac{L_{int} + C_f R_{int}^2}{C_f R_{int} L_{int}} \right) + \frac{2}{L_{int} C_f}} = \frac{1}{2} \cdot \frac{\omega_{na}^2}{s^2 + 2\zeta_a \omega_{na} s + \omega_{na}^2} \tag{40}$$

The Sepic converter is a fifth order system, and the ideal system dynamics should be of fourth order, but expression (40) denotes a second order system. The fourth order dynamics consists of two decoupled pairs of complex conjugated poles that can be factorized in two second order functions as seen in (41). The converter is stable, since all its coefficients of $D_a(s)$ and $D_b(s)$ exist and are positive. In a clear contrast to $H_{CfP}(s)$, where a pair of complex conjugated poles is hidden, in other system transfer functions the whole dynamics are visible. One of these cases is the $H_{C1P}(s)$ transfer function, where according to expressions (42) and (43), the full dynamics is shown. Realize that s expected, at DC conditions $V_{Cf} = V_{C1}$.

$$P(s) = \left(s^2 + 2\zeta_a\omega_{na}s + \omega_{na}^2\right) \cdot \left(s^2 + 2\zeta_b\omega_{nb}s + \omega_{nb}^2\right) = D_a(s) \cdot D_b(s) \tag{41}$$

$$H_{C_1 P}(s) = \frac{\hat{V}_{C_1}(s)}{\hat{V}_P(s)} = \frac{\hat{V}_{C_1}(s)}{\hat{V}_{C_f}(s)} \cdot \frac{\hat{V}_{C_f}(s)}{\hat{V}_P(s)} = H_{C1Cf}(s) \cdot H_{C_f P}(s) \tag{42}$$

$$H_{C1Cf}(s) = \frac{\hat{V}_{C_1}(s)}{\hat{V}_{C_f}(s)} = \frac{-s^2 L_2\left(V_P^*+2V_{Bat}\right)+\left[4V_{Bat}^2 L_1+2V_P^* L_2 V_{Bat}+V_P^* L_2\left(V_P^*+2V_{Bat}\right)\right]R_{int}\cdot s+4R_{int}^2 V_{Bat}^2}{2s^2\left[R_{int}^2 L_2 C_1 V_{Bat}\left(V_P^*+2V_{Bat}\right)\right]+s\cdot R_{int}L_2 V_P^{*2}+4R_{int}^2 V_{Bat}^2} \tag{43}$$

6. Experimental Circuits of the Impedance Adaptors

Figures 15 and 16 depict the Sepic converter and its control board, respectively.

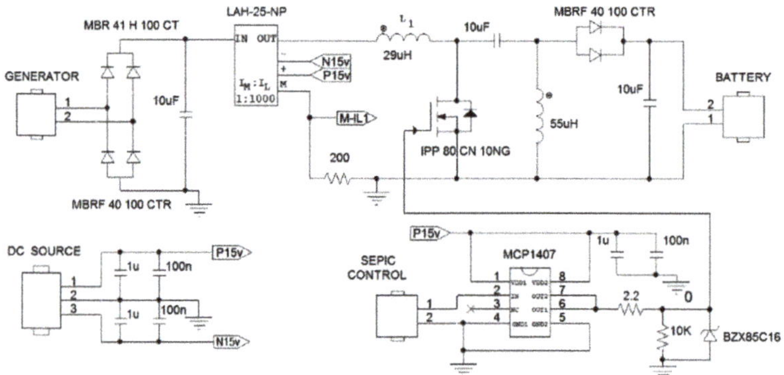

Figure 15. Sepic converter power stage.

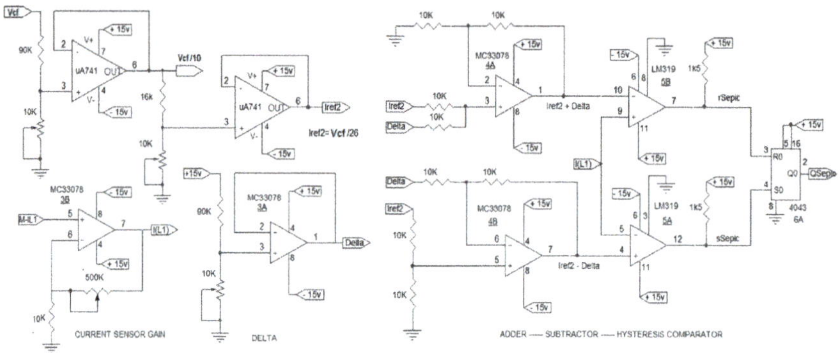

Figure 16. Sepic control board.

Figures 17 and 18 depict the HBB power stage and its control board. The boost mode HBB and the Sepic control circuits mode are equal, because the control surface is the same, see (17) and (33).

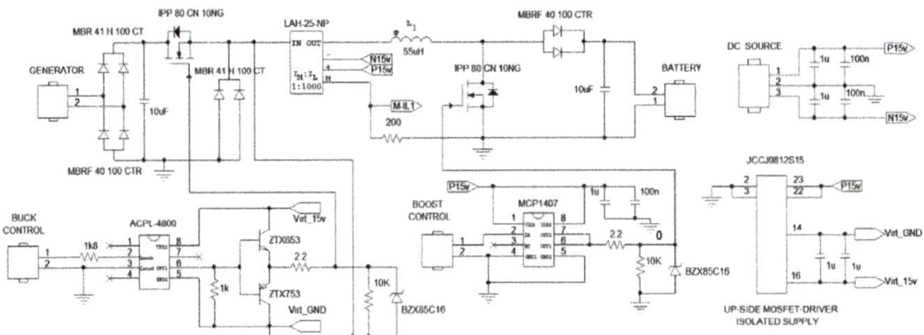

Figure 17. HBB converter power stage.

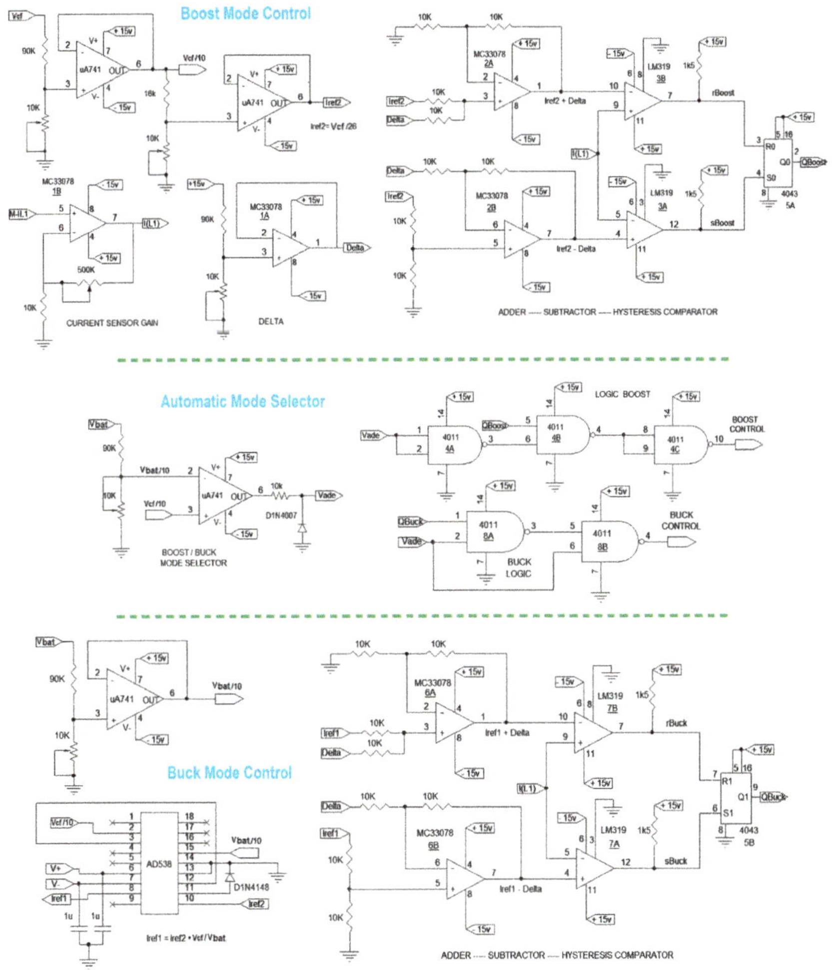

Figure 18. HBB control board.

7. Experimental Verification

The performance of both converters, the hybrid and the Sepic, have been simulated and verified experimentally with two types of input signals. The first signal $v_p(t)$ is the pulse waveform provided by the harvesting device of Figure 8. The simulated results are shown in Figure 19a–f. The second signal $v_p(t)$ is a sinusoidal waveform, and the simulated and experimental results are given in Figure 20a–f. Experimental results with a third pulse waveform are in Figure 21a–d.

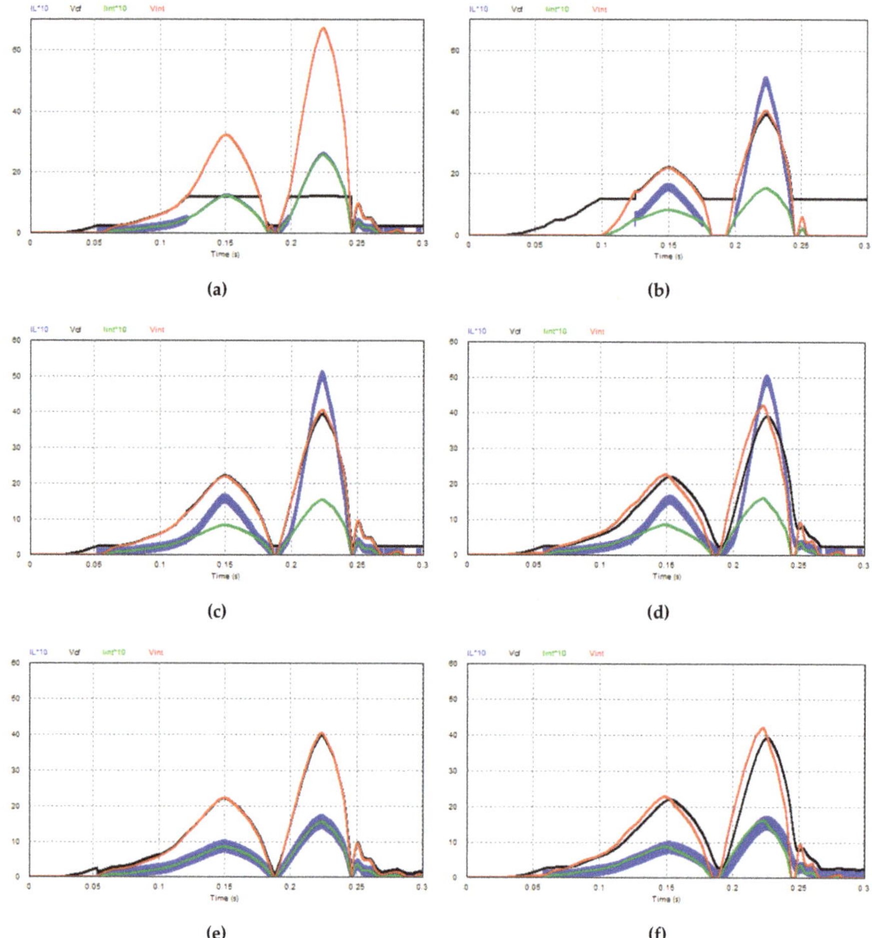

Figure 19. PSIM simulations with a pulse input: (**a**) HBB boost mode C_f = 10 μF, (**b**) HBB buck mode C_f = 10 μF, (**c**) HBB C_f = 10 μF, (**d**) HBB C_f = 200 μF, (**e**) Sepic C_f = 10 μF, (**f**) Sepic C_f = 200 μF.

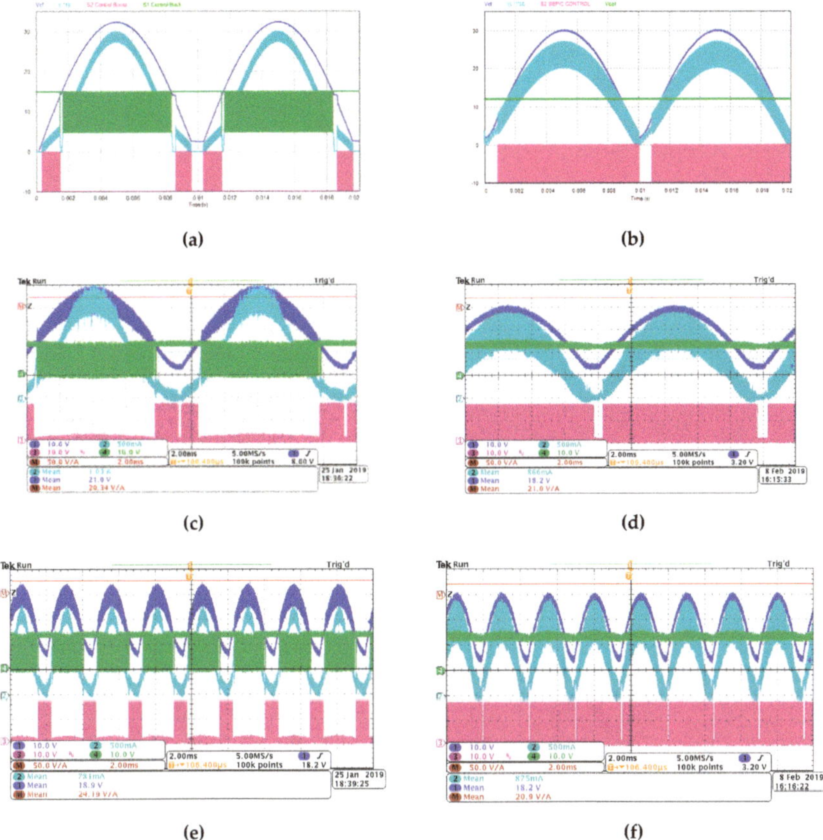

Figure 20. Simulation and experimental tests with $C_f = 10$ μF, and a sinusoidal input of $v_p(t) = 60 \cdot sin(2\pi f_o \cdot t)$: (**a**) HBB simulation with $f_o = 50$ Hz, (**b**) Sepic simulation with $f_o = 50$ Hz, (**c**) HBB prototype with $f_o = 50$ Hz, (**d**) Sepic prototype with $f_o = 50$ Hz, (**e**) HBB prototype with $f_o = 200$ Hz, and (**f**) Sepic prototype with $f_o = 200$ Hz.

In the simulation captions of Figure 19, the signals shown are: $i_L(t)$ in blue, $v_{cf}(t)$ in black, $i_{int}(t)$ in light green, and finally $v_{int}(t)$ in red. The currents are shown amplified by a factor 10. Current $i_{int}(t)$ is the adaptor input current, this is, the absolute value of the harvesting device current. Signal $v_{int}(t)$ is the absolute value of the harvesting device internal voltage drop, this is $v_{int}(t) = R_{int} \cdot i_{int}(t)$. If impedance matching occurs, then $v_{int}(t)$ must coincide with $v_{cf}(t)$, the adaptor input voltage.

The simulated cases are the following: Figure 19a HBB with only the boost mode enabled; Figure 19b HBB only enabling the buck mode; in cases Figure 19c,d the HBB adaptor can work with both modes, but the capacitor value is $C_f = 10$ μF in case Figure 19c, and $C_f = 200$ μF in case Figure 19d; and finally cases Figure 19e,f depict the Sepic converter, using $C_f = 10$ μF in Figure 19e and with $C_f = 200$ μF in Figure 19f.

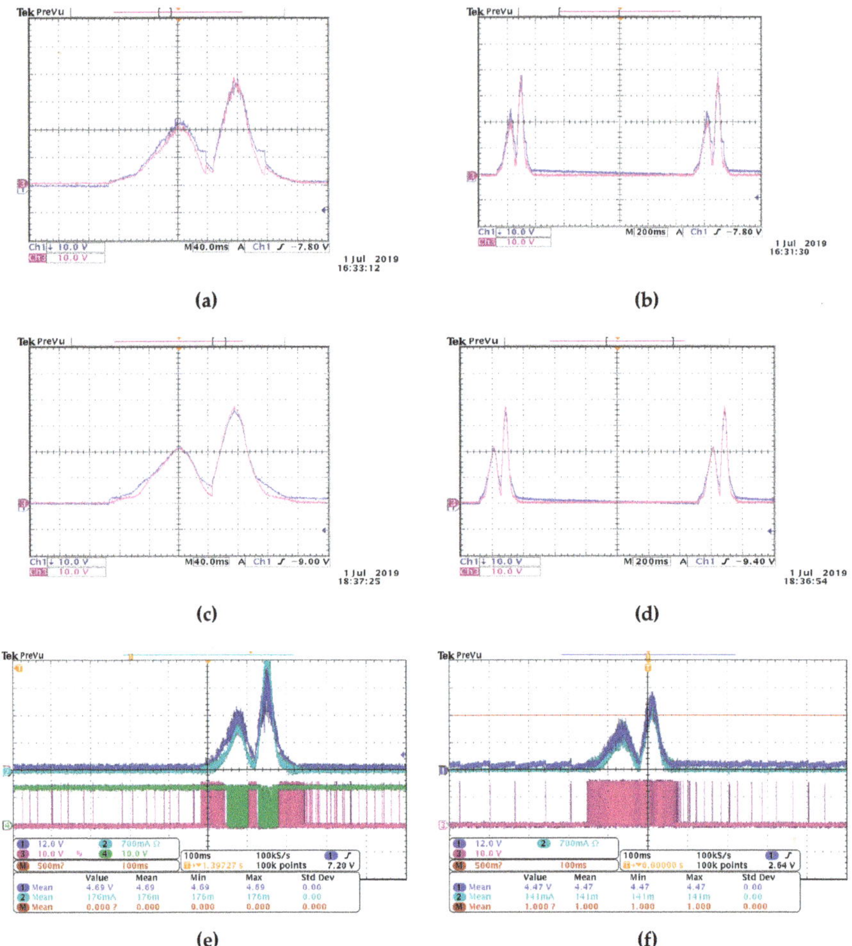

Figure 21. Experimental tests with pulse-train, $C_f = 10\ \mu F$: (**a**) HBB pulse detail, (**b**) HBB pulse-train, (**c**) Sepic pulse detail, (**d**) Sepic pulse-train, (**e**) HBB with control signals, and (**f**) Sepic with control signals.

As expected, and verified in Figure 19, in the Sepic converter, the adaptor input current $i_{int}(t)$ and the inductor current $i_{L1}(t)$ are equal, but in the HBB converter both currents only coincide in the boost mode, in the buck mode both currents are different. The effect of the C_f filter capacitor can also be observed. When its value is 200 μF, $v_{cf}(t)$ and $v_{int}(t)$ are different. There is a phase shift between them, and therefore the impedance matching is not perfect. Finally, in Figure 19a,b it can be appreciated that impedance matching occurs only when a converter is switching. In the HBB boost mode, the impedance matching only happens when $v_{cf}(t) < Vbat$, see Figure 19a, whereas in the buck mode, impedance matching takes place only when $v_{cf}(t) > Vbat$, as can be seen in Figure 19b.

Figure 20 depicts the simulated and experimental results obtained with a sinusoidal waveform for $v_p(t)$. Both converters have been tested at different frequencies from 50 Hz to 200 Hz. Realize that in a sea wave-train, waves are separated 10–20 seconds, implying a wave-train frequency of 0.05 Hz–0.1 Hz. In this way, a wave pulse can stand for 0.25–2.5 seconds, that corresponds to a range of fundamental frequencies of 0.4 Hz–4 Hz. To assure a good behavior with sea waves, the tests have been made at unrealistically high frequencies for the harvesting device. If the tests are satisfactory, the $v_{Cf}(t)$ tracking behavior of a real sea-wave pulse voltage $v_p(t)$ will deliver even better results.

The signals shown at Figure 20 are the following: The input adaptor voltage $v_{Cf}(t)$ in shown in dark blue, the inductor current $i_{L1}(t)$ in cyan color, the HBB buck mode switch gate signal $S_{buck}(t)$ is shown light green, and finally the Sepic switch, and the HBB boost mode switch gate signals $S_{sepic}(t)$ and $S_{boost}(t)$ are shown in pink.

The oscilloscope captions at Figure 20c,d corresponding to the real experimental test at 50 Hz show a good agreement with the corresponding simulated results, also at 50 Hz. The experimental results at 200 Hz, show that at his frequency the filtering effect of $C_f = 10~\mu F$ does not allow the adaptor input voltage $v_{Cf}(t)$ to drop at zero volts. Figure 20a shows clearly the dead-zone effect around $V_{bat} = 12$ V, where the HBB operating in boost mode changes to buck operation mode and vice-versa. Comparing the HBB and the Sepic converter experimental results, the adaptor input voltage $v_{Cf}(t)$ exhibit a small ripple in the Sepic case and in the HBB operation mode because the adaptor input current $i_{int}(t)$ is the inductor one $i_{L1}(t)$. Conversely, in the HBB buck mode, the adaptor input current $i_{L1}(t)$ is a pulsating one, and the ripple is higher.

Figure 21 depicts the experimental results using a laboratory programmable power supply that delivers a train of voltage pulses imitating a sea wave-train. Each voltage pulse reproduces the harvesting-device pulse given in Figure 8. This power supply supplies the impedance matching circuit through a series resistor of 26 Ω reproducing the harvesting device output impedance.

The captions in Figure 21a–d show two waveforms. The voltage-drop in the 26 Ω resistance $v_{int}(t)$ is shown in pink color, whereas the impedance matching circuit input voltage $v_{Cf}(t)$ is shown in dark blue color. As can be seen in these figures, there is a good agreement between both curves proving the correct behavior of the impedance matching circuit, for both converters: The HBB and the Sepic converter. The remaining captions, Figure 21e,f, show various signals from the HBB and the Sepic converters, respectively. The signals shown in these captions are: $v_{int}(t)$ in cyan color, $v_{Cf}(t)$ in dark blue, $S_{boost}(t)$ and $S_{sepic}(t)$ in pink color, and finally $S_{buck}(t)$ in light grey.

The input impedance Z_{in}, and the η_M matching efficiency (44) for both converters are depicted in Figure 22a–d. The data shown in these graphs have been obtained along with the sinusoidal experiments of Figure 20. Three different frequencies are used: 50 Hz (red), 100 Hz (blue), and 200 Hz (green). The HBB and the Sepic converter input impedances Z_{in} are plotted, respectively, in Figure 22a,b.

The impedance matching efficiency η_M (44) of both converters is shown Figure 22c,d. That matching efficiency has been defined as the ratio between the power absorbed by the converter, and the power absorbed by the converter in case of an ideal matching, when $Z_{in} = R_{int} = 26~\Omega$.

$$\eta_M = \frac{Z_{in} \cdot V_P^2}{(26 + Z_{in})^2} \cdot \frac{(26 + 26)^2}{26 \cdot V_P^2} = 104 \cdot \frac{Z_{in}}{(26 + Z_{in})^2} \tag{44}$$

This experimental section ends with two photographs given in Figure 23. Thus, Figure 23a shows the experimental workbench testing the HBB converter, whereas Figure 23b show the same environment with the Sepic converter.

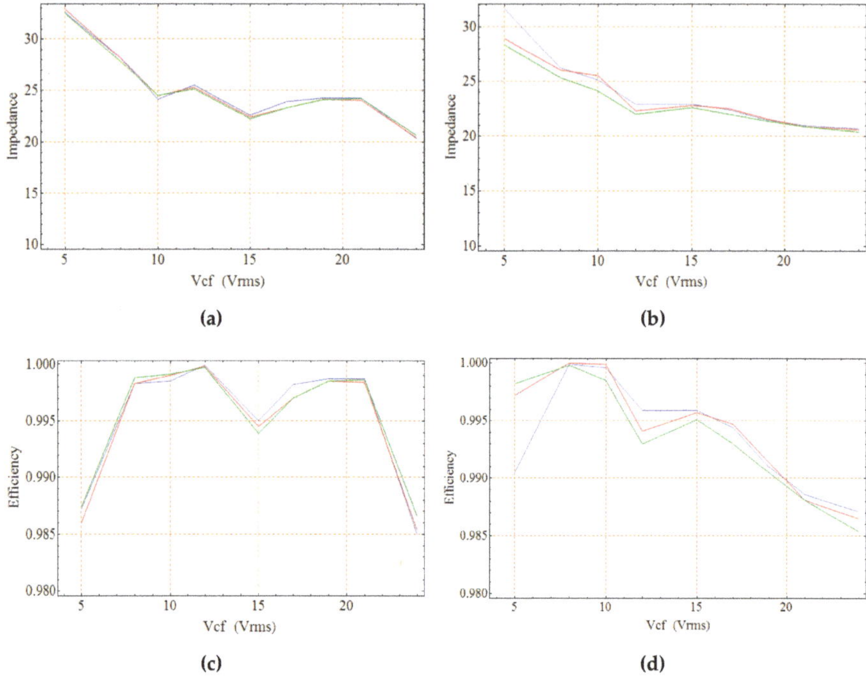

Figure 22. Experimental tests with sinusoidal signals, $C_f = 10$ µF: (**a**) HBB input impedance, (**b**) Sepic input impedance, (**c**) HBB matching efficiency, and (**d**) Sepic matching efficiency.

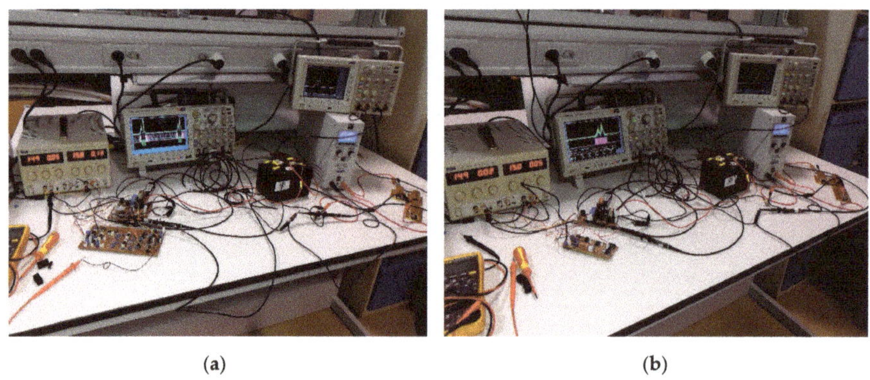

Figure 23. Experimental workbench during the test with the eperiment of Figure 22: (**a**) HBB adaptor and (**b**) Sepic adaptor.

8. Conclusions

Although extracting electric power from sea-waves at great scale is worldwide developing, the extraction of small amounts of energy from the waves, conveniently stored in a battery, can be solution to supply small stand-alone equipment, with occasional and short-lived power consumption peaks: Weather stations, telecom relays, and similar equipment.

The design, simulation, realization, and verification of an inductive harvesting generator that profits the differential position of two sea-surface points has been presented. This device supposes a novelty compared to sea-wave energy buoys and harvesting devices based on the vertical oscillation of a single sea point.

The proposed harvesting generator is made with three transducers, that are based on a coil-magnet arrangement, where a permanent magnet moves linearly inside the coil. The generator operation has also been simulated and verified experimentally.

Next, two switching converters, a hybrid buck-boost cell and a Sepic converter have been proposed, analyzed, and verified experimentally as matching circuits to transfer the energy from the harvesting device to the storage 12 V battery. Two different sliding mode control surfaces are required to guarantee the impedance matching using the HBB converter. Thus, during the boost mode the corresponding surface controls the input current, whereas in the buck mode, the surface controls the output current. Conversely, a clear advantage of the Sepic converter is that only a single surface is needed for the same purpose, because in this case the adaptor circuit input current is always the controllable inductor current. Indeed, the Sepic converter control law is the same that has been used for the boost mode operation in the HBB converter.

The main difference between both adaptor circuits is their behavior when the adaptor input voltage $v_{Cf}(t)$ coincides with the battery voltage. While the hybrid converter must change its operation mode, nothing happens with the Sepic converter. In the hybrid converter case, there is an input voltage dead-zone where the converter is not switching. Although the switching losses disappear, as the converter becomes uncontrolled, the impedance matching is lost during those instants.

Both converters behave reasonably well, and show good results matching the generator output impedance and the converter input impedance, as shown in the extensive experimental and simulation results. In both cases, as can be seen in Figure 23a,b there is a slight variation of the adaptor input resistance, due the current sensor lack of linearity working in a wide range of current values, the variable delay caused by the mosfet drivers, and finally the converter efficiency, which is not obviously constant along the input voltage range. Anyway, the matching efficiency is always over 95% whatever be the input voltage, as can be seen in Figure 22c,d.

To reduce the input impedance variation in terms of the input voltage, the hysteresis width used to implement the sliding surfaces by means of hysteretic comparators must be reduced. Nevertheless, although this would increase the tracking precision, leading to a slight increment of the matching efficiency, maybe the switching losses would be greatly increased leading to a global reduction of the energy delivered to the storage battery.

In this paper two switching converters operating as impedance matching circuits have been studied. To achieve this, they must be forced to behave like a loss free resistor, and here sliding mode control is introduced as an easy technique to assure this behavior in a wide range of input and output voltage variations. Buck-boost and similar converters in DCM operation are used in the literature for this proposal, in harvesting applications, but the input voltage variation range is usually small. The solution proposed here allows impedance matching and loss free resistor behavior with harvesting generators with a wide input voltage variation range. The simplicity of the sliding mode control technique, its inherent robustness against parametric variation, and input and output disturbances, are the main advantages of the technique proposed here.

These issues, as other related to a different configuration of the harvesting device coils will be the subject of a future research.

Author Contributions: Conceptualization, J.A.G.-C., and H.V.-B.; methodology, H.V.-B., J.A.G.-C., J.A.B.-R. and À.C.-P.; validation, J.A.G.-C.; formal analysis, H.V.-B. and J.A.G.-C.; investigation, H.V.-B., J.A.G.-C. and J.A.B.-R.; resources, H.V.-B.; writing—original draft preparation, J.A.G.-C., and H.V.-B.; writing—review and editing, H.V.-B., J.A.G.-C., J.A.B.-R, and À.C.-P.; visualization, J.A.G.-C. and À.C.-P.; supervision, H.V.-B.; project administration, H.V.-B.; funding acquisition, H.V.-B.

Funding: This research was funded by the Spanish Agencia Estatal de Investigación (AEI) and the Fondo Europeo de Desarrollo Regional (FEDER) under research projects DPI2015-67292-R (AEI/FEDER, UE).

Conflicts of Interest: The authors declare no conflict of interest.

References

1. Falcao, A.F. Wave energy utilization: A review of the technologies. *Renew. Sustain. Energy Rev.* **2010**, *14*, 900–915. [CrossRef]
2. Faiz, J.; Nematsaberi, A. Linear electrical generator topologies for direct-drive marine wave energy conversion—An overview. *IET Renew. Power Gener.* **2017**, *11*, 1163–1176. [CrossRef]
3. Phillips, O.M. *The Dynamics of the Upper Ocean*; Cambridge University Press: Cambridge, UK, 1977; ISBN 0-521-29801-6.
4. Herbich, J.B. *Handbook of Coastal Engineering*; McGraw-Hill: New York, NY, USA, 2000; ISBN 978-0-07-134402-9.
5. Bou-Mosleh, C.; Rahme, P.; Beaino, P.; Mattar, R.; Nassif, E.A. Contribution to Clean Energy Production using a Novel Wave Energy Converter. In Proceedings of the 2nd Renewable Energy for Developing Countries—REDEC 2014, Beirut, Lebanon, 26–27 November 2014; pp. 108–111.
6. Karayaka, H.; Mahlke, H.; Bogucki, D.; Mehrubeoglu, M. A Rotational Wave Energy Conversion System Development and Validation with Real Ocean Wave Data. In Proceedings of the 2011 IEEE Power and Energy Society General Meeting, Detroit, MI, USA, 24–28 July 2011.
7. Liang, H.; Zhang, D.; Yang, J.; Tan, M.; Huang, C.; Wang, J.; Chen, Y.; Xu, C.; Sun, K. Hydrodynamic research of a novel floating type pendulum wave energy converter based on simulations and experiments. In Proceedings of the OCEANS 2016, Shanghai, China, 10–13 April 2016.
8. Tasneem, N.; Suri, S.; Mahbub, I. A low-power CMOS voltage boosting rectifier for wireless power transfer applications. In Proceedings of the 2018 Texas Symposium on Wireless and Microwave Circuits and Systems (WMCS), Waco, TX, USA, 5–6 April 2018.
9. Belal, E.; Mostafa, H.; Ismail, Y.; Said, M. A Voltage Multiplying AC/DC Converter for Energy Harvesting Applications. In Proceedings of the 2016 28th International Conference on Microelectronics (ICM), Giza, Egypt, 17–20 December 2016; pp. 229–232.
10. Cheng, S.; Jin, Y.; Rao, Y.; Arnold, D. An Active Voltage Doubling AC/DC Converter for Low-Voltage Energy Harvesting Applications. *IEEE Trans. Power Electron.* **2011**, *26*, 2258–2265. [CrossRef]
11. Haoyu, W.; Yichao, T.; Khaligh, A. A Bridgeless Boost Rectifier for Low-Voltage Energy Harvesting Applications. *IEEE Trans. Power Electron.* **2013**, *28*, 5206–5214.
12. Sun, Q.; Patil, S.; Stoute, S.; Sun, N.; Lehman, B. Optimum design of magnetic inductive energy harvester and its AC-DC converter. In Proceedings of the 2012 IEEE Energy Conversion Congress and Exposition (ECCE), Raleigh, NC, USA, 15–20 September 2012; pp. 394–400.
13. Lefeuvre, E.; Audigier, D.; Richard, C.; Guyomar, D. Buck-Boost Converter for Sensorless Power Optimization of Piezoelectric Energy Harvester. *IEEE Trans. Power Electron.* **2007**, *22*, 2018–2025. [CrossRef]
14. Guo, T.; Lerley, R.; Ha, D.S. Development of a power conditioning circuit for railcar energy harvesting. In Proceedings of the MWSCAS—The 56th international Midwest Symposium on Circuits and Systems, Columbus, OH, USA, 4–7 August 2013; pp. 513–516.
15. Chen, N.; Wei, T.; Dong, H.; Jung, H.J.; Lee, S. Alternating Resistive Impedance Matching for an Impact-Type Micro-Wind Piezoelectric Energy Harvester. *IEEE Trans. Ind. Electron.* **2018**, *65*, 7374–7382. [CrossRef]
16. Sun, Q.; Patil, S.; Sun, N.; Lehman, B. Phase/RMS Maximum Power Point Tracking for Inductive Energy Harvesting System. In Proceedings of the 2015 IEEE Energy Conversion Congress and Exposition (ECCE), Montreal, QC, Canada, 20–24 September 2015; pp. 408–413.
17. Heo, S.; Yang, Y.; Lee, J.; Lee, S.; Kim, J. Micro Energy Management for Energy Harvesting at Maximum Power Point. In Proceedings of the 2011 International Symposium on Integrated Circuits, Singapore, 12–14 December 2011; pp. 136–139.
18. Shousha, M.; Dinulovic, D.; Haug, M. A universal topology based on buck-boost converter with optimal resistive impedance tracking for energy harvesters in battery powered applications. In Proceedings of the 2017 IEEE Applied Power Electronics Conference and Exposition (APEC), Tampa, FL, USA, 26–30 March 2017; p. 2111.

19. Cid-Pastor, A.; Martínez-Salamero, L.; el Aroudi, A.; Giral, R.; Calvente, L.; Leyva, R. Synthesis of Loss-Free Resistors based on sliding-mode control and its applications in power processing. *Control Eng. Pract.* **2013**, *21*, 689–699. [CrossRef]
20. Griffith, D. *Introduction to Electrodynamics*, 3rd ed.; Pearson: London, UK, 2013; ISBN 978-0-321-85656-2.
21. Singer, S.; Ozeri, S.; Shmilovitz, D. A Pure Realization of Loss-Free Resistor. *IEEE Trans. Circuits Syst. I Regul. Pap.* **2004**, *51*, 1639–1647. [CrossRef]

Article

Control of Output and Circulating Current of Modular Multilevel Converter Using a Sliding Mode Approach

Waqar Uddin [1], Kamran Zeb [1,2], Muhammad Adil Khan [3], Muhammad Ishfaq [1], Imran Khan [4], Saif ul Islam [1], Hee-Je Kim [1,*], Gwan Soo Park [1] and Cheewoo Lee [1]

[1] School of Electrical and Computer Engineering, Pusan National University, Busan 46241, Korea; waqudn@pusan.ac.kr (W.U.); kami_zeb@yahoo.com (K.Z.); engrishfaq1994@gmail.com (M.I.); shaheen_575@yahoo.com (S.u.I.); gspark@pusan.ac.kr (G.S.P.); cwlee1014@pusan.ac.kr (C.L.)

[2] Department of Electrical Engineering, National University of Science and Technology, Islamabad 44000, Pakistan

[3] Department of Electrical and Computer Science, Air University, Islamabad 44000, Pakistan; adil.khan@mail.au.edu.pk

[4] C2N, University of Paris Sud, University of Paris Saclay, 10 Boulevard Thomas Gobert, 91120 Palaiseau, France; imran.khan@c2n.upsaclay.fr

* Correspondence: heeje@pusan.ac.kr; Tel.: +82-10-3492-9677

Received: 27 September 2019; Accepted: 24 October 2019; Published: 25 October 2019

Abstract: The modular multilevel converter (MMC) has been prominently used in medium- and high-power applications. This paper presents the control of output and circulating current of MMC using sliding mode control (SMC). The design of the proposed controller and the relation between control parameters and validity condition are based on the system dynamics. The proposed designed controller enables the system to track its reference values. The controller is designed to control both output current and circulating current along with suppression of second harmonics contents in circulating current. Furthermore, the capacitor voltage and energy of the converter are also regulated. The control of output current is carried out in dq-axis as well as in $\alpha\beta - axis$ with first-order switching law. However, a second-order switching law-based super twisting algorithm is used for controlling circulating current and suppression of its second harmonics contents. The stability of the controlled system is numerically calculated and verified by Lyapunov stability conditions. Moreover, the simulation results of the proposed controller are critically compared with the conventional proportional resonant (PR) controller to verify the effectiveness of the proposed control strategy. The proposed controller attains faster dynamic response and minimizes steady-state error comparatively. The simulation of the MMC model is carried out in MATLAB/Simulink.

Keywords: modular multilevel converter; sliding mode control; Lyapunov stability

1. Introduction

The modular multilevel converter has been widely used in medium and high-voltage applications [1,2], integration of renewable energy [3,4], medium-voltage drives [5,6], and battery energy management system [7,8] due to its versatile and promising features i.e., modularity, redundancy, excellent harmonic performances, and transformerless configuration [1,9–11]. The structure of modular multilevel converter (MMC) is composed of different modules connected in different configuration i.e., half-bridge and full-bridge, as depicted in Figure 1.

The control of MMC is categorized into internal current control and output current control. Output current control is used to control the active and reactive power of converters [9,12]. Along with the control of output power, the internal dynamic of MMC i.e., circulating current and submodule capacitor voltage, is also in focus. In order to achieve multiple objectives i.e., energy balancing, circulating current control, and output current control, various control schemes have been introduced in the literature.

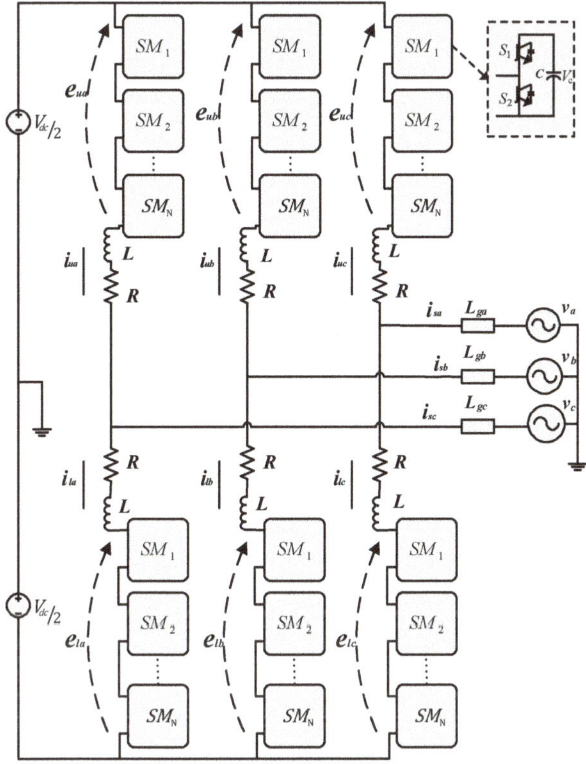

Figure 1. 3 Phase circuit diagram of modular multilevel converter (MMC).

The control scheme introduced by Hagiwara and Akagi [13] is mostly adopted in the literature. The circulating current control is used to suppress the harmonic contents produced due to capacitor ripples. The magnitude of circulating current has a high impact on the arm current. It increases the root mean square (RMS) value of arm current, which will increase power losses and second-order harmonics will result in further generation of other higher-order harmonics that will increase power loss and current stresses on switching devices [14,15]. However, no impact of circulating current is seen on the alternating current (AC) network.

A different current control method is proposed, such as $dq - axis$ control using proportional-integral controller [16,17]; the output current is controlled in $dq - axis$ with grid frequency while the control of circulating current is carried out in double frequency $dq - axis$ frame. But the controller fails to completely suppress harmonics in circulating current due to limited gain at harmonic frequencies. This will result in the insufficient performance of MMC in a steady state. PR controller is also proposed in [18,19] to control the circulating current. However, the design and tuning of multi-resonant control is difficult and complicated. Moreover, any small deviation in frequency will lead to a larger deviation of the control variable from its reference. In [20–23], the authors proposed a harmonic repetitive controller and plug-in repetitive controller for suppression of second harmonic in circulating current, but achieving a stable operation and excellent control performance may be a trivial task. Furthermore, these controllers are tuned at a specific frequency, and a small variation in frequency can lead to the failure of the controller. A combination of spatial comb and the spatial repetitive controller is proposed in [24] for suppression of second harmonic current and controller of output current, but the author has not discussed the balancing of capacitor voltage and energy of converter. A sliding mode controller is proposed in [25] to control the current of the converter, but the author does not consider the store energy and energy difference between arms. Moreover, the circulating current is controlled in a double-frequency $dq-axis$ frame. Interconversion of different frames will cause a computation burden for a large system. In [26], sliding mode control is proposed only for controlling the output current, although the circulating current and energy balance of the converter is not considered. Moreover, the circulating current is controlled in the dq-axis frame. Hence, interconversion of the different frames will increase the computational burden. The arm inductor also has a prominent effect on the circulating current studied in [27], but the large value of inductor will result in bulkiness of the converter. A backstepping controller is introduced in [28] to eradicate the harmonic content of circulating current. Although the amplitude of second harmonic current is reduced, it is not eliminated completely. A feedback linearization [29], model predictive approach [30–32], and langrage optimization-based control [33] was proposed to control parameters of MMC but the computational complexity, a well-developed mathematical model, and variable frequency operation limit the effectiveness of these strategies.

In order to cope with the problem and complexities of different control strategies, SMC [34] is proposed for controlling output current, circulating current, capacitor voltage, and energy balance of MMC. The indirect SMC control can be easily adapted for power converters. The remarkable characteristic of SMC is its easy and simple implementation and tuning. Moreover, the response of a closed-loop system becomes insensitive to some uncertainties due to the use of SMC. This principle will cope with the intrinsic model variations. Furthermore, the implementation of circulating current control in ABC frame will reduce the computation burden of different transformation ($T_{abc \Leftrightarrow dq}$) needed during desigthe n of control schemes. The two terms of control law i.e., equivalent term and attraction term, will combinedly move the system trajectory to the sliding surface and keep it steered on it. The condition of attraction is satisfied by choosing a proper derivative. A first-order SMC is designed for output current control in $dq-axis$. However, a second-order SMC based on the super-twisting algorithm is designed to suppress the second-order harmonic current and steer the value of circulating current on the DC reference.

The paper is organized as follows. Section 2 describes the modeling and operation of MMC. Control structure design and stability analysis is represented in Section 3. Section 4 contains the results and discussion, while the paper is concluded in Section 5.

2. Modeling of MMC

The equivalent single-phase circuit of MMC is depicted in Figure 2. The different currents flowing in arms of MMC are defined as [35]

$$i_{u,j} = i_{c,j} + \frac{i_{o,j}}{2} \tag{1}$$

$$i_{l,j} = i_{c,j} - \frac{i_{o,j}}{2} \tag{2}$$

$$i_{o,j} = i_{u,j} - i_{l,j} \tag{3}$$

$$i_{c,j} = \frac{1}{2}(i_{u,j} + i_{l,j}). \tag{4}$$

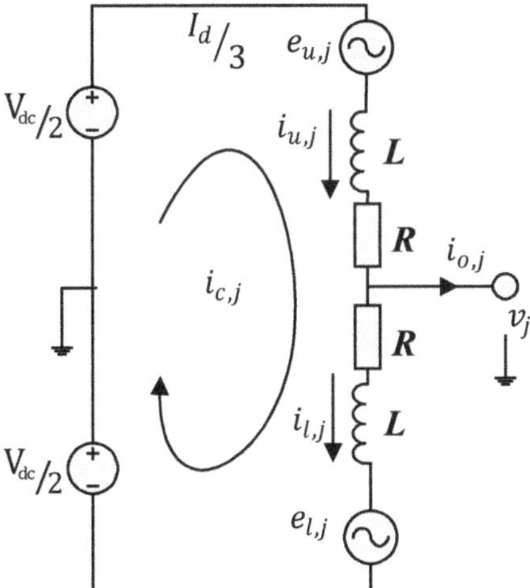

Figure 2. Phase equivalent circuit of MMC.

By applying Kirchhoff's voltage law on the upper and lower arm of the circuit in Figure 2, we get the following voltage equations

$$-\frac{1}{2}V_{dc} + e_{u,j} + i_{u,j}R + L\frac{di_{u,j}}{dt} + v_j = 0, \tag{5}$$

$$-\frac{1}{2}V_{dc} + e_{l,j} + i_{l,j}R + L\frac{di_{l,j}}{dt} - v_j = 0. \tag{6}$$

Adding and subtracting Equations (5) and (6), along with the substitution of Equations (3) and (4), will result in a new set of equations given as

$$\frac{L}{2}\frac{di_{o,j}}{dt} = \underbrace{\left(\frac{-e_{u,j} + e_{l,j}}{2}\right)}_{v_{s,j}} - \frac{R}{2}i_{o,j} - v_j, \tag{7}$$

$$L\frac{di_{c,j}}{dt} = -\underbrace{\left(\frac{e_{u,j} + e_{l,j}}{2}\right)}_{v_{c,j}} - Ri_{c,j} + \frac{V_{dc}}{2}. \tag{8}$$

Equations (7) and (8) characterize the dynamics of MMC. By analyzing Equation (7), it is concluded that as v_j is AC bus voltage, only $v_{s,j}$ can be manipulated to control outputhe t current $i_{o,j}$. Similarly, Equation (8) is used to control the circulating current dynamics of MMC. Likewise, in Equation (8), $i_{c,j}$ can be controlled by manipulation of the internal voltage $v_{c,j}$. Moreover, the internal dynamics of the SM capacitor also has a great effect on the circulating current. It is represented in terms of arm voltages and arm currents as

$$\frac{C}{N}\frac{dv^{\Sigma}_{cu,j}}{dt} = n_{u,j}i_{u,j}, \tag{9}$$

$$\frac{C}{N}\frac{dv^{\Sigma}_{cl,j}}{dt} = n_{l,j}i_{l,j}. \tag{10}$$

$(n^i_{u,j}, n^i_{u,j}) = 1$ means the capacitor is inserted while $(n^i_{u,j}, n^i_{u,j}) = 0$ means the capacitor is bypassed in respective arms. Equations (8)–(10) combinedly represent the internal dynamic of MMC.

Equations (1) and (2) are substituted in Equations (9) and (10) to represent capacitor dynamics in terms of output current, circulating current, and insertion indices (n_u, n_l).

$$\frac{1}{N}\frac{Cdv^{\Sigma}_{cu,j}}{dt} = n_u\left(i_{c,j} + \frac{i_{o,j}}{2}\right),$$ (11)

$$\frac{1}{N}\frac{Cdv^{\Sigma}_{cl,j}}{dt} = n_l\left(i_{c,j} - \frac{i_{o,j}}{2}\right),$$ (12)

While $n_u = \frac{v^*_c - v^*_s}{v^{\Sigma}_{cu}}$ and $n_l = \frac{v^*_c + v^*_s}{v^{\Sigma}_{cl}}$, v^*_c and v^*_s are the controlled inputs to the plant. The internal dynamic of MMC is controlled through the convergence of $i_{c,j}$ to a DC component $\frac{P}{MV_{dc}}$. In control of MMC, n_u and n_l are available for manipulation. Also, v^{Σ}_{cu} and v^{Σ}_{cl} are forced to converge to V_{dc} for controlling the internal dynamics of MMC. Equations (1)–(12) are modeled in MATLAB/Simulink as depicted in Figure 3.

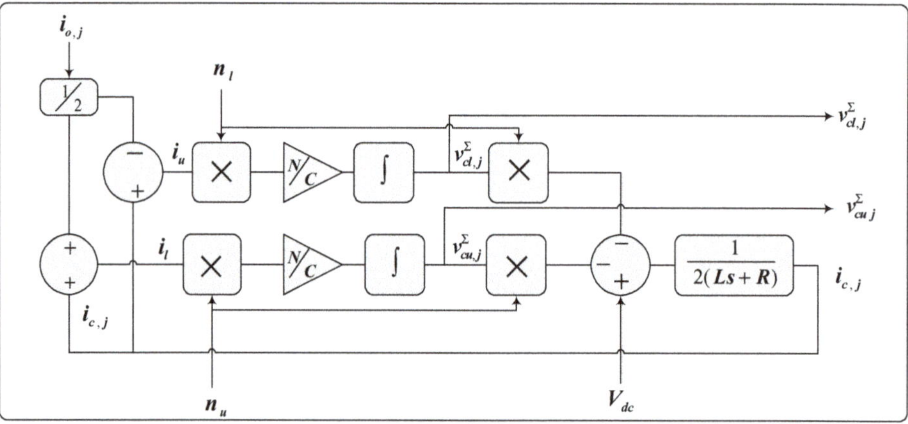

Figure 3. Simulink model.

3. Control Structure of MMC

3.1. SMC Design Overview

SMC has the remarkable properties of easy tuning, robustness, accuracy, and easy implementation. These features allow for the use of SMC in various domains for controlling nonlinear processes. Hence, SMC is used in control of different applications such as electrical drives [36–38], power converter converters [39], microgrid control [40,41], wind energy control [42], and many more.

The SMC system is designed to steer the states of a system onto a specific surface in state space, called the sliding surface. The two conditions, invariance and attractivity, made the trajectory steer on the sliding surface. The conditions are given as

$$\begin{cases} \dot{S}_s(x) = 0 & if\ S_s(x) = 0 \\ \dot{S}_s(x) < 0 & if\ S_s(x) > 0 \\ \dot{S}_s(x) > 0 & if\ S_s(x) < 0 \end{cases}.$$ (13)

The condition given in Equation (13) defines the control law for a given system. The control law is composed of both invariance and attractive condition. The mathematical expression for control law is given as

$$u^* = u^e + u^{att}. \tag{14}$$

u^* represent control input to the plant. u^e, u^{att} represents invariance and attractive terms respectively. The two terms in control law have distinct features.

1. The first term is equivalent control, which keeps the system trajectory on a sliding surface.
2. The second term is used for attractivity; it brings the system from outside to the sliding surface. Usually, it determines system dynamics from initial point to sliding surface.

Both the above conditions are graphically illustrated in Figure 4.

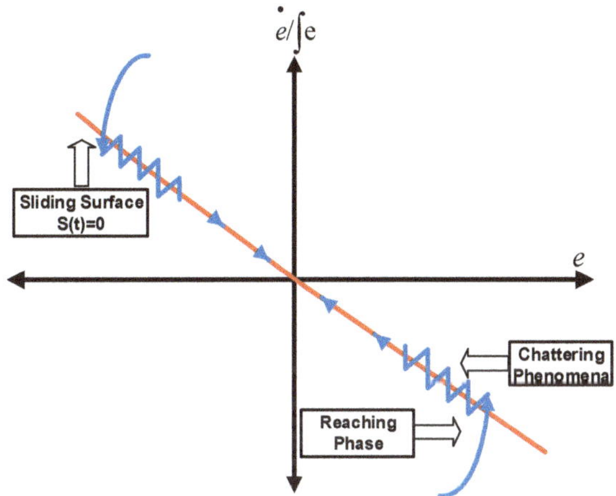

Figure 4. Sliding surface.

Consider a nonlinear system described as in [43] for designing of control law

$$\dot{x} = f(x) + g(x)u, \tag{15}$$

where $x \in R^n$ is the system state vector and $f(x)$ and $g(x)$ are a nonlinear function of the system state vector (x) and input vector (u). The switching function for a system is defined as

$$S_s(x) = C^T x = \sum_{i=1}^{n-1} c_i x_i + x_n, \tag{16}$$

where $C^T = [c_1, c_2 \dots c_{n-1}, 1]$ are the coefficients of the sliding surface. The derivative of the sliding surface is given as

$$\dot{S}_s(x) = C^T \dot{x}. \tag{17}$$

Substitute the value of $\dot{x}$ in Equation (17)

$$\dot{S}_s(x) = C^T f(x) + C^T g(x)u, \tag{18}$$

by assuming the reaching law

$$\dot{S}_s(x) = -Qsgn(S_s(x)) - KS_s(x), \tag{19}$$

while Q and K are positive real numbers. Combining Equations (15)–(19) we get control law as

$$u = -(C^T g(x))^{-1} C^T f(x) - (C^T g(x)t)^{-1} Q(sgn(S(x)) - KS(x)). \tag{20}$$

Equation (20) gives the control input to the plant. By choosing a proper value for Q and K, the system output parameters will follow the desired performance.

SMC is very effective in the control domain, but an undesirable chattering problem is associated with it, as depicted in Figure 4. This should be eliminated as it gives rise to oscillations and appears in the output. Different techniques are used to solves chattering issues. In our control scheme, we use the saturation function for output current loop as described in [40,41], while the super twisting algorithm is used for control of circulating current to cope with this issue effectively. The second-order algorithm makes the law continuous; hence, it handles the chattering issue in a very efficient way.

3.2. Output Current Control

The dynamics of the output current are shown in Equation (7). As it is clear from the equation that $v_{s,j}$ is available for manipulation to control the output current, Equation (7) can be also be presented in the below form

$$\begin{bmatrix} v_{s,a} \\ v_{s,b} \\ v_{s,c} \end{bmatrix} = \frac{L}{2} \begin{bmatrix} \frac{di_{o,a}}{dt} \\ \frac{di_{o,b}}{dt} \\ \frac{di_{o,c}}{dt} \end{bmatrix} + \frac{R}{2} \begin{bmatrix} i_{o,a} \\ i_{o,b} \\ i_{o,c} \end{bmatrix} + \begin{bmatrix} v_a \\ v_b \\ v_c \end{bmatrix}. \tag{21}$$

Equation (21) is transformed from *abc* to *dqo* using Park's transformation ($T_{abc \Rightarrow dq}$) [16].

$$T_{abc \Rightarrow dq} = \frac{2}{3} \begin{bmatrix} \cos \omega_s t & \cos\left(\omega_s t - \frac{2\pi}{3}\right) & \cos\left(\omega_s t + \frac{2\pi}{3}\right) \\ -\sin \omega_s t & -\sin(\omega_s t - \frac{2\pi}{3}) & -\sin(\omega_s t + \frac{2\pi}{3}) \\ \frac{1}{2} & \frac{1}{2} & \frac{1}{2} \end{bmatrix} \tag{22}$$

For balanced three-phase system, Equation (21) can be represented in $dq - axis$ as

$$\begin{bmatrix} v_{ds} \\ v_{qs} \end{bmatrix} = \frac{L}{2} \begin{bmatrix} \frac{di_{d,o}}{dt} \\ \frac{di_{q,o}}{dt} \end{bmatrix} + \begin{bmatrix} \frac{R}{2} & -\omega_s L \\ \omega_s L & \frac{R}{2} \end{bmatrix} \begin{bmatrix} i_{d,o} \\ i_{q,o} \end{bmatrix} + \begin{bmatrix} v_d \\ v_q \end{bmatrix}. \tag{23}$$

Equation (23) is used to formulate the control law. The control law is formulated by considering both invariance and attraction condition as

$$v_{ds}^* = v_{ds}^e + v_{ds}^{att}, \tag{24}$$

$$v_{qs}^* = v_{qs}^e + v_{qs}^{att}. \tag{25}$$

The first terms in Equations (24) and (25) represent the equivalent voltage vector. The equivalent voltage vector is active during steady-state, and it will steer the system trajectory onto the sliding manifold, while later terms in Equations (24) and (25) represent attractive voltage that is active in the transient state. The attractive term forces the system trajectory to the sliding surface. The error for both d-axis and q-axis current is defined as

$$e = i_{dq}^* - i_{dq}. \tag{26}$$

The relative degree of the system is $r = 1$. Hence, the sliding surfaces selected for I_{ds} and I_{qs} are as follows

$$S_{sd} = i_{d,o}^* - i_{d,o} = 0 \tag{27}$$

$$S_{sq} = i_{q,o}^* - i_{q,o} = 0. \tag{28}$$

The v_{ds}^e and v_{qs}^e terms that are calculated from Equations (27) and (28) given as

$$\dot{S}_{sd} = -\frac{di_{d,o}}{dt} = -\frac{1}{L}\left(2v_{ds}^e - Ri_{d,o} + \omega_s Li_{q,o} - 2v_d\right) = 0 \tag{29}$$

$$\dot{S}_{sq} = -\frac{di_{q,o}}{dt} = -\frac{1}{L}\left(2v_{qs}^e - Ri_{q,o} - \omega_s Li_{d,o} - 2v_q\right) = 0. \tag{30}$$

By simplification of Equations (29) and (30), we can get an equation for the invariance condition of the design controller.

$$v_{ds}^e = v_d + \frac{1}{2}\left(Ri_{d,o} - \omega_s Li_{q,o}\right) \tag{31}$$

$$v_{qs}^e = v_q + \frac{1}{2}\left(Ri_{q,o} + \omega_s Li_{d,o}\right) \tag{32}$$

From Equations (26) and (27), $i_{d,o}^* = i_{do}^m$ and $i_{q,o}^* = i_{q,o}^m$ replace the value in Equations (29) and (30), and solve for the value v_{ds}^* and v_{qs}^* given as

$$v_{ds}^* = \underbrace{v_d + \frac{1}{2}\left(Ri_{d,o}^* - \omega_s Li_{q,o}^*\right)}_{v_{ds}^e} - \underbrace{\frac{L}{2}\frac{dS_{sd}}{dt}}_{v_{ds}^{att}} \tag{33}$$

$$v_{qs}^* = \underbrace{v_q + \frac{1}{2}\left(Ri_{q,o}^* + \omega_s Li_{d,o}^*\right)}_{v_{qs}^e} - \underbrace{\frac{L}{2}\frac{dS_{sq}}{dt}}_{v_{qs}^{att}}. \tag{34}$$

The switching functions S_{sd} and S_{sq} used in control law are given as

$$v_{ds}^{att} = -\frac{L}{2}\frac{dS_{sd}}{dt} = -\frac{L}{2}\left(-Q_d sgn(S_{sd}) - K_d S_{sd}\right) \tag{35}$$

$$v_{qs}^{att} = -\frac{L}{2}\frac{dS_{sq}}{dt} = -\frac{L}{2}\left(-Q_q sgn(S_{sq}) - K_q S_{sq}\right). \tag{36}$$

The $-K_d S_{sd}$ and $-K_q S_{sq}$ terms force the state trajectory to approach the sliding surface. The rise time will reduce with a larger value of $K_{d,q}$ while a small value of $Q_{d,q}$ will minimize the oscillation. The fully controlled law is given as

$$\begin{bmatrix} v_{ds}^* \\ v_{qs}^* \end{bmatrix} = \begin{bmatrix} v_{ds}^e \\ v_{qs}^e \end{bmatrix} + \begin{bmatrix} v_{ds}^{att} \\ v_{qs}^{att} \end{bmatrix} \tag{37}$$

while

$$\begin{bmatrix} v_{ds}^e \\ v_{qs}^e \end{bmatrix} = \begin{bmatrix} \frac{R}{2} & -\omega_s L \\ \omega_s L & \frac{R}{2} \end{bmatrix} \begin{bmatrix} i_{d,o}^* \\ i_{q,o}^* \end{bmatrix} + 2 \begin{bmatrix} v_d \\ v_q \end{bmatrix} \tag{38}$$

$$\begin{bmatrix} v_{ds}^{att} \\ v_{qs}^{att} \end{bmatrix} = L\left(\begin{bmatrix} Q_d & 0 \\ 0 & Q_q \end{bmatrix} \begin{bmatrix} sgn(S_{sd}) \\ sgn(S_{sq}) \end{bmatrix} + \begin{bmatrix} K_d & 0 \\ 0 & K_q \end{bmatrix} \begin{bmatrix} S_{sd} \\ S_{sq} \end{bmatrix}\right). \tag{39}$$

The implemented control structure for output current control is depicted in Figure 5.

Theorem 1. *The proposed designed control scheme based on SMC is asymptotically stable and ensures boundedness of tracking error of output current if the reference dq-axis voltage v_{dqs}^* is chosen as in Equation (37).*

Proof of Theorem 1. The stability of the system can be checked using Lyapunov stability analysis [43].

A Lyapunov function is defined as

$$V = \frac{1}{2}S_s^2.$$ (40)

In order to ensure the stability of Equations (33) and (34), the following two conditions need to be satisfied

$$\dot{V} < 0 \text{ for } S_s \neq 0$$
$$\lim_{|S|\to\infty} V = \infty.$$ (41)

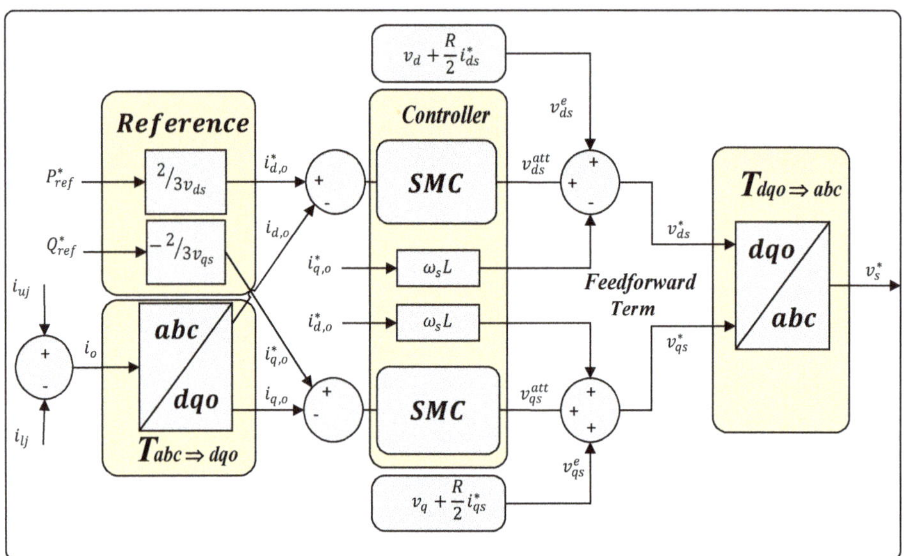

Figure 5. Output current control structure.

In order to check the first condition of Lyapunov stability, the derivative of Lyapunov is given as

$$\dot{V} = S_{sdq}\dot{S}_{sdq}.$$ (42)

The value of $\dot{S}_{sdq}$ is substituted from Equations (29) and (30)

$$S_{sdq}\dot{S}_{sdq} = -S_{sdq}\frac{1}{L}(2v_{dqs}^* - 2v_{dq} - Ri_{dq,o} \pm \omega_s Li_{qd,o}).$$ (43)

Similarly, from Equations (33) and (34)

$$v_{dqs}^* = v_{dq} + \frac{1}{2}(Ri_{dq,o}^* \mp \omega_s Li_{qd,o}^*) - \frac{L}{2}\dot{S}_{sdq}.$$ (44)

The value of v_{dqs}^* is substituted into Equation (43). The equation is modified as

$$S_{sdq}\dot{S}_{sdq} = -\frac{S_{sdq}}{L}\left(R(i_{dq,o}^* - i_{dq,o}) \mp \omega_s L(i_{qd,o}^* - i_{qd,o}) - \frac{L}{2}\dot{S}_{sdq}\right).$$ (45)

Substituting the value of Equation s(27) and (28) and Equations (35) and (36)

$$S_{sdq}\dot{S}_{sdq} = -\frac{R}{L}S_{sdq}^2 \pm \omega_s S_{sd}S_{sq} - \frac{1}{2}Q_{dq}S_{sdq}sgn(S_{sdq}) - \frac{1}{2}K_d S_{sdq}^2.$$ (46)

This equation for q-axis current returns a negative value regardless of the sign of sliding surface S_{sd} and S_{sq} that ensures the stability of the system. However, for d-axis, $\omega_s S_{sq} < \frac{R}{L} S_{sd} + \frac{S_{sd}}{2}(Q_d sgn(S_{sd}) + K_d)$ Equation (44) shows that for any value of switching function, the derivative of the Lyapunov equation is negative. Similarly, the second condition of stability is also fulfilled. $\square$

3.3. Circulating Current Control

The circulating current is generated due to the inner voltage imbalances between the arms. It flows in each phase leg of the converter. Circulating current is in the form of a negative sequence current with double grid frequency [16]. It does not affect the output current or voltage, except it increases the losses of converter. The circulating current has two parts is given as

$$i_{c,j} = \frac{I_{dc}}{3} + I_{2h,j}. \tag{47}$$

The value of $i_{c,j}$ is forced to follow DC current $(\frac{I_{dc}}{3})$, which will result in the suppression of second harmonic current. The switching law for the circulating current control is composed of equivalent voltage and attractive voltage.

$$v_{c,j}^* = v_{c,j}^e + v_{c,j}^{att} \tag{48}$$

Similarly, the switching function for circulating current control is given as

$$S_{sc} = i_{c,j}^* - i_{c,j} = 0. \tag{49}$$

By taking the derivative of Equation (49) and substituting the value of $\frac{di_{c,j}}{dt}$ from Equation (8), we get

$$\frac{dS_{sc}}{dt} = -\frac{di_{c,j}}{dt} = -\frac{1}{L}\left(-v_{c,j}^e - Ri_{c,j} + \frac{V_{dc}}{2}\right) = 0. \tag{50}$$

By solving Equation (50) for equivalent voltage '$v_{c,j}^e$',

$$v_{c,j}^e = -Ri_{c,j} + \frac{V_{dc}}{2}. \tag{51}$$

Substituting Equationa (51) and (49) in Equation (48), we get

$$v_{c,j}^* = \underbrace{-Ri_{c,j}^* + \frac{V_{dc}}{2}}_{v_{c,j}^e} + \underbrace{L\frac{dS_{sc}}{dt}}_{v_{c,j}^{att}}. \tag{52}$$

In order to converge switching function to zero, the super-twisting algorithm is used for circulating current. Due to its continuous nature and integral term, it will compensate for the high-frequency disturbances i.e., second harmonic current. It will retain the continuity of function along with attenuation of disturbance. The super-twisting control algorithm is given as

$$v_{c,j}^{att} = L\frac{dS_{sc}}{dt} = L\left(-\sqrt{K}|S_{sc}|^\alpha sgn(S_{sc}) - 1.1K\int sgn(S_{sc})\right) \tag{53}$$

where $\alpha = \frac{1}{2}$. The larger value of K will result in a good performance of the closed-loop system. This second-order controller will reduce the oscillatory contents of circulating current and will track the DC reference effectively. Equations (51) and (53) combinedly givesthe control law for circulating current control. The overall equation is given as

$$v^*_{c,j} = -Ri_{c,j} + \frac{V_{dc}}{2} - L\left(\sqrt{K}|S_{sc}|^\alpha sgn(S_{sc}) + 1.1K \int sgn(S_{sc}) \right). \tag{54}$$

$v^*_{c,j}$ will provide a controlled input to the plant. The reference value of circulating ($i^*_{c,j}$) current is one-third of DC current. Along with the DC component, an increment (Δi^*_c) derived from the energy equation is added to the reference to keep the converter's energy balanced. The energy balancing technique used in [44] is adopted in this paper. The time derivative of energy sum and energy difference is given as

$$\frac{dW_\Sigma}{dt} = 2v^*_{c,j}i_c - v^*_s i_o \tag{55}$$

$$\frac{dW_\Delta}{dt} = v^*_{c,j}i_o - 2v^*_s i_c. \tag{56}$$

$W_\Sigma = W_u + W_l$ and $W_\Delta = W_u - W_l$. In order to enhance the performance of circulating current controller, the mean value of $v^\Sigma_{cu,j}$ and $v^\Sigma_{lu,j}$ is made equal to V_{dc} by converging W_Σ to $W_{\Sigma o}$ and W_Δ to zero. Equations (55) and (56) are integrated to get W_Σ and W_Δ as in [35]. Hence, the increment term Δi^*_c is obtained

$$\Delta i^*_c = K_{sum}(W_{\Sigma o} - LPF(W_\Sigma)) - K_\Delta LPF(W_\Delta)cos\omega t. \tag{57}$$

LPF represents low pass filter, and LPF and K_Δ are positive constants. The control structure for circulating current control is depicted in Figure 6. However, the overall control scheme implemented for control of MMC is depicted in Figure 7.

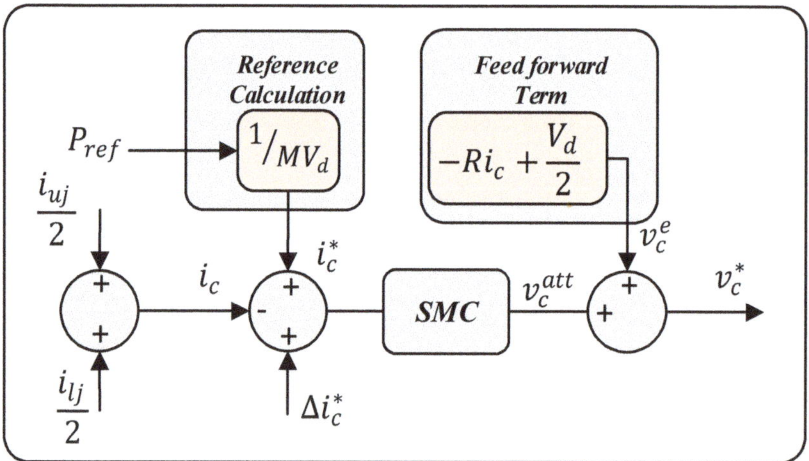

Figure 6. Current control structure.

Theorem 2. *The proposed designed SMC-based control scheme for circulating current is asymptotically stable and ensures boundedness of tracking error if the reference internal voltage $v^*_{c,j}$ is chosen as in Equation (54).*

Proof of Theorem 2. Like the output current controller, the stability of the circulating current controller is also ensured. From Equation (42), in the case of the circulating current controller, the value of the switching function is substituted from Equation (49)

$$S_{sc}\dot{S}_{sc} = -S_{sc}\frac{di_{c,j}}{dt} = -S_{sc}\frac{1}{L}\left(-v_{c,j}^* - Ri_{c,j} + \frac{V_{dc}}{2}\right).$$ (58)

$v_{c,j}^*$ is calculated as

$$v_{c,j}^* = -Ri_{c,j}^* + \frac{V_{dc}}{2} + L\dot{S}_{sc}.$$ (59)

The value of $v_{c,j}^*$ is substituted into Equation (59), and simplifying the equation,

$$S_{sc}\dot{S}_{sc} = -S_{sc}\frac{1}{L}\left(R\left(i_{c,j}^* - i_{c,j}\right) + L\dot{S}_{sc}\right).$$ (60)

Finally, the value of $\dot{S}_{sc}$ is substituted with Equation (53)

$$S_{sc}\dot{S}_{sc} = -\frac{R}{L}S^2 + \sqrt{K}S_{sc}|S_{sc}|^{\frac{1}{2}}sgn(S_{sc}) + 1.1K\int S_{sc}sgn(S_{sc}).$$ (61)

Hence, $S_{sc}\dot{S}_{sc} < 0$ if

$$\left|\frac{R}{L}S^2\right| > \left(\sqrt{K}S|S|^{\frac{1}{2}}sgn(S) + 1.1K\int Ssgn(S)\right).$$ (62)

In the case of the circulating current controller, Equation (62) should be satisfied for the stability of the system. However, the second condition of Lyapunov is satisfied.

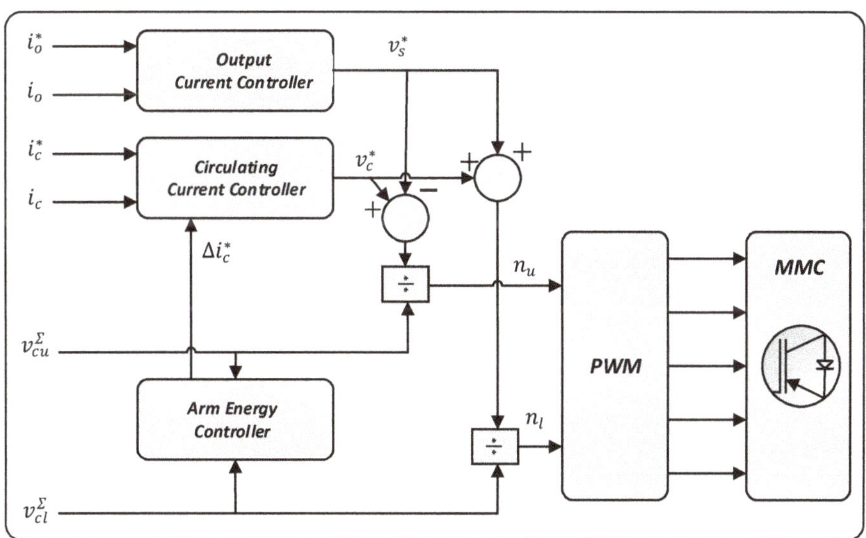

Figure 7. Structure of MMC.

4. Results and Discussion

A three-phase model of MMC is simulated in MATLAB/Simulink to analyze the performance of the design control scheme. The value of different parameters is presented in Appendix B. The simulation was carried out for 1 s, but to show the transient response, the time axis is scaled to 0.3 s. Moreover, the scale of the y-axis is kept the same for easy visual comparison.

The output current of MMC is transformed and controlled in *dq*- axis. However, for the sake of comparison with the PR controller, the output current is also controlled in a stationary frame ($\alpha\beta - axis$). The reference value of output current is generated from the desired active power and reactive power. The measured values of output current (I_d, I_q) track their respective references perfectly as depicted in Figure 8a,b. The d-axis and q-axis currents are attracted to the reference values and perfectly follow it. The designed controlled attractive term and equivalent term contribute well in achieving control goals. The system reaches steady-state at $t = 0.02$ s. The controlled value of these currents ensures that MMC is delivering the desired active and reactive power to the grid. The response of proposed SMC is fast and efficient as the controller works well in dynamic as well as in steady-state.

Figure 9 shows the control of output current in the stationary reference frame. The results of the proposed controller are compared with PR. The control law developed in the case of a stationary reference frame is given in Appendix A. Figure 9a shows output current controlled through SMC while the control of output current through PR is depicted in Figure 9b. The comparison of both controllers shows that the SMC is capable of reaching the respective reference value quickly, which shows the fast response of the proposed controller and perfect tracking of the desired value with almost zero steady-state error. As an initial condition for the sliding surface is zero, β-axis current catches its reference values right from the start and the attractive term of controller helps α-axis current to reach its respective reference. After reaching the reference, the equivalent term perfectly keeps the measured value on the track to minimize steady-state error, while in the case of the PR controller, the transient response of the controller is slow. It takes too much time to track the reference value. However, the response gets better in steady-state; but the steady-state error still exists. The error of both controllers is represented in Figure 10, and provides a clear comparison of both controllers. Figure 10a shows that SMC has an error of 5 A both in transient and steady-state, but the PR has a peak of almost 100 A in transient while 20 A in steady-state. This shows the crystal-clear effectiveness of the proposed controller.

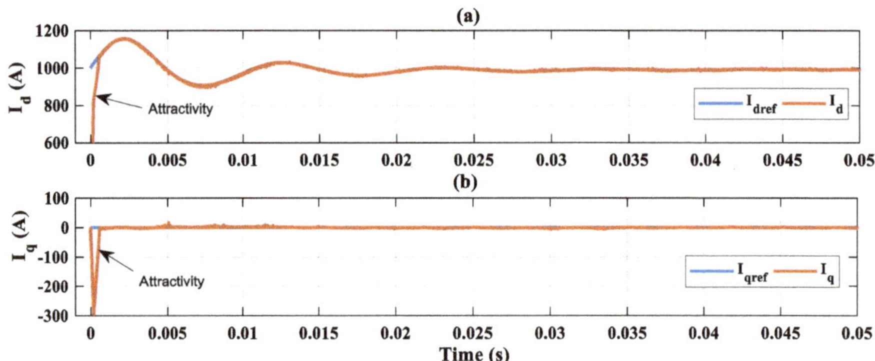

Figure 8. Output current; (**a**) d-axis current; (**b**) q-axis current.

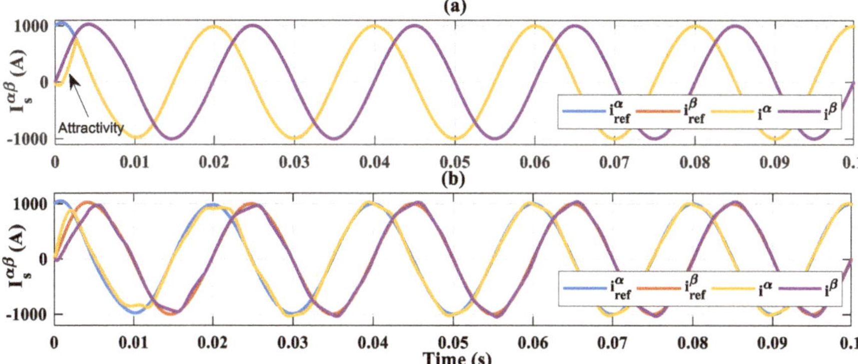

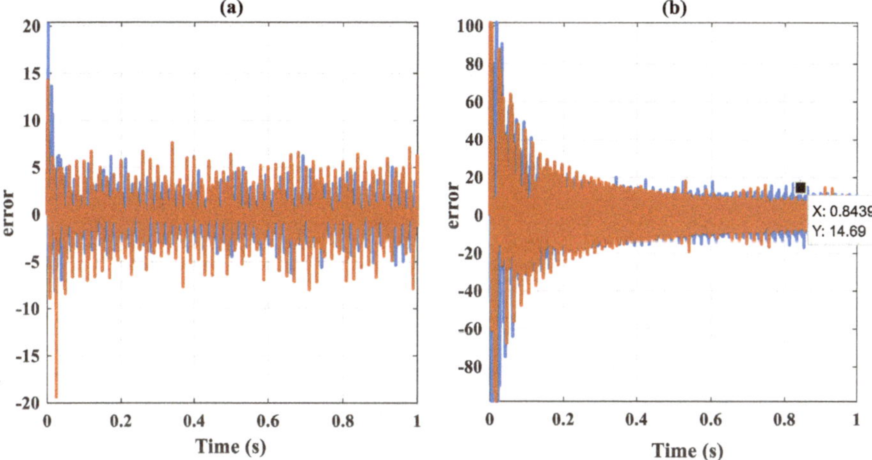

Figure 9. Output current in $\alpha\beta$-axis and its references; (**a**) control using sliding mode control (SMC); (**b**) control using proportional resonant (PR).

Figure 10. Error signal of output current; (**a**) SMC; (**b**) PR.

Figure 11 shows the reference and measured values of circulating current ($i_{c,j}$). The reference of $i_{c,j}$ is set according to value $\frac{I_{dc}}{3} = \frac{P}{MV_{dc}}$ i.e., 250 A. The convergence of $i_{c,j}$ ensures that the second harmonic current flowing in each leg of the converter is suppressed and only the DC component of the current is flowing in each leg. Consequently, it will result in the reduction of converter losses and reduces current stresses on the switches. Figure 10a,c shows the dynamic response of circulating current using SMC and PR controller, respectively. The response of SMC is fast and efficient as it reaches a steady-state in 0.05 s while PR takes almost 0.15 s to reach a steady state. This shows the response of SMC is three times faster than PR in case of circulating current control. The responses in steady-state are given in Figure 11b,d for SMC and PR controllers, respectively. The effectivity of SMC is persistent in steady-state as well. The current is converged perfectly at reference value in the case of SMC while the response of PR is subjected to large amplitude oscillation and different current is flowing in different legs of MMC. Notably, phase C has a large deviation from the desired value. The difference between the measured value and the reference value is almost 20 A. Besides this difference, the second harmonic contents of circulating current still exist. The PR controller fails to fully suppress the second harmonic current. This will lead to more converter losses and current stress in devices for the same size and rating of converter.

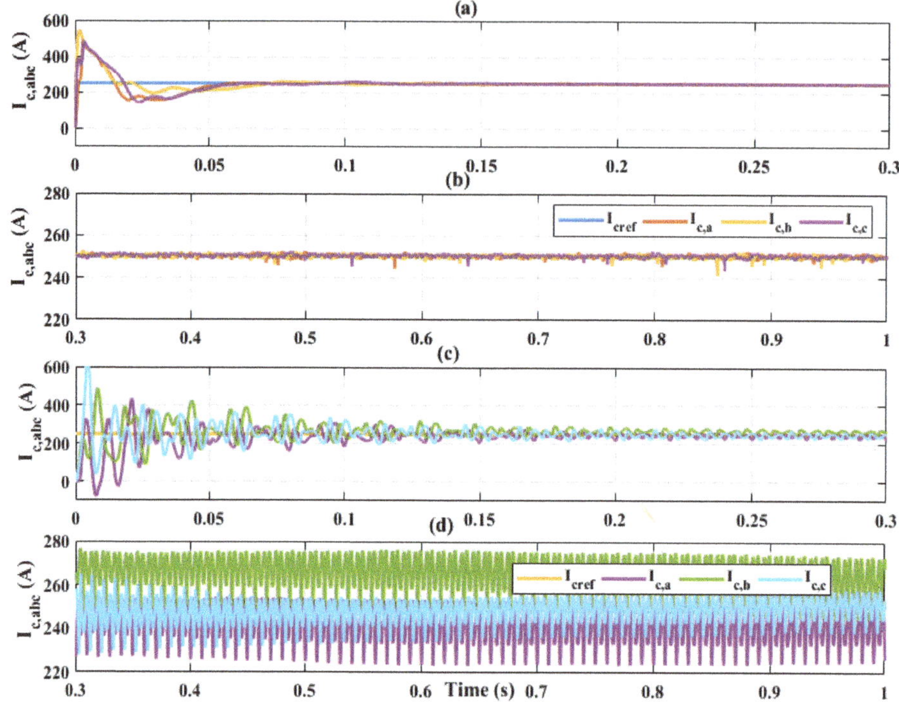

Figure 11. Current with references (**a**,**b**) control with SMC; (**c**,**d**) control with PR.

Figure 12a,b shows the error of circulating current. As it is clear from Figure 11, the error has an amplitude of about 200 A up to t = 0.05 s while after the dynamic response, still there is a large steady-state error that exists in the case of the PR controller. However, the SMC has a good contribution toward achieving the control goal. The error in the case of SMC has an amplitude of maximum of 5 A in steady-state.

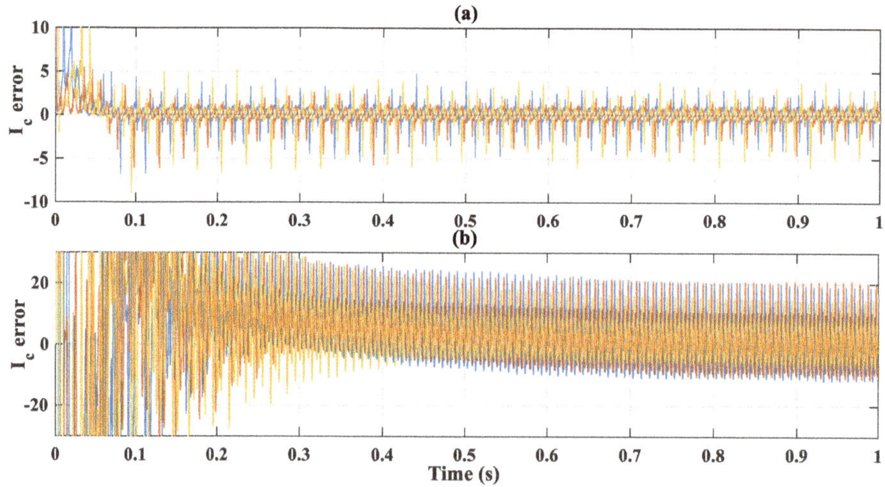

Figure 12. Circulating current error; (**a**) SMC; (**b**) PR.

The response of the upper arm capacitor voltage is depicted in Figure 13. This response of the lower arm capacitor voltage is not shown as both the upper arm and lower arm have almost the same response. Figure 13a shows the capacitor voltage response in the case of SMC while Figure 13b shows the capacitor voltage in the case of the PR controller. As both controllers are using the same energy-balancing control based on only proportional control, the ripple magnitude and capacitor voltage of the upper arm is almost the same. The uses of SMC guarantee a well-regulated capacitor voltage for all three-phase arms. The smoothness in the capacitor voltage ripple reflects the output voltage of the converter.

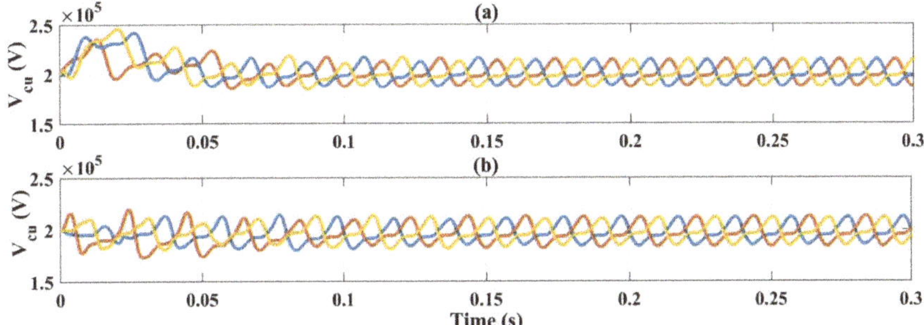

Figure 13. Arm capacitor voltage; (**a**) control with SMC; (**b**) control with PR.

The energy sum and difference between the upper and lower arm are shown in Figure 14a–d. Figure 14a,b shows the responses of during use of the proposed controller while Figure 14c,d show the response of the PR controller.

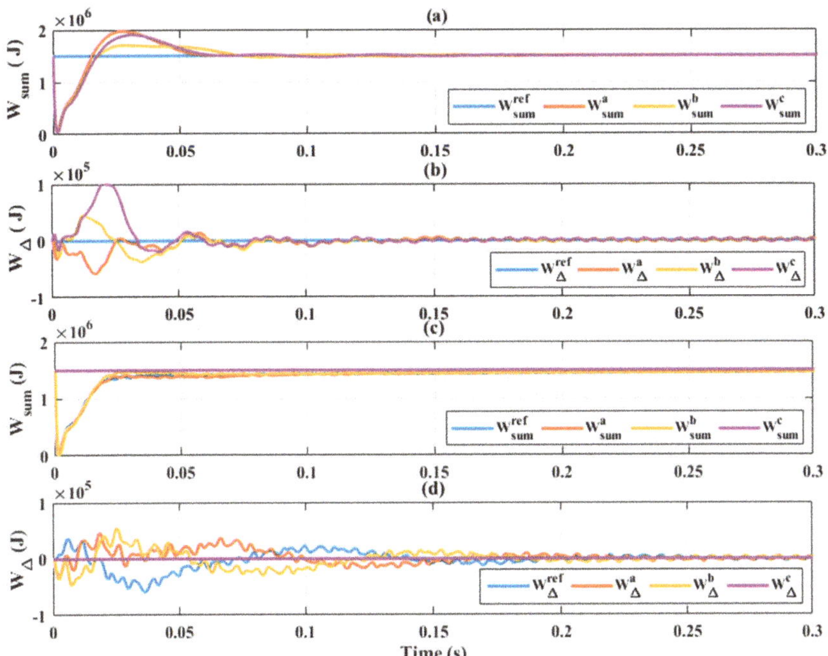

Figure 14. (**a**) Energy sum control with SMC; (**b**) energy difference with SMC; (**c**) energy sum with PR; (**d**) energy difference with PR.

The energy sum is equal to the phase leg energy of MMC. The reference of W_{sum} is calculated from the equation $NC(v_{sm})^2$. v_{sm} is SM voltage. The measured value of each arm is stabilized and follow the respective reference perfectly after $t = 0.06$ s. As SM capacitor voltage is responsible for energy storage, the regulated capacitor voltage results in a good and fast response of leg energy. After the initial charging of a capacitor, the response becomes well-regulated and follows the respective reference value. Similarly, Figure 14b depicts the energy difference between arms. Initially, due to the charging of the capacitor, there is a difference between an upper and lower arm of phase C but the controller tackles the problem and energy difference becomes zero afterward. The energy difference also ensures to track its reference value. The difference becomes zero at $t = 0.04$ s. This will cause both arms to reach equipotential and, hence, now charges will flow from one arm to another. However, in the case of PR, the measured leg energy reaches its reference value at $t = 0.3$ s, which shows a very slow response as compared to the proposed controller. Similarly, the energy difference becomes zero at $t = 0.2$ s, which also indicates a slow response. However, in steady-state, the PR controller tracks the reference energy difference signal.

Figure 15 depicts the output voltage of MMC. The output voltage has a good response to the proposed controller. The good response of output voltage is due to the well-regulated capacitor voltages of the upper arm and lower arm.

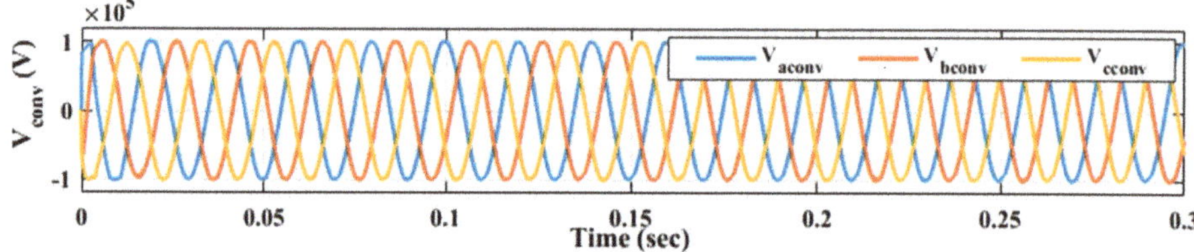

Figure 15. Output voltage.

The results of the proposed controller and PR controller are shown in this section. SMC shows good encourageable performance as compared to PR. The attractive term in control law shows a good response, which is reflected in the result. The result shows that SMC has a fast dynamic response compared to PR. Moreover, in steady-state, it also minimizes the error, while in the case of the circulating current controller, the super-twisting algorithm has effectively suppressed second-order content in circulating current, resulting in a very small error as compared to PR.

Moreover, to have a clear picture and conclusion, the performance indices of both the controller are measured and represented in Tables 1 and 2. The lower value of performance indices indicates good and stable control response.

Table 1. Performance indices of output current controllers.

		$i_{o,\alpha}$			$i_{o,\beta}$	
PR	1.	ISE:	905.6	1.	ISE:	509.6
	2.	IAE:	9.652	2.	IAE:	9.44
	3.	ITAE:	2.679	3.	ITAE:	2.831
Proposed Controller	1.	ISE:	1914	1.	ISE:	0.01842
	2.	IAE:	2.161	2.	IAE:	0.09404
	3.	ITAE:	0.03498	3.	ITAE:	0.03941

ISE: Integral square error, IAE: Integral absolute error, ITAE: Integral time absolute error.

Table 2. Performance indices of circulating current controllers.

		Phase A ($i_{c,a}$)			Phase B ($i_{c,b}$)			Phase C ($i_{c,c}$)	
	1.	ISE:	1353	1.	ISE:	1288	1.	ISE:	1456
PR	2.	IAE:	16.13	2.	IAE:	16.13	2.	IAE:	16.69
	3.	ITAE:	4.344	3.	ITAE:	4.125	3.	ITAE:	3.815
	1.	ISE:	220	1.	ISE:	7.42	1.	ISE:	75.41
Proposed Controller	2.	IAE:	1.493	2.	IAE:	0.6611	2.	IAE:	1.155
	3.	ITAE:	0.2748	3.	ITAE:	0.2711	3.	ITAE:	0.2905

ISE: Integral square error, IAE: Integral absolute error, ITAE: Integral time absolute error.

Table 1 shows the performance indices of output current controller. In the case of "$i_{o,\alpha}$", the integral square error (ISE) value of SMC controller is higher than PR controller. This is because the initial value of the sliding surface is zero while the initial value of "$i_{o,\alpha}$" is 1000 A. However, as the controller approaches its desired surface, it gives lower integral absolute error (IAE) and integral time absolute error (ITAE) values than the PR controller for $i_{o,\alpha}$. However, in the case of $\beta-$ component of output current, SMC has lower values of ISE, IAE, and ITAE, which show the good response of controller.

Similarly, the performance indices for circulating current controllers are given in Table 2. By analyzing the value of all three-phase current, values of ISE, IAE, and ITAE are much smaller in the case of SMC compared to PR. As small values of performance indices shows best performance, it is concluded that SMC has a better response than PR in the case of circulating current.

5. Conclusions

This work is based on the current control of MMC using SMC design. Output and circulating current are controlled using separate controllers. The controller ensures the reference tracking of both output current and circulating current, and the second harmonics contents is well suppressed. Moreover, the proposed control scheme regulates capacitor arm voltages and the energy of system. The design of SMC-based control shows fast response in a transient state as well as in a steady state. The attractive term and equivalent terms in control law work very well. The measure value is attracted and steered on the reference value effectively. The design shows optimal performance of MMC parameters. The stability and robustness of controller is proved using Lyapunov analysis.

Author Contributions: W.U., K.Z., and M.I., proposed the main idea of the paper. W.U. implemented the mathematical derivations, simulation verifications, and analyses. The paper was written by W.U., and was revised by K.Z., M.I., S.u.I., M.A.K., and H.-J.K. All the authors were involved in preparing the final version of this manuscript. Furthermore, this whole work was supervised by H.-J.K., G.S.P., and C.L.

Funding: This research was supported by Basic Research Laboratory through the National Research Foundations of Korea funded by the Ministry of Science, ICT and Future Planning (NRF-2015R1A4A1041584).

Nomenclature

Acronym	Meaning	Units
*	Reference values	–
e, att	Equivalent and attractive terms	–
$j = a, b, c$	Three phases (*abc*)	–
$i_{u,j}$, $i_{l,j}$	upper arm and lower arm currents of j phase	A
$i_{o,j}$, $i_{c,j}$	output and circulating currents of j phase	A
V_{dc}, I_{dc}	DC link voltage and current respectively	V, A
$I_{2h,j}$	2nd harmonic of circulating current	A
$e_{u,j}$, $e_{l,j}$	internal voltage of upper arm and lower arm	V
R, L	Phase leg resistance and Inductance	Ω, H
v_j	Grid voltage	V
$v_{s,j}$, $v_{c,j}$	Converter output and internal voltages respectively	V
C	Submodule capacitance	F
N	Number of submodules in one arm	–
v_j	Submodule capacitor voltage	V
$v^{\Sigma}_{cu,j}$, $v^{\Sigma}_{cl,j}$	sum of capacitor voltage of upper arm and lower arm	V
$n_{u,j}$, $n_{l,j}$	insertion indices	–
P	active power	W
$M = 3$	number of phases	–
S_s	Sliding surface	–
u	control input	–
abc / dq / $\alpha\beta$	Quantity in three phase (*abc*), *dq* and $\alpha\beta$ systems respectively	–
ω_s	angular frequency of grid voltage	$\frac{rad}{s}$
v_{ds}, v_{qs}	Output voltage in *dq-axis*	V
v_d, v_q	Grid voltage in *dq-axis*	V
$i_{d,o}$, $i_{q,o}$	output current in *dq-axis*	A
$Q_{d,q}$, $K_{d,q}$, K, $K_{\Delta,sum}$	Positive constant	–
V	Lyapunov function	–
W_{Σ}, W_{Δ}	Energy sum and difference of both arms	J
sgn	Signum function	–
x	State vector	–
$f(x)$, $g(x)$	nonlinear function of system state vector	–
$C^T = [c_1, c_2 \ldots c_{n-1}, 1]$	co-efficient of sliding function	
LPF	Low pass filter	–

Appendix A

Transforming Equation (21) to stationary reference frame ($\alpha\beta$) by using Clarke's Transformation ($T_{abc \Rightarrow \alpha\beta o}$), we get

$$\begin{bmatrix} v_{s,\alpha} \\ v_{s,\beta} \end{bmatrix} = \frac{L}{2} \begin{bmatrix} \frac{di_{o,\alpha}}{dt} \\ \frac{di_{o,\beta}}{dt} \end{bmatrix} + \frac{R}{2} \begin{bmatrix} i_{o,\alpha} \\ i_{o,\beta} \end{bmatrix} + \begin{bmatrix} v_\alpha \\ v_\beta \end{bmatrix}. \tag{A1}$$

Equation (A1) is solved for value $v^*_{s,\alpha\beta}$ given as

$$v^*_{s,\alpha\beta} = \underbrace{v_{\alpha\beta} + \frac{1}{2} R i^*_{o,\alpha\beta}}_{v^e_{s,\alpha\beta}} - \underbrace{\frac{L}{2} \frac{dS_{s,\alpha\beta}}{dt}}_{v^{att}_{s,\alpha\beta}}. \tag{A2}$$

The switching function $S_{s,\alpha\beta}$ used in control law is given as

$$v^{att}_{s,\alpha\beta} = -\frac{L}{2} \frac{dS_{s,\alpha\beta}}{dt} = \frac{L}{2} \left(Q_{\alpha\beta} sgn\left(S_{s,\alpha\beta}\right) + K_{\alpha\beta} S_{s,\alpha\beta} \right). \tag{A3}$$

$Q_{\alpha\beta}$, and $K_{\alpha\beta}$ are positive and real numbers. The ful control law is given as

$$\begin{bmatrix} v_s^* \\ v_{qs}^* \end{bmatrix} = \begin{bmatrix} v_{ds}^e \\ v_{qs}^e \end{bmatrix} + \begin{bmatrix} v_{ds}^{att} \\ v_{qs}^{att} \end{bmatrix}$$
$$v_{s,\alpha\beta}^* = v_{s,\alpha\beta}^e + v_{s,\alpha\beta}^{att} \tag{A4}$$

while

$$v_{s,\alpha\beta}^* = v_{\alpha\beta} + \frac{1}{2}Ri_{o,\alpha\beta}^* + \frac{L}{2}\left(Q_{\alpha\beta}sgn\left(S_{s,\alpha\beta}\right) + K_\alpha S_{s,\alpha\beta}\right). \tag{A5}$$

Appendix B

Table A1. Converter Speciation.

Parameters	Value	Symbols	Units
D.C Voltage	200	V_d	kV
Grid Voltage	100	V_s	kV
Output Peak Current	1	I_s	kA
Frequency	50	f	Hz
Number of level	12	N	-
Inducatnce	50	L	mH
Resistance	1.57	R	Ω
Capcitance	0.45	C	mF

References

1. Kouro, S.; Malinowski, M.; Gopakumar, K.; Pou, J.; Franquelo, L.G.; Bin, W.; Rodriguez, J.; Pérez, M.A.; Leon, J.I. Recent Advances and Industrial Applications of Multilevel Converters. *IEEE Trans. Ind. Electron.* **2010**, *57*, 2553–2580. [CrossRef]
2. Perez, M.A.; Bernet, S.; Rodriguez, J.; Kouro, S.; Lizana, R. Circuit Topologies, Modeling, Control Schemes, and Applications of Modular Multilevel Converters. *IEEE Trans. Power Electron.* **2015**, *30*, 4–17. [CrossRef]
3. Kolluri, S.; Gorla, N.B.Y.; Sapkota, R.; Panda, S.K. A new control approach for improved dynamic performance of MMC based HVDC subsea power transmission system. In Proceedings of the 2017 IEEE Innovative Smart Grid Technologies-Asia: Smart Grid for Smart Community (ISGT-Asia 2017), Auckland, New Zealand, 4–7 December 2017.
4. Debnath, S.; Saeedifard, M. A New Hybrid Modular Multilevel Converter for Grid Connection of Large Wind Turbines. *IEEE Trans. Sustain. Energy* **2013**, *4*, 1051–1064. [CrossRef]
5. Hagiwara, M.; Nishimura, K.; Akagi, H. A Medium-Voltage Motor Drive With a Modular Multilevel PWM Inverter. *IEEE Trans. Power Electron.* **2010**, *25*, 1786–1799. [CrossRef]
6. Li, B.; Zhou, S.; Xu, D.; Yang, R.; Xu, D.; Buccella, C.; Cecati, C. An Improved Circulating Current Injection Method for Modular Multilevel Converters in Variable-Speed Drives. *IEEE Trans. Ind. Electron.* **2016**, *63*, 7215–7225. [CrossRef]
7. Vasiladiotis, M.; Rufer, A. Analysis and control of modular multilevel converters with integrated battery energy storage. *IEEE Trans. Power Electron.* **2015**, *30*, 163–175. [CrossRef]
8. Li, Y.; Han, Y. A Module-Integrated Distributed Battery Energy Storage and Management System. *IEEE Trans. Power Electron.* **2016**, *31*, 8260–8270. [CrossRef]
9. Zeb, K.; Uddin, W.; Khan, M.A.; Ali, Z.; Ali, M.U.; Christofides, N.; Kim, H.J. A comprehensive review on inverter topologies and control strategies for grid connected photovoltaic system. *Renew. Sustain. Energy Rev.* **2018**, *94*, 1120–1141. [CrossRef]
10. Nami, A.; Liang, J.; Dijkhuizen, F.; Demetriades, G.D. Modular Multilevel Converters for HVDC Applications: Review on Converter Cells and Functionalities. *IEEE Trans. Power Electron.* **2015**, *30*, 18–36. [CrossRef]
11. Zeb, K.; Khan, I.; Uddin, W.; Khan, M.A.; Sathishkumar, P.; Busarello, T.D.C.; Ahmad, I.; Kim, H.J. A review on recent advances and future trends of transformerless inverter structures for single-phase grid-connected photovoltaic systems. *Energies* **2018**, *11*, 1968. [CrossRef]

12. Ishfaq, M.; Uddin, W.; Zeb, K.; Khan, I.; Islam, S.U.; Khan, M.A.; Kim, H.J. A new adaptive approach to control circulating and output current of modular multilevel converter. *Energies* **2019**, *12*, 1118. [CrossRef]

13. Hagiwara, M.; Akagi, H. Control and Experiment of Pulsewidth-Modulated Modular Multilevel Converters. *IEEE Trans. Power Electron.* **2009**, *24*, 1737–1746. [CrossRef]

14. Lizana, R.; Perez, M.A.; Arancibia, D.; Espinoza, J.R.; Rodriguez, J. Decoupled Current Model and Control of Modular Multilevel Converters. *IEEE Trans. Ind. Electron.* **2015**, *62*, 5382–5392. [CrossRef]

15. Riar, B.S.; Madawala, U.K. Decoupled control of modular multilevel converters using voltage correcting modules. *IEEE Trans. Power Electron.* **2015**, *30*, 690–698. [CrossRef]

16. Bahrani, B.; Debnath, S.; Saeedifard, M. Circulating Current Suppression of the Modular Multilevel Converter in a Double-Frequency Rotating Reference Frame. *IEEE Trans. Power Electron.* **2016**, *31*, 783–792. [CrossRef]

17. Qingrui, T.; Zheng, X.; Lie, X. Reduced Switching-Frequency Modulation and Circulating Current Suppression for Modular Multilevel Converters. *IEEE Trans. Power Deliv.* **2011**, *26*, 2009–2017. [CrossRef]

18. Lacerda, V.A.; Coury, D.V.; Monaro, R.M. Proportional-Resonant controller applied to modular multilevel converter for HVDC systems. In Proceedings of the 2018 Simposio Brasileiro de Sistemas Eletricos (SBSE), Niteroi, Brazil, 12–16 May 2018; pp. 1–6.

19. Li, S.; Wang, X.; Yao, Z.; Li, T.; Peng, Z. Circulating Current Suppressing Strategy for MMC-HVDC Based on Nonideal Proportional Resonant Controllers Under Unbalanced Grid Conditions. *IEEE Trans. Power Electron.* **2015**, *30*, 387–397. [CrossRef]

20. He, L.; Zhang, K.; Xiong, J.; Fan, S. A Repetitive Control Scheme for Harmonic Suppression of Circulating Current in Modular Multilevel Converters. *IEEE Trans. Power Electron.* **2015**, *30*, 471–481. [CrossRef]

21. Yang, S.; Wang, P.; Tang, Y.; Zagrodnik, M.; Hu, X.; Tseng, K.J. Circulating Current Suppression in Modular Multilevel Converters With Even-Harmonic Repetitive Control. *IEEE Trans. Ind. Appl.* **2018**, *54*, 298–309. [CrossRef]

22. Yang, S.; Wang, P.; Tang, Y.; Zhang, L. Explicit Phase Lead Filter Design in Repetitive Control for Voltage Harmonic Mitigation of VSI-Based Islanded Microgrids. *IEEE Trans. Ind. Electron.* **2017**, *64*, 817–826. [CrossRef]

23. Zhang, M.; Huang, L.; Yao, W.; Lu, Z. Circulating Harmonic Current Elimination of a CPS-PWM-Based Modular Multilevel Converter With a Plug-In Repetitive Controller. *IEEE Trans. Power Electron.* **2014**, *29*, 2083–2097. [CrossRef]

24. Kolluri, S.; Gorla, N.B.Y.; Sapkota, R.; Panda, S.K. A New Control Architecture with Spatial Comb Filter and Spatial Repetitive Controller for Circulating Current Harmonics Elimination in a Droop Regulated Modular Multilevel Converter for Wind Farm Application. *IEEE Trans. Power Electron.* **2019**, *34*, 10509–10523. [CrossRef]

25. Yang, Q.; Saeedifard, M.; Perez, M.A. Sliding Mode Control of the Modular Multilevel Converter. *IEEE Trans. Ind. Electron.* **2019**, *66*, 887–897. [CrossRef]

26. Ishfaq, M.; Uddin, W.; Zeb, K.; Islam, S.U.; Hussain, S.; Khan, I.; Kim, H.J. Active and reactive power control of modular multilevel converter using sliding mode controller. In Proceedings of the 2019 2nd International Conference on Computing, Mathematics and Engineering Technologies (iCoMET 2019), Sukkur, Pakistan, 30–31 January 2019.

27. Uddin, W.; Hussain, S.; Zeb, K.; Khalil, I.U.; Ullah, Z.; Dildar, M.A.; Adil, M.; Ishfaq, M.; Khan, I.; Kim, H.J. Effect of Arm Inductor on Harmonic Reduction in Modular Multilevel Converter. In Proceedings of the 4th International Conference on Power Generation Systems and Renewable Energy Technologies (PGSRET 2018), Islamabad, Pakistan, 10–12 September 2018.

28. Din, W.U.; Zeb, K.; Ishfaq, M.; Islam, S.U.; Khan, I.; Kim, H.J. Control of internal dynamics of grid-connected modular multilevel converter using an integral backstepping controller. *Electronics* **2019**, *8*, 456. [CrossRef]

29. Yang, S.; Wang, P.; Tang, Y. Feedback Linearization-Based Current Control Strategy for Modular Multilevel Converters. *IEEE Trans. Power Electron.* **2018**, *33*, 161–174. [CrossRef]

30. Riar, B.S.; Geyer, T.; Madawala, U.K. Model Predictive Direct Current Control of Modular Multilevel Converters: Modeling, Analysis, and Experimental Evaluation. *IEEE Trans. Power Electron.* **2015**, *30*, 431–439. [CrossRef]

31. Vatani, M.; Bahrani, B.; Saeedifard, M.; Hovd, M. Indirect Finite Control Set Model Predictive Control of Modular Multilevel Converters. *IEEE Trans. Smart Grid* **2015**, *6*, 1520–1529. [CrossRef]

32. Dekka, A.; Wu, B.; Yaramasu, V.; Fuentes, R.L.; Zargari, N.R. Model Predictive Control of High-Power Modular Multilevel Converters—An Overview. *IEEE J. Emerg. Sel. Top. Power Electron.* **2019**, *7*, 168–183. [CrossRef]

33. Bergna, G.; Garces, A.; Berne, E.; Egrot, P.; Arzande, A.; Vannier, J.-C.; Molinas, M. A Generalized Power Control Approach in ABC Frame for Modular Multilevel Converter HVDC Links Based on Mathematical Optimization. *IEEE Trans. Power Deliv.* **2014**, *29*, 386–394. [CrossRef]

34. Gao, W.; Hung, J.C. Variable structure control of nonlinear systems: A new approach. *IEEE Trans. Ind. Electron.* **1993**, *40*, 45–55.

35. Sharifabadi, K.; Harnefors, L.; Nee, H.-P.; Norrga, S.; Teodorescu, R. Dynamics and Control. In *Design, Control and Application of Modular Multilevel Converters for HVDC Transmission Systems*; John Wiley & Sons, Ltd.: Chichester, UK, 2016; pp. 133–213.

36. Zeb, K.; Ayesha; Haider, A.; Uddin, W.; Qureshi, M.B.; Mehmood, C.A.; Jazlan, A.; Sreeram, V. Indirect Vector Control of Induction Motor using Adaptive Sliding Mode Controller. In Proceedings of the 2016 Australian Control Conference (AuCC), Newcastle, Australia, 3–4 November 2016; pp. 358–363.

37. Zeb, K.; Uddin, W.; Khan, M.A.; Ayesha; Haider, A.; Kim, H.J. A comparative assessment of scalar controlled induction motor using PI, adaptive sliding mode, and FLC based on SD controllers. In Proceedings of the 2017 1st International Conference on Latest Trends in Electrical Engineering and Computing Technologies (INTELLECT 2017), Karachi, Pakistan, 15–16 November 2017.

38. Zeb, K.; Din, W.; Khan, M.; Khan, A.; Younas, U.; Busarello, T.; Kim, H. Dynamic Simulations of Adaptive Design Approaches to Control the Speed of an Induction Machine Considering Parameter Uncertainties and External Perturbations. *Energies* **2018**, *11*, 2339. [CrossRef]

39. Zeb, K.; Islam, S.U.; Din, W.U.; Khan, I.; Ishfaq, M.; Busarello, T.D.C.; Ahmad, I.; Kim, H.J. Design of fuzzy-PI and fuzzy-sliding mode controllers for single-phase two-stages grid- connected transformerless photovoltaic inverter. *Electronics* **2019**, *8*, 520. [CrossRef]

40. Baghaee, H.R.; Mirsalim, M.; Gharehpetian, G.B.; Talebi, H.A. A Decentralized Power Management and Sliding Mode Control Strategy for Hybrid AC/DC Microgrids including Renewable Energy Resources. *IEEE Trans. Ind. Inform.* **2017**. [CrossRef]

41. Baghaee, H.R.; Mirsalim, M.; Gharehpetian, G.B.; Talebi, H.A. Decentralized Sliding Mode Control of WG/PV/FC Microgrids under Unbalanced and Nonlinear Load Conditions for On- and Off-Grid Modes. *IEEE Syst. J.* **2018**, *12*, 3108–3119. [CrossRef]

42. Beltran, B.; Benbouzid, M.E.H.; Ahmed-Ali, T. Second-Order sliding mode control of a doubly fed induction generator driven wind turbine. *IEEE Trans. Energy Convers.* **2012**, *27*, 261–269. [CrossRef]

43. Asad, M.; Ashraf, M.; Iqbal, S.; Bhatti, A.I. Chattering and stability analysis of the sliding mode control using inverse hyperbolic function. *Int. J. Control. Autom. Syst.* **2017**, *15*, 2608–2618. [CrossRef]

44. Antonopoulos, A.; Angquist, L.; Nee, H.-P.P. On dynamics and voltage control of the Modular Multilevel Converter. In Proceedings of the 2009 13th European Conference on Power Electronics and Applications, Barcelona, Spain, 8–10 September 2009.

Article

Digital Control of a Buck Converter Based on Input-Output Linearization. An Interpretation Using Discrete-Time Sliding Control Theory

Enric Vidal-Idiarte [1], Carlos Restrepo [2], Abdelali El Aroudi [1], Javier Calvente [1] and Roberto Giral [1,*]

[1] Departament d'Enginyeria Electrònica, Elèctrica i Automàtica, Escola Tècnica Superior d'Enginyeria, Universitat Rovira i Virgili, 43007 Tarragona, Spain

[2] Department of Electromechanics and Energy Conversion, Universidad de Talca, Curicó 3340000, Chile

* Correspondence: roberto.giral@urv.cat; Tel.: +34-977559620

Received: 14 June 2019; Accepted: 16 July 2019; Published: 17 July 2019

Abstract: This paper presents the analysis and design of a PWM nonlinear digital control of a buck converter based on input-output linearization. The control employs a discrete-time bilinear model of the power converter for continuous conduction mode operation (CCM) to create an internal current control loop wherein the inductor current error with respect to its reference decreases to zero in geometric progression. This internal loop is as a constant frequency discrete-time sliding mode control loop with a parameter that allows adjusting how fast the error is driven to zero. Subsequently, an outer voltage loop designed by linear techniques provides the reference of the inner current loop to regulate the converter output voltage. The two-loop control offers a fast transient response and a high regulation degree of the output voltage in front of reference changes and disturbances in the input voltage and output load. The experimental results are in good agreement with both theoretical predictions and PSIM simulations.

Keywords: two-loop digital control; buck converter; input-output linearization; PWM; sliding mode

---

1. Introduction

Since breaking into the regulation of dc-dc switching converters at the end of the 1990s, the digital control has undergone successive phases of expansion and stagnation. Destined to be a new standard for the switching converters at the moment of its appearance, the reality, however, has shown that its adoption by the industry has been so far practically inexistent. Despite the inherent advantages of the digital control such as insensitivity to noise, design reuse and reprogram capability, the user's mistrust, on the one hand, and the difficulty of design, on the other hand, have contributed to this absence of industrial penetration [1].

From the user's perspective, adopting a new technology more expensive than the present one and not working so well in terms of accuracy in steady-state and dynamic performance was a risk that the manufacturers did not want to take. Time-delay and numerical limit cycles are intrinsic drawbacks associated with the digital design that eventually degrade the controller's response. The time-delay problem is introduced mainly by the analog-to-digital converter in the cycle by cycle realization of the sampling/conversion process [2,3], which provokes a controller's delayed reaction of one switching cycle. To mitigate this problem, the sampling/conversion process is performed at large sampling rates up to 32 times the switching frequency [4,5] and small quantization steps are often employed, which results in an increasing of cost, hardware complexity and consumption of the controller. The limit cycle issue is caused by the digital pulse width modulator and the quantization error of the analog-to-digital converter, this yielding an oscillation in the converter output voltage [6].

From the designer's point of view, the realization of the digital controller is less intuitive than the corresponding analog implementation. In addition to this fact, it has to be pointed out that the first designs required silicon areas considerably bigger than those needed presently and their power consumption was significantly larger than that of their corresponding analog counterparts. However, the exponential decrease of cost and size of digital circuits, which in the period elapsed since 1998 has dramatically incremented the capacity of integrating electronic functions, has revived the idea of digital control as the definitive substitute of the analog equivalent in dc-dc switching converters. This idea is being supported by the fact that recent digital research has been focused on non-realizable aspects in the analog domain such as integrating communications, controller's self-tuning, converter's efficiency monitoring and complex nonlinear control techniques implementation. An example of this are the works reported in [5,7], where the dynamic response to perturbations in the power converter is improved by a specific nonlinear control with self-tuning and the overall performance is ameliorated by the action of an active efficiency monitoring system. Other contributions have dealt with new modulation schemes [8] to optimize the converter's efficiency and extend its input range. Some recent reports are also found when testing new complex control strategies either deterministic [9] or random modulation-based [10], and in introducing new on-line self-tuning protocols [11].

Most digital control realizations are of hybrid type, which means that they combine the action of a nonlinear controller to facilitate a fast transient response with the performance of a linear controller to obtain a precise regulation in steady-state. This is the case of [12], wherein a fuzzy logic-based controller implements a PI algorithm to eliminate the output voltage error in steady-state. However, when that error or its time derivative are relatively high, the duty cycle changes rapidly by means of a nonlinear action. A similar situation is described in [5], where a PID algorithm for steady-state regulation is combined with a sliding mode control for minimum time transient operation. A posterior attempt to solve the problem of minimum time transient recovery is reported in [13] together with a concise description of different strategies dealing with this problem. Nevertheless, the controller does not operate on-line because it only uses memory accesses and comparisons of previously stored data. Another approach on minimization concerns the output deviation as reported in [14], where a specific integrated circuit for a digital control of a single and a two-phase buck converters is introduced showing that the deviation of the output is four times smaller than that of a fast PID compensator. Other cases exhibiting this hybrid nature are found in the optimal-time control whose goal is to lead the converter's dynamics to the steady-state in minimum time [15], in the near-optimal voltage deviation and recovery time [16,17], and finally in the use of a hysteretic analog modulator together with a digital linear control loop and a digital frequency regulation loop [18].

On the other hand, most linear digital controllers have been designed from linear analog controllers by mapping the s-plane into the z-plane and employing either frequency domain methods or pole-zero assignments in the z-plane. The first exception is the work reported in [19], where the controller design is based on a discrete-time linear model of the plant, which is obtained with the method presented in [20] for switching converters operating in continuous conduction mode with constant switching frequency. A second exception is found in [21], where a time-domain design method is used to fit a digital PID template to the desired response. In [22], after a review of previous contributions, a methodology to design non-linear digital controllers based on discrete-time sliding mode control is presented where a dead-beat response is achieved [23]. Also, in [24] this methodology is applied to regulate the output voltage of a boost converter with constant power load, providing a comparison with one of the most referenced works in digital control of power converters [25].

The aim of this work is to develop a digital control that combines a good transient response and a good steady-state regulation in only one algorithm. This has the advantage of simplifying the control implementation with respect to the hybrid control realizations found extensively in the literature. The starting point of the controller proposed in this article is the bilinear recurrence developed in [26] for PWM converters in CCM, which is reviewed in Section 2. The proposed nonlinear controller is obtained by input-output linearization and presented in Section 3, in which a discrete-time sliding mode interpretation is also provided. Output voltage regulation and experimental results are reported in Sections 4 and 5 respectively. Conclusions are presented in Section 6.

2. Nonlinear Recurrence

The dynamic behaviour of a dc-to-dc switching converter operating in CCM with constant switching frequency can be described by the following pair of state equations

$$\dot{X} = A_1 X + b_1 V_g \qquad nT \leq t < (n + d_n)T \tag{1}$$

$$\dot{X} = A_2 X + b_2 V_g \qquad (n + d_n)T \leq t < (n+1)T \tag{2}$$

where X is the converter state vector, A_1, A_2, b_1, and b_2 are constant-coefficient matrices, T is the switching period, nT is the instant at the beginning of the conduction state ("ON" state) in n^{th} cycle, and $d_n T$ is the duration of "ON" state. Let d_n the duty cycle during the switching cycle $(nT, (n+1)T)$ and $\bar{d}_n = 1 - d_n$.

The solution of the first equation with initial condition $X(nT)$ is given by

$$X((n + d_n)T) = e^{A_1 d_n T} X(nT) + \int_{nT}^{(n+d_n)T} e^{A_1((n+d_n)T - \zeta)} b_1 V_g d\zeta. \tag{3}$$

Taking into account that the pair $[A_i, e^{A_i t}]$ commutes, i.e., $A_i^{-1} e^{A_i t} = e^{A_i t} A_i^{-1}$, Equation (3) becomes as follows

$$X((n + d_n)T) = e^{A_1 d_n T} X(nT) + \left(e^{A_1 d_n T} - I \right) A_1^{-1} b_1 V_g \tag{4}$$

where I is a unitary matrix with appropriate dimensions. Similarly, the solution of Equation (2) with initial condition $X((n + d_n)T)$ will be given by Equation (5),

$$X((n + 1)T) = e^{A_2 \bar{d}_n T} X((n + d_n)T) + \int_{(n+d_n)T}^{(n+1)T} e^{A_2((n+1)T - \zeta)} b_2 V_g d\zeta. \tag{5}$$

By inserting Equation (4) in (5) we obtain

$$X((n + 1)T) = e^{A_2 \bar{d}_n T} [e^{A_1 d_n T} X(nT) + (e^{A_1 d_n T} - I) A_1^{-1} b_1 V_g] + (e^{A_2 \bar{d}_n T} - I) A_2^{-1} b_2 V_g. \tag{6}$$

Please note that terms in the form $(e^{AT} - I)A^{-1}$ have a matrix series expansion even if A is singular

$$(e^{AT} - I)A^{-1} = T \sum_{n=0}^{\infty} \frac{(AT)^n}{(n + 1)!}. \tag{7}$$

In practice, the switching frequency is much larger than the natural frequencies of the power converters and therefore the exponential matrices containing T can be approximated by their respective first terms of a Taylor's series at $T = 0$, so that Equation (6) can be written as follows in (8),

$$X((n + 1)T) \approx [I + (A_1 d_n + A_2 \bar{d}_n)T] X(nT) + (b_1 V_g d_n + b_2 V_g \bar{d}_n)T]. \tag{8}$$

The recurrence equation defined in Equation (8) can be expressed as

$$X((n + 1)T) = HX(nT) + FX(nT)d_n T + Ed_n T + G \tag{9}$$

where

$$E = (b_1 - b_2)V_g, \quad F = (A_1 - A_2), \quad G = b_2 V_g T, \quad H = I + A_2 T. \tag{10}$$

The nonlinear nature of recurrence Equation (9) is observed in the second and forth terms by showing the respective multiplication of the duty cycle (control) by the state vector and by an additive term depending on the input voltage (energy). In the particular case of the buck converter $A_1 = A_2$, which implies $F = 0$ and reveals that the nonlinear behaviour is an afine characteristic. In the case of a boost converter $b_1 = b_2$, which results in $E = 0$ and shows that the nonlinear characteristic is produced by the product of the control and state vector.

Although a constant value has been considered for V_g, the analysis developed could be applied also to slowly varying input voltages. In this case there will be also a nonlinearity produced by the product of the control and the input voltage.

3. Current Control Loop Based on Input-Output Linearization

In the particular case of a buck converter (Figure 1), the state vector is $X = [v_C \ i_L]^T$ and therefore the nonlinear recurrence can be simplified as follows

$$X((n+1)T) = HX(nT) + Ed_nT \tag{11}$$

since F and G are zero because $A_1 = A_2$ and $b_2 = 0$ in such converter.

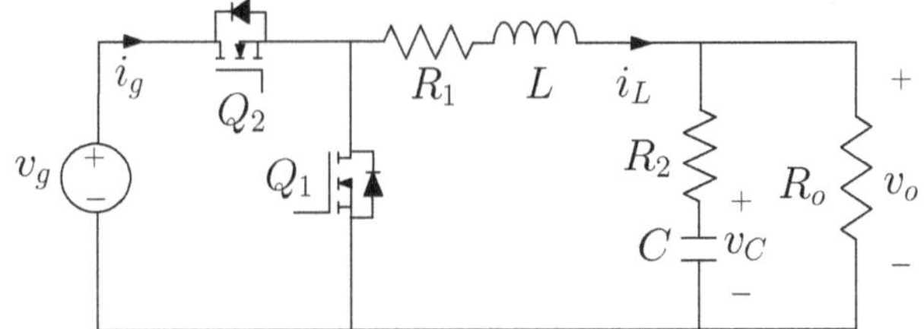

Figure 1. Buck converter configuration with parasitic resistances in the reactive elements.

Matrix H can be expressed as

$$H = \left(h_{ij}\right)_{2\times2} \tag{12}$$

where

$$h_{11} = 1 - \frac{R_aT}{L}, \quad h_{12} = -\frac{(1-\varepsilon)T}{L}, \quad h_{21} = \frac{(1-\varepsilon)T}{C}, \quad h_{22} = 1 - \frac{T}{CR_b} \tag{13}$$

and

$$R_a = R_1 + \frac{R_oR_2}{R_o+R_2}, \quad \varepsilon = \frac{R_2}{R_o+R_2}, \quad \text{and} \quad R_b = R_o+R_2. \tag{14}$$

Hence, recurrence Equation (11) can be expressed as

$$i_L((n+1)T) = h_{11}i_L(nT) + h_{12}v_C(nT) + \frac{V_g}{L}d_nT \tag{15}$$

$$v_C((n+1)T) = h_{21}i_L(nT) + h_{22}v_C(nT). \tag{16}$$

Now, let's impose that the current sample at instant $(n+1)T$ reaches a reference $i_{REF}(nT)$ with an approximation error dynamics that decreases in geometric progression. The error dynamics can be expressed as follows

$$i_L((n+1)T) - i_{REF}(nT) = w\left(i_L(nT) - i_{REF}(nT)\right) \tag{17}$$

where

$$-1 < w < 1 \tag{18}$$

and w is the ratio of the decreasing geometric progression. Taking into account Equation (15), imposing the current error dynamics given by Equation (17) requires a control law given by

$$d_n \;=\; \frac{L}{V_g T}\big((1-w)i_{REF}(nT) - h_{12}v_C(nT) - (h_{11}-w)i_L(nT)\big). \tag{19}$$

Assuming that the current has tracked its digitally generated reference of constant value $i_{REF}(nT) = I_{REF}u(nT) = i_{REF}[n]$ for the sake of simplicity, the remaining dynamics corresponding to the capacitor voltage will be described by

$$v_C((n+1)T) = h_{21}i_{REF}[n] + h_{22}v_C(nT). \tag{20}$$

Besides, the converter parameters have the following bounds

$$0 < h_{11} < 1, \quad 0 < h_{22} < 1, \quad h_{12} < 0, \quad \text{and} \quad h_{21} > 0. \tag{21}$$

Therefore the coefficient of $v_C(nT)$ in recurrence Equation (20) is positive and smaller than one. Hence, the recurrence will always exhibit a stable behaviour around the converter steady-state operating point given by

$$I_L = I_{REF} \tag{22}$$

$$V_C = \frac{I_{REF}h_{21}}{1 - h_{22}} \tag{23}$$

$$D = \frac{I_{REF}L}{V_g T}\frac{(1-h_{22})(1-h_{11}) - h_{12}h_{21}}{(1-h_{22})}. \tag{24}$$

It has to be pointed out that the recurrence based on the current valley is accurate to describe the inductor current dynamics but it is less exact to explain the capacitor voltage behaviour. This discrepancy is due to the ripple existence in both inductor current and capacitor voltage. Thus, if no losses are assumed, the steady-state mean value of the capacitor voltage will be given by the product of the corresponding mean value of the inductor current and the load resistance. This is not the case in Equation (23), which predicts a capacitor voltage given by the product of the current valley and the load resistance $V_C = R_o\,I_{REF}$. Hence, there is a prediction error in each recurrence period given by

$$\langle V_C \rangle - V_C = \frac{I_{MAX} - I_{REF}}{2}R_o \tag{25}$$

where $\langle V_C \rangle$ and I_{MAX} are the mean and maximum values of capacitor voltage and inductor current in steady-state respectively.

For this reason, the recurrence expressing the voltage behaviour is now modified with the introduction of some additional terms corresponding to a trapezoidal approximation in the calculation of the inductor current mean value. For the sake of simplicity, no losses are considered ($R_1 = 0$, $R_2 = 0$) so that the new recurrence for the capacitor voltage can be expressed as follows:

$$v_C((n+1)T) = h'_{21}i_{REF}[n] + h'_{22}v_C(nT) + (-d_n^2 + 2d_n)\frac{T^2}{2LC}V_g \tag{26}$$

where d_n was given in Equation (19), $h'_{21} = \frac{T}{C}$, and $h'_{22} = 1 - \frac{T}{R_oC} - \frac{T^2}{2LC}$.

The new equilibrium point is

$$I_L = I_{REF} \tag{27}$$

$$V_C = \frac{V_g}{2}\left[(1 - \frac{2L}{R_o T}) + \sqrt{\left(1 - \frac{2L}{R_o T}\right)^2 + \frac{8LI_{REF}}{TV_g}}\right] \tag{28}$$

$$D = \frac{V_C}{V_g}, \tag{29}$$

which is coincident with the reported in [22] using an equivalent discrete-time sliding approach to obtain the recurrence for the output capacitor voltage. In fact, both approaches are completely equivalent if $w = 0$. It is possible to provide a sliding mode interpretation of our approach Equation (17) considering that it proposes a more general switching surface s_e than the one in [22], which was the current error with negative sign. The new surface Equation (31) includes a dynamical term of the current error with the same decreasing geometric progression of the input-output linearization and becomes Equation (30) for $w = 0$.

$$
\begin{aligned}
s(nT) &= i_L(nT) - I_{REF}(nT) \tag{30}\\
s_e(nT) &= s(nT) - w\,s((n-1)T) \tag{31}\\
s_e((n+1)T) &= 0. \tag{32}
\end{aligned}
$$

Adding a dynamical term in the discrete recurrence increases the order of the closed loop reference-to-current transfer function in the z domain Equation (34) that, since it exhibits the order reduction associated with sliding mode ideal dynamics, is of first order and can be directly determined from the recurrence equation of the current.

$$
\begin{aligned}
i_L((n+1)T) &= w\,i_L(nT) + (1-w)I_{REF}(nT) \tag{33}\\
G_I(z) &= \frac{I_L(z)}{I_{REF}(z)} = \frac{1-w}{z-w}. \tag{34}
\end{aligned}
$$

Please note that with $w = 0$ the pole of the I_{REF}-to-I_L current loop transfer function is located at the origin like in a dead-beat control of a first order discrete-time system [23].

4. Voltage Regulation

An outer loop establishing the reference of the inner current loop is added now in order to regulate the output voltage to a desired level V_{REF}, the reference being the sum of two terms which are respectively proportional to the output voltage error and to the integral of the error.

In the descriptive equations of the system, Equations (17), (19), and (26), the discrete current reference, and proportional and integral errors can be expressed as follows

$$
\begin{aligned}
i_{REF}[n] &= k_1 e[n] + k_2 inte[n]\\
e[n] &= V_{REF}u(nT) - v_C(nT)\\
inte[n] &= inte[n-1] + e[n] \tag{35}
\end{aligned}
$$

where V_{REF} is the desired output voltage and k_1, k_2 are respectively the proportional and integral coefficients of the voltage regulation loop. Equivalently, Equation (35) can be written in compact form as

$$i_{REF}[n] = i_{REF}[n-1] + \frac{k_n}{k_{VI}}(e[n] - \beta\,z_P\,e[n-1]) \tag{36}$$

where the control parameters have been normalized with respect to Equation (13) as $k_n = (k_1 + k_2)k_{VI}$ and $\beta = \frac{k_1}{(k_1+k_2)z_P}$.

It can be observed that, although the equation corresponding with the voltage regulator is linear, Equations (17), (19), (26) should be linearized around the equilibrium point so that the choice of coefficients k_1, k_2 or k_n, β can be carried out by means of linear control techniques. The analysis of the linearized equations together with Equation (35) in the z domain results in the following transfer equations

$$G_P(z) = \frac{V_C(z)}{I_{REF}(z)} = \frac{k_{VI}(1-w)(z-z_D)}{(z-w)(z-z_P)} = G_I(z)\frac{k_{VI}(z-z_D)}{(z-z_P)} \tag{37}$$

$$G_C(z) = \frac{I_{REF}(z)}{E(z)} = \frac{k_n}{h_{VI}}\frac{(z-\beta z_P)}{(z-1)} \tag{38}$$

where coefficients k_{VI}, z_N and z_P are

$$k_{VI} = \frac{T(V_g - V_C)}{CVg} \geq 0$$

$$z_D = -\frac{V_C}{V_g - V_C} < 0$$

$$z_P = 1 - \frac{2LT + R_oT^2(2V_C/V_g - 1)}{2LR_oC}. \tag{39}$$

It should be noted that the transfer function in Equation (37) presents a zero depending on the operating point associated with the delay inherent to the digitally PWM controlled converter [22].

The closed loop gain will be

$$\mathcal{T}(z) = \frac{k_n(1-w)(z-z_D)(z-\beta z_P)}{(z-w)(z-z_P)(z-1)} \tag{40}$$

where $\beta = 1$ will fix a closed loop pole in its open loop position. A sensible design of the control parameters will usually consider the ranges $0 < \beta < 1$ and $0 < k_n$.

Let us consider the converter coefficients and operation point described in [22]. It can be shown that $z_D = -1$ for $D = 0.5$ ($V_C = V_g/2$) and the pole z_P in the plant Equation (37) corresponds to the converter coefficient h_{22}. For the given voltage operation point, while the previous zero and pole of the plant are already determined from the point of view of the controller design, the pole at w corresponding to the input-output linearization in the current loop can be adjusted to improve the performance of the voltage controller. Applying the Jury criterion to the characteristic equation $1 + \mathcal{T}(z) = 0$ the following expression can be obtained.

$$-\frac{1+w}{(1-w)h_{22}} < k_n\beta < \frac{1}{h_{22}}. \tag{41}$$

In the most restrictive case ($w \to -1$) Equation (41) becomes

$$0 < k_n < \frac{1}{\beta h_{22}}. \tag{42}$$

Other necessary conditions provided by the Jury criterion have been omitted because they are more complex and not so useful for the control design.

5. Experimental Results and Numerical Simulations

To verify the performance of the proposed digital control and, in particular, the effects of the inner current loop parameter w, a buck converter power stage with synchronous rectification was built and its control implemented on a Texas Instruments TMS320F28335 Digital Signal Controller (DSC) as shown in Figure 2.

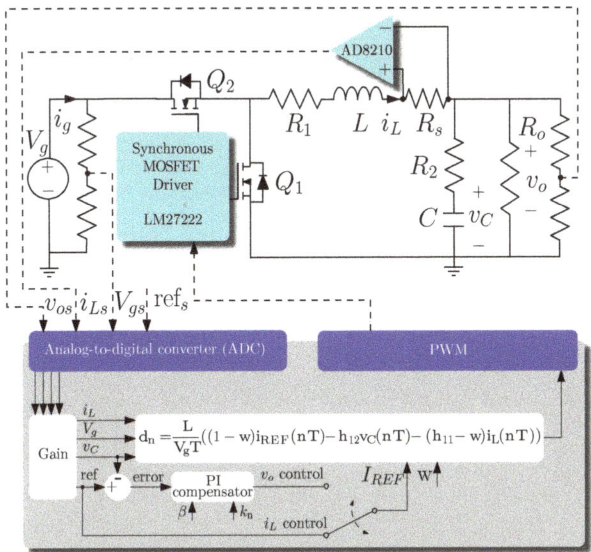

Figure 2. Schematic of the buck regulator.

As mentioned previously, the power stage, whose components are described in Table 1, is the same used in [22]. Since the parameters of the X7R dielectric vary with the frequency, temperature, current ripple and dc applied voltage, an approximated value of 350 µF for the output capacitor bank was experimentally determined from the ac voltage ripple at the operating point. It has been assumed that R_2 is negligible.

On the other hand, the variables required by the control (V_g, v_o, and i_L) are sensed by means of resistive voltage dividers for input and output voltages and by means of an AD8210 difference amplifier for the inductor current. For simplicity reasons, instead of providing an external analog signal, the references are variables within the DSC code. All the analog signals are converted into digital values by means of a 12-bit Analog-to-Digital Converter (ADC) and processed to calculate the control law according to Equation (19) as illustrated in Figure 2. The core of the control law (19) is the same for both the current control (only internal loop) and the voltage control (two loops), the current reference variable i_{REF} being an independent variable in the case of the current control implementation, while in the case of output voltage regulation, the value of i_{REF} is given by the PI compensator output.

The control coefficients k_n and β are only required for voltage control and their selection will be discussed later, whereas three different values have been considered for the ratio of the decreasing geometric progression $w \in [-0.5, 0.0, 0.5]$ of the current loop.

In addition, the PWM signal of Q_2 with duty cycle d_n corresponding to the control law is obtained by using a Digital Pulse Width Modulator (DPWM) of the DSC and sent to the LM27222 integrated circuit to drive the switch formed by the complementary action of Q_1 and Q_2 (IRF3708 MOSFETs).

Figure 3 shows a picture of the experimental setup with the buck power stage and the digital signal controller in front and the main dc power source, an electronic load, the oscilloscope, and an auxiliary power supply that provides 5 V to the digital board at the rear.

Table 1. Components and parameters of the synchronous buck converter

Component/Parameter	Description	Type/Value
Q_1, Q_2	Power MOSFET	IRF3798
C	Ceramic Capacitor X7R dielectric	C5750X7R1C476M230KB $11 \times 47\,\mu F$ 5-V estimated $C \approx 350\,\mu F$
L	Power Inductor Ferrite core	WE-HCC 7443320330 L at 100 kHz: $3.3\,\mu H$
$R_1 + R_s$	Series Resistor Current Sensing + Inductor DCR	Total Resistance: $\approx 6.6\,m\Omega$ SMD $2.2\,m\Omega$ $4.4\,m\Omega$
R_o	Load Power Resistor	Aluminium Housed $1\,\Omega$
T	Switching period	$10\,\mu s$
$f = 1/T$	Switching frequency	100 kHz
Δi_L	Inductor peak-to-peak current ripple ($V_g = 10$ V, $v_o = 5$ V and $R_o = 1\ \Omega$)	7.6 A
I_L	Inductor dc current ($V_g = 10$ V, $v_o = 5$ V and $R_o = 1\ \Omega$)	5 A
Δv_o	Output peak-to-peak voltage ripple ($V_g = 10$ V, $v_o = 5$ V and $R_o = 1\ \Omega$)	27 mV (0.54%)
MOSFET Driver	High-speed synchronous MOSFET driver Texas Instruments	LM27222
Current monitor	Bidirectional current shunt monitor Analog Devices	AD8210

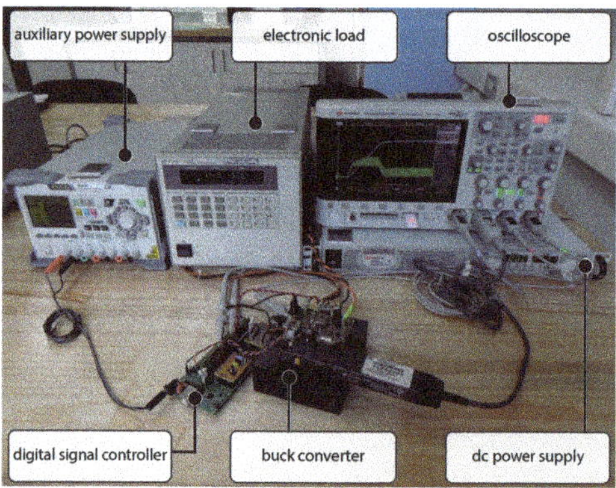

Figure 3. Picture of the experimental setup.

5.1. Inner Current Loop

A sequential sampling of the input and output voltages and the current signal has been synchronized with the PWM sawtooth signal and adjusted so that all variables are available to start performing the duty cycle calculations at the beginning of each cycle. It has been verified that a typical duty cycle calculation requires about 1.5 µs, well below the nominal ON time of 5 µs of the trailing-edge modulation considered. Despite configuring the current to be the last sampled variable, the analog-to-digital sampling, conversion, and latency times, makes impossible to mimic exactly the theoretical procedure. Therefore, instead of sampling the inductor current at its minimum value, it is sampled at about 200 ns before the end of a switching cycle, still at the OFF subinterval where the current slope is negative.

Figure 4 show simulations (left) and experimental results (right) of the inductor current response to a step increase from 3 A to 5 A of the valley current reference, where the three different values of the geometric convergence factor $w \in \{-0.5, 0.0, 0.5\}$ have been considered. These values have been selected so that three different qualitative responses to the same step reference change can be discussed.

The ADC 200 ns-delay, together with the effects of the 350-kHz bandwidth of the implemented current sensor, and the first order RC antialiasing filter of the control card are the main causes for the differences between the minimum current values and their references observed in the PSIM simulations. As it can be seen, since all these factors have been considered, the simulated currents are in a remarkable good agreement with the corresponding experimental waveforms. Being digitally generated, the reference waveform is not shown at the oscilloscope captures.

In Figure 4a,b the convergence factor is $w = 0.5$. The error between the current minimum values and its final steady-state value is reduced in half every switching cycle and an exponential envelope linking the minimums can be easily visualized. In Figure 4c,d the convergence factor has been reduced to $w = 0.0$ so that, disregarding the delay associated with the modulation, the steady-state zero-error is reached in one cycle. As delay associated with the modulation we mean that, assuming ideally no delays and instantaneous calculation times, all reference changes at any point between two consecutive sampling points are seen by the control as a change at the beginning of the cycle, and result in the same response. The $w = 0$ response seems optimal from the point of view of transients in the current loop but, since our objective is to regulate the output voltage, we have decided to analyze also the possibility of using a negative convergence factor. The effects of $w = -0.5$ can be seen in Figure 4e,f, where the transient duration is the same as in the opposite sign case, $w = 0.5$, but the alternation in the error sign makes it difficult to imagine the two exponential envelopes, one increasing and one decreasing, linking every other minimum point of the current. The negative convergence factor effect is similar than in analog peak and valley current-mode control which also could result in a negative discrete-time pole. The difference being that, in our case, the pole is imposed by the value of w while in analog control it is a consequence of many parameters.

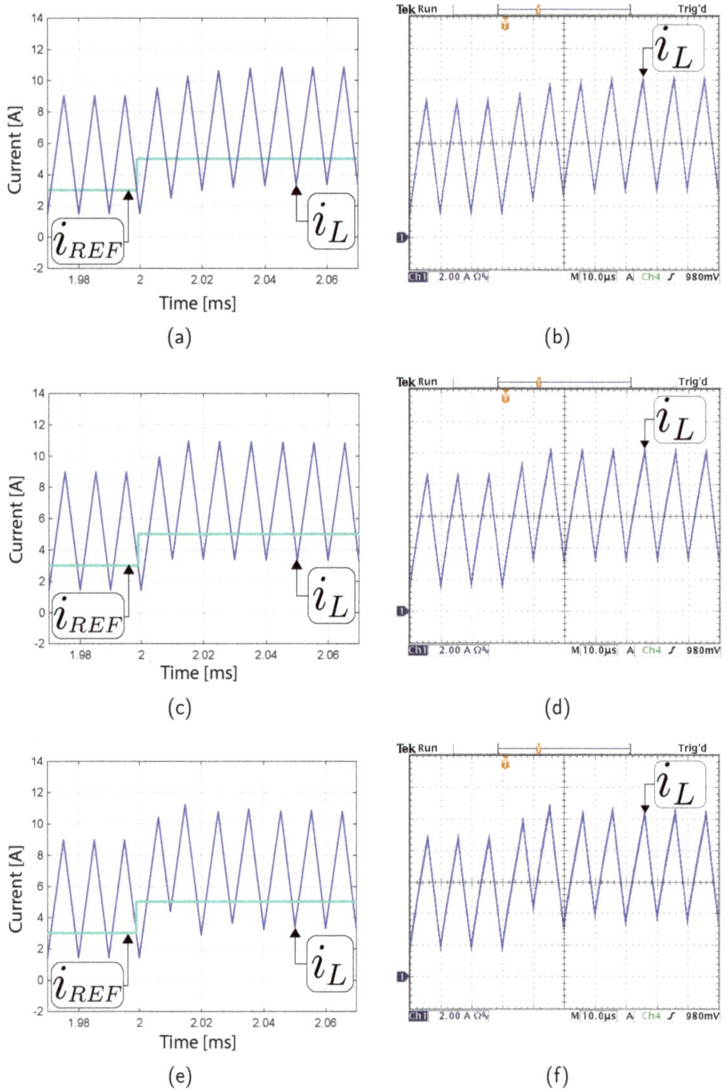

Figure 4. Inner current loop step responses, at t = 2 ms the current reference is increased from 3 A to 5 A: (**a**) PSIM simulation, $w = 0.5$; (**b**) experimental measurement, $w = 0.5$; (**c**) PSIM simulation, $w = 0.0$; (**d**) experimental measurement, $w = 0.0$; (**e**) PSIM simulation, $w = -0.5$; and (**f**) experimental measurement, $w = -0.5$..

5.2. Voltage Regulation Loop

The current reference-to-output voltage discrete transfer in Equation (44) corresponding to $w = 0.0$ has been used to design the parameters of a PI compensator in MATLAB's Control System Designer Toolbox. The PI parameters have been slightly rounded to final values of $k_n = 0.275$ and $\beta = 0.85$ that provide a crossover frequency (CF) of about 8 kHz and a phase margin (PM) of about $45°$. The exact CF and PM figures are provided in Table 2 where the theoretical values are compared with the results obtained from PSIM closed loop ac-sweep simulations and the experimental measurements obtained using a Venable 3120 frequency response analyzer (FRA). The table provides also data obtained with $w = 0.5$ and $w = -0.5$ using the same PI parameters. To help readers to reproduce the theoretical results, all three numerical current reference-to-output voltage discrete transfer functions are provided next.

$$G_P(z)|_{w=0.5} = \frac{z+1}{140z^2 - 206z + 68} \tag{43}$$

$$G_P(z)|_{w=0.0} = \frac{z+1}{70z^2 - 68z} \tag{44}$$

$$G_P(z)|_{w=-0.5} = \frac{3(z+1)}{140z^2 - 66z - 68}. \tag{45}$$

The common PI discrete controller is

$$G_C(z) = 19.3 \, \frac{z - 0.8257}{z - 1}. \tag{46}$$

Table 2. Crossover frequency (CF) and phase margin (PM) for different values of w (with $\beta = 0.85$ and $k_n = 0.55$).

w	Theoretical		Simulated		Experimental	
	CF	PM	CF	PM	CF	PM
$(-)$	(kHz)	(deg)	(kHz)	(deg)	(kHz)	(deg)
0.5	7.3	23.3	6.1	34.0	5.2	36.9
0.0	8.4	43.4	7.3	44.2	6.2	46.4
-0.5	8.6	53.2	7.6	48.7	7.3	47.4

Simulation and experimental results are in good agreement despite a relatively ideal simulation circuit, which has no switching losses and a very simple model of the switching delays. As expected, since the theoretical transfer functions have been obtained after linearizing the approximate model (9), theoretical predictions differ more than simulations from the experimental results. The differences are more significant in the cases of $w = 0.5$ and $w = -0.5$ than for $w = 0.0$. The maximum error is between the theoretical and the experimental values of PM for $w = 0.5$. For the three kinds of data origins (theoretical predictions, simulations, and experimental measurements), selecting $w = -0.5$ provides wider loop bandwidths (CFs) and higher PMs with the same computational efforts in the experimental setup.

Figure 5 depicts the three theoretical discrete root locus plots corresponding to the previous transfer Equations (43), (44), and (45). All the root locus have been obtained for the same PI compensator in Equation (46) (with $\beta = 0.85$ and $k_n = 0.275$), and only the inner current loop convergence factor w is different among them. Root locus in Figure 5b with $w = 0.0$ is the reference diagram predicting a closed loop positive discrete pole and a pair of complex conjugated poles with a damping factor of 0.741, close to $1/\sqrt{2}$. Root locus (a) corresponding to $w = 0.5$ depicts a similar real pole but has a pair of less damped complex poles (damping factor 0.26). Finally, the effect of $w = -0.5$ is depicted in plot (c) where the damping factor of the dominant complex pole pair is 0.94 and the real pole is negative $(-0.305\ \mathrm{s}^{-1})$.

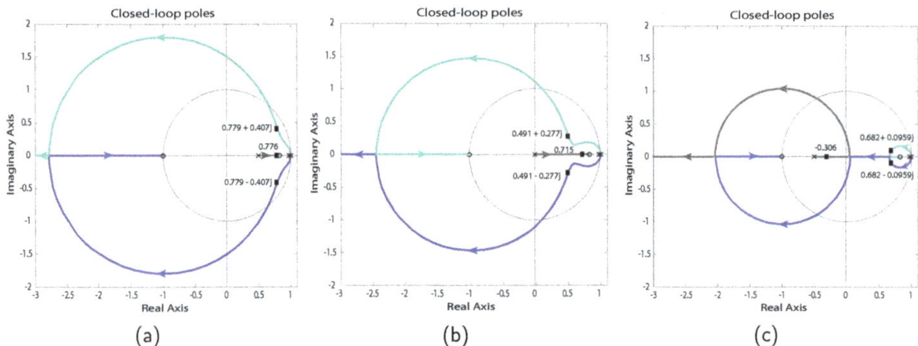

Figure 5. Root locus plots for different current loop convergence factors (w) using the same PI as a voltage loop compensator: (**a**) $w = 0.5$; (**b**) $w = -0.0$; (**c**) $w = -0.5$.

Closed voltage loop simulated and experimental responses to ±2-A step changes in the nominal load are provided in Figure 6. Left plots correspond to PSIM simulations and right plots to experimental oscillograms. Again, the PI compensator is the same in Equation (46) for all cases shown, being w the only parameter that is different among the lines of subplots. The less damped voltage responses in top Figure 6a,b that correspond to $w = 0.5$ are qualitatively in good agreement between them and also with the theoretical underdamped closed loop poles of Figure 5a root locus. The amplitudes of the over- and undershoots and frequency of the ringings are similar although the experimental plot seems slightly more damped, which is to be expected because the simulation considers only conduction losses. Middle Figure 6c,d depict the simulated and experimental load regulation responses when $w = 0.0$. Both voltage responses are a bit more damped than the one expected from a system with a theoretical damping factor slightly larger than $1/\sqrt{2}$. Finally, the dynamics of the voltage responses in bottom Figure 6e,f are coincident with those of an almost critically damped system (dominant complex poles with a theoretically damping factor of 0.94) with settling times of about 120 µs – 140 µs. As expected from the similar PMs in Table 2 there are no large differences between the dynamics with $w = 0.0$ and $w = -0.5$. The main difference is in the over and under shoot absolute amplitudes that are 20 mV smaller for $w = -0.5$ than for $w = 0.0$.

Since experimental load regulation results with $w = -0.5$ are slightly better, the following simulations and experiments are focused on this case. Figure 7 depicts the system responses to relatively large amplitude ±1-V reference changes. In both simulated and experimental cases the valley current reference has been limited between -5 A and 8 A. It is important to limit the maximum current, although indirectly, to avoid saturating the power stage inductor and the oscilloscope current probe. Establishing saturation limits to the current reference is also useful to avoid exceeding the current sensor linear operation range. In both simulation and experiments reference changes from 5 V to 6 V and back from 6 V to 5 V have been considered. In the positive changes, the output voltage reaches the 6-V steady state in about 140 µs. It is worth noting that the 8-A valley current reference limitation acts for about 5 switching periods. The response to the negative reference change is slightly faster, about 120 µs, because in the experiment there is no saturation of the negative current reference and in the simulation it is of less than one switching period, and it is required more duration of the saturation to have a significant effect on the voltage dynamics.

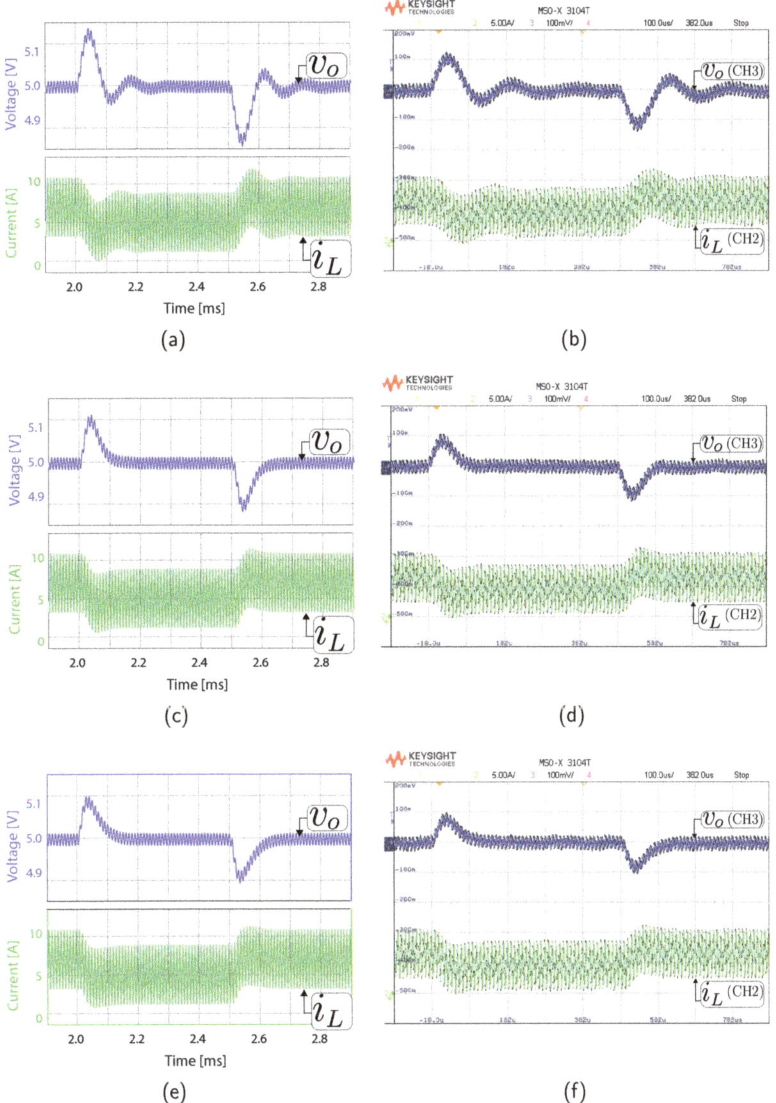

Figure 6. Output voltage and inductor current load regulation responses in closed voltage loop. The system is at steady-state with a load current of 7-A that is suddenly decreased to its nominal value of 5-A (1 Ω resistor) and, after voltage stabilization, is increased back to 7 A: (**a**) PSIM simulation, $w = 0.5$; (**b**) experimental measurement, $w = 0.5$; (**c**) PSIM simulation, $w = 0.0$; (**d**) experimental measurement, $w = 0.0$; (**e**) PSIM simulation, $w = -0.5$; and (**f**) experimental measurement, $w = -0.5$. In the oscillograms, CH1: output voltage v_o (ac coupling, 100 mV/div), CH2: inductor current i_L (dc coupling, 5 A/div).

Figure 8 shows the inductor current and output voltage at the system start-up from zero initial conditions. In this large-signal voltage situation, in addition to the current reference limit of 8 A, a limitation of the minimum duty cycle to 15% ensures that duty cycles smaller than the calculation time do not cause problems with the DPWM. The DSC DPWM comparator generates an interrupt when the duty cycle registry is equal to that of the digital ramp. Therefore, the interrupt is not generated if the duty cycle registry is written with a value smaller than the digital ramp of the DPWM, which causes the control signal to be at high state for the full period and, usually in start-up, severe current spikes. As it can be seen, although the mentioned saturations, the output voltage reaches its desired steady state in about 400 µs in the simulation and in about 600 µs in the experiment without noticeable overvoltages. The difference in start-up times is due to the fact that the X7R dielectric of the 11 small-footprint ceramic capacitors connected in parallel in the experimental power stage is highly nonlinear with respect to the operating voltages and temperatures. From a theoretical value of 517 µF around 0 V the capacitance at 5 V, estimated from current and voltage ripple measurements, derated to about 350 µF.

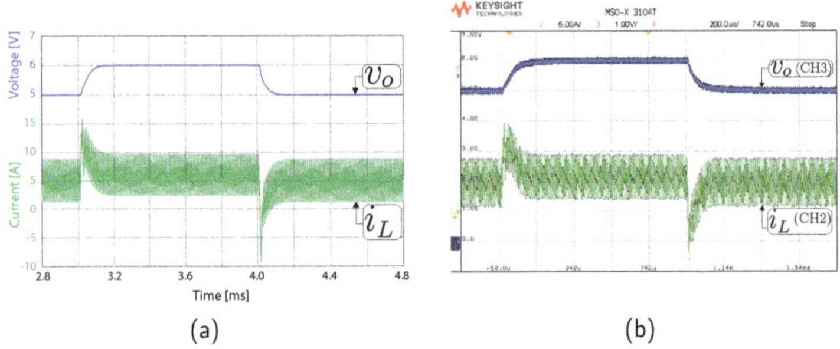

Figure 7. Voltage regulation waveforms corresponding to reference changes from 5 V to 6 V and again to 5 V ($w = -0.5$): (**a**) Simulation; (**b**) Experimental measurement.

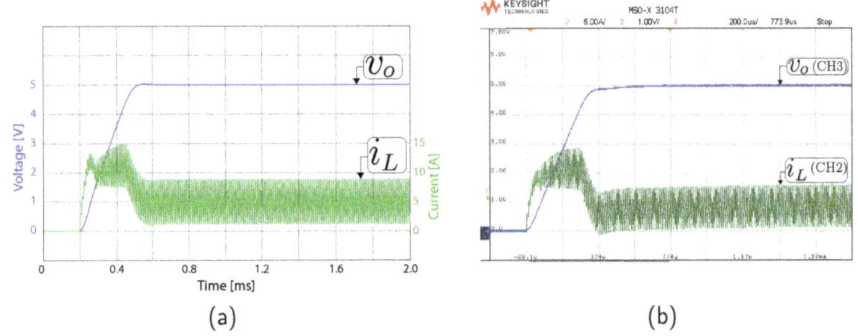

Figure 8. System start-up from zero initial conditions. (**a**) PSIM simulation; (**b**) Experimental measurements.

To avoid having a discontinuous input current, the buck power stage has a 1-mF capacitive input filter that makes difficult to apply high slew-rate perturbations in the input voltage. The experimental results in Figure 9b shows the input voltage obtained when programming at the dc power supply an ideal 25-Hz square voltage between 10 V and 14 V. As it can be seen, after a voltage change the real input voltage converges exponentially to the new steady-state value. The simulation in Figure 9a has been adjusted so that is shows the same input voltage pattern as in the experiment. As expected, the line regulation is excellent, with just small changes in the ripple amplitudes of the output voltage that are caused by the peak-to-peak amplitude modulations in the inductor current.

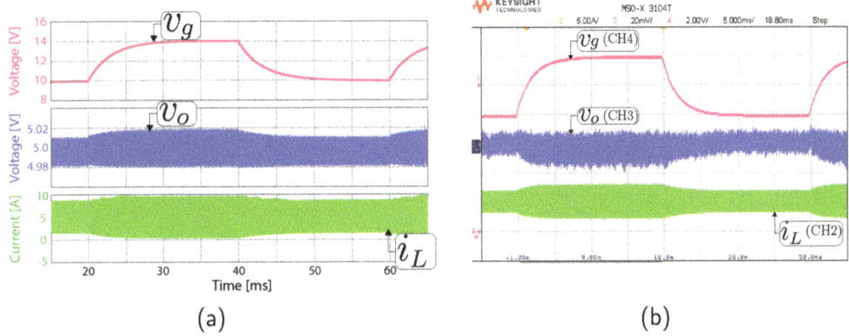

Figure 9. System line regulation. (**a**) PSIM simulation; (**b**) Experimental measurements.

6. Conclusions

This work develops a nonlinear digital control in a PWM buck converter with steady-state voltage regulation and fast transient response. Uses a discrete-time bilinear model of the converter in CCM and applies input-output linearization of the inductor current dynamics to obtain the control law. The current control algorithm employs the samples of capacitor voltage and inductor current at the beginning of the ON interval to determine the duty cycle in the same switching cycle. The main limitation occurs at low voltage values because the calculation time of the particular experimental implementation imposes a minimum duty cycle of about 15%. The internal current loop is stable for all the permitted range of duty cycle values when it operates either alone as an inductor current regulator or in cooperation with an outer loop for output voltage regulation. It has been determined that selecting a negative value for the convergence factor w in the current loop instead of zero, as in the case of a sliding control, can provide a slightly better load regulation performance. The start-up large-signal transient is fast and together with line regulation simulations and experiments verify the excellent regulation of the output voltage.

Author Contributions: Where it is not specified, all authors contributed equally. Conceptualization, E.V.-I. and J.C.; methodology, E.V.-I. and J.C.; software, E.V.-I., C.R. and R.G.; validation, E.V.-I. and R.G.; data curation, C.R.; writing—review and editing, E.V.-I., J.C., A.E.A., C.R. and R.G.; visualization, C.R.; supervision, R.G.

Funding: This research was funded by the Spanish Agencia Estatal de Investigación (AEI) and the Fondo Europeo de Desarrollo Regional (FEDER) under research projects DPI2016-80491-R (AEI/FEDER, UE) and DPI2017-84572-C2-1-R (AEI/FEDER, UE). The work was also supported by the the Chilean Government under Project CONICYT/FONDECYT 1191680 and by SERC Chile (CONICYT/FONDAP/15110019).

Conflicts of Interest: The authors declare no conflict of interest.

References

1. Liu, Y.F.; Meyer, E.; Liu, X. Recent Developments in Digital Control Strategies for DC/DC Switching Power Converters. *IEEE Trans. Power Electron.* **2009**, *24*, 2567–2577, doi:10.1109/TPEL.2009.2030809. [CrossRef]
2. Patella, B.; Prodic, A.; Zirger, A.; Maksimovic, D. High-frequency digital PWM controller IC for DC-DC converters. *IEEE Trans. Power Electron.* **2003**, *18*, 438–446, doi:10.1109/TPEL.2002.807121. [CrossRef]
3. Peterchev, A.; Sanders, S. Quantization resolution and limit cycling in digitally controlled PWM converters. *IEEE Trans. Ind. Electron.* **2003**, *18*, 301–308, doi:10.1109/TPEL.2002.807092. [CrossRef]
4. Costabeber, A.; Corradini, L.; Mattavelli, P.; Saggini, S. Time optimal, parameters-insensitive digital controller for DC-DC buck converters. In Proceedings of the IEEE Power Electronics Specialists Conference, Rhodes, Greece, 15–19 June 2008; pp. 1243–1249, doi:10.1109/PESC.2008.4592101. [CrossRef]
5. Yousefzadeh, V.; Babazadeh, A.; Ramachandran, B.; Alarcon, E.; Pao, L.; Maksimovic, D. Proximate Time-Optimal Digital Control for Synchronous Buck DC-DC Converters. *IEEE Trans. Power Electron.* **2008**, *23*, 2018–2026, doi:10.1109/TPEL.2008.924843. [CrossRef]

6. Yeh, C.A.; Lai, Y.S. Digital Pulsewidth Modulation Technique for a Synchronous Buck DC/DC Converter to Reduce Switching Frequency. *IEEE Trans. Ind. Electron.* **2012**, *59*, 550–561, doi:10.1109/TIE.2011.2143381. [CrossRef]

7. Vidal-Idiarte, E.; Carrejo, C.; Calvente, J.; Martínez-Salamero, L. Two-Loop Digital Sliding Mode Control of DC-DC Power Converters Based on Predictive Interpolation. *IEEE Trans. Ind. Electron.* **2011**, *58*, 2491–2501, doi:10.1109/TIE.2010.2069071. [CrossRef]

8. York, B.; Yu, W.; Lai, J.S. Hybrid-Frequency Modulation for PWM-Integrated Resonant Converters. *IEEE Trans. Power Electron.* **2013**, *28*, 985–994, doi:10.1109/TPEL.2012.2201960. [CrossRef]

9. Kapat, S.; Krein, P. Formulation of PID Control for DC-DC Converters Based on Capacitor Current: A Geometric Context. *IEEE Trans. Power Electron.* **2012**, *27*, 1424–1432, doi:10.1109/TPEL.2011.2164423. [CrossRef]

10. Tsai, C.; Yang, C.; Wu, J. A Digitally Controlled Switching Regulator with Reduced Conductive EMI Spectra. *IEEE Trans. Ind. Electron.* **2012**, *60*, 3938–3947, doi:10.1109/TIE.2012.2207650. [CrossRef]

11. Abu Qahouq, J.A.; Arikatla, V. Online Closed-Loop Autotuning Digital Controller for Switching Power Converters. *IEEE Trans. Ind. Electron.* **2013**, *60*, 1747–1758, doi:10.1109/TIE.2012.2190373. [CrossRef]

12. Perry, A.; Feng, G.; Liu, Y.F.; Sen, P. A Design Method for PI-like Fuzzy Logic Controllers for DC-DC Converter. *IEEE Trans. Ind. Electron.* **2007**, *54*, 2688–2696, doi:10.1109/TIE.2007.899858. [CrossRef]

13. Pitel, G.E.; Krein, P.T. Minimum-Time Transient Recovery for DC-DC Converters Using Raster Control Surfaces. *IEEE Trans. Power Electron.* **2009**, *24*, 2692–2703, doi:10.1109/TPEL.2009.2030805. [CrossRef]

14. Radić, A.; Lukić, Z.; Prodić, A.; de Nie, R.H. Minimum-Deviation Digital Controller IC for DC-DC Switch-Mode Power Supplies. *IEEE Trans. Power Electron.* **2013**, *28*, 4281–4298, doi:10.1109/TPEL.2012.2227503. [CrossRef]

15. Soto, A.; de Castro, A.; Alou, P.; Cobos, J.; Uceda, J.; Lotfi, A. Analysis of the Buck Converter for Scaling the Supply Voltage of Digital Circuits. *IEEE Trans. Power Electron.* **2007**, *22*, 2432–2443. [CrossRef]

16. Meyer, E.; Zhang, Z.; Liu, Y.F. Digital Charge Balance Controller to Improve the Loading/Unloading Transient Response of Buck Converters. *IEEE Trans. Power Electron.* **2012**, *27*, 1314–1326. [CrossRef]

17. Meyer, E.; Liu, Y.F. Digital Charge Balance Controller With an Auxiliary Circuit for Improved Unloading Transient Performance of Buck Converters. *IEEE Trans. Power Electron.* **2013**, *28*, 357–370, doi:10.1109/TPEL.2012.2198923. [CrossRef]

18. Huerta, S.; Alou, P.; Garcia, O.; Oliver, J.; Prieto, R.; Cobos, J. Hysteretic Mixed-Signal Controller for High-Frequency DC-DC Converters Operating at Constant Switching Frequency. *IEEE Trans. Power Electron.* **2012**, *27*, 2690–2696, doi:10.1109/TPEL.2012.2184770. [CrossRef]

19. Maksimovic, D.; Zane, R. Small-Signal Discrete-Time Modeling of Digitally Controlled PWM Converters. *IEEE Trans. Power Electron.* **2007**, *22*, 2552–2556, doi:10.1109/TPEL.2007.909776. [CrossRef]

20. Brown, A.R.; Middlebrook, R.D. Sampled-data modeling of switching regulators. In Proceedings of the IEEE Power Electronics Specialists Conference, Boulder, CO, USA, 29 June–3 July 1981; pp. 349–369.

21. Peretz, M.; Ben-Yaakov, S. Time-Domain Design of Digital Compensators for PWM DC-DC Converters. *IEEE Trans. Power Electron.* **2012**, *27*, 284–293, doi:10.1109/TPEL.2011.2160358. [CrossRef]

22. Vidal-Idiarte, E.; Marcos-Pastor, A.; Giral, R.; Calvente, J.; Martinez-Salamero, L. Direct digital design of a sliding mode-based control of a PWM synchronous buck converter. *IET Power Electronics* **2017**, *10*, 1714–1720, doi:10.1049/iet-pel.2016.0975. [CrossRef]

23. Ogata, K. *Discrete-Time Control Systems*; Prentice-Hall, Inc.: Upper Saddle River, NJ, USA, 1995; pp. 242–257.

24. El Aroudi, A.; Martínez-Treviño, B.; Vidal-Idiarte, E.; Cid-Pastor, A. Fixed Switching Frequency Digital Sliding-Mode Control of DC-DC Power Supplies Loaded by Constant Power Loads with Inrush Current Limitation Capability. *Energies* **2019**, *12*, 1055. [CrossRef]

25. Chen, J.; Prodic, A.; Erickson, R.W.; Maksimovic, D. Predictive digital current programmed control. *IEEE Trans. Power Electron.* **2003**, *18*, 411–419, doi:10.1109/TPEL.2002.807140. [CrossRef]

26. Majo, J.; Martinez, L.; Poveda, A.; de Vicuna, L.; Guinjoan, F.; Sanchez, A.; Valentin, M.; Marpinard, J. Large-signal feedback control of a bidirectional coupled-inductor Cuk converter. *IEEE Trans. Ind. Electron.* **1992**, *39*, 429–436, doi:10.1109/41.161474. [CrossRef]

Article

Sliding Mode Based Control of Dual Boost Inverter for Grid Connection

Diana Lopez-Caiza [1,†,*], **Freddy Flores-Bahamonde** [1,†], **Samir Kouro** [1,†], **Victor Santana** [2,†], **Nicolás Müller** [1,†] **and Andrii Chub** [3,†]

[1] Electronics Engineering Department, Universidad Técnica Federico Santa María, Valparaíso 2390123, Chile; freddy.flores@usm.cl (F.F.-B.); samir.kouro@usm.cl (S.K.); nicolas.muller@alumnos.usm.cl (N.M.)

[2] Advanced Center for Electrical and Electronic Engineering, Valparaíso 2390212, Chile; victor.santana@usm.cl

[3] Department of Electrical Power Engineering and Mechatronics, Tallinn University of Technology, 19086 Tallinn, Estonia; andrusha.chub@gmail.com

* Correspondence: diana.lopez.5@sansano.usm.cl

† These authors contributed equally to this work.

Received: 27 September 2019; Accepted: 4 November 2019; Published: 7 November 2019

Abstract: Single-stage voltage step-up inverters, such as the Dual Boost Inverter (DBI), have a large operating range imposed by the high step-up voltage ratio, which together with the converter of non-linearities, makes them a challenge to control. This is particularly the case for grid-connected applications, where several cascaded and independent control loops are necessary for each converter of the DBI. This paper presents a global current control method based on a combination of a linear proportional resonant controller and a non-linear sliding mode controller that simplifies the controller design and implementation. The proposed control method is validated using a grid-connected laboratory prototype. Experimental results show the correct performance of the controller and compliance with power quality standards.

Keywords: sliding mode control; dual boost inverter; step-up inverter; grid connection

---

1. Introduction

Two-stage power converters are generally used for connecting low-voltage DC sources, such as photovoltaic modules, batteries, fuel cells, and super-capacitors, to AC grids. The input voltage is boosted beyond the peak voltage of the grid by the first stage, a DC–DC converter, which is then converted to AC by the second stage, the grid–tie inverter [1]. However, the efficiency of a two-stage conversion system, particularly when a high step-up voltage DC–DC stage is required, is the main disadvantage of such configuration. The size, cost, and reliability are also factors to take into account in two-stage conversion systems. In this context, single-stage power converters have been proposed to improve the overall efficiency, by reducing the numbers of elements in the system. One of these topologies is the Dual Boost Inverter (DBI) originally introduced in [2].

The DBI consists of two bidirectional DC–DC boost converters, connected in parallel at the DC input and differential mode at the AC output. To obtain a sinusoidal output voltage, each DC–DC boost converter generates a sinusoidal output (with opposite phase between each other), relative to a substantial DC-bias (equal for both converters) which is canceled through the differential connection, leaving only the AC component at the output. Thus, each converter works around an operating point (the DC-bias) but with a large variable output voltage range (the AC component). Additionally, the product between the control and state variables, present in the averaged model of the DC–DC boost converter, shows the high non-linearity of the system [3]. Consequently, the main challenge of DBI is in design of a control system.

From the introduction of the DBI, several control techniques have been proposed in the literature, which can be classified into two main groups of control strategies: independent and global. In the first group, each boost converter is controlled individually to generate its respective sinusoidal output voltage employing linear and non-linear methods. In most cases, a cascaded linear control scheme is used, with a slower outer control loop for the capacitor voltage, and a faster (higher bandwidth) control loop for inductor current. Examples of this control strategy can be found in [4–6], where proportional-integral (PI) controllers are employed in both loops. However, PI controllers can lead to steady-state errors and phase shifts when used to control sinusoidal signals. For this reason, proportional resonant (PR) controllers have been proposed in [7,8], as an alternative to overcome these issues. One common condition for these control strategies is to ensure that the minimum DC-bias is composed of the input DC voltage and half of the amplitude of the output AC voltage to achieve the proper operation of each DC–DC converter.

Also, some non-linear techniques can be found in the group of independent control strategies. Among them: sliding mode control, with a switching surface composed by the error in the voltage of the capacitor and the inductor current is presented in [9]; a dynamic linearizing modulator used to control the capacitor voltage is presented in [10]; the differential flatness propriety, as shown in [11], where the individual control of the output voltage is indirectly accomplished through the regulation of energy stored in each boost converter; and finite control set model predictive control, where a non-linear discrete model of the DBI is used to predict and optimize the behavior of each converter, as introduced in [12].

In contrast, in the global control strategy group, the differential output AC voltage of the DBI is considered to be the main control objective. This type of control was introduced for the first time in [3], where a cascaded control diagram based on the sliding mode approach is applied to achieve the sinusoidal output voltage. The external control loop regulates the output voltage error of the inverter using a PI controller. The inner control loop corresponds to a switching surface, synthesized from the difference between the current of the inductors and the external controller output. One advantage of this strategy is the reduction of control loops, which leads to a decrease in the number of required sensors. An extended analysis of the equilibrium point for this control strategy is presented in [13], where the DC component of the capacitor voltages is automatically adjusted to the two-fold of the input voltage.

In most cases, the control strategies have been tested for passive loads (R and RL loads). Although good performances under perturbations have been achieved, the grid connection has not been thoroughly analyzed. This is mainly because it is difficult to find a relationship between the output current and the control variables of the inverter. Nevertheless, experimental validations of grid-connected DBIs can be found in [7,8]. In both cases, the cascaded linear strategy is used to control individually each boost converter, including an additional control loop based on active and reactive power. Therefore, five control loops are necessary to connect the DBI to the grid, making the design and implementation of the control system a complex process. This is particularly an issue for the DBI, which is intended for low power applications, such as grid-connected photovoltaic microinverters, for which low cost control platform are usually used. Other high performance contributions regarding DC–DC converter control have been successfully proposed such as Robust Time-Delay Control for a boost converter [14], as well as adaptive SMC [15], and higher order SMC techniques [16]. However, these have only been proposed for DC–DC converters (not for a DBI with the generation of an AC waveform), and while their extension to current control for DBI may be interesting, they are inherently more complex to implement and require high-end control platforms.

The main contributions of this work are the development of a simple and low computational control system based on SMC with only two control loops. One external linear control loop that regulates the grid current through a PR controller, and an internal non-linear control loop that is composed of a switching surface to control the difference between the current of the DBI inductors. This is feasible due to the symmetry of the DBI allowing the control of both DC–DC converters as a single system by means of a unique control signal, based on an extension of the theoretical derivation of the SMC presented in [13]. However, in this paper the system model and controller derivation has been modified to control the output current instead of voltage. Furthermore, this paper is the first time this principle has been applied to a grid-connected system, with an AC current output, and evaluated experimentally. In addition, experimental performance under grid perturbations and dynamic behavior of the current controller are included. The proposed control system can perform in such circumstances while complying with IEEE standard 1547. The DC-bias achieved by the proposed method is double the input voltage, which is lower than the DC-bias required by traditional methods [4–8], which impacts the size of the capacitors and blocking voltage of the devices.

This paper is organized as follows, a detailed description of the DBI topology is presented in Section 2, the control strategy proposed in this work is introduced in Section 3, the experimental validation and main results of the grid-connected DBI are presented in Section 4, and Section 5 presents the main accomplishments and conclusions of this work.

2. Topology Description

The concept of a generic step-up voltage single-stage differential inverter is shown in Figure 1. The inverter is composed of two bidirectional DC–DC converters, which share the same input source, while their output voltages are connected in differential mode. Each DC–DC converter generates a sinusoidal output voltage with a DC-bias (V_{dc}), as shown below

$$v_{an}(t) = V_{dc} + \frac{v_{ac}(t)}{2} \tag{1}$$

$$v_{bn}(t) = V_{dc} - \frac{v_{ac}(t)}{2} \tag{2}$$

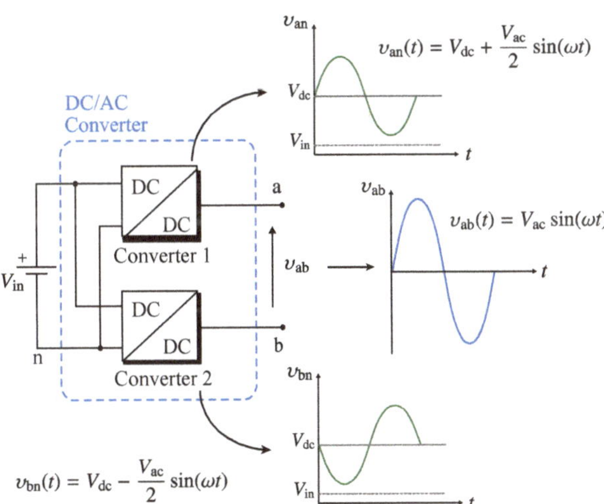

Figure 1. The generic concept of a single-stage step-up differential mode inverter.

The AC component of the output signal of each converter is in the opposite phase regarding the other converter. Thus, considering the same DC component for both converters, the output voltage of the inverter is given by

$$v_{ab}(t) = v_{an}(t) - v_{bn}(t) = V_{ac} \cdot \sin(\omega t) \tag{3}$$

where V_{ac} is the amplitude of the output voltage of the inverter. By generating opposite phase AC signals, the total converter output voltage doubles the individual converter AC amplitude. Therefore, this configuration can achieve a high step-up voltage ratio conversion, one provided by the DC–DC converter boost ratio and one that doubles voltage due to the differential connection. Please note that the DBI fulfills two functions with a single-stage conversion: voltage step-up the DC to AC conversion, to accomplish the grid connection.

In the literature, several bidirectional DC–DC converters have been used, e.g., flyback [17,18], cuk [19,20], and boost [2–8]. The latter topology, also known as Dual Boost Inverter (DBI), is the one under analysis in this work. The power circuit of the DBI consists of two bidirectional DC–DC boost converters as shown in Figure 2 for a grid-connected application. Please note that the outputs of the two DC–DC converters are connected to the grid through a symmetrically divided inductive filter L_s. The grid resistance R_s is shown for modeling purposes.

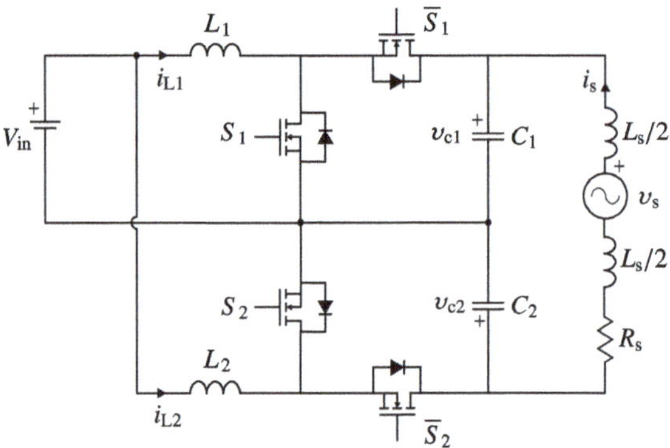

Figure 2. Dual boost inverter topology.

To obtain a single averaged switched model of the whole system, two complementary control signals are considered to be in [13], defining the global operation of the system. This idea signifies the difference regarding other works, where the model of the inverter is obtained for each boost converter. Considering the signals $u(t) = S_1$ and $1 - u(t) = S_2$, the averaged switched model of the DBI is described by

$$L_1 \frac{di_{L1}(t)}{dt} = V_{in} - v_{c1}(t) \cdot (1 - u(t)) \tag{4}$$

$$L_2 \frac{di_{L2}(t)}{dt} = V_{in} - v_{c2}(t) \cdot u(t) \tag{5}$$

$$C_1 \frac{dv_{c1}(t)}{dt} = (1 - u(t)) \cdot i_{L1}(t) + i_s(t) \tag{6}$$

$$C_2 \frac{dv_{c2}(t)}{dt} = u(t) \cdot i_{L2}(t) - i_s(t) \tag{7}$$

where i_{L1} and i_{L2} are the currents through the inductors L_1 and L_2, i_s is the grid current, v_{c1} and v_{c2} are the voltage of the capacitors, V_{in} is the input voltage, $u(t)$ and $1 - u(t)$ are the duty cycles, and S_1 and S_2 are the switching signals.

The product between the state variables and control input (bilinear term) in Equations (4)–(7) shows that the non-linearity of the inverter model is preserved. Moreover, the DBI is integrated to the grid through an inductive filter L_s (R_s represents the resistance of the inductive filter and grid), as shown in the equivalent circuit of Figure 3, and the voltage equation can be obtained by

$$L_s \frac{di_s(t)}{dt} = v_o(t) - i_s(t)R_s - v_s(t) \tag{8}$$

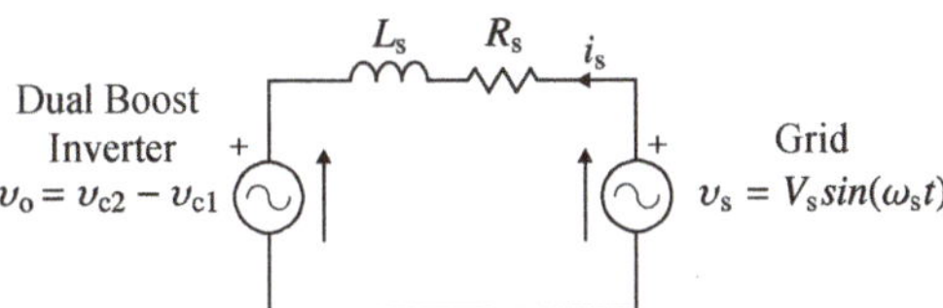

Figure 3. Equivalent model of the DBI with grid connection.

3. Control Strategy

The proposed control of the DBI is shown in Figure 4, which consists of cascaded control loops. The fast non-linear inner control loop, based on sliding mode control, regulates the difference of the current in the inductors of DC–DC boost converters, while the linear and slower outer control loop, manages through a PR the current injected to the grid.

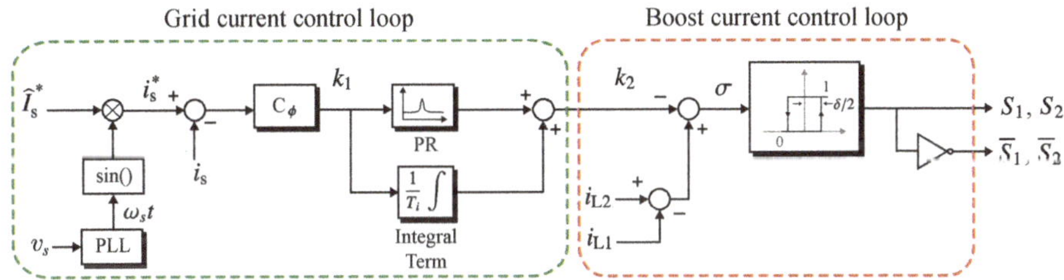

Figure 4. The cascaded control scheme of the grid-connected DBI, with external PR controller and internal sliding mode controller.

3.1. Inner Current Control Loop

To control the output current of the DBI, the sliding mode approach is proposed in the present work for the non-linear inner control loop, due to its inherent properties guaranteeing stability and robustness against variation of parameters with high regulation dynamics, as shown in [21,22]. The analysis here developed has its foundation in the state variables behavior of the dual boost inverter under the presence of a sinusoidal reference introduced in [13]. However, this was solved in [13] for a voltage control loop with a linear load, which cannot be directly extended for a current control for grid-connected applications. This adds a new state variable to the system defined in Equation (8). Thus, it is necessary to adapt the SMC law to fulfill this new control objective. To accomplish this, the analysis is based on the Filippov's method [23] and its corresponding equivalent control approach.

3.1.1. Sliding Surface Selection

Considering that the output voltage of the inverter is obtained from subtracting the voltage of the capacitors (v_{c1} and v_{c2}), and that the capacitor voltage control is related to the inductor current, it is possible to establish that the difference between the current of the inductors controls indirectly the output voltage of the dual boost inverter [3]. Therefore, the sliding surface ($\sigma(t)$) can be defined by

$$\sigma(t) = -k_2(t) + i_{L2}(t) - i_{L1}(t) \tag{9}$$

where k_2 is the output of external control loop.

Please note that by considering the current derivatives of Equations (4) and (5), the sliding surface can be rewritten as

$$\sigma(t) = -k_2(t) + \int_{t_0}^{t} \left[\frac{V_{in}}{L_2} - \frac{v_{c2}(t)}{L_2} \cdot u(t) \right] dt - \int_{t_0}^{t} \left[\frac{V_{in}}{L_1} - \frac{v_{c1}(t)}{L_1} \cdot (1 - u(t)) \right] dt \tag{10}$$

3.1.2. Equivalent Control

In other to guarantee that the sliding mode is maintained on the selected surface, it is necessary to find the equivalent control (u_{eq}) through the invariance condition given by

$$\left. \frac{d\sigma}{dt} \right|_{\substack{\sigma=0 \\ u=u_{eq}}} = 0 \tag{11}$$

Hence, the derivative of the sliding surface evaluated in $\sigma = 0$ and $u = u_{eq}$ is

$$\frac{d\sigma}{dt} = -\frac{dk_2(t)}{dt} + \frac{V_{in}}{L_2} - \frac{v_{c2}(t)}{L_2} \cdot u_{eq}(t) - \frac{V_{in}}{L_1} + \frac{v_{c1}(t)}{L_1} \cdot (1 - u_{eq}(t)) = 0 \tag{12}$$

From (12) and due to the symmetry of the inverter ($L = L_1 = L_2$), the equivalent control is defined as

$$u_{eq}(t) = \left(-\frac{dk_2(t)}{dt} + \frac{v_{c1}(t)}{L} \right) \frac{L}{v_{c1}(t) + v_{c2}(t)} \tag{13}$$

3.1.3. Existence Condition

With the expression of equivalent control in Equation (13), the next step is to prove the existence condition, which can be determined by

$$\sigma(x, t) \cdot \frac{d\sigma(x, t)}{dt} < 0 \tag{14}$$

Thus, the derivative of the surface is replaced in Equation (14), expressing the existence condition as

$$\sigma \left[-\frac{dk_2(t)}{dt} + \frac{V_{in}}{L} - \frac{v_{c2}(t)}{L} \cdot u(t) - \frac{V_{in}}{L} + \frac{v_{c1}(t)}{L} \cdot (1 - u(t)) \right] < 0 \tag{15}$$

To establish a relationship between $u(t)$ and $u_{eq}(t)$, the expression $-u_{eq}(t) + u_{eq}(t) = 0$ is added in Equation (15), resulting in

$$\sigma \left[-\frac{dk_2(t)}{dt} + \frac{V_{in}}{L} - \frac{v_{c2}(t)}{L} \cdot \left(u(t) \underbrace{-u_{eq}(t) + u_{eq}(t)}_{=0} \right) - \frac{V_{in}}{L} + \frac{v_{c1}(t)}{L} \cdot \left(1 - \left(u(t) \underbrace{-u_{eq}(t) + u_{eq}(t)}_{=0} \right) \right) \right] < 0 \tag{16}$$

It is possible to reduce (16) by considering Equation (12), which leads to

$$\sigma \left[\frac{v_{c1}(t) + v_{c2}(t)}{L} \cdot (-u(t) + u_{eq}(t)) \right] < 0 \tag{17}$$

Finally, evaluating (17), and considering that $v_{c1} > 0$, $v_{c2} > 0$ and $L > 0$, it can be determined that if the switching surface is positive, the term $(-u(t) + u_{eq}(t))$ should be negative to accomplish the existence condition, which implies $u = u^+ = 1$. Otherwise, if the switching surface is negative, the term

$(-u(t) + u_{eq}(t))$ should be positive and the action control takes the minimum value ($u = u^- = 0$), which can be expressed as

$$\begin{cases} \sigma > 0 \text{ and } u > u_{eq} & \longrightarrow u = u^+ = 1 \\ \sigma < 0 \text{ and } u < u_{eq} & \longrightarrow u = u^- = 0 \end{cases} \tag{18}$$

The control action defined by Equation (18) leads to the state trajectory to slide on the switching surface and eventually reach the intersection of the switching surface and the equilibrium point converging in a finite time, as demonstrated in [13].

3.2. Outer Current Loop

The main goal of this loop is to regulate the grid current of the DBI. The angle of i_s^* is obtained from a PLL, which is used to reconstruct a sinusoidal waveform enabling synchronization with the grid [24,25]. The amplitude for the current reference will be considered to be a given value, provided externally to fulfill purposes of the application. Since the current reference is sinusoidal, a proportional resonant controller is used, which is tuned to the grid angular frequency ω_s.

3.2.1. Linearization

To design the controller of the outer loop, it is necessary to find the transfer function between the grid current i_s and the output of the external control loop k_2. The equivalent control approach (Equation (13)) is used to introduce the variable k_2 in the averaged switched model of the inverter. Therefore, the equivalent control of $u_{eq}(t)$ and its complement $(1 - u_{eq}(t))$ are redefined as

$$u_{eq}(t) = \frac{v_{c1}(t)}{v_{c1}(t) + v_{c2}(t)} - \frac{L}{v_{c1}(t) + v_{c2}(t)} \cdot \frac{dk_2(t)}{dt} \tag{19}$$

$$1 - u_{eq}(t) = \frac{v_{c2}(t)}{v_{c1}(t) + v_{c2}(t)} + \frac{L}{v_{c1}(t) + v_{c2}(t)} \cdot \frac{dk_2(t)}{dt} \tag{20}$$

The control signals $u(t)$ and $1 - u(t)$ are substituted by the equivalent control $u_{eq}(t)$ and $1 - u_{eq}(t)$ in Equations (4)–(7), where it is possible to identify that the derivative of current through L_1 presents the same behavior of the derivative of i_{L2}. For this reason, i_{L1} was omitted and the state variables of the non-linear model are defined by

$$\begin{aligned} \dot{x}(t) &= f\left[x(t)\right] + \mathbf{B}u(t) \\ y(t) &= \mathbf{C}x(t) \end{aligned} \tag{21}$$

where

$$f(x) = \begin{bmatrix} f_1 \\ f_2 \\ f_3 \\ f_4 \end{bmatrix} = \begin{bmatrix} \dot{x}_1 \\ \dot{x}_2 \\ \dot{x}_3 \\ \dot{x}_4 \end{bmatrix} = \begin{bmatrix} \dfrac{di_{L2}(t)}{dt} \\ \dfrac{dv_{c1}(t)}{dt} \\ \dfrac{dv_{c2}(t)}{dt} \\ \dfrac{di_s(t)}{dt} \end{bmatrix} \tag{22}$$

$$y = \begin{bmatrix} i_s \end{bmatrix} = \begin{bmatrix} x_4 \end{bmatrix}$$

$$u = \begin{bmatrix} k_2 \end{bmatrix}$$

Taking into account that $\sigma(t) = 0$, the current through inductor L_1 can be obtained as,

$$i_{L1}(t) = i_{L2}(t) - k_2(t) \tag{23}$$

Considering (23), $v_o = v_{c2} - v_{c1}$, and that the value of the capacitors are the same $C_1 = C_2 = C$, the linear model of the system can be expressed as

$$\Delta \dot{x} = \begin{bmatrix} \Delta \dot{x}_1 \\ \Delta \dot{x}_2 \\ \Delta \dot{x}_3 \\ \Delta \dot{x}_4 \end{bmatrix} = \underbrace{\begin{bmatrix} 0 & -\dfrac{1}{4L} & -\dfrac{1}{4L} & 0 \\ \dfrac{1}{2C} & 0 & 0 & \dfrac{1}{C} \\ \dfrac{1}{2C} & 0 & 0 & -\dfrac{1}{C} \\ 0 & -\dfrac{1}{L_s} & \dfrac{1}{L_s} & \dfrac{R_s}{L_s} \end{bmatrix}}_{\mathbf{A}} \cdot \begin{bmatrix} \Delta x_1 \\ \Delta x_2 \\ \Delta x_3 \\ \Delta x_4 \end{bmatrix} + \underbrace{\begin{bmatrix} 0 \\ -\dfrac{1}{2C} \\ 0 \\ 0 \end{bmatrix}}_{\mathbf{B}} \cdot \Delta u + \underbrace{\begin{bmatrix} \dfrac{1}{L} \\ 0 \\ 0 \\ 0 \end{bmatrix}}_{\mathbf{E}} \cdot \Delta p \quad (24)$$

$$\Delta y = \underbrace{\begin{bmatrix} 0 & 0 & 0 & 1 \end{bmatrix}}_{\mathbf{C}} \cdot \begin{bmatrix} \Delta x_1 \\ \Delta x_2 \\ \Delta x_3 \\ \Delta x_4 \end{bmatrix} \quad (25)$$

Please note that the same equilibrium point shown in [13] was used in this analysis, which is defined as $[x_{10}, x_{20}, x_{30}, x_{40}, k_{20}] = [0, 2 \cdot V_{in}, 2 \cdot V_{in}, 0, 0]$.

Considering the equations in Laplace domain, the transfer function can be defined as

$$G(s) = \frac{\Delta y(s)}{\Delta u(s)} = \mathbf{C} \cdot (s\mathbf{I} - \mathbf{A})^{-1} \cdot \mathbf{B} \quad (26)$$

Replacing the matrices and performing some algebraic operations, the relation between the grid current i_s and k_2 can be derived as

$$G(s) = \frac{\Delta y(s)}{\Delta u(s)} = \frac{i_s(s)}{k_2(s)} = \frac{1}{2CL_s} \cdot \frac{1}{s^2 + \dfrac{R_s}{L_s}s + \dfrac{2}{CL_s}} \quad (27)$$

Please note that the transfer function is of second order and depends only on the capacitor value, the grid filter, and the grid resistance.

3.2.2. Outer Control Design

The plant $G(s)$ is critically stable because it presents a complex conjugate pole pair in the left half-plane close to the imaginary axis. To better illustrate this issue, the frequency response of the plant is shown in Figure 5a, where a significant resonant peak located at 1 kHz can be appreciated. To compensate this peak, to assure a zero steady-state error at 60 Hz (grid frequency) and to regulate the grid current of the inverter, a PR controller is used. The transfer function of the PR controller in the Laplace domain is given by

$$C_{PR}(s) = k_p + \frac{2k_i\omega_c s}{s^2 + 2\omega_c s + \omega_o{}^2} \quad (28)$$

where k_p is the proportional gain, k_i is the resonant gain, ω_c is the cut-off frequency, and ω_o is the fundamental frequency. Please note that to deal with the sensitivity issue of the ideal PR controller, a bandwidth around the resonant frequency of the controller is added through the cut-off frequency, obtaining a non-ideal PR controller with finite gain [26]. The parameters applied to calculate the PR controller are shown in Table 1.

The closed-loop Bode diagram of the system (T_1) is shown in Figure 5b. Although a finite gain at grid frequency is introduced to obtain a zero-state error in the tracking of the grid current reference, a sufficient degree of the relative stability is not achieved, since the phase margin is equal to zero [27].

Therefore, a phase compensator C_ϕ is included to increase the phase margin. The transfer function of this compensator is given by

$$C_\phi(s) = k \cdot \frac{s+a}{s+b} \tag{29}$$

This compensator is composed of a pole and a zero to incorporate phase in the system [28]. As a result, the effect of the cascaded phase compensator is shown in the bode diagram of closed-loop $T_2(s)$ of Figure 5b, where a phase margin of 31.1° is achieved. The parameters of the phase compensator are shown in Table 1.

In addition, to avoid an offset in the grid current of the inverter, due to the fact that in a practical implementation both dc–dc converters will not be exactly the same, an integration term is incorporated in the control scheme, as shown in Figure 4, to force the steady-state error to zero at $\omega = 0$.

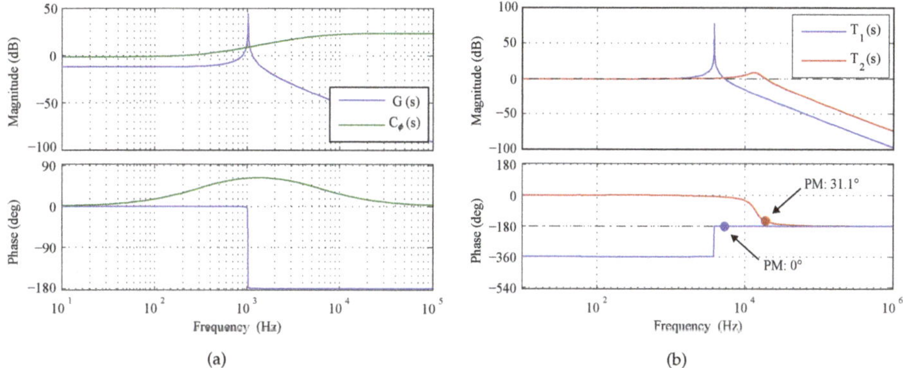

Figure 5. Bode diagrams: (**a**) plant and phase compensator, (**b**) closed-loop without and with the phase compensator.

Table 1. Main parameters of the experimental setup.

Symbol	Parameter	Experimental Value
Grid Parameters		
v_s	Grid voltage	110 [V_{rms}]
f_s	Grid frequency	60 [Hz]
L_s	Grid filter inductance	10 [mH]
Converter Parameters		
V_{in}	Input voltage	70 [V]
L_1, L_2	Inverter inductors	55 [μH]
C_1, C_2	Inverter capacitors	5 [μF]
Control Parameters		
k	Gain of C_ϕ	1
k_p	Proportional gain of PR	50
k_i	Resonant gain of PR	700
ω_c	Cut-off frequency of PR	5 [rad/s]
a	Zero of phase compensator	2000
b	Pole of phase compensator	35,000

4. Experimental Results

The proposed control for the DBI is validated using the experimental setup shown in Figure 6. The experimental prototype is composed of the power and control parts. In the power part, two dc–dc boost converters have been connected in differential mode, the differential output is connected to the grid through a line filter. The nominal parameters of the setup are summarized in Table 1.

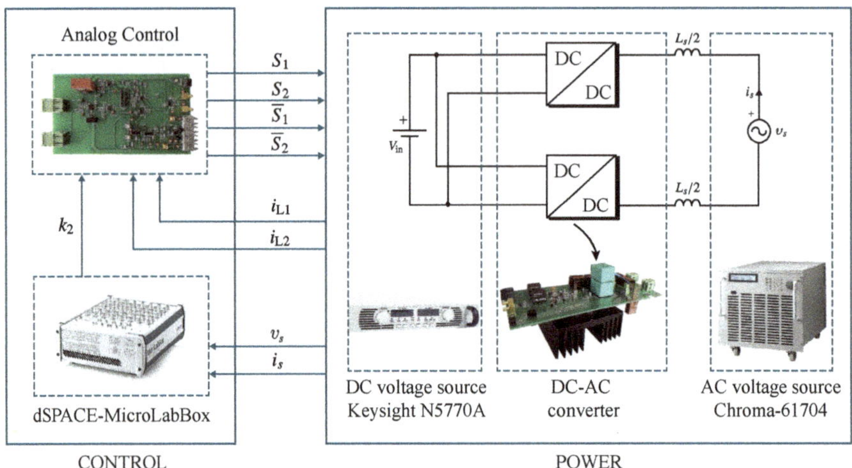

Figure 6. Experimental diagram of dual boost inverter.

Conventional control systems for power converters are implemented using digital platforms such as DSP and FPGA, due to their fast computational times. On the other hand, sliding mode control is commonly implemented with analog circuits because a hysteresis comparator is used to achieve a finite switching frequency. In this work, a hybrid implementation is proposed taking the advantages of both types of implementation. The phase compensator and PR controller are implemented in a DSpace MicroLabBox platform using Matlab/Simulink, while an analog circuit contains the sliding mode inner control loop, which controls the current in the boost converters. Figure 7 shows the schematic diagram of the analog implementation of the sliding mode controller, where two operational amplifiers are used to generate the switching surface, while the hysteresis is achieved through a comparator LM319 and a J-K Flip-Flop (MC14027B integrated circuit). The hysteresis boundaries given by voltage signals are regulated through variable resistors.

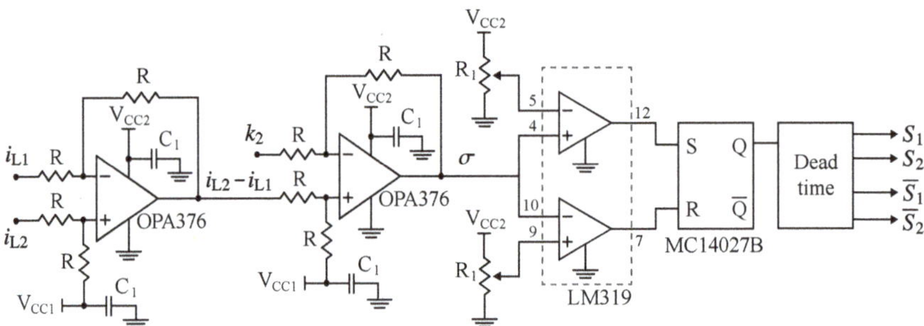

Figure 7. Analog circuit for control stage, where R = 20 kΩ, R_1 = 10 kΩ, V_{CC1} = 2.5 V, V_{CC2} = 5 V, and C_1 = 0.1 μF.

To show the operation of the proposed control, the behavior of the converter is tested when it is connected to the grid. For this purpose, the input has been emulated using a dc voltage source (Keysight N5770A) operating at 70 V_{dc}, while the grid has been emulated using an AC source (Chroma 61704) operating at 110 $V_{ac,rms}$/60 Hz. The experimental results are shown in Figure 8.

The grid current reference i_s^* and measured current i_s of the DBI are presented in Figure 8a, where it is possible to verify that accurate tracking of the current reference is achieved by the PR controller. The angle of the reference was extracted from the grid voltage v_s, which is shown in Figure 8d. The output of this external control loop $k2$ is the reference for the difference between the inductor currents, which can be seen in Figure 8a.

Figure 8b shows the voltage of the capacitors and currents in the inductors, which are balanced despite not being directly controlled. Please note that the magnitudes of the dc and AC components of the output voltages and inductor currents of each dc–dc converter are the same. In the case of the voltages, the dc component of the capacitor voltage is 140 V, which is the double of the input voltage (V_{in}). The maximum value of the amplitude of AC component is around 110 V, and the phases of these voltages are shifted by $180°$.

From Figure 8c it can be seen that the output voltage of the inverter is sinusoidal, despite the voltage of each capacitor is not purely sinusoidal. Additionally, the condition of proper operation of the inverter ($v_o > v_s$) is fulfilled, because the output voltage of the inverter is around 111 $V_{ac,rms}$, which is higher than the grid voltage (110 $V_{ac,rms}$) .

Figure 8d presents the grid current (i_s) and voltage (v_s), together with the output voltage and current through the inductance of one of the dc–dc converters. The grid current is very close to a sinusoidal waveform and is always in phase with the grid voltage to ensure the power factor close to unity.

The spectrum and total harmonic distortion (THD) of the grid current in the steady state were obtained to analyze the power quality. These results are presented together with the limits of the IEEE standard 1547 in Figure 9. Note how the harmonics present in the grid current comply with the standard. The total harmonic distortion obtained for the grid current is 4.47%. This is a very good result, considering this is the first iteration of a laboratory prototype (stray inductances and other circuit components have not been optimized), and a simple inductive filter was used for grid connection. The harmonic content could be improved further with higher order filters (such as LC or LCL), typically used in such applications.

Two tests were performed to evaluate the dynamic performance of the proposed control method. The first test consists of a step-down and a step-up in the output (grid) current reference (i_s^*) to assess the tracking performance. The second test consists on applying a voltage dip in the grid voltage (v_s) to assess the performance under system perturbations. Figure 10a shows the experimental results associated with the step-down (1.0 to 0.8 A) in the grid current reference, while Figure 10b presents the results obtained for the step-up (0.8 to 1.0 A) in the grid current reference. As illustrated by both figures, a fast dynamic behavior is achieved by the proposed control method, where the tracking of the grid current reference is promptly accomplished. As a result, the grid current variations are reflected in the amplitude of the current through the inductors i_{L1} (and i_{L2}); however the voltage of the capacitor v_{c1} (and v_{c2}) in both cases is kept constant. Both step changes were introduced at the peak value of the current reference, to evaluate the most demanding dynamic scenario for the controller.

For the second dynamic test, shown in in Figure 11, a voltage dip of 20% was introduced in the grid voltage (only the transition from voltage dip to nominal voltage is shown). The grid voltage amplitude transitions from 88 to 110 $V_{ac,rms}$. This generates an increase of the AC component in the voltage of both output capacitors (v_{c1} and v_{c2}), while the grid current (i_s) remains controlled without reflecting any change caused by this perturbation, highlighting the robustness of the proposed control method. However, since the inductor current i_{L1} (and i_{L2}) depends on the difference between the input voltage v_{in} and the voltage in each output capacitor, a small variation is experienced by i_{L1} and i_{L2}. Please note that during this test the output (grid) current reference was kept constant.

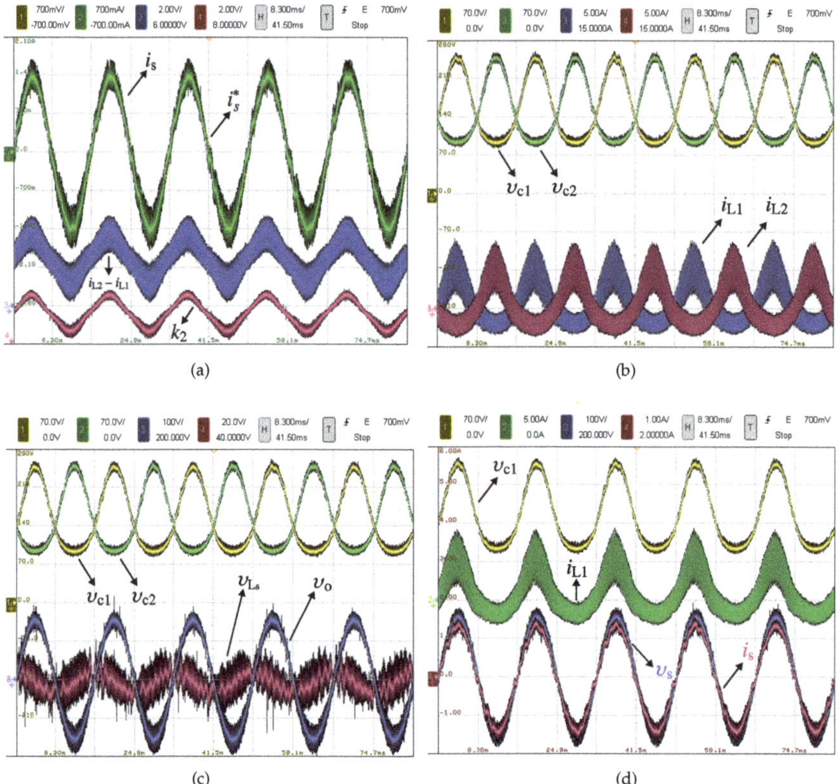

Figure 8. Experimental results with grid connection and input voltage of 70 V: (**a**) reference and measurement of grid current, difference of the inductor currents and output of outer control loop (k_2), (**b**) voltage of the capacitors (v_{c1}, v_{c2}) and current through the inductors (i_{L1}, i_{L2}), (**c**) voltage of the capacitors (v_{c1}, v_{c2}), voltage of the line filter, output voltage (v_o) and (**d**) v_{c1}, i_{L1}, grid voltage and current.

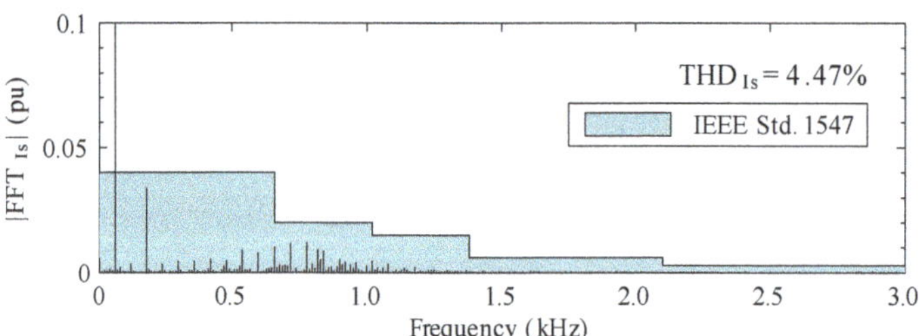

Figure 9. Grid current spectrum with 70 V of input voltage.

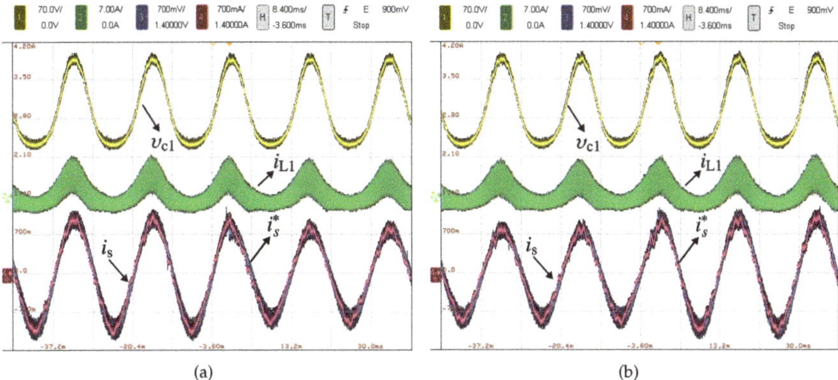

(a)   (b)

Figure 10. Experimental results under variations in the output current reference (voltage of the capacitor C_1 (v_{c1}), current through the inductor L_1 (i_{L1}), reference (i_s^*) and measurement (i_s) of grid current): (**a**) Step-down in the output current reference, and (**b**) Step-up in the output current reference.

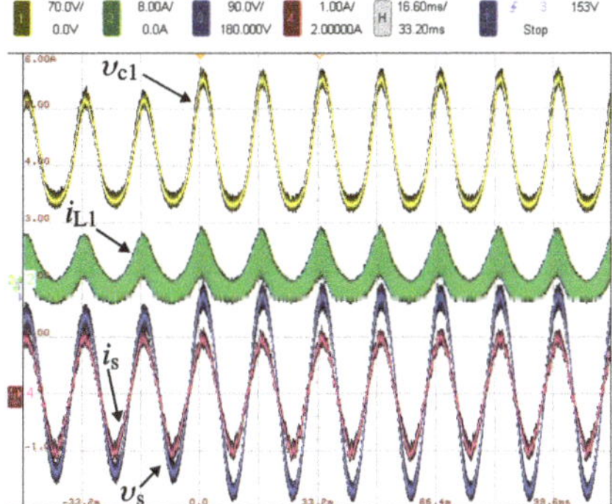

Figure 11. Experimental dynamic performance under grid perturbation (from 20% voltage dip to nominal voltage): voltage of the capacitor C_1 (v_{c1}), current through the inductor L_1 (i_{L1}), measurement of grid current (i_s), and grid voltage (v_s).

5. Conclusions

In this paper, a global sliding mode current control scheme for a grid-connected DBI is presented. Two control loops compose the proposed method: the linear PR outer control loop regulates the output current of the inverter (grid current), while the non-linear sliding mode inner control loop regulates the difference between the current of the inductors of the DBI, which allows control of the output voltages of the capacitors indirectly. Hence, fewer control loops are obtained compared to previous control schemes presented for the grid-connected DBI, because the inverter plus grid connection is analyzed globally.

Experimental results show the performance of the proposed control method for a grid-connected DBI. Both reference tracking and power quality show good performance despite using only an inductive filter for grid connection. Hence more sophisticated filters, such as LC or LCL, commonly used in grid-connected applications, could further improve the power quality. In addition, the dc component in the capacitor voltages is double the input voltage, which allows reducing the elevation ratio of the converters, the blocking voltage of the devices, and the capacitor size, compared to traditional methods used for this topology with voltage control loops.

The proposed control method was also tested under dynamic conditions in the current reference and under grid voltage perturbations, achieving good performance in both cases.

Author Contributions: D.L.-C. conceived, designed and performed the experimental evaluations and wrote the original draft; F.F.-B. and S.K. provided practical insight, resources, and supervision and reviewed the manuscript; V.S. was part of the team to design and develop the experimental prototype; N.M. and A.C. reviewed the manuscript and provided practical insight.

Funding: The authors acknowledge the financial support provided by FONDECYT 1191532, AC3E (CONICYT/BASAL/FB0008), SERC Chile (CONICYT/FONDAP/15110019), and CONICYT-PFCHA/Doctorado Nacional/2018-21181781.

Conflicts of Interest: The authors declare no conflict of interest.

References

1. Kouro, S.; Leon, J.I.; Vinnikov, D.; Franquelo, L.G. Grid-Connected Photovoltaic Systems: An Overview of Recent Research and Emerging PV Converter Technology. *IEEE Ind. Electron. Mag.* **2015**, *9*, 47–61. [CrossRef]
2. Caceres, R.; Barbi, I. A boost DC-AC converter: Operation, analysis, control and experimentation. In Proceedings of the 21st Annual Conference on IEEE Industrial Electronics (IECON '95), Orlando, FL, USA, 6–10 November 1995; Volume 1, pp. 546–551. [CrossRef]
3. Cortes, D.; Vazquez, N.; Alvarez-Gallegos, J. Dynamical Sliding-Mode Control of the Boost Inverter. *IEEE Trans. Ind. Electron.* **2009**, *56*, 3467–3476. [CrossRef]
4. Jang, M.; Agelidis, V.G. A Minimum Power-Processing-Stage Fuel-Cell Energy System Based on a Boost-Inverter with a Bidirectional Backup Battery Storage. *IEEE Trans. Power Electron.* **2011**, *26*, 1568–1577. [CrossRef]
5. Sanchis, P.; Ursaea, A.; Gubia, E.; Marroyo, L. Boost DC-AC inverter: A new control strategy. *IEEE Trans. Power Electron.* **2005**, *20*, 343–353. [CrossRef]
6. Zhu, G.; Tan, S.; Chen, Y.; Tse, C.K. Mitigation of Low-Frequency Current Ripple in Fuel-Cell Inverter Systems through Waveform Control. *IEEE Trans. Power Electron.* **2013**, *28*, 779–792. [CrossRef]
7. Jang, M.; Ciobotaru, M.; Agelidis, V.G. A Single-Phase Grid-Connected Fuel Cell System Based on a Boost-Inverter. *IEEE Trans. Power Electron.* **2013**, *28*, 279–288. [CrossRef]
8. Abeywardana, D.B.W.; Hredzak, B.; Agelidis, V.G. A Rule-Based Controller to Mitigate DC-Side Second-Order Harmonic Current in a Single-Phase Boost Inverter. *IEEE Trans. Power Electron.* **2016**, *31*, 1665–1679. [CrossRef]
9. Caceres, R.O.; Barbi, I. A boost DC-AC converter: Analysis, design, and experimentation. *IEEE Trans. Power Electron.* **1999**, *14*, 134–141. [CrossRef]
10. Jha, K.; Mishra, S.; Joshi, A. High-Quality Sine Wave Generation Using a Differential Boost Inverter at Higher Operating Frequency. *IEEE Trans. Ind. Appl.* **2015**, *51*, 373–384. [CrossRef]
11. Renaudineau, H.; Lopez, D.; Flores-Bahamonde, F.; Kouro, S. Flatness-based control of a boost inverter for PV microinverter application. In Proceedings of the 2017 IEEE 8th International Symposium on Power Electronics for Distributed Generation Systems (PEDG), Florianopolis, Brazil, 17–20 April 2017; pp. 1–6. [CrossRef]
12. Lopez, D.; Flores-Bahamonde, F.; Kouro, S.; Perez, M.A.; Llor, A.; Martínez-Salamero, L. Predictive control of a single-stage boost DC-AC photovoltaic microinverter. In Proceedings of the IECON 2016, 42nd Annual Conference of the IEEE Industrial Electronics Society, Florence, Italy, 23–26 October 2016; pp. 6746–6751. [CrossRef]
13. Flores-Bahamonde, F.; Valderrama-Blavi, H.; Bosque-Moncusi, J.M.; García, G.; Martínez-Salamero, L. Using the sliding-mode control approach for analysis and design of the boost inverter. *IET Power Electron.* **2016**, *9*, 1625–1634. [CrossRef]
14. Wang, Y.; Yu, D.; Kim, Y. Robust Time-Delay Control for the DC–DC Boost Converter. *IEEE Trans. Ind. Electron.* **2014**, *61*, 4829–4837. [CrossRef]
15. Pandey, S.K.; Patil, S.L.; Phadke, S.B. Comment on "PWM-Based Adaptive Sliding-Mode Control for Boost DC–DC Converters" [Aug 13 3291-3294]. *IEEE Trans. Ind. Electron.* **2018**, *65*, 5078–5080. [CrossRef]

16. Ma, R.; Wu, Y.; Breaz, E.; Huangfu, Y.; Briois, P.; Gao, F. High-order Sliding Mode Control of DC-DC Converter for PEM Fuel Cell Applications. In Proceedings of the 2018 IEEE Industry Applications Society Annual Meeting (IAS), Portland, OR, USA, 23–27 September 2018; pp. 1–7.

17. Liu, Y.; Huang, M.; Sun, J.; Zha, X. Active power decoupling method for isolated micro-inverters. In Proceedings of the 2014 International Power Electronics and Application Conference and Exposition, Shanghai, China, 5–8 November 2014; pp. 1222–1225. [CrossRef]

18. Kjær, S.B.; Blaabjerg, F. A novel single-stage inverter for the AC-module with reduced low-frequency ripple penetration. In Proceedings of the Epe'2003, Toulouse, France, 2–4 September 2003.

19. Mazumder, S.K.; Mehrnami, S. A low-device-count single-stage direct-power-conversion solar microinverter for microgrid. In Proceedings of the 2012 3rd IEEE International Symposium on Power Electronics for Distributed Generation Systems (PEDG), Aalborg, Denmark, 25–28 June 2012; pp. 725–730. [CrossRef]

20. Mehrnami, S.; Mazumder, S.K. Discontinuous Modulation Scheme for a Differential-Mode Ćuk Inverter. *IEEE Trans. Power Electron.* **2015**, *30*, 1242–1254. [CrossRef]

21. Skvarenina, T.L. *The Power Electronics Handbook*; CRC Press: Boca Raton, FL, USA, 2002.

22. Vecchio, C. Sliding Mode Control: Theoretical Developments and Applications to Uncertain Mechanical Systems. Ph.D. Thesis, The University of Pavia, Pavia, Italy, 2008.

23. Arscott, F.; Filippov, A. *Differential Equations with Discontinuous Righthand Sides: Control Systems*; Mathematics and Its Applications; Springer: Dordrecht, The Netherlands, 1988.

24. Teodorescu, R.; Liserre, M.; Rodriguez, P. *Grid Converters for Photovoltaic and Wind Power Systems*; John Wiley & Sons: Hoboken, NJ, USA, 2011; Volume 29.

25. Rodríguez, P.; Luna, A.; Muñoz-Aguilar, R.S.; Etxeberria-Otadui, I.; Teodorescu, R.; Blaabjerg, F. A Stationary Reference Frame Grid Synchronization System for Three-Phase Grid-Connected Power Converters Under Adverse Grid Conditions. *IEEE Trans. Power Electron.* **2012**, *27*, 99–112. [CrossRef]

26. Teodorescu, R.; Blaabjerg, F.; Liserre, M.; Loh, P.C. Proportional-resonant controllers and filters for grid-connected voltage-source converters. *IEE Proc. Electr. Power Appl.* **2006**, *153*, 750–762. [CrossRef]

27. Kazimierczuk, M. *Pulse-Width Modulated DC-DC Power Converters*; Wiley: Hoboken, NJ, USA, 2008.

28. Corradini, L.; Maksimovic, D.; Mattavelli, P.; Zane, R. *Digital Control of High-Frequency Switched-Mode Power Converters*; IEEE Press Series on Power Engineering; Wiley: Hoboken, NJ, USA, 2015.

Article

Sliding-Mode Approaches to Control a Microinverter Based on a Quadratic Boost Converter

Hugo Valderrama-Blavi [1,*], **Ezequiel Rodríguez-Ramos** [2], **Carlos Olalla** [1] **and Xavier Genaro-Muñoz** [1]

[1] Department of Electrical, Electronic, and Automatic Control Engineering, Universitat Rovira i Virgili, 43007 Tarragona, Spain; carlos.olalla@urv.cat (C.O.); xavier.genaro@urv.cat (X.G.-M.)
[2] School of Electrical and Electronic Engineering, Nanyang Technological University, Singapore 639798, Singapore; ezequiel001@e.ntu.edu.sg
* Correspondence: hugo.valderrama@urv.cat (H.V.-B.); Tel.: +34-977-558-523 (H.V.-B.)

Received: 28 August 2019; Accepted: 19 September 2019; Published: 27 September 2019

Abstract: A comparative analysis of the dynamic features of a step-up microinverter based on the cascade connection of two synchronized boost stages and a full-bridge is presented in this work. In the conventional approach the output of the cascaded boost converter is a 350–400 DC voltage that supplies the full-bridge that makes the DC-AC conversion. Differently from the classical approach, in this work, the cascaded boost converter delivers a sinusoidal rectified voltage of 230 Vrms to the full-bridge converter that operates as unfolding stage. This stage changes the voltage sign of one of every two periods of the rectified sinusoidal signal providing the final output AC waveform. In contrast to a classical full-bridge inverter, the unfolding stage lacks output filter, and has zero order dynamics. Thus, the approach presented here implies a second order dynamics reduction that will be increased applying sliding motions to control the system. After introducing the inverter circuit, two sliding control alternatives, input current mode and pseudo-oscillating mode, are presented. Both alternatives are analyzed, simulated, and verified experimentally. Furthermore, detailed description of the microinverter power stage and control circuits are also given.

Keywords: microinverter; sliding mode control (SMC), self-oscillating system; two cascaded-boosts converters

---

1. Introduction

Practical applications for microinverters are gradually increasing, from autonomous systems based on renewable energies, to supply small electrical appliances in embedded systems. Essentially, a microinverter is a low-power converter employed to obtain a sinusoidal AC waveform from a low DC voltage value, such as the one provided by a car battery, a photovoltaic module, or a fuel cell stack. The power rating of a microinverter is usually below 200 W, and the output voltage and frequency are those of the electrical grid.

Microinverters can be classified into two groups. The first group comprises all converter circuits behaving as an AC current source [1,2]. These are typically used in PV grid-tie applications, such as in the AC-module concept [2]. The second group of microinverter topologies operate as AC voltage sources. Consequently, these circuits are particularly suited for stand-alone.

Normally, a microinverter has two stages, see Figure 1. The first-stage is usually a step-up converter type, and the second stage is commonly a full-bridge inverter. Galvanic isolation [3] can be incorporated either in the first stage, or in the second one.

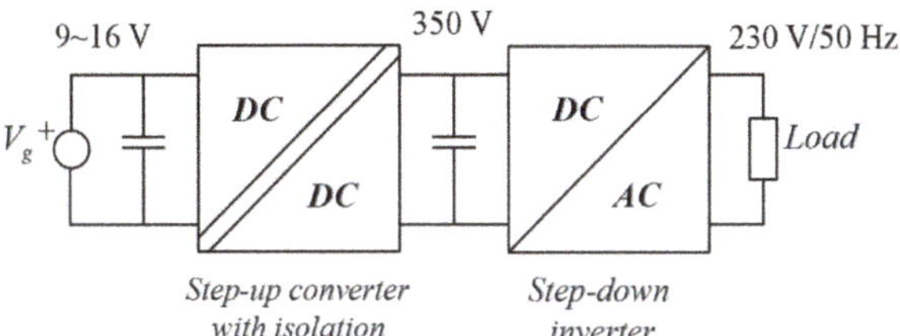

Figure 1. General diagram of the conventional microinverter configuration.

Many different two-stage microinverter topologies have been proposed in order to provide a large AC voltage (230 Vrms, 50 Hz) from a low DC supply in the range of 9 to 16 V [3–5].

A common approach consists of boosting the DC voltage above the required AC peak voltage in the first stage, whereas the second one realizes the inverting function (Figure 1). Sometimes, a high-frequency isolation transformer is used in the DC-DC stage to achieve a large conversion ratio avoiding extreme duty ratios. This can be achieved by using, for example, two interleaved flyback converters [5–7]. Nevertheless, if galvanic isolation is not required, those approaches are far from being optimal in power density, weight, size, and cost. Figure 2 depicts diverse converter static gains.

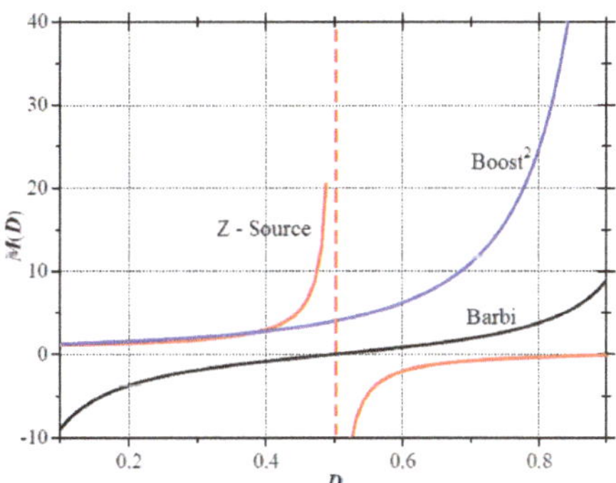

Figure 2. Ideal static transfer function M(D) of different converters.

Since there is growing interest in developing transformerless alternatives [3], several single-stage approaches have been proposed. Among the most reported configurations are the boost inverter (also named Barbi) [8–10], shown in Figure 3a, and the Z-source inverter [8], in Figure 3b. Although these last approaches avoid the use of a transformer, they exhibit a high nonlinear behavior and offer a very limited gain conversion ratio (typically below 10), and a poor efficiency [5–12], resulting in non-competitive solutions. This is especially important, in the Z-source inverter, where the shoot-through states management complicates the converter control, and the high-stress of the switching devices reduces the efficiency.

Two-stage microinverters, based on the combination of a step-up DC-DC converter and a step-down DC-AC inverter can avoid the use of a transformer and achieve a large conversion ratio at the same time [5]. In this context, topologies derived from the quadratic boost converter [12–14] are advantageous. Quadratic boost converters can provide larger gain conversion ratios at lower duty ratios, exceeding the voltage gain capabilities of the Z-source and Barbi converters, as shown in Figure 3. The efficiency analysis of four quadratic boost topologies presented in [4,13] shows that the quadratic voltage gain topology with the highest efficiency is the cascade connection of two boost converters in synchronous operation.

To summarize, this work employs the same quadratic boost topology under an advantageous operational mode in which the first stage of the microinverter is responsible for generating the AC waveform. In order to regulate the converter, two control methods providing robustness and tight performances are proposed and compared.

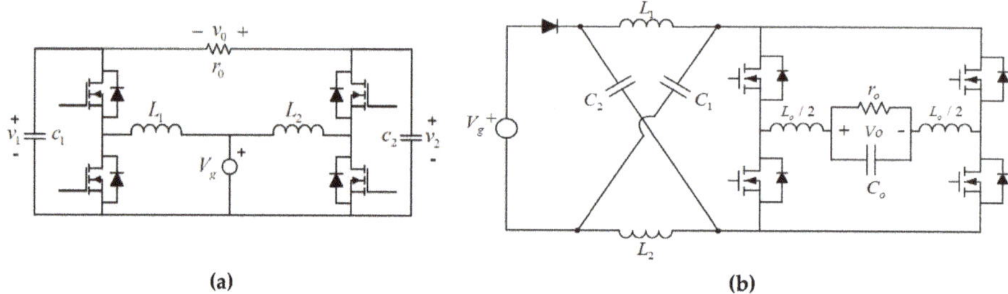

Figure 3. Single-stage transformerless inverters with small voltage gain: (**a**) Boost inverter (Barbi), and (**b**) Z-source single-phase inverter.

The remainder of this paper is structured as follows. In Section 2, the proposed operational mode is presented and compared with the conventional approach, showing their advantages and drawbacks. In Section 3, the dynamics of the proposed operational mode are analyzed. Next, in Sections 4 and 5, two sliding-mode control methods are proposed to regulate the converter. The first controller, based on the control of the input current of the microinverter, is introduced in Section 4. The second controller, which exhibits a self-oscillating behavior, is proposed in Section 5. Section 6 shows the inverter power stage and the control circuits for the two proposed controllers. The experimental results verifying the proposed operational mode and comparing the results of the two controllers are given in Section 7. Finally, Section 8 shows the work conclusions and some research guidelines.

2. Comparison of Microinverter Operation Strategies

Figure 4 depicts the two different operation strategies. The conventional approach depicted in Figure 4a is named "slow-fast" approach in this work. The proposed operational mode, shown in Figure 4b, is named "fast-slow".

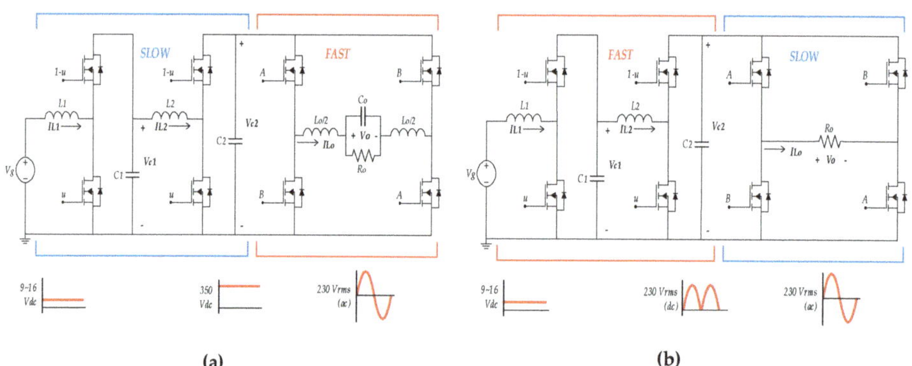

Figure 4. Strategies to perform a step-up inverter based on a two cascaded-boost converter: (**a**) conventional approach (slow-fast), (**b**) proposed approach (fast-slow).

In the "slow-fast" operation, the first stage provides a large DC intermediate voltage V_{c2} in the range of 350–400 V whereas the full-bridge inverter realizes the DC-AC conversion. The output voltage is obtained with a bipolar or unipolar sinusoidal PWM modulation (SPWM). Although at equal switching frequency the unipolar modulation exhibits lower switching losses than the bipolar one, the unipolar modulation requires a more complicated switching logic.

Oppositely, in the "fast-slow" approach, the first stage converter provides a rectified sinusoidal voltage of amplitude 230 $\sqrt{2}$ and the unfolding stage makes the sign inversion. In this case, as the full-bridge is switching at 50 Hz and the switching losses are negligible.

The only drawback of this strategy is that the quadratic boost minimum output voltage is Vg instead of zero, causing a zero-cross distortion problem in the inverter output voltage that is addressed in Section 6.

According to Figure 4, no matter what the selected strategy is, "slow-fast" or "fast-slow", the first stage is always a quadratic boost converter, and consequently, a fourth order plant. Nevertheless, in the "slow-fast" strategy, the second stage is a buck inverter, and has a second order dynamics; whereas in the "fast-slow" approach, the full-bridge is an unfolding stage, lacks an output filter, and has zero-order dynamics. Figure 5 compares both microinverter operation strategies.

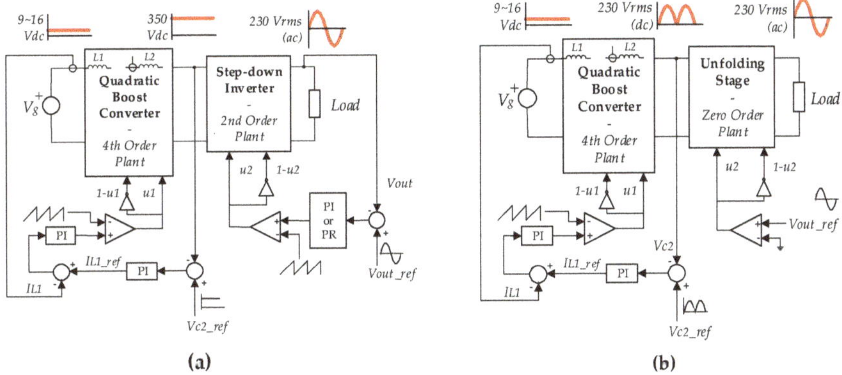

Figure 5. Common and proposed approaches with PWM control: (**a**) common case (slow-fast): 6th order plant, 9th -10th order system depending on inverter controller; (**b**) fast-slow strategy: 4th order plant, 6th order system.

Assuming PWM control, Figure 5 shows the control loops required for each approach. Both strategies require a two-loop controller to regulate the first stage output voltage according to a given voltage reference (constant in case of Figure 5a, variable in case of Figure 5b). The main dynamical difference comes from the full-bridge stage operation. In the conventional approach the full bridge is a buck inverter and requires a PR (proportional resonant) or PI (proportional integral) controller to track a sinusoidal reference to obtain a sinusoidal output voltage. Although PR is normally preferred because it can totally cancel the tracking error, differently from the PI; the PR is a second order controller whereas the PI has a single pole at the origin. In the fast-slow approach, the full-bridge is used to alternately change the output voltage sign, and only a comparator (a zero dynamics controller) is required.

The pole position of the quadratic boost small-signal output-to-control transfer function depends on the converter duty ratio. Whereas in the conventional approach the first stage output voltage V_{C2} is constant (350 V), and the duty ratio only depends on the input voltage variations, in the "fast-slow" strategy the poles variability is enhanced because $v_{C2}(t)$ is ranging from Vg to the output voltage peak 230 $\sqrt{2} = 325$ V. This fact could complicate the control design and constitutes an additional issue that justifies the use of sliding mode control.

Among Sliding Mode Control (SMC) advantages, the most known are the following: simplicity of implementation, inherent robustness against perturbations and parametric variations, and dynamic order reduction. As the microinverter has two stages, a control surface per converter-stage, this means two surfaces, could be considered.

Applying sliding motions implies a system-dynamics order reduction per surface. In the conventional "slow-fast" approach the PWM-controlled 6th order plant, when sliding motions are applied, becomes a 4th order plant. Thus, the quadratic boost dynamics is reduced to a third-order plant, and then, the second-order full-bridge inverter stage becomes a first-order plant.

Conversely, in the "fast-slow" strategy, as the second microinverter stage has no dynamics, no control surface is required. As a result, the whole "fast-slow" microinverter plant becomes a third-order system, one order less than in the conventional "slow-fast" approach.

In addition, if sliding motions are applied in the "fast-slow" approach, to assure that the quadratic boost output tracks the reference voltage, a PI regulator is required, obtaining finally a 4th order system. Finally, it is worth mentioning that the order of the system would be much higher if regular PWM regulators would be used in these plants. With the "fast-slow" approach, the system would exhibit 6th order dynamics. With the "slow-fast" operation, the order would even be higher, adding up to 9th order.

In this work, two different sliding surfaces are applied to a quadratic-boost "fast-slow" microinverter operating with the "fast-slow" strategy. The first proposed controller (Figure 6a) is based on input current regulation [12], whereas the second proposed controller (Figure 6b) is based on a self-oscillating behavior [15,16]. After steady state analysis in Section 3, the input current surface is analyzed in Section 4, and the self-oscillating surface in Section 5.

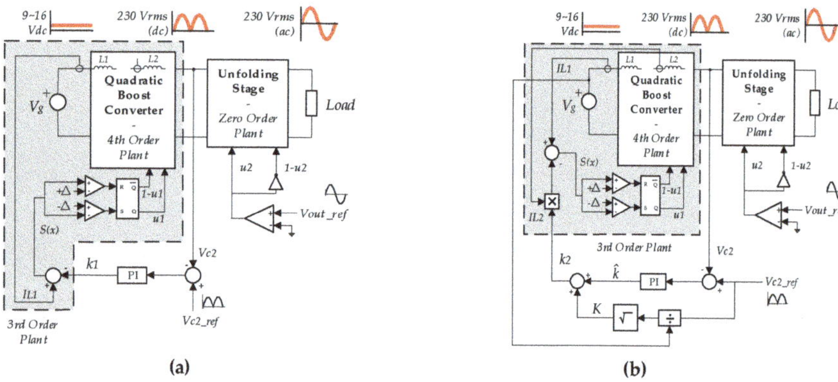

Figure 6. Output voltage regulation. Proposed sliding surfaces: (**a**) using the microinverter input current $S_1(x,t) = i_{L1}(t) - k_1(t) = 0$, (**b**) inducing a self-oscillating motion $S_2(x,t) = i_{L1}(t) - k_2(t) \cdot i_{L2}(t) = 0$.

3. "Fast-Slow" Microinverter Analysis

Figure 4 shows the schematic of the proposed "fast-slow" step-up inverter based on the cascade connection of two boost circuits. The goal is to achieve a sinusoidal voltage signal, $v_0(t)$, of a given RMS value and frequency from a low DC input voltage, $v_g(t)$. To do that, the voltage at the output of the quadratic boost stage, $v_{c2}(t)$, must be kept at a given reference value $v_{ref}(t)$ despite disturbances, see Equation (1).

$$\begin{cases} v_{C2}(t) = V_{ref} \cdot |\sin(\omega_0 t)| \\ v_0(t) = v_{C2}(t) \cdot \text{sign}\left(\sin(\omega_0 t)\right) \end{cases} \tag{1}$$

Only the quadratic boost stage is analyzed due to the zero dynamics of the unfolding stage. The fast stage is a fourth-order variable structure system with its two commutation cells in synchronous operation. Continuous conduction mode (CCM) operation was assumed. As shown in Figure 7, this converter has two equivalent circuit configurations for CCM operation during a switching period.

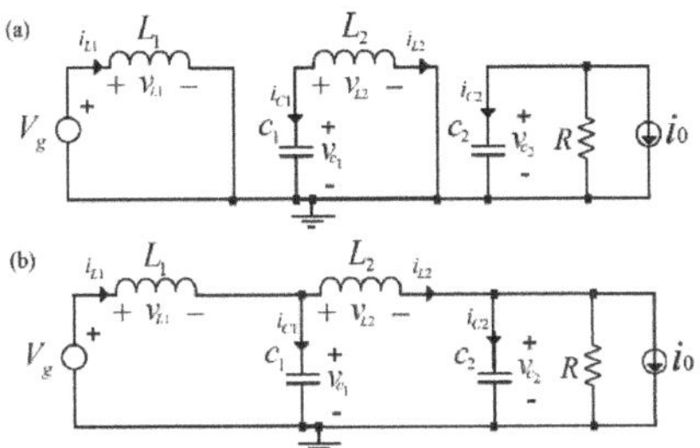

Figure 7. Cascade-boost converter topologies: (**a**) ON, and (**b**) OFF.

In the equation set, current source $i_0(t)$ models load disturbances, R models the converter nominal load, and L_1, L_2, C_1, and C_2, stand for the inductances and capacitances values, respectively. Thus, the dynamic expressions in state-space at each position of the switch are characterized as,

$$\begin{cases} \dot{x}(t) = A_1 x(t) + B_1 & \text{during Ton}\,(u(t) = 1) \\ \dot{x}(t) = A_2 x(t) + B_2 & \text{during Toff}\,(u(t) = 0) \end{cases}, \tag{2}$$

where A_i and B_i ($i \in \{1, 2\}$) are the state matrices and the input vector respectively, and for each subinterval (*Ton* and *Toff*), and where $x(t) = [i_{L1}, i_{L2}, v_{C1}, v_{C2}]^T$ is the state vector, which groups the inductor currents and capacitor voltages.

Thus, Equation (2) can be compacted into a single bilinear form, see Equation (3):

$$\begin{cases} \dot{x}(t) = Ax(t) + \delta + [Bx(t) + \gamma] \cdot u(t) \\ A = A_2,\ \delta = B_2,\ B = (A_1 - A_2),\ \gamma = (B_1 - B_2) \end{cases}, \tag{3}$$

where the discrete control variable $u(t) \in \{0, 1\}$ stands for the gate signal of the low-side controlled switches of the two-cascade-boost circuits, which force the commutation between *Ton* and *Toff*.

In summary, the converter switching dynamics can be written as

$$\dot{x}(t) = \begin{pmatrix} 0 & 0 & -\frac{1-u(t)}{L_1} & 0 \\ 0 & 0 & \frac{1}{L_2} & -\frac{1-u(t)}{L_2} \\ \frac{1-u(t)}{C_1} & -\frac{1}{C_1} & 0 & 0 \\ 0 & \frac{1-u(t)}{C_2} & 0 & -\frac{1}{RC_2} \end{pmatrix} x(t) + \begin{pmatrix} \frac{v_g(t)}{L_1} \\ 0 \\ 0 \\ -\frac{i_0(t)}{C_2} \end{pmatrix}. \tag{4}$$

Replacing the control variable $u(t)$ in Equation (4) by its averaged value, the duty cycle D(t), the ideal quadratic-boost converter quasi-static gain M[D(t)], operating in CCM, can be obtained using Equation (5). The corresponding permanent-regime averaged state vector $x(t) = [i_{L1}(t), i_{L2}(t), v_{C1}(t), v_{C2}(t)]$ is given in Equation (6). Realize that switching frequency is at least three orders of magnitude higher than the variation frequency of the sinusoidal averaged state variables.

$$M[D(t)] = \frac{v_{C1}}{V_g} \cdot \frac{v_{C2}}{v_{C1}} = \frac{1}{D'(t)} \cdot \frac{1}{D'(t)} = \frac{1}{D'^2(t)}\ \text{where } D'(t) = 1 - D(t), \tag{5}$$

$$x(t) = \left[\frac{V_{ref}^2}{V_g R} \sin^2(\omega_o t),\ \frac{1}{R}\sqrt{\frac{V_{ref}^3}{V_g}}\left|\sin(\omega_o t)\right|,\ \sqrt{V_g \cdot V_{ref}}\left|\sin(\omega_o t)\right|,\ V_{ref}\left|\sin(\omega_o t)\right| \right]^T \tag{6}$$

$$\text{where } x(t) = \left[\ i_{L1}(t),\ i_{L2}(t),\ v_{C1}(t),\ v_{C2}(t)\ \right]^T \text{and } V_{ref} = 230\sqrt{2}$$

Finally, to obtain a regulated output voltage, rejecting input voltage perturbations and load variations, a two-loop system is required. The outer-loop regulates the output voltage with a PI controller. For the inner-loop two sliding mode surfaces are proposed: an input current surface (Equation (7)) analyzed in Section 4, and a self-oscillating one (Equation (8)) studied in Section 5.

$$\left.\begin{array}{l} S_1(x,t) = i_{L1}(t) - k_1(t) = 0 \\ e(t) = V_{ref}\cdot\left|\sin(\omega_0 t)\right| - v_{C2}(t) \\ k_1(t) = k_{P,1}\cdot e(t) + k_{I,1}\int_0^t e(\tau)d\tau \end{array}\right\}, \tag{7}$$

$$\left.\begin{array}{l} S_2(x,t) = i_{L1}(t) - k_2(t)i_{L2}(t) = 0 \\ e(t) = V_{ref}\left|\sin(\omega_0 t)\right| - v_{C2}(t) \\ k_2(t) = \sqrt{\dfrac{V_{ref}\left|\sin(\omega_0 t)\right|}{V_g}} + k_{P,2}e(t) + k_{I,2}\int_0^t e(\tau)d\tau \end{array}\right\}. \tag{8}$$

4. Microinverter Control with Input Current Approach

In this mode of operation, the control law forces the input current,

$i_{L1}(t)$, to track a signal of the form $k(t) = K\sin^2(\omega_0 t)$. Realize that this fact is in perfect agreement with the state vector in permanent regime (Equation (6)), the sliding mode control-law described in Equation (7), and the *PoPi* behavior of any converter. Figure 8 depicts the block diagram of input current control strategy.

$$P_i(t) = P_o(t) \Rightarrow \left\{\begin{array}{c} P_i(t) = V_g \cdot i_{L1}(t) \\ P_o(t) = \dfrac{v_o^2}{R}(t) = \dfrac{V_{ref}^2}{R}\sin^2(\omega_0 t) \end{array}\right\} \Rightarrow i_{L1}(t) = K_1\sin^2(\omega_0 t), \tag{9}$$

$$\text{where } K_1 = \frac{V_{ref}^2}{RV_g}, \text{ and } V_{ref} = 230\sqrt{2} = 325\,V. \tag{10}$$

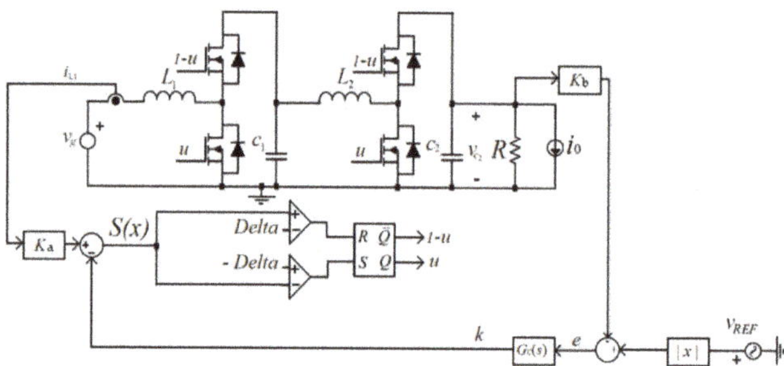

Figure 8. Input current control strategy.

The outer-loop establishes the reference of the inner loop by means of a PI compensator $G_C(s)$ processing the voltage error e(t). Figure 8 illustrates the hysteresis-based two-loop control, where the inner current loop is defined by means of a sliding surface (Equation (7)), which drives the converter to a stable desired point.

Assuming the frequency ω_0 is considerably lower than the switching frequency, i.e., the evolution of the sliding surface is much slower than the transients among states, it can be considered that the input inductor current $i_{L1}(t)$ follows its reference $k_1(t)$.

The discontinuous switching function u(t) that yields sliding motion is given in Equation (11).

$$u(t) = \left\{\begin{array}{l} 1 \text{ when } S_1(x,t) < 0 \\ 0 \text{ when } S_1(x,t) > 0 \end{array}\right.. \tag{11}$$

Based on the sliding surface (Equation (7)) and applying the invariance conditions ($S_1(x,t) = 0$ and $dS_1(x,t)/dt = 0$) to the converter equation set (Equation (4)), the equivalent control (Equation (12)) is obtained. The control will exist unconditionally since $v_{C1}(t) > 0$, $\forall t > 0$.

$$0 \leq u_{eq}(x,t) = 1 - \frac{v_g(t) - L_1 \frac{dk_1(t)}{dt}}{v_{C1}(t)} \leq 1. \tag{12}$$

Now, by replacing the expression of the equivalent control (Equation (12)) in the converter equation set (Equation (4)) and considering the surface (Equation (7)), $i_{L1}(t) = k_1(t)$, the resulting ideal sliding dynamics is obtained. As the input current $i_{L1}(t)$ dynamics is given by the external reference $k_1(t)$, the remaining system dynamics is reduced one order once the surface is reached.

$$\begin{cases} \frac{di_{L2}}{dt} = -\frac{1}{L_2} \cdot \frac{v_{C2}}{v_{C1}}\left(v_g - L_1 \frac{dk_1}{dt}\right) + \frac{v_{C1}}{L_2} = g_1(x,t) \\ \frac{dv_{C1}}{dt} = \frac{1}{C_1} \cdot \frac{k_1}{v_{C1}}\left(v_g - L_1 \frac{dk_1}{dt}\right) - \frac{i_{L2}}{C_1} = g_2(x,t) \\ \frac{dv_{C2}}{dt} = \frac{1}{C_2} \cdot \frac{i_{L2}}{v_{C1}}\left(v_g - L_1 \frac{dk_1}{dt}\right) - \frac{v_{C2}}{RC_2} - \frac{i_o}{C_2} = g_3(x,t) \end{cases} \tag{13}$$

The nonlinear behavior of the ideal sliding dynamics is depicted in Equation (13). To synthesize an appropriate controller using a small-signal model, a linearization around the equilibrium point is required. By means of a quasi-static approximation [16] the state vector $x(t)$ evolution can be modelled as a succession of equilibrium points that evolve according the output voltage sinusoid value, that it $f(\theta) = |\sin(\theta)| = |\sin(\omega_o t)|$. By nullifying the derivatives in Equation (13), the succession of equilibrium points (Equation (14)) is obtained.

$$X^*(\theta) = \left[K_1 f^2(\theta), \ \left(\frac{K_1^3 V_g}{R}\right)^{\frac{1}{4}} f^{\frac{3}{2}}(\theta), \ (K_1 R V_g^3)^{\frac{1}{4}} f^{\frac{1}{2}}(\theta), \ \sqrt{K_1 R V_g}\, f(\theta) \right]^T, \ f(\theta) = |\sin(\theta)| \tag{14}$$

$$\text{where } X^*(\theta) = \left[I_{L1}^*(\theta), \ I_{L2}^*(\theta), \ V_{C1}^*(\theta), \ V_{C2}^*(\theta) \right]^T \text{and } K_1 = \frac{V_{ref}^2}{R V_g}$$

Once the equilibrium points are calculated, the linearization of the ideal sliding dynamics (Equation (13)) to get a small-signal linear model around a given equilibrium point (Equation (14)) is a well-known procedure. The equation set of Equation (15) is the linearized version of Equation (13).

$$\begin{aligned} \frac{d\hat{i}_{L2}}{dt} &\approx \mathbf{a} \cdot \hat{v}_{C1}(t) + \mathbf{b} \cdot \hat{v}_{C2}(t) + \mathbf{c} \cdot \hat{v}_g(t) + \mathbf{d} \cdot \frac{d\hat{k}_1(t)}{dt} \\ \frac{d\hat{v}_{C1}}{dt} &\approx \mathbf{e} \cdot \hat{i}_{L2}(t) + \mathbf{f} \cdot \hat{k}_1(t) + \mathbf{g} \cdot \hat{v}_{C1}(t) + \mathbf{h} \cdot \hat{v}_g(t) + \mathbf{x} \cdot \frac{d\hat{k}_1(t)}{dt} \\ \frac{d\hat{v}_{C2}}{dt} &\approx \mathbf{z} \cdot \hat{v}_{C2}(t) + \mathbf{y} \cdot \hat{i}_{L2}(t) + \mathbf{m} \cdot \hat{v}_{C1}(t) + \mathbf{n} \cdot \hat{v}_g(t) + \mathbf{p} \cdot \frac{d\hat{k}_1(t)}{dt} + \mathbf{q} \cdot \hat{i}_o(t) \end{aligned} \tag{15}$$

The corresponding coefficients of the linearized model (Equation (15)) are calculated by means of the Jacobian matrix and given in Equation (16).

$$\begin{aligned} \mathbf{a} &= \left.\frac{dg_1}{dv_{C1}}\right|_{X*} = \frac{2}{L_2} & \mathbf{b} &= \left.\frac{dg_1}{dv_{C2}}\right|_{X*} = \frac{-V_g}{L_2 V_{C1}^*(\theta)} & \mathbf{c} &= \left.\frac{dg_1}{dv_g}\right|_{X*} = \frac{-V_{C2}^*(\theta)}{L_2 V_{C1}^*(\theta)} & \mathbf{d} &= \left.\frac{dg_1}{d\left(\frac{dk_1}{dt}\right)}\right|_{X*} = \frac{L_1 V_{C2}^*(\theta)}{L_2 V_{C1}^*(\theta)} \\[2ex] \mathbf{e} &= \left.\frac{dg_2}{di_{L2}}\right|_{X*} = \frac{-1}{C_1} & \mathbf{f} &= \left.\frac{dg_2}{dk_1}\right|_{X*} = \frac{V_g}{C_1 V_{C1}^*(\theta)} & \mathbf{g} &= \left.\frac{dg_2}{dv_{C1}}\right|_{X*} = \frac{-K_1 \cdot f^2(\theta)}{C_1 V_{C2}^*(\theta)} & \mathbf{x} &= \left.\frac{dg_2}{d\left(\frac{dk_1}{dt}\right)}\right|_{X*} = \frac{-L_1 K_1 \cdot f^2(\theta)}{C_1 V_{C1}^*(\theta)} \\[2ex] \mathbf{h} &= \left.\frac{dg_2}{dv_g}\right|_{X*} = \frac{K_1 \cdot f^2(\theta)}{C_1 V_{C1}^*(\theta)} & \mathbf{z} &= \left.\frac{dg_3}{dv_{C2}}\right|_{X*} = \frac{-1}{RC_2} & \mathbf{y} &= \left.\frac{dg_3}{di_{L2}}\right|_{X*} = \frac{V_g}{C_2 V_{C1}^*(\theta)} & \mathbf{p} &= \left.\frac{dg_3}{d\left(\frac{dk_1}{dt}\right)}\right|_{X*} = \frac{-L_1 I_{L2}^*(\theta)}{C_2 V_{C1}^*(\theta)} \\[2ex] \mathbf{m} &= \left.\frac{dg_3}{dv_{C1}}\right|_{X*} = \frac{-I_{L2}^*(\theta)}{C_2 V_{C2}^*(\theta)} & \mathbf{n} &= \left.\frac{dg_3}{dv_g}\right|_{X*} = \frac{I_{L2}^*(\theta)}{C_2 V_{C1}^*(\theta)} & \mathbf{q} &= \left.\frac{dg_3}{di_o}\right|_{X*} = \frac{-1}{C_2} \end{aligned} \tag{16}$$

Placing the coefficient values (Equation (16)) in the linearized model (Equation (15)) and taking the Laplace transform the corresponding small-signal model transfer functions are obtained. Figure 9 shows the closed-loop small-signal diagram block. The controller box $G_C(s)$ included in Figure 9 accounts for the PI integrator that closes the voltage loop.

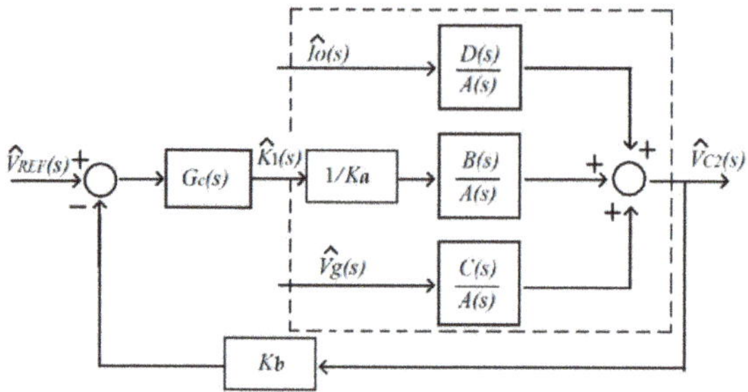

Figure 9. Block diagram of the output voltage regulation loop.

Then, considering the PI controller transfer function is $G_C(s) = k_P + k_{In}/s$, the closed-loop transfer function is derived (Equation (17)), where K_a and K_b represent the input current and output voltage sensor gains, respectively. The model polynomials A(s), B(s), C(s) and D(s) are given in Equation (18). As can be deduced from Equation (18) the converter poles, zeros, and gains are found to vary along the AC output sinusoid.

$$\hat{g}_{ref}(s) = \frac{\hat{v}_{C2}(s)}{\hat{v}_{ref}(s)} = \frac{1}{K_a}\frac{G_C(s)B(s)}{A(s) + \frac{K_b}{K_a}G_C(s)B(s)}, \tag{17}$$

$$\left.\begin{aligned}
A(s) &= s^3 + s^2 \cdot \left[\frac{\beta(\theta)}{C_1} + \frac{1}{RC_2}\right] + \frac{s}{L_2C_2} \cdot \left[\frac{1}{R\beta(\theta)} + \frac{2C_2}{C_1} + \frac{L_2\beta(\theta)}{RC_1}\right] + \frac{4}{RL_2C_1C_2} \\
B(s) &= -s^3 \cdot \left[\frac{L_1\beta(\theta)}{C_2}\right] + s^2 \cdot \left[\frac{L_1}{L_2C_2}\right] - \frac{s}{C_1C_2} \cdot \left[\frac{2L_1\beta(\theta)}{L_2} + \frac{1}{R}\right] + \frac{2}{L_2C_1C_2R\beta(\theta)} \\
C(s) &= s^2 \cdot \left[\frac{\beta(\theta)}{C_2}\right] - \frac{s}{L_2C_2} + \frac{2\beta(\theta)}{L_2C_1C_2} \\
D(s) &= -\frac{s^2}{C_2} - s \cdot \left[\frac{\beta(\theta)}{C_1C_2}\right] - \frac{2}{L_2C_1C_2} \quad \text{where } \beta(\theta) = \frac{V_{ref}}{RV_g}|\sin\theta| = \frac{V_{ref}}{RV_g}f(\theta)
\end{aligned}\right\}. \tag{18}$$

4.1. Parameter Variation Analysis

To appropriately design the PI coefficients k_P and k_{In}, it is convenient to realize a parametric study of the root-locus of the open-loop control-to-output transfer function $\hat{g}_k(s) = \hat{v}_{C2}(s)/\hat{k}_1(s)$. Two different average output powers are considered: the nominal case (100 W) and twice that power.

The main converter parameters are R = {250, 1000} Ω, V_g = 13 V, L_1 = 8 µH, L_2 = 800 µH, C_1 = 0.5 µF, C_2 = 0.8 µF, ω_o = 100π rad/s, K_a = 0.2 V/A, K_b = 0.01 V/V, 2.5° < θ < 90°. Figure 10a charts the poles and zeros placement for 100 W (K_1 = 16.67 A) and Figure 10b shows it for twice the power (K_1 = 33.33 A).

By inspection of Figure 10a, it is clear the existence of a real dominant pole on the left half-plane for θ > 7.5°, and a pair of complex conjugate poles whose damping increases when the output voltage or the angular parameter θ increase. The dominant pole location is proportional to the mean output power. Thus, for 100 W the pole is at s = −5·10^3 rad/s, whereas for 200 W it is close to s = −10^4 rad/s.

The behavior for 100 and 200 W is similar. In both cases there are real right half-plane zeros and all the poles are in the left half-plane. Nevertheless, as depicted in Figure 10b, in the 200 W case, there is a dominant pole while θ < 60°, but for θ > 60°, the system exhibits is a second order dominance with a pair of complex conjugate poles with a damping factor close to 1.

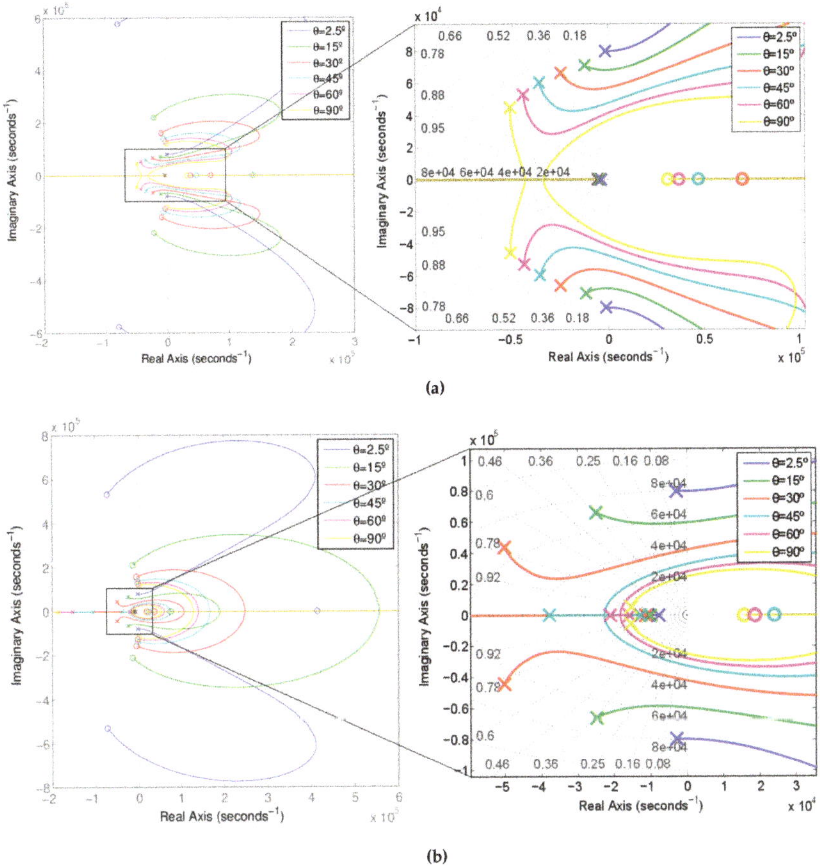

Figure 10. Root locus of $\hat{g}_k(s)$ (**a**) for 100 W, ($R = 484\ \Omega$, $K_1 = 16.67$ A); (**b**) for 200 W, ($R = 242\ \Omega$, $K_1 = 33.33$ A)

4.2. Synthesis of the Voltage Regulation Loop

To assure at least local closed-loop stability for all operation ranges $\omega_0 t = \theta \in [0,\ \pi/2]$, the Routh–Hurwitz criterion is applied to the closed-loop characteristic polynomial (Equation (19)) of the voltage regulation transfer function (Equation (17)).

$$P(s) = s\cdot\left[A(s) + \frac{K_b}{K_a}G_C(s)B(s)\right] = p_4\cdot s^4 + p_3\cdot s^3 + p_2\cdot s^2 + p_1\cdot s + p_0, \tag{19}$$

$$p_0 = \frac{2k_{In}}{L_2C_1C_2R\beta(\theta)}\frac{K_b}{K_a} \quad p_1 = \frac{4}{RL_2C_1C_2} - \frac{k_{In}}{C_1C_2}\frac{K_b}{K_a}\left[\frac{2L_1\beta(\theta)}{L_2} + \frac{1}{R}\right] + \frac{2k_P}{L_2C_1C_2R\beta(\theta)}\frac{K_b}{K_a}$$

$$p_2 = \frac{1}{L_2C_2}\left[\frac{1}{R\beta(\theta)} + \frac{2C_2}{C_1} + \frac{L_2\beta(\theta)}{RC_1}\right] + \frac{L_1k_{In}}{L_2C_2}\frac{K_b}{K_a} - \frac{k_P}{C_1C_2}\frac{K_b}{K_a}\left[\frac{2L_1\beta(\theta)}{L_2} + \frac{1}{R}\right] \tag{20}$$

$$p_3 = \frac{\beta(\theta)}{C_1} + \frac{1}{RC_2} + \frac{L_1}{C_2}\left(-\beta(\theta)k_{In} + \frac{k_P}{L_2}\right)\frac{K_b}{K_a} \quad p_4 = 1 - \frac{L_1\beta(\theta)k_P}{C_2}\frac{K_b}{K_a}$$

According to the Routh criterium, the small-signal stability of the system can be ensured if the following set of constraints is satisfied:

$$\begin{aligned} p_4 > 0,\ p_3 > 0,\ (p_2p_3 - p_1p_4) > 0, \\ (p_1p_2p_3 - p_1{}^2p_4 - p_0p_3{}^2) > 0,\ p_0 > 0 \end{aligned} \tag{21}$$

From the aforesaid procedure, a stable region in the $k_P - k_{In}$ plane can be derived. Given the nominal input voltage, output voltage, and load R, Figure 11 charts the PI controller constants to assure stability (Equation (21)) for a representative set of θ values. In the nominal case, stability is assured if $k_P \in [0, 1.23]$ and $k_{In} \in [0, 4\cdot10^5]$. Equivalent plots were considered for other load and input voltage values. Finally, to enhance the transient performance, and assure stability, the following gains were selected $k_P = 1$, and $k_{In} = 2\cdot10^4$.

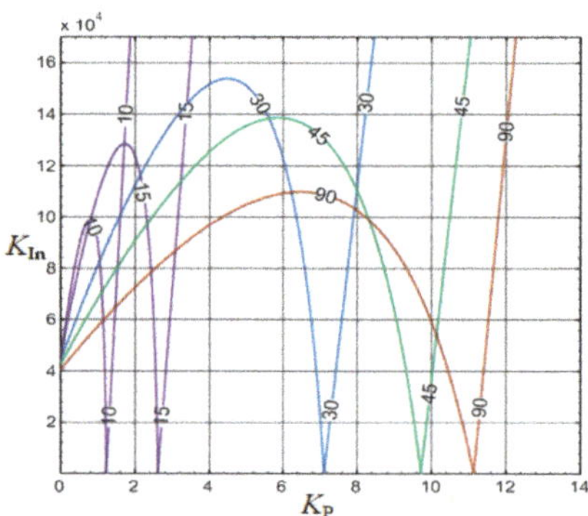

Figure 11. Evaluation of the stability region in the k_P - k_{In} plane, for the nominal case.

5. Microinverter Control with Self-Oscillating Approach

As in the previous case, an invariant rectified sine signal of 230 Vrms and 100 Hz at the output of the two cascaded-boosts stage is required. Unlike in the previous input current control, the reference output voltage is achieved establishing a proportional relationship $k_2(t)$ among both inductor currents, $i_{L1}(t)$ and $i_{L2}(t)$. In this case, differently from previous works [15–17], $k_2(t)$ is a time-variant signal modulated by an external reference signal.

The self-oscillation approach implies that a system is self-switching, with any external reference causing the converter state changes, and the resulting output voltage gain depends only on the value of the proportionality constant. However, in this particular case, as the quadratic-boost output voltage must track a given reference voltage, in a quasi-static approximation, the proportionality constant $k_2(t)$ must also change slowly to assure that the output voltage is a sinusoidal signal. To find the required steady-state link between the two inductor currents $K_2(\omega_0 t)$, the following conditions need to be considered: (i) the converter exhibits a *PoPi* behavior ($V_g\cdot I_{L1} = V_{C1}\cdot I_{L2}$), and (ii) the medium voltage capacitor is the geometric mean of the input and output voltage (Equation (6)).

$$K_2(\omega_0 t) = K_2(\theta) = \frac{I_{L1}}{I_{L2}} = \frac{V_{C1}}{V_g} = \frac{\sqrt{V_{C2}V_g}}{V_g} = \sqrt{\frac{V_{ref}|\sin(\theta)|}{V_g}}. \tag{22}$$

It is worth mentioning that load perturbations will not affect the proportionality constant $K_2(\omega_0 t)$. As a result, this control-law should assure ideal load regulation, and by means of the implicit nonlinear V_g feedforward (Equation (22)), an ideal line regulation could also be expected. Nevertheless, as converter losses exist, and intending to compensate those losses or any other perturbation, a PI controller is used, resulting in the aforesaid sliding surface (Equation (8)).

The main difference between Equations (7) and (8) is that the voltage regulation PI controller of the input current approach is in charge of computing the reference current completely by itself, whereas in the self-oscillating approach we establish the equilibrium point of the converter by means of $K_2(\theta)$ (Equation (22)), and therefore the output of the PI compensator $\hat{k}_2$ will be zero unless a disturbance arises. Furthermore, to perform an effective line regulation in Equation (8), a feedforward control loop is required to update the steady-state relationship among inductor currents (Equation (22)) effectively and rapidly.

The main interest of this control method is that it is possible to fix the output voltage through a correct selection of the transformation ratio (Equation (23)), despite output power requirements (transformer characteristic). Hence, no feedback control loop would be required if there were no conversion losses. Figure 12 depicts the control block diagram of the self-oscillating strategy.

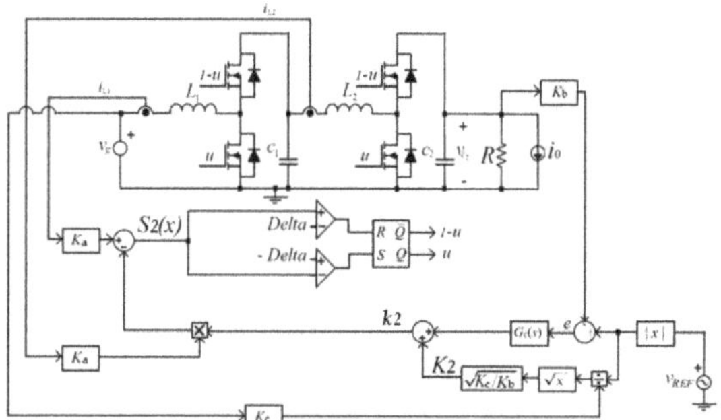

Figure 12. Self-oscillating control strategy.

Based on the sliding surface (Equation (8)) and applying the invariance conditions in the dynamics of the converter (Equation (4)), leads to the equivalent control (Equation (23)).

$$u_{eq}(x,t) = 1 - \frac{L_2 v_g(t) - L_1 k_2(t) v_{C1}(t) - L_1 L_2 \frac{dk_2(t)}{dt} i_{L2}(t)}{L_2 v_{C1}(t) - L_1 k_2(t) v_{C2}(t)} \tag{23}$$
$$u_{eq}(x,t) \text{ exists if } k_2(t) \neq \sqrt{L_2/L_1}$$

Unlike in the input current control, where there is an unconditional transversality, the self-oscillating surface has a conditional transversality, and the existence of an equivalent control must be ensured through a correct design of both inductors. A possible solution for that is given in Equation (24).

$$0 < K_2(\theta) < K_{2max} = \sqrt{\frac{V_{ref}}{V_{g\_min}}} = \sqrt{\frac{325}{9}} \approx 6 \text{ and } \sqrt{\frac{L_2}{L_1}} = \sqrt{\frac{800\,\mu H}{8\,\mu H}} = 10. \tag{24}$$

From the equivalent control expression (Equation (23)), the converter equation set (Equation (4)), and the surface (Equation (8)), the resulting ideal sliding-mode dynamics of the converter (Equation (25)) can be derived.

$$\begin{cases} \dfrac{di_{L2}}{dt} = \dfrac{-v_{C2}}{L_2} \cdot \dfrac{-L_1 L_2 \frac{dk_2}{dt} i_{L2} - L_1 k_2 v_{C1} + L_2 v_g}{L_2 v_{C1} - L_1 k_2 v_{C2}} + \dfrac{v_{C1}}{L_2} = g_1(x,t) \\[2mm] \dfrac{dv_{C1}}{dt} = \dfrac{k_2 i_{L2}}{C_1} \cdot \dfrac{-L_1 L_2 \frac{dk_2}{dt} i_{L2} - L_1 k_2 v_{C1} + L_2 v_g}{L_2 v_{C1} - L_1 k_2 v_{C2}} - \dfrac{i_{L2}}{C_1} = g_2(x,t) \\[2mm] \dfrac{dv_{C2}}{dt} = \dfrac{i_{L2}}{C_2} \cdot \dfrac{-L_1 L_2 \frac{dk_2}{dt} i_{L2} - L_1 k_2 v_{C1} + L_2 v_g}{L_2 v_{C1} - L_1 k_2 v_{C2}} - \dfrac{v_{C2}}{RC_2} - \dfrac{i_o}{C_2} = g_3(x,t) \end{cases} \tag{25}$$

The equilibrium point (Equation (26)) is derived by nullifying the derivatives in the nonlinear dynamics (Equation (25)). As shown in Equation (26) it has a variable transformer characteristic, with turns ratio n = $V_{C2}/V_g = K^2(\theta)$. The small signal control block diagram is shown in Figure 13.

$$X^*(\theta) = \left[\frac{V_g}{R} K_2^4(\theta), \ \frac{V_g}{R} K_2^3(\theta)^3, \ V_g \cdot K_2(\theta), \ V_g \cdot K_2^2(\theta)\right]^T \quad K_2(\theta) = \sqrt{\frac{V_{C2}}{V_g}} = \sqrt{\frac{V_{ref}|\sin(\theta)|}{V_g}}. \tag{26}$$

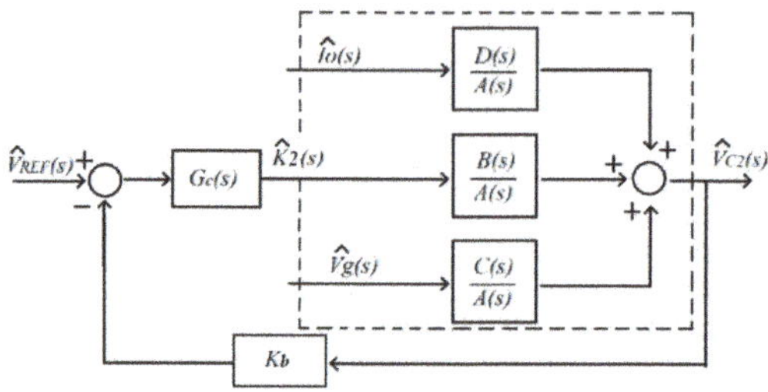

Figure 13. Block diagram of the output voltage regulation loop.

By linearizing the ideal dynamics (Equation (25)) around the equilibrium point (Equation (26)), it is possible to obtain a small signal model (Equation (27)),

$$
\begin{aligned}
\frac{d\hat{i}_{L2}}{dt} &\approx \mathbf{a}\cdot\hat{v}_{C1}(t) + \mathbf{b}\cdot\hat{v}_{C2}(t) + \mathbf{c}\cdot\hat{v}_g(t) + \mathbf{d}\cdot\frac{d\hat{k}_2(t)}{dt} \\
\frac{d\hat{v}_{C1}}{dt} &\approx \mathbf{e}\cdot\hat{k}_2(t) + \mathbf{f}\cdot\hat{v}_{C1}(t) + \mathbf{g}\cdot\hat{v}_g(t) + \mathbf{h}\cdot\hat{v}_{C2}(t) + t\cdot\frac{d\hat{k}_2(t)}{dt} \\
\frac{d\hat{v}_{C2}}{dt} &\approx \mathbf{z}\cdot\hat{v}_{C2}(t) + \mathbf{y}\cdot\hat{i}_{L2}(t) + \mathbf{m}\cdot\hat{v}_{C1}(t) + \mathbf{n}\cdot\hat{v}_g(t) + \mathbf{q}\cdot\hat{i}_o(t) + \mathbf{w}\cdot\frac{d\hat{k}_2(t)}{dt}
\end{aligned}
\tag{27}
$$

where coefficients are defined in Equation (28).

$$
\begin{aligned}
\mathbf{a} &= \left.\frac{dg_1}{dv_{C1}}\right|_{X_*} = \frac{-2}{K^2(\theta)L_1 - L_2} &
\mathbf{b} &= \left.\frac{dg_1}{dv_{C2}}\right|_{X_*} = \frac{K^{-1}(\theta)}{K^2(\theta)L_1 - L_2} &
\mathbf{n} &= \left.\frac{dg_3}{dv_g}\right|_{X_*} = \frac{-L_2 K^2(\theta)}{RC_2(K^2(\theta)L_1 - L_2)} \\[2mm]
\mathbf{h} &= \left.\frac{dg_2}{dv_{C2}}\right|_{X_*} = \frac{-L_1 K_2^3(\theta)}{RC_1(K_2^2(\theta)L_1 - L_2)} &
\mathbf{e} &= \left.\frac{dg_2}{dk_2}\right|_{X_*} = \frac{V_g K_2^2(\theta)}{RC_1} &
\mathbf{f} &= \left.\frac{dg_2}{dv_{C1}}\right|_{X_*} = \frac{(K_2^2(\theta)L_1 + L_2)K_2^2(\theta)}{RC_1(K_2^2(\theta)L_1 - L_2)} \\[2mm]
\mathbf{g} &= \left.\frac{dg_2}{dv_g}\right|_{X_*} = \frac{-L_2 K_2^3(\theta)}{RC_1(K^2(\theta)L_1 - L_2)} &
\mathbf{q} &= \left.\frac{dg_3}{di_o}\right|_{X_*} = \frac{-1}{C_2} &
\mathbf{t} &= \left.\frac{dg_2}{d\left(\frac{dk_2}{dt}\right)}\right|_{X_*} = \frac{V_g L_1 L_2 K_2^6(\theta)}{R^2 C_1(K_2^2(\theta)L_1 - L_2)} \\[2mm]
\mathbf{z} &= \left.\frac{dg_3}{dv_{C2}}\right|_{X_*} = \frac{-2L_1 K_2^2(\theta) + L_2}{RC_2(K_2^2(\theta)L_1 - L_2)} &
\mathbf{y} &= \left.\frac{dg_3}{di_{L2}}\right|_{X_*} = \frac{1}{C_2 K_2(\theta)} &
\mathbf{m} &= \left.\frac{dg_3}{dv_{C1}}\right|_{X} = \frac{(K_2^2(\theta)L_1 + L_2)K_2(\theta)}{RC_2(K_2^2(\theta)L_1 - L_2)} \\[2mm]
\mathbf{n} &= \left.\frac{dg_3}{dv_g}\right|_{X_*} = \frac{-L_2 K_2^2(\theta)}{RC_2(K_2^2(\theta)L_1 - L_2)} &
\mathbf{c} &= \left.\frac{dg_1}{dv_g}\right|_{X_*} = \frac{K_2(\theta)}{K_2^2(\theta)L_1 - L_2} &
\mathbf{d} &= \left.\frac{dg_1}{d\left(\frac{dk_2}{dt}\right)}\right|_{X_*} = \frac{-L_1 V_g K_2^4(\theta)}{R(K_2^2(\theta)L_1 - L_2)}
\end{aligned}
\tag{28}
$$

As in the previous case, the linearized dynamics (Equation (27)) allows to determine the open-loop transfer functions of the small signal model depicted in Figure 13. Note that in this case, the output of the controller is understood as the variation of the ratio among inductor currents instead of the variation of the input current reference amplitude, as in the input current surface.

The polynomials $A(s)$, $B(s)$, $C(s)$, and $D(s)$, defining the small signal model transfer functions are shown in Equation (29), and as in the previous case, the poles and zeros have also a clear dependence of K_2, R, V_g, and $sin(\theta)$. The characteristic polynomial of the closed-loop transfer functions is defined in Equation (30).

$$
\begin{aligned}
A(s) &= s^3\cdot\left[K_2^2(\theta)L_1 - L_2\right] + s^2\cdot\left[\frac{2L_1 K_2^2(\theta) - L_2}{RC_2} - \frac{K_2^2(\theta)\left[K_2^2(\theta)L_1 + L_2\right]}{RC_1}\right] - s\cdot\left[\frac{K_2^2(\theta)\left[K_2^2(\theta)L_1 + L_2\right]}{R^2 C_1 C_2} + \frac{1}{K_2^2(\theta)C_2}\right] + \frac{-1}{RC_1 C_2} \\[2mm]
B(s) &= s^3\cdot\left[\frac{L_1 L_2 K_2^5(\theta)V_g}{R^2 C_2}\right] - s^2\cdot\left[\frac{L_1 K_2^3(\theta)V_g}{RC_2}\right] + s\cdot\left[\frac{K_2^3(\theta)V_g\left[2K_2^2(\theta)L_1 + L_2\right]}{R^2 C_1 C_2}\right] + \frac{-2K_2(\theta)V_g}{RC_1 C_2} \\[2mm]
C(s) &= -s^2\cdot\left[\frac{K_2^2(\theta)L_2}{RC_2}\right] + \frac{s}{C_2} + \frac{-K_2^2(\theta)}{RC_1 C_2} \qquad
D(s) = -s^2\cdot\left[\frac{K_2^2(\theta)L_1 - L_2}{C_2}\right] + s\cdot\left[\frac{K_2^2(\theta)\left[K_2^2(\theta)L_1 + L_2\right]}{RC_1 C_2}\right].
\end{aligned}
\tag{29}
$$

$$
P(s) = s\cdot[A(s) + K_b G_c(s)B(s)] = p_4 s^4 + p_3 s^3 + p_2 s^2 + p_1 s + p_0
\tag{30}
$$

5.1. Parameter Variation Analysis

Figure 14a,b depict the root locus of the open-loop control-to-output (plant) transfer function = $B(s)/A(s)$ in the same test conditions and with the converter component parameters used in the previous case, the input current surface.

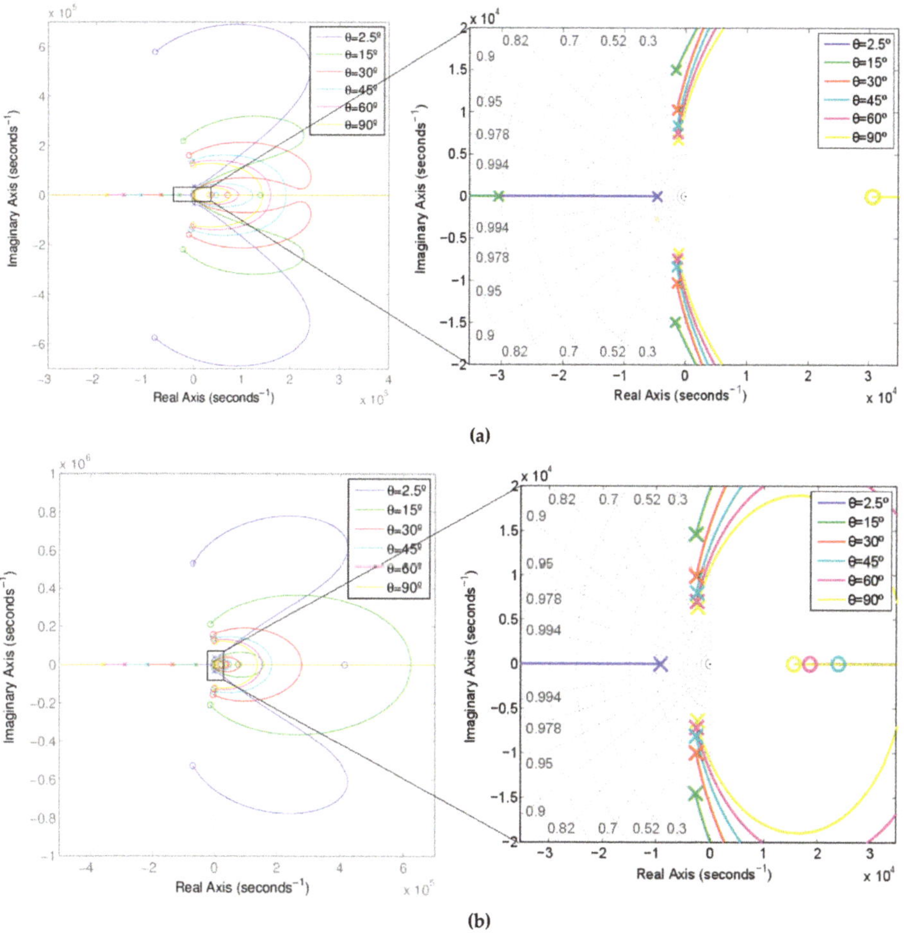

Figure 14. Root locus of for (**a**) 100 W, ($R = 484\ \Omega$), (**b**) for 200 W, ($R = 242\ \Omega$).

The charts of Figure 14 show the existence of a pair of complex conjugate dominant poles on the left half-plane. Just as in the input current control case, the real part of the dominant complex conjugate poles is proportional to the output power. Thus, for 100 W the real part is $\text{Re}[s] = -1380$ rad/s and for the 200 W case $\text{Re}[s] = -2640$ rad/s. As in the preceding case, the damping factor ζ of the dominant poles increases with the load instantaneous power, this means when $\omega_0 t = \theta$ increases from 0° to 90°. On the contrary, in the preceding case, the damping factor could reach $\zeta = 1$, and here the damping never exceeds 0.38, and the dynamics is more oscillatory.

Nevertheless, and this is the most important, with the first control technique, the plant of the converter could be approximated by means of a first order system whose dynamics are four times faster than in the self-oscillating approach. In addition, the proximity of the dominant complex poles to the imaginary axis in the self-oscillating approach in conjunction with the non-minimum phase behavior further complicates the design of the control loop due to the lower bandwidth.

5.2. Synthesis of the Voltage Regulation Loop

In a similar way to the previous case, the Routh–Hurwitz criterion is used to assure, at least, local stability in closed-loop. Once the Routh criterion is applied to the characteristic polynomial $P(s)$ (Equation (30)), the corresponding stability constraints are given in Equation (31), and the k_P-k_{In} chart is also obtained.

$$\begin{aligned}
\left(p_2 p_3 - p_1 p_4\right) &> 0, \\
\left(p_1 p_2 p_3 - p_1^2 p_4 - p_0 p_3^2\right) &< 0, \; p_0 < 0
\end{aligned} \tag{31}$$

Similarly, as is done with the input current control surface, from the aforesaid procedure, a stable region in the k_P - k_{In} plane can be derived. Given the nominal input voltage, output voltage, and load R, Figure 15 charts the PI controller constants to assure stability (Equation (31)) for a representative set of θ values. In the nominal case, stability is assured if $k_P \in [0, 0.72]$ and $k_{In} \in [0, 2 \cdot 10^3]$. Equivalent plots were considered for other load and input voltage values. Finally, to enhance the transient performance, and assure stability, the following gains were selected $k_P = 0.5$, and $k_{In} = 10^3$.

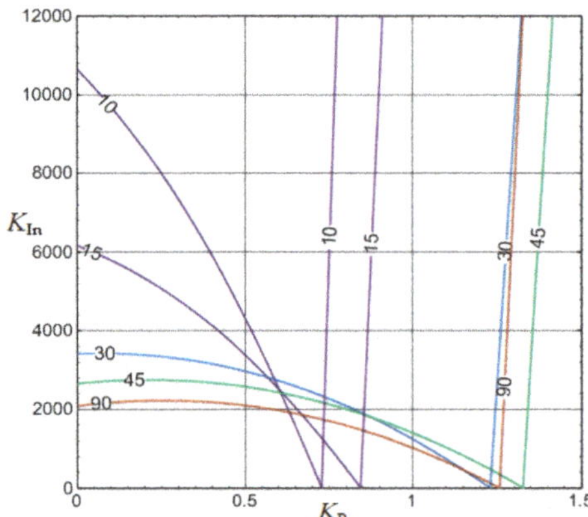

Figure 15. Evaluation of the stability region in the K_P - K_{In} plane for the nominal case.

6. Inverter Realization

In this section, the realization of a 100 W a "fast-slow" inverter prototype using two cascaded-boost circuits and an unfolding-stage is discussed, and the resulting power stage and control circuit schematics will be given. Nevertheless, the quadratic step-up nature of the microinverter first stage, makes the minimum output voltage of this stage equal to the input voltage V_g. Then, as this voltage cannot reach zero, the inverter output voltage changes instantaneously from V_g to $-V_g$ and/or vice-versa each time where $\omega_0 t = k \cdot \pi$, causing a zero-cross distortion. This is the first issue to solve.

6.1. Solving the Zero-Cross Distortion of the "Fast-Slow" Approach

The first option could be a hybrid operation of the microinverter second stage. During most of the sinusoidal cycle the full-bridge can operate as unfolding stage. However, at the vicinity of the zero-crossing instants $\omega_0 t = k \cdot \pi$, the full-bridge output could be PWM modulated to smooth the output voltage in the zero-cross region. However, to be operative a PWM requires that a second-order low-pass output filter is required. If the filter operates the whole sinusoidal cycle, the positive effects of the "fast-slow" approach are eliminated, that is, the zero-order dynamics of the unfolding stage. If the filter only operates in the zero-cross region, the complexity of the control circuit increases.

A second solution consists of connecting the unfolding stage in differential mode to the microinverter first stage. In other words, the ground of the unfolding stage is connected to the input voltage terminal Vg. This solution is depicted in Figure 16a.

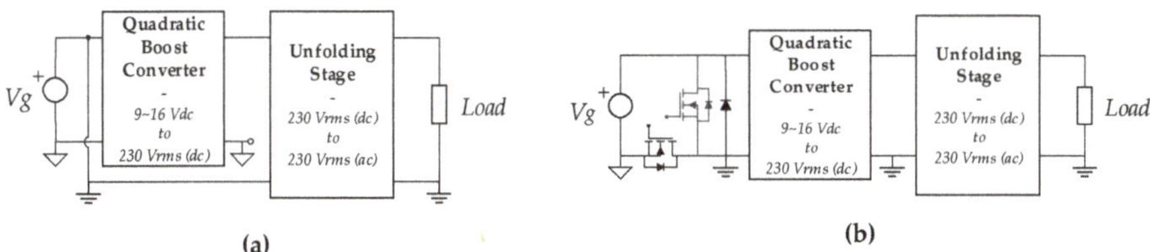

Figure 16. Solutions for the output voltage distortion in the "fast-slow" approach: (**a**) Unfolding stage differential connection, and (**b**) introducing a Buck-switch cell.

Figure 16b depicts the chosen solution. A buck-switch stage is connected between the input source and the microinverter first stage. This buck-switch disconnects the first stage from the input source voltage some hundreds of microseconds around the zero-crossing instants $\omega_0 t = k \cdot \pi$. During these disconnection periods, the quadratic-boost input voltage is zero, the output voltage can reach zero voltage, and the zero-cross distortion is eliminated.

6.2. Power-Stage Circuits and Implementation

The inverter was designed to deliver an output regulated voltage of 230 Vrms/50 Hz for an input voltage range from 9 to 16 V, and an output load range from 50 to 200 W. The converter component values are $L_1 = 8$ µH, $L_2 = 800$ µH, $C_1 = 0.5$ µF, and $C_2 = 0.8$ µF. Capacitor C_1 is implemented with a set of 10 parallel *47 nF* multilayer ceramic capacitors (MLCC) and capacitor C_2 with a set of eight parallel *100 nF* tantalum electrolytic type ones. Both types are characterised by a low series resistance and high voltage capability, appropriate for switching applications.

The following MOSFET transistors were selected: (a) the IRFB4110PBF (180 A, 100 V, 4 mΩ) for the Buck-switch cell, (b) the IPP110N20NA (88 A, 200 V, 9.9 mΩ) for the high-side and low-side switches of the first boost cell, (c) the SiC MOSFETs C3M0120090D (23 A, 900 V, 120 mΩ) for the second boost cell, and finally (d) for the unfolding stage switches, the SPW47N60C3 device (47 A, 650 V, 70 mΩ) was used.

Several driver ICs are used: two IR2110 are used in the quadratic boost, two IR21834 are used in the unfolding stage, and a low-side only driver MCP1407P is employed in the buck-switch stage. The free-wheeling MOSFET of the buck-switch stage is driven by a complementary bipolar transistor pair (ZTX653/ZTX753), for the free-wheeling MOSFET. Figures 17–20 depict the power stage circuits.

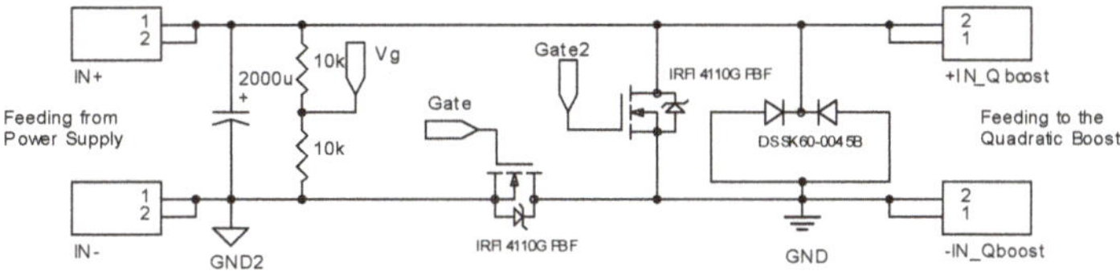

Figure 17. Power stage circuit of the Buck-switch cell.

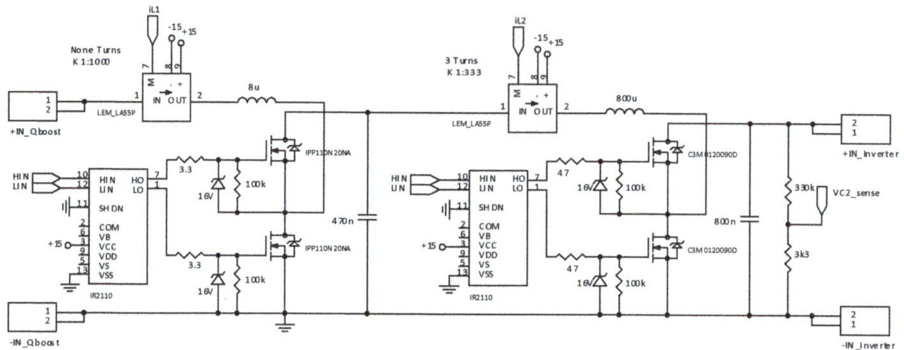

Figure 18. Power stage circuit of the proposed Quadratic Boost.

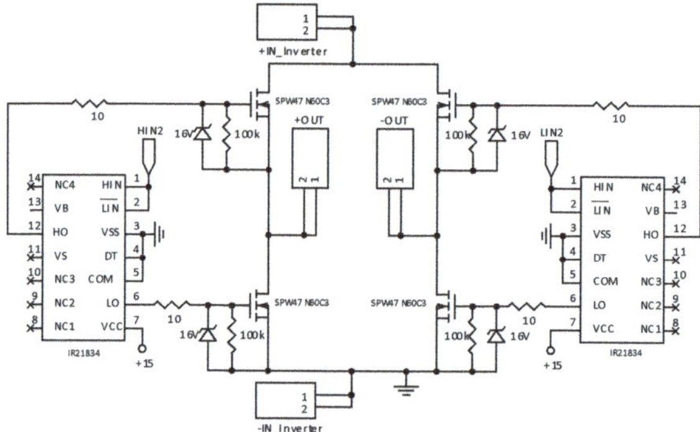

Figure 19. Power stage circuit and drivers of the unfolding/inverter stage.

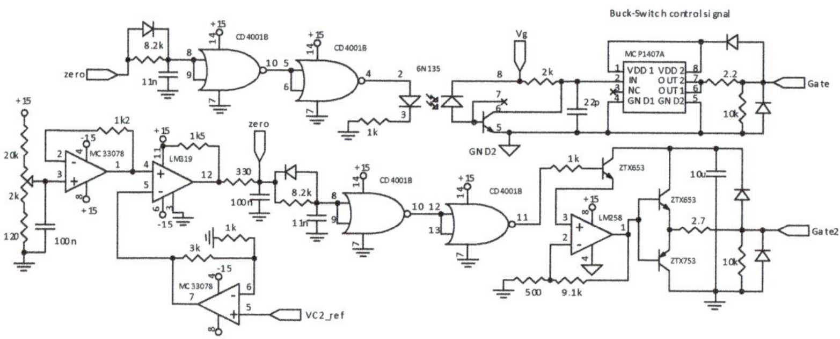

Figure 20. Zero-crossing detection circuit and Buck-switch control drivers.

6.3. Control-Surfaces and Voltage-Loop Implementation

To implement the proposed sliding mode control surfaces, a hysteretic comparator is required. For practical reasons to be explained in Section 7, and related to the experimental results, two different type of hysteresis are available for test purposes: constant and variable. The inverter will be able to operate with any of them. The constant hysteresis width Δdc, can be adjusted in the range $\Delta dc = [0.1, 0.3]$. In the variable hysteresis type $\Delta(t) = \Delta dc + \Delta v \cdot |\sin(\theta)|$, the offset Δdc and the amplitude Δv can be adjusted, where $\Delta v = [0.1, 0.5]$. The hysteresis parameters have a considerable impact in the microinverter switching losses and conversion efficiency. Figures 21–23 depict the control circuits.

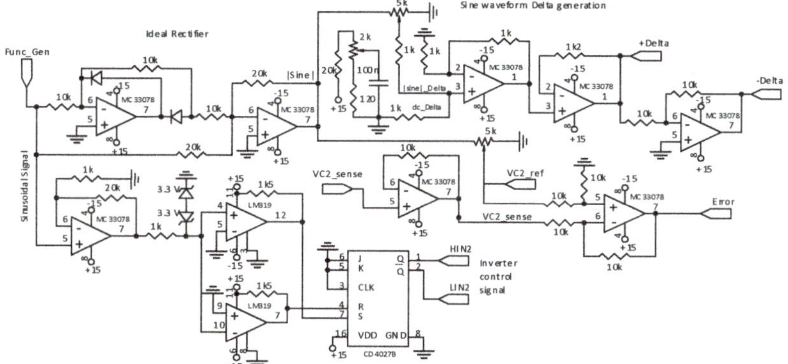

Figure 21. Generator of ancillary signals.

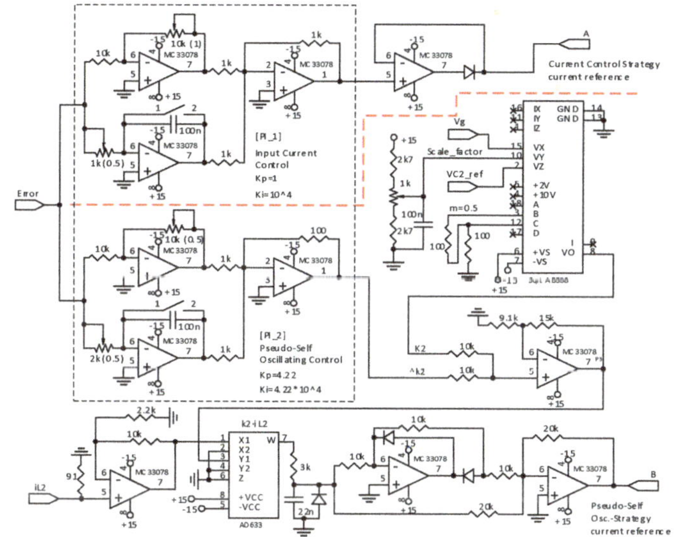

Figure 22. Control-loop and surface generator.

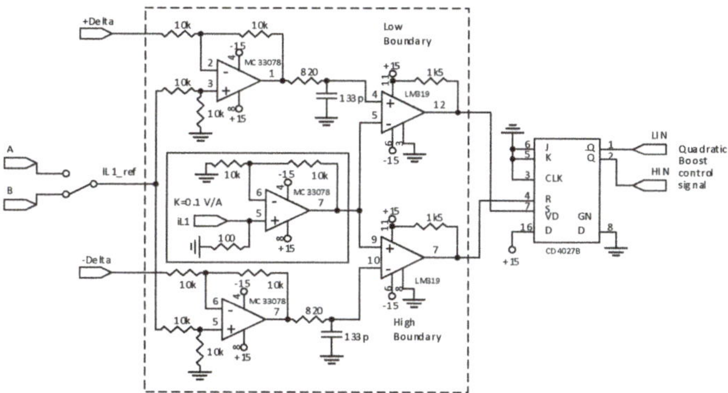

Figure 23. Sliding surface selection and hysteretic comparator.

Figure 21 depicts diverse ancillary functions. Taking a sinusoidal signal from a function generator, this circuit board generates (a) the output voltage reference signal $v_{ref}(\theta) = |\sin(\theta)|$ by means of an ideal rectifier, (b) the output voltage error $e(t) = v_{ref}(t)$-$v_{C2}(t)$, (c) the unfolding-stage switching signals H_{IN2} and L_{IN2}, and finally (d) both hysteresis widths, the constant one and the variable.

Figure 22 depicts the two PI controllers, one per surface, that were used to close the voltage regulation loop for both control surfaces. The output signals "A" and "B" are the references used by the hysteretic comparator circuit in Figure 23 to implement the sliding surfaces (Equation (32)).

$$\begin{aligned} S(x,t) = i_{L1}(t) - A = 0 \;\; where\, A = K_1(t) \quad input\, current\, surface \\ S(x,t) = i_{L1}(t) - B = 0 \;\; where\, B = K_2(t)\cdot i_{L2}(t) \;\; self-oscillating\, surface \end{aligned} \tag{32}$$

7. Experimental Results

Figure 24 shows a photograph of the microinverter prototype and the experimental set-up. This test environment includes a Delta Elektronika SM70-45-D (3 kW, 70 V, 45 A) power supply for the power stage, a TENMA 72-10505 (90 W, 30 V, 3 A) to supply the control circuit, a Yokogawa DLM4038 (2.5 GS/s, 350 MHz) oscilloscope with Tektronix TCP2020 (20 Arms) current probes and Yokogawa 700924 (±1400 V) differential probes to measure and capture the electrical variables.

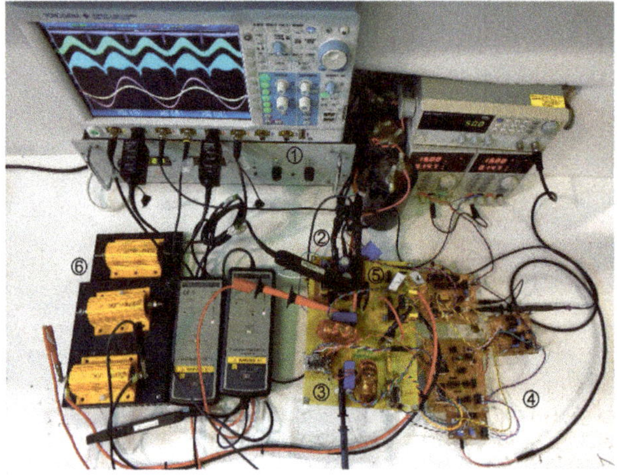

Figure 24. Experimental setup and microinverter prototype.

The reference $v_{ref}(t)$ is a rectified sinusoidal signal, provided by the ancillary circuit of Figure 22 that rectifies a sinusoidal voltage given by an external function generator.

The items displayed in Figure 24 are (1) the power supply SM70-45-D, (2) the buck-switch stage, (3) the two cascaded-boosts, (4) the ancillary and control PCBs, including both surfaces; (5) the unfolding stage; and finally the (6) load resistances (50–200 W). The oscilloscope, the function generator, and the power supply for the control boards are also shown.

Concerning the microinverter output waveforms, the results are very similar despite the control surface used or the hysteresis type. As both control surfaces are designed to do the same, a proof of their good performance is that the experimental results are practically identical.

The effect of the buck-switch to solve zero-cross distortion can be seen in Figure 25. The dark-blue color signal V_g is the quadratic-boost input voltage, the light-green signal is V_{c1}, the cyan signal is V_{c2}, the red one is the unfolding output voltage V_{out}, and finally the pink color signal is the reference voltage V_{ref}. Realize that the introduction of the quadratic-boost supply allows that all voltage variables go to zero at crossing-instants reducing the distortion effects.

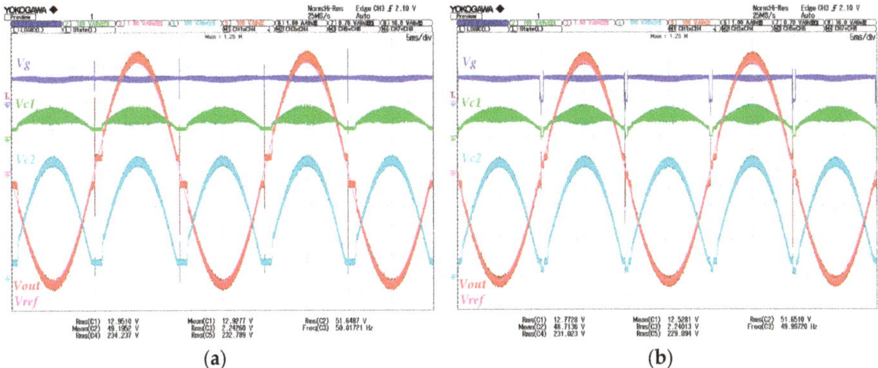

(a)   (b)

Figure 25. Input current surface with constant-width hysteresis. Microinverter variables traces, (**a**) without the Buck-switch, and (**b**) including the Buck-switch.

A load transient from 50 to 150 W was carried out to see the dynamic response, and similar results were obtained for both surfaces. Constant or variable hysteresis width have no influence in dynamic response for input current surface case (Equation (7)). Nevertheless, the self-oscillating surface requires a very small hysteresis width when both current variables i_{L1} and i_{L2}, are near to zero to start oscillating, and to avoid excessive switching losses a variable hysteresis width is used.

In the caption of Figure 26, the load current is the orange color signal, and the microinverter output voltage is depicted in red.

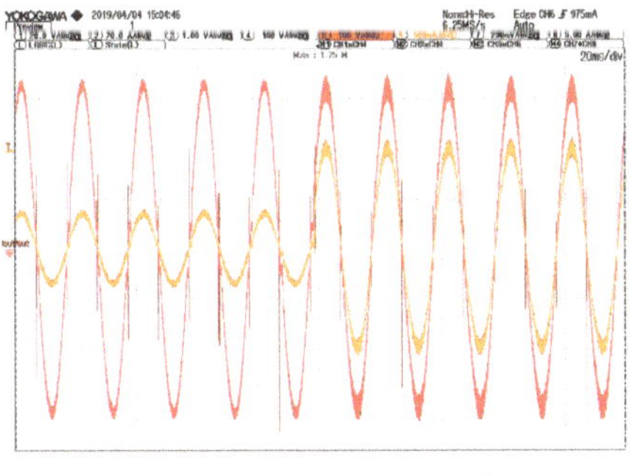

Figure 26. Load transition of 50 to 150 W.

Next oscilloscope captions, see Figure 27, depict the experimental inverter waveforms at 100 W, for both surfaces and different hysteresis types to illustrate their effect on the inverter output. Captions are given in two types. On the left side, there is always the oscilloscope captions in sample mode. In the right side, we include the same experiment when the oscilloscope is averaging 256 samples. This allows to see the real averaged value of all the variables, neglecting their high-frequency ripple.

As can be seen in all the figures, the output voltage and current, shown in red and orange colors respectively, are always equal, despite the selected hysteresis and control surface type.

The input current $i_{L1}(t)$, depicted in light-green color, and the intermediate capacitor voltage $v_{C1}(t)$, in cyan color, show clearly the effect of the hysteresis type, as can be observed comparing the different ripple amplitudes in the left captions (sample mode). On the contrary, in the right side captions, where digital oscilloscope averaging is applied, the input current $i_{L1}(t)$ shows clearly a $sin^2(\omega_0 t)$ waveform type, whereas the $v_{C1}(t)$ exhibits a waveform of type $|sin(\omega_o t)|^{1/2}$, as expected from the state vector (Equation (6)) reproduced here in Equation (33).

$$x(t) = \left[\begin{array}{cccc} \dfrac{V_{ref}^2}{V_g R} \sin^2(\omega_o t), & \dfrac{1}{R}\sqrt{\dfrac{V_{ref}^3}{V_g}}|\sin(\omega_o t)|, & \sqrt{V_g \cdot V_{ref}|\sin(\omega_o t)|}, & V_{ref}\,|\sin(\omega_o t)| \end{array} \right]^T$$
$$where \; x(t) = \left[\begin{array}{cccc} i_{L1}(t), & i_{L2}(t), & v_{C1}(t), & v_{C2}(t) \end{array} \right]^T and \; V_{ref} = 230\sqrt{2} \tag{33}$$

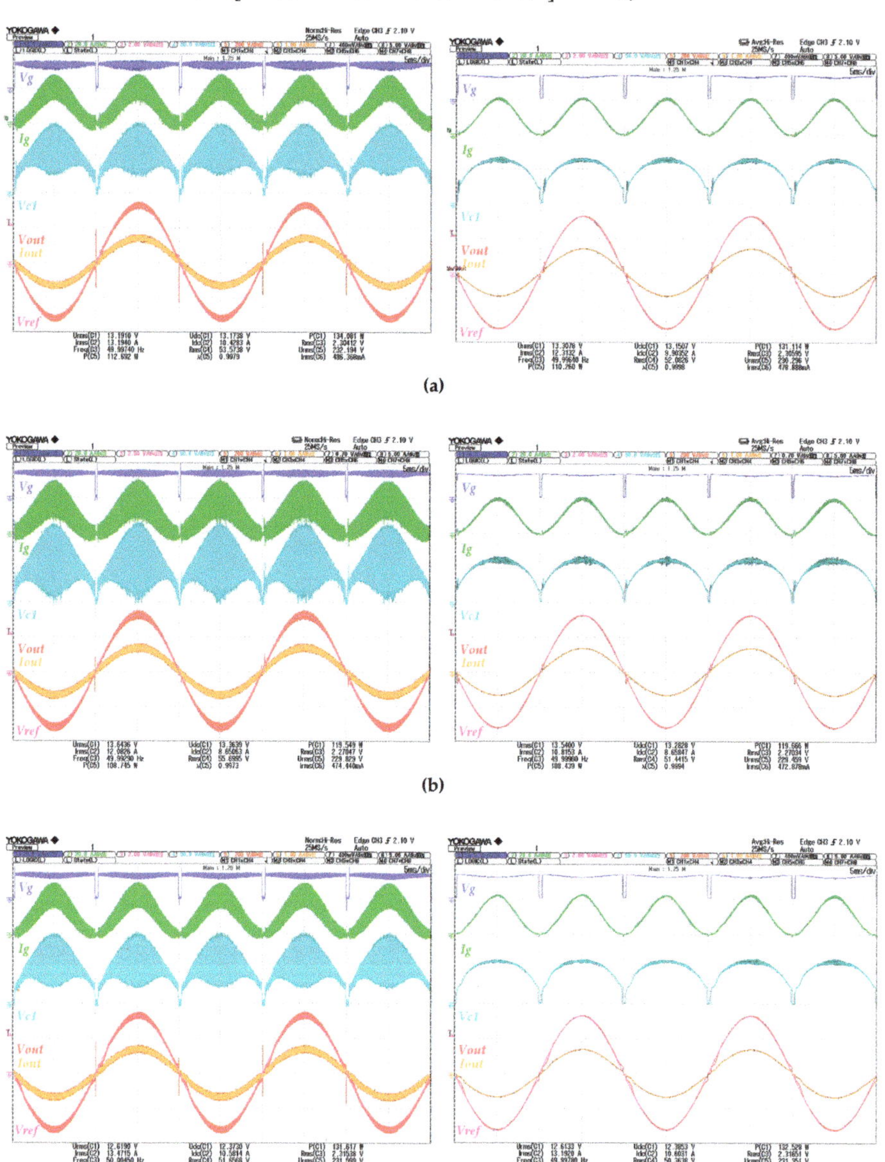

Figure 27. Microinverter captions at 100 W under different conditions. (**a**) Input current control and constant hysteresis. (**b**) Input current surface and variable hysteresis, and (**c**) self-oscillating surface and variable hysteresis.

Figures 28 and 29, Figures 30 and 31 give, respectively, the inverter efficiency, and the minimum and maximum switching frequency, and the average switching frequency for diverse output powers P(W) = {25, 50, 75, 100, 125, 150}. Different control surfaces and hysteresis types are used. The input current surface with variable hysteresis exhibits the best efficiency, around 90%.

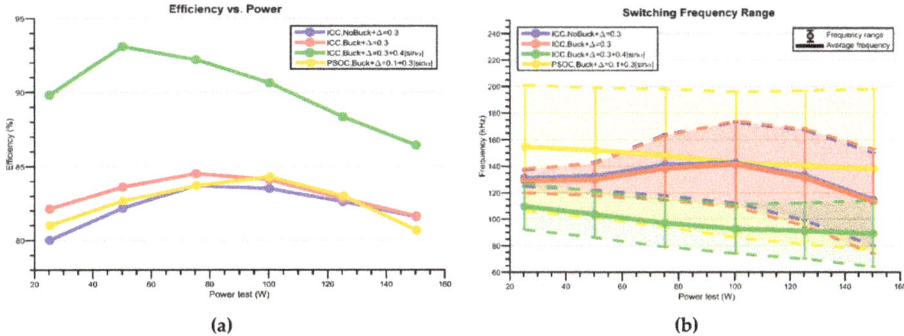

Figure 28. Microinverter results at different output powers, (**a**) efficiency, and (**b**) switching frequency.

Control Type	Buck Switch	Hysteresis Type	25 W	50 W	75 W	100 W	125 W	150 W
Input Current	*OFF*	$\Delta = 0.3$	80.31	82.16	83.70	83.51	82.60	81.59
Input Current	*ON*	$\Delta = 0.3$	82.10	83.61	84.52	84.10	82.82	81.62
Input Current	*ON*	$\Delta = 0.3 + 0.4|\sin\alpha|$	89.83	93.01	92.20	90.62	88.33	86.42
Self − Oscillating	*ON*	$\Delta = 0.1 + 0.3|\sin\alpha|$	81.05	82.66	83.76	84.29	83.10	80.67

Figure 29. Microinverter efficiency in %.

Control Type	Buck Switch	Hysteresis Type	25 W	50 W	75 W	100 W	125 W	150 W
Input Current	*OFF*	$\Delta = 0.3$	125-137	122-143	118-164	112-173	99-167	80-150
Input Current	*ON*	$\Delta = 0.3$	120-138	118-142	114-163	109-174	95-168	74-153
Input Current	*ON*	$\Delta = 0.3 + 0.4|\sin\alpha|$	92-127	86-121	79-115	74-111	70-112	64-114
Self − Oscillating	*ON*	$\Delta = 0.1 + 0.3|\sin\alpha|$	108-201	105-199	98-198	89-196	84-197	78-198

Figure 30. Microinverter maximum and minimum switching frequency in kHz.

Control Type	Buck Switch	Hysteresis Type	25 W	50 W	75 W	100 W	125 W	150 W
Input Current	*OFF*	$\Delta = 0.3$	131	132.5	141	142.5	133	115
Input Current	*ON*	$\Delta = 0.3$	129	130	138.5	141.5	131.5	113.5
Input Current	*ON*	$\Delta = 0.3 + 0.4|\sin\alpha|$	109.5	103.5	97	92.5	91	89
Self − Oscillating	*ON*	$\Delta = 0.1 + 0.3|\sin\alpha|$	154.5	152	148	142.5	140.5	138

Figure 31. Microinverter average switching frequency in kHz.

8. Conclusions

A quadratic-boot based microinverter is presented in this work. The first stage is a cascaded boost converter, and the second stage is a full-bridge. In the classical approach (slow-fast), the cascaded boost converter output is a 350–400 DC voltage that supplies a full-bridge inverter. This work explores a new alternative, the "fast-slow" strategy. In this approach, the first stage delivers a sinusoidal rectified voltage of 230 Vrms to the full-bridge unfolding stage to get the AC output.

Comparing both strategies, in the classical approach, the sinusoidal voltage is obtained by PWM modulation. Thus, the full-bridge stage must switch at high frequency to deliver a low THD sinusoidal output. In contrast, in the "fast-slow" approach the full-bridge is switching at 50 Hz, as operates as unfolding stage, and the switching losses of this stage are negligible.

If the first stage output voltage in both strategies is compared, in the classical approach this voltage is 350–400V, but for the approach presented here, the peak value is 325 V and its rms value is 230 Vrms. This means that the stress of the switching devices is lower, and therefore, at equal switching frequency, the switching losses will be lower.

Now, if the dynamics involved in both approaches are compared, in the classical approach, the microinverter plant has 6th order dynamics, whereas in the "slow-fast" approach, as the unfolding stage lacks the second order filter, and the plant has only 4th order dynamics.

Next, if PWM control is applied in the "slow-fast" case, three PI regulators are required to operate the inverter, two for the first stage (inner current loop and voltage loop) and the third PI for the full-bridge inverter voltage loop. This implies that the whole microinverter would have 9th order dynamics. Applying PWM control to the "fast-slow" strategy means two PI regulators for the first stage, the whole system becoming dynamics of the 6th order.

To conclude, it is important to remark that the dynamics of the cascaded-boost is highly nonlinear, and the small-signal output voltage to control transfer function, and therefore, the plant poles, depend strongly on the complementary converter duty cycle (D′ = 1-D). As a result, in the "slow-fast" approach the pole position is quite constant, and only depends on the input voltage variations. Oppositely, in the "fast-slow" approach, as the first stage output voltage is following a sinusoidal waveform, and the plant poles are continuously changing making practically impossible a classical control approach using PWM techniques.

Sliding mode control is a non-linear control technique with inherent robustness to parametric variations and perturbations, very simple to implement (only a hysteretic comparator is required), and it produces a system-order reduction. When this control technique is applied to the "fast-slow" microinverter, the 4th order plant, becomes a 3rd order one. Besides, to regulate the inverter output voltage, only one external loop is required, the first-stage PI controller. Thus, the closed-loop microinverter becomes finally a 4th order system.

Two different sliding control surfaces were proposed. An input current surface, and a self-oscillating surface. Both surfaces were analyzed, and the corresponding PI regulators were designed. The input current surface is relatively classic and is based on controlling the converter input energy. The converter with the self-oscillating approach behaves like a variable transformer, with ideal rejection of load and line disturbances, and directly establishes the sinusoidal output voltage. In fact, the voltage regulation loop is only required to compensate the converter losses.

The input current surface has a first order dominant pole at low output voltage that evolves to a pair of critically damped ($\zeta = 1$) real poles, whereas the self-oscillating surface exhibits a second order dominance with a pair of undamped complex conjugated poles. At both surfaces the pole position is practically proportional to the mean load power, whereas the damping ratio increases with the output voltage. In addition, as the dominant pole ($5 \cdot 10^3$–10^4 rad/s) of the input current surface is five times faster than the real part of the self-oscillating surface dominant poles (10^3–$2 \cdot 10^3$ rad/s), the input current surface bandwidth is higher, and the closed-loop dynamic response is faster.

The validity of the proposed "fast-slow" approach with any of the proposed surfaces was demonstrated with several oscilloscope captions. Two types of hysteresis were considered, constant and variable. According to the experimental results, the input current surface with a variable hysteresis width exhibits the better efficiency results.

Author Contributions: Conceptualization, H.V.-B.; Data curation, E.R.-R. and X.G.-M.; Formal analysis, H.V.-B., E.R.-R. and C.O.; Funding acquisition, H.V.-B.; Investigation, H.V.-B., E.R.-R., C.O. and X.G.-M.; Methodology,

H.V.-B. and C.O.; Project administration, H.V.-B.; Resources, H.V.-B.; Supervision, H.V.-B. and C.O.; Validation, H.V.-B., E.R.-R. and X.G.-M.; Visualization, H.V.-B., E.R.-R. and X.G.-M.; Writing—original draft, H.V.-B., E.R.-R., C.O. and X.G.-M.; Writing—review & editing, H.V.-B. and C.O.

Funding: This research was funded by the Spanish Agencia Estatal de Investigación (AEI) and the Fondo Europeo de Desarrollo Regional (FEDER) under research projects DPI2015-67292-R (AEI/FEDER, UE).

Conflicts of Interest: The authors declare no conflict of interest. The funders had no role in the design of the study; in the collection, analyses, or interpretation of data; in the writing of the manuscript, or in the decision to publish the results.

References

1. Zhao, Z.; Xu, M.; Chen, Q.; Lai, J.-H.; Cho, Y. Derivation, analysis, and imlemention of a boost-buck coverter-based high-efficiency PV inverter. *IEEE Trans. Power Electron.* **2012**, *27*, 1304–1311. [CrossRef]

2. Oldenkamp, H.; de Jong, I. The Return of the AC Module Inverter. In Proceedings of the 24th European Photovoltaic Solar Energy Conference, Hamburg, Germany, 21–25 September 2009; pp. 3101–3104.

3. Hasan, R.; Mekhilef, S.; Seyedmahmoudian, M.; Horan, B. Grid-connected isolated PV microinverters: A review. *Renew. Sustain. Energy Rev.* **2017**, *67*, 1065–1080. [CrossRef]

4. Choudhury, T.R.; Nayak, B. Comparison and analysis of cascaded and quadratic boost converter. In Proceedings of the IEEE Power, Communication and Information Technology Conference (PCITC), Bhubaneswar, India, 15–17 October 2015; pp. 1065–1080.

5. Lopez, D.; Flores-Bahamonde, F.; Renaudineau, H.; Kouro, S. Double voltage step-up photovoltaic microinverter. In Proceedings of the 2017 IEEE International Conference on Industrial Technology, Toronto, ON, Canada, 22–25 March 2017; pp. 406–411.

6. Fornage, M. Method and Apparatus for Converting Direct Current to Alternating Current. US Patent 7,796,412, 14 September 2010.

7. Yuhao Luo, J. Alternating Parallel Fly Back Converter with Alternated Master-Slave Branch Circuits. US Patent App. 13/807,0532, 9 June 2011.

8. Cáceres, R.O.; Barbi, I. A Boost DC-AC Converter: Analysis, Design, and Experimentation. *IEEE Trans. Power Electron.* **1999**, *14*, 134–141. [CrossRef]

9. Sanchis, P.; Ursæa, A.; Gubía, E.; Marroyo, L. Boost DC-AC inverter: A new control strategy. *IEEE Trans. Power Electron.* **2005**, *20*, 343–353. [CrossRef]

10. Flores-Bahamonde, F.; Valderrama Blavi, H.; Garcia, G.; Martinez Salamero, L. Análisis de un Ondulador Boost basado en Control en Modo Deslizamiento para Energías Renovables. In Proceedings of the SAAEI's 14, Tangiers, Morocco, 25–27 June 2014.

11. Galigekere, V.P.; Kazimierczuk, M.K. Analysis of PWM Z-Source DC-DC Converter in CCM for Steady State. *IEEE Trans. Circuits Syst. I Regul. Pap.* **2012**, *59*, 854–863. [CrossRef]

12. Lopez Santos, O.; Martinez Salamero, L.; Garcia, G.; Valderrama Blavi, H. Robust Sliding-Mode Control Design for a Voltage Regulated Quadratic Boost Converter. *IEEE Trans. Power Electron.* **2015**, *30*, 2313–2327. [CrossRef]

13. Lopez Santos, O.; Martinez Salamero, L.; Garcia, G.; Valderrama Blavi, H.; Sierra Polanco, T. Comparison of the Quadratic Boost Topologies Operating under Sliding-Mode Control. In Proceedings of the Brazilian Power Electronics Conference (COBEP), Gramado, Brazil, 27–31 October 2013.

14. López Santos, O.; Martínez Salamero, L.; García, G.; Valderrama Blavi, H.; Mercuri, D.O. Efficiency Analysis of a Sliding-Mode controlled Quadratic Boost Converter. *IET Power Electron.* **2013**, *6*, 364–373. [CrossRef]

15. Erickson, R.W.; Maksimovic, D. *Fundamentals of Power Electronics*, 2nd ed.; Kluwer Academic: Dordrecht, The Netherlands.

16. Martinez-salamero, L.; Valderrama-blavi, H.; Giral, R. Self-Oscillating DC-to-DC Switching Converters with Transformer Characteristics. *IEEE Trans. Aerosp. Electron. Syst.* **2005**, *41*, 710–716. [CrossRef]

17. Kocher, M.J.; Steigerwald, R.L. An AC-to-DC Converter with High Quality Input Waveforms. *IEEE Trans. Ind. Appl.* **1983**, *IA-19*, 586–599. [CrossRef]

Article

Sliding Mode Output Regulation for a Boost Power Converter [†]

Jorge Rivera [1,*], Susana Ortega-Cisneros [2] and Florentino Chavira [3]

[1]   CONACYT–Advanced Studies and Research Center (CINVESTAV), National Polytechnic Institute (IPN), Guadalajara Campus, Zapopan 45015, Mexico

[2]   Advanced Studies and Research Center (CINVESTAV), National Polytechnic Institute (IPN), Guadalajara Campus, Zapopan 45015, Mexico; sortega@gdl.cinvestav.mx

[3]   Ceti Unidad Colomos, Calle Nueva Escocia 1885, Providencia 5a Sección, Guadalajara 44638, Mexico; jose196868@yahoo.com.mx

*   Correspondence: riveraj@gdl.cinvestav.mx; Tel.: +52-33-3777-3600

†   This paper is an extended version of our paper published in CCE 2012, Mexico city, Mexico, 26–28 September 2012.

Received: 21 November 2018; Accepted: 21 February 2019; Published: 6 March 2019

Abstract: This work deals with the novel application of the sliding mode (discontinuous) output regulation theory to a nonlinear electrical circuit, the so-called boost power converter. This theory has excelled due to the fact that trajectory tracking plays a central role. The control of a boost power converter for the output tracking of a DC biased sinusoidal signal is a challenging problem for control engineers. The main difficulties are the computation of a proper reference signal for the inductor current, and the stabilization of the inductor current dynamics or to guarantee the correct output tracking of the capacitor voltage. With the application of the discontinuous output regulation these problems are solved in this work. Simulations and real time experiments were carried out with an unknown variation of the DC input voltage, where the good output tracking of the capacitor voltage was verified along with the stabilization of the inductor current. The discontinuous output regulation theory has proven to be a suitable tool in the output tracking for the boost power converter.

Keywords: output regulation; state feedback; sliding mode control; DC-DC power converter

---

1. Introduction

Switched mode DC-DC power converters [1], are mainly used as constant current sources for LEDs, industry lighting, mobile phones, and photovoltaic systems [2,3]. Among the well known DC-DC converter topologies as buck, boost, buck-boost and cúk converters, the last three mentioned topologies result in being nonminimum-phase when directly controlling the output capacitor voltage variable [4]. Therefore, these topologies constitute a challenging area in the nonlinear control design point of view, attracting attention from researchers. Hence, several control techniques, either linear or nonlinear, to regulate these converters have been proposed such as input-output feedback linearization [5], boundary-conduction mode [6], linear designs [7], sliding mode control [8–10], current-mode-control [11], artificial neural networks [12], fuzzy logic control [13], passivity-based control [14], discontinuous conduction mode [15,16], among others.

With respect to the DC-AC power conversion using the traditional switched DC-DC power converter topologies, this has also attracted attention from researchers. In particular, in the work presented in [17], two closed-loop boost power converters were proposed for the DC-AC power conversion problem. One boost converter was controlled for the tracking of a DC biased sinusoidal signal, producing at the output a unipolar voltage. The other boost converter was controlled to track at the output the same signal, but with a phase value of 180°. The load was connected differentially across the converters. Thus, whereas a DC bias appears at each end of the load with respect to ground, the differential DC voltage over the load is zero, obtaining an AC voltage across the load. Two main advantages are noted in this work: the peak value of the AC voltage is larger than the DC input voltage due to the voltage raise property of the boost power converter, the other advantage is the possibility of harmonic reduction that will depend on the accuracy of the controllers for the generation of DC biased sinusoidal voltages at the outputs of each boost converter. Also, in the work [18], a cascade control strategy based on the sliding mode control was presented for the boost inverter. An analytical approach to describe the sliding mode motion was presented, and the generation of a sinusoidal signal with grid frequency and very low distortion was experimentally verified.

After the work presented in [17], the control design for the output tracking of a DC biased sinusoidal signal for a single boost power converter, became popular among researchers, where two control methods are distinguished: the direct and indirect methods. These control methods are characterized by the following problems:

- In the direct control method, the output capacitor voltage is directly controlled for tracking a proposed reference signal, yielding to a nonminimum-phase system, i.e., the residual inductor current dynamics is unstable. For a given capacitor voltage reference signal, the computation of a proper reference signal for the inductor current is a difficult task;
- In the indirect control method, the inductor current is directly controlled for tracking a proposed reference signal, yielding to a minimum-phase system, i.e., the residual capacitor voltage dynamics is stable, but the proposal of the inductor current reference signal that yields the desired behaviour at the output capacitor voltage is also a difficult task.

In the effort to solve these problems, one can find in the literature several works as the one in [8] where the flatness property of the system is exploited with an indirect control approach. This work is characterized by determining the inductor current in an iteratively fashion, and by a residual dynamics of the output capacitor voltage where its convergence to its desired reference signal results difficult to be demonstrated. In [19,20] direct tracking sliding mode controllers were proposed. In particular, in [19], the reference signal for the inductor current is provided as a solution of the linearized internal dynamics, restricting the validity of this solution to the vicinity of the linearization point; and in [20], based on preliminary results provided in [19], the work is improved by considering a dynamic sliding manifold, nevertheless, local results are provided. In the work by [21] the inductor current was obtained by means of a uniformly convergence sequence of Galerkin approximations, but the effectiveness of the control scheme depends on several hypotheses for which sufficient conditions are not provided. The approximations of periodic solutions with constant sign for Abel ordinary differential equation of the 2nd kind in the normal form is developed by [22] using an iterative scheme, where the tracking control is developed by means of a stable inversion-based approach. In the work presented in [23], the author avoids the stability problem when directly controlling the output voltage by approximating linearly the boost converter with a transfer function. Although a quasi–sliding mode control technique is applied, the solution based on this model is restricted to the vicinity of an operating point. In [24] a sliding mode controller based on the equivalent control method is presented, where perturbations are restricted to be constant, moreover, this work lacks of sliding mode stability analysis. Another work based on the sliding mode control technique is presented in [25], where the advantage of the order reduction property of the sliding mode dynamics is not taken into account.

Despite the effort made by researchers, they have overlooked the output regulation theory, which can perfectly match with the tracking problem of a DC biased sinusoidal signal along with the reckoning of a reference signal for the inductor current. Moreover, a discontinuous regulator can tackle the problem of nonminimum-phase system as in [26].

Tracking control and perturbation rejection problem for nonlinear systems is a challenging task. When the reference signals and perturbation are generated by an autonomous system known as exosystem, the problem is well known as output regulation [27]. The output regulation problem has been solved for the linear setting in [28], where the solution relies on the existence of a solution for a set of algebraic matrix equations. For nonlinear systems, the problem was solved based on the solvability of a set of nonlinear differential equations known as the Francis-Byrnes-Isidori (FBI) equations. The main idea behind the solution to the output regulation problem is to design an attractive and invariant steady state.

The output regulation theory has been successfully combined with popular control techniques such as sliding modes [29], fuzzy control [30], and artificial neural networks [31].

Now we present the results of a particular interest in the combination with sliding modes, which yields to the well known discontinuous output regulator strategy. Sliding modes add to the closed–loop system robustness properties to matched perturbations [32,33]. The main idea behind the discontinuous output regulator is to design a sliding surface on which the dynamics of the system are constrained to evolve by means of a discontinuous control law. The sliding surface contains the output steady state, where the dynamics of the system tends asymptotically along the sliding manifold to the steady-state behaviour.

Some advances were already provided in [34], where a discontinuous output regulator was designed for the tracking of a DC biased sinusoidal signal for a single boost power converter. In the mentioned work, a classical sliding surface and control action based on the sign function were designed. Simulations results were only reported.

Therefore, based on the direct control method principle, the aims of this work are the application of an improved state feedback discontinuous output regulator for the output tracking problem of a DC biased sinusoidal signal on a DC-DC boost power converter, and the validation of the proposed controller by means of real time experimentation. There is a sliding surface improvement that relies on the addition of an integral action for compensating a constant disturbance in the sliding mode dynamics (closed–loop inductor current unstable dynamics), that helps to improve the stability of such dynamics; and for the control action, the addition to the sign function of the equivalent control term that renders invariant the sliding manifold. Although the sign function can also render invariant the sliding manifold, its functionality is limited by the switching frequency of transistors. There is a clear advantage of using a direct method based on the discontinuous output regulator over existing techniques that consist of the framework provided for the relatively easy calculation of the steady–state (inductor current reference signal), and of the stabilization of the inductor current dynamics by means of a proper sliding mode function.

The rest of this work is organized as follows: in Section 2 the discontinuous output regulation theory is briefly revisited. In Section 3, the discontinuous output regulator for the boost power converter circuit is designed. Section 4 deals with a simulations study, and Section 5 deals with the presentation of experimental results, finally some comments conclude the work in Section 6.

2. Recalls on Discontinuous Output Regulation Theory

In this section, the main ideas behind classical and discontinuous output regulation theory are briefly revisited, as in [29].

Discontinuous Regulator for Nonlinear Systems in Regular Form

Consider a nonlinear system in the regular form as presented in [35]:

$$\dot{x}_1 = f_1(x_1, x_2) + d_1(x_1, x_2)\omega \tag{1}$$

$$\dot{x}_2 = f_2(x_1, x_2) + g_2(x_1, x_2)u + d_2(x_1, x_2)\omega \tag{2}$$

$$\dot{\omega} = s(\omega) \tag{3}$$

$$e = h(x_1, x_2) - q(\omega) \tag{4}$$

where $x_1 \in \mathbb{X}_1 \subset \mathbb{R}^{n-m}$, $x_2 \in \mathbb{X}_2 \subset \mathbb{R}^m$, $u \in \mathbb{R}^m$, and $rank[g_2(x_1, x_2)] = m \; \forall x \in \mathbb{X} \subset \mathbb{R}^n$. The vectors $f_1(x_1, x_2), f_2(x_1, x_2), h(x_1, x_2), q(\omega), s(\omega)$, and the columns of $g_2(x_1, x_2), d_1(x_1, x_2)$ and $d_2(x_1, x_2)$ are smooth vector fields of class $\mathbb{C}^\infty_{[t,\infty)}$, with $f_1(0,0) = 0$, $f_2(0,0) = 0$ and $h(0) = 0$.

The State Feedback Sliding Mode Output Regulator Problem (SFSMORP) [36] is defined as the problem of finding a sliding manifold

$$\sigma(x, \omega) = 0, \quad \sigma = (\sigma_1, \ldots \sigma_m)^T \tag{5}$$

and a discontinuous controller

$$u_i = \begin{cases} u_i^+ & if \quad \sigma_i(x, \omega) > 0 \\ u_i^- & if \quad \sigma_i(x, \omega) < 0 \end{cases} \quad i = 1, \ldots, m \tag{6}$$

where $u = (u_1, \ldots, u_m)^T$. Here u_i^+, u_i^-, and the sliding manifold (5) are chosen such that, the following conditions are met:

(SMS_{SF}) (Sliding Mode Stability) the control (6) is designed to induce sliding mode motion on the sliding manifold (5) in finite time,

(S_{SF}) the equilibrium $x = 0$ of the sliding mode dynamics

$$\dot{x} = f(x) + g(x)u_{eq} \Big|_{\sigma(x, \omega) = 0} \tag{7}$$

is asymptotically stable, where u_{eq} is the equivalent control defined as a solution of $\dot{\sigma} = 0$,

(R_{SF}) there exists a neighborhood $\mathbb{V} \subset \mathbb{X} \times \mathbb{W}$ of $(0, 0)$ such that, for each initial condition $(x(0), \omega(0)) \in \mathbb{V}$, the output tracking error (4) goes asymptotically to zero, i.e., $\lim_{t \to \infty} e(t) = 0$.

The following assumptions for system (1)–(4) will be instrumental for the solving the SFSMORP:

Assumption 1. *The Jacobian matrix* $S = \left[\dfrac{\partial s}{\partial \omega}\right]_{\omega=0}$ *at the equilibrium point* $\omega = 0$ *has all its eigenvalues on the imaginary axis.*

Assumption 2. *The pair* (A_{11}, A_{12}) *is stabilizable with* $A_{1i} = \left[\dfrac{\partial f_1}{\partial x_i}\right]_{x=0}$, *with* $i = \{1, 2\}$.

The steady state for x_1 and x_2 is introduced as $\pi_1(\omega)$ and $\pi_2(\omega)$, respectively. Then, defining the steady state error:

$$z = x - \pi(\omega) = \begin{pmatrix} z_1 \\ z_2 \end{pmatrix} = \begin{pmatrix} x_1 \\ x_2 \end{pmatrix} - \begin{pmatrix} \pi_1(\omega) \\ \pi_2(\omega) \end{pmatrix}. \tag{8}$$

The dynamics for (8), and the output tracking error as functions of the steady state error are calculated using (1)–(4) of the following form:

$$\dot{z}_1 = f_1(z_1 + \pi_1(\omega), z_2 + \pi_2(\omega)) + d_1(z_1 + \pi_1(\omega), z_2 + \pi_2(\omega))\omega - \frac{\partial \pi_1(\omega)}{\partial \omega}s(\omega) \tag{9}$$

$$\dot{z}_2 = f_2(z_1 + \pi_1(\omega), z_2 + \pi_2(\omega)) + g_2(z_1 + \pi_1(\omega), z_2 + \pi_2(\omega))u$$

$$+ d_2(z_1 + \pi_1(\omega), z_2 + \pi_2(\omega))\omega - \frac{\partial \pi_2(\omega)}{\partial \omega}s(\omega) \tag{10}$$

$$e = h(z_1 + \pi_1(\omega), z_2 + \pi_2(\omega)) - q(\omega). \tag{11}$$

The sliding surface is proposed as follows:

$$\sigma = z_2 + \sigma_1(z_1) = 0, \quad \sigma_1(0) = 0, \quad \left[\frac{\partial\sigma_1}{\partial z_1}\right]_{(0)} = \Sigma_1 \tag{12}$$

where its corresponding sliding regime equation of order $(n-m)$th is similar to that in (7), and is given by

$$\dot{z}_1 = f_1(z_1 + \pi_1(\omega), \sigma_1(z_1) + \pi_2(\omega)) + d_1(z_1 + \pi_1(\omega), \sigma_1(z_1) + \pi_2(\omega))\omega - \frac{\partial\pi_1(\omega)}{\partial\omega}s(\omega) \tag{13}$$

The linear approximation of (9)–(11) and (3) is useful for analyzing the stability under the sliding regime:

$$\begin{pmatrix} \dot{z}_1 \\ \dot{z}_2 \end{pmatrix} = \begin{pmatrix} A_{11} & A_{12} \\ A_{21} & A_{22} \end{pmatrix} \begin{pmatrix} z_1 \\ z_2 \end{pmatrix} + \begin{pmatrix} 0 \\ B_2 \end{pmatrix} u + \begin{pmatrix} R_1 \\ R_2 \end{pmatrix} \omega + \begin{pmatrix} \phi_1(z,\omega) \\ \phi_2(z,\omega) \end{pmatrix} \tag{14}$$

$$\dot{\omega} = S\omega + \phi_\omega(\omega) \tag{15}$$

$$e = C_1 z_1 + C_2 z_2 + (C_1\Pi_1 + C_2\Pi_2 - Q)\omega + \Phi_e(z,\omega) \tag{16}$$

hence by using (12) in (14)–(16), the corresponding linear approximation of the sliding mode Equation (13) can be described as follows:

$$\dot{z}_1 = (A_{11} - A_{12}\Sigma_1)z_1 + R_1\omega + \phi_{1,s}(z,\omega) \tag{17}$$

$$\dot{\omega} = S\omega + \phi_\omega(\omega) \tag{18}$$

$$e = (C_1 - C_2\Sigma_1)z_1 + (C_1\Pi_1 + C_2\Pi_2 - Q)\omega + \Phi_e(z,\omega) \tag{19}$$

where $R_1 = A_{11}\Pi_1 + A_{12}\Pi_2 - \Pi_1 S + D_1$, $R_2 = A_{21}\Pi_1 + A_{22}\Pi_2 - \Pi_2 S + D_2$, $A_{2j} = (\partial f_2/\partial z_j)_{(0,0)}$, $B_2 = g_2(0,0)$, $C_i = (\partial h/\partial z_i)_{(0,0)}$, $D_i = d_i(0,0)$, $\Pi_i = (\partial\pi_i/\partial\omega)_{(0)}$, $Q = (\partial q/\partial\omega)_{(0)}$, and the higher order terms (H.O.T.) $\phi_i(z,\omega)$, $\phi_\omega(\omega)$, $\phi_e(z,\omega)$, $\phi_{1,s}(z,\omega)$ vanish at the origin with its first derivatives; with $i,j = \{1,2\}$, and the constant matrix S defined in assumption 1. Before defining the sliding manifold and discontinuous control, the conditions that will solve the SFSMORP for the nonlinear system in Regular form will be established.

Proposition 1. *Under assumptions 1 and 2, if there exists C^k ($k \geq 2$) mappings $x_1 = \pi_1(\omega)$ and $x_2 = \pi_2(\omega)$, with $\pi_1(0) = 0$ and $\pi_2(0) = 0$, defined in a neighborhood $\mathbb{W}$ of 0 that satisfy the following conditions:*

$$f_1(\pi_1(\omega), \pi_2(\omega)) + d_1(\pi_1(\omega), \pi_2(\omega))\omega = \frac{\partial\pi_1(\omega)}{\partial\omega}s(\omega) \tag{20}$$

$$h(\pi_1(\omega), \pi_2(\omega)) - q(\omega) = 0 \tag{21}$$

at $(x_1, x_2, \omega, e) = (0,0,0,0)$ the SFSMORP for nonlinear systems in Regular form is solvable.

Proof. Given the sliding manifold (12), the discontinuous control law is defined as follows:

$$u = -MB_2^{-1}sign(\sigma).$$

If the control gain M is chosen such that $M > \|g_2(z_1 + \pi_1(\omega), z_2 + \pi_2(\omega))u_{eq}(z,\omega)\|$ with u_{eq} as a solution of $\dot{\sigma} = 0$, then, the condition (SMS_{SF}) holds. After the sliding mode occurs, the relation $z_2 = -\sigma_1(z_1)$ is true (the corresponding linear approximation is $z_2 = -\Sigma z_1$), and the motion in the closed–loop system will be governed by (17)–(19). Recalling that for the linear system in regular form (14), the matrix $(A_{11} - A_{12}\Sigma_1)$ is Hurwitz by a proper choice of Σ_1, and if condition (20) holds, then

$$R_1\omega + \phi_{1,s}(z,\omega) = f_1(\pi_1(\omega), \pi_2(\omega)) + d_1(\pi_1(\omega), \pi_2(\omega))\omega - \frac{\partial \pi_1(\omega)}{\partial \omega}s(\omega) = 0. \qquad (22)$$

It is clear that under relation (22), and under the property of centre manifold results in the following relation: $z_1(t) \to 0 \Rightarrow x_1(t) \to \pi_1(\omega(t))$ as $t \to \infty$. Thus, the requirement (S_{SF}) is fulfilled. Finally, if condition (21) holds too, then

$$(C_1\Pi_1 + C_2\Pi_2 - Q)\omega + \Phi_e(z,\omega) = h(\pi_1(\omega), \pi_2(\omega)) - q(\omega) = 0 \qquad (23)$$

So, by continuity, if relation (23) holds, then the output tracking error (19) converges to zero and condition (R_{SF}) holds too. $\square$

3. Discontinuous Output Regulation for a Boost Power Converter

In the following, the mathematical model of a boost converter and the problem formulation are presented, then the discontinuous output regulation technique is designed in order to solve the posed control problem.

3.1. Mathematical Model and Problem Formulation for the Boost Power Converter

Figure 1 shows an electric diagram of a boost power converter under the assumption of ideal switches.

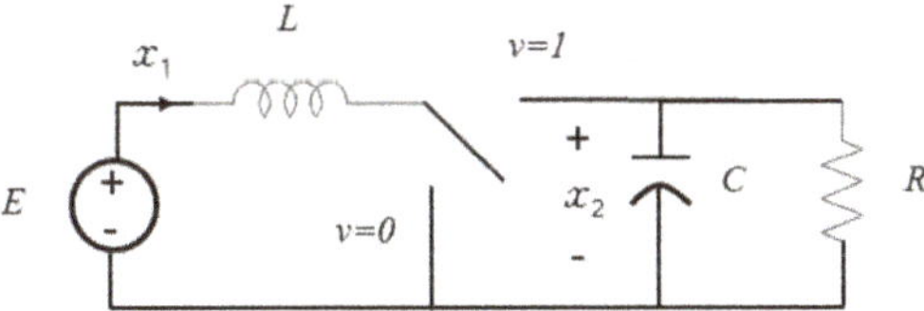

Figure 1. Boost power converter circuit.

The state-space average mathematical model in continuous current mode of the Boost converter is given by the following equations as in [8]:

$$\dot{x}_1 = -\frac{vx_2}{L} + \frac{E}{L} \qquad (24)$$

$$\dot{x}_2 = \frac{vx_1}{C} + \frac{x_2}{RC} \qquad (25)$$

$$y = x_2 \qquad (26)$$

with x_1 as the inductor current, x_2 as the output voltage capacitor, the control input v represents the switch position and can only take the value of 0 or 1, moreover, E is the DC input voltage. To account for variations in the input voltage E, it is considered that

$$E = E_\circ + \Delta E \qquad (27)$$

with $E_\circ$ as the nominal DC input voltage, and ΔE as an unknown and bounded constant deviation from the nominal DC input voltage, i.e., $|\Delta E| \leq d_1$, where d_1 is a positive constant upper bound. The constant parameters are the resistance R, the inductance denoted by L, and the capacitance denoted by C, where the storage elements are considered ideally lossless.

In general, the control problem consists of designing a direct controller for the boost power converter, that in order to deal with the nonminimum-phase stability problem of the inductor current, a sliding mode controller is proposed, and for dealing with the computation of a proper inductor current reference signal, output regulation theory is also proposed, yielding to a discontinuous output regulator design.

In particular, the control problem consists of forcing the output (26) to track a given reference signal $x_{2,r}$ and at the same time to reject the unknown perturbation $\Delta E/L$. Therefore one can consider the following output tracking error

$$e = x_2 - x_{2,r} \tag{28}$$

with $x_{2,r}$ as a reference signal for the output voltage capacitor. The reference signal is supposed to be generated by an autonomous exosystem (3) given by

$$
\begin{aligned}
\dot{w}_1 &= -\alpha w_2 & (29)\\
\dot{w}_2 &= \alpha w_1 & (30)\\
\dot{w}_3 &= 0 & (31)\\
\dot{w}_4 &= 0 & (32)\\
x_{2,r} &= q(w) = w_1 + w_3 & (33)
\end{aligned}
$$

with initial conditions $w_1(0) = w_2(0) = a$, $w_3(0) = b$, $w_4(0) = c$, with a, b, c, and α as positive constants. Equations (29) and (30) correspond to a harmonic oscillator, where its solution w_1 and w_2 will be sinusoidal shape signals with an amplitude of $\sqrt{2}a$ and frequency value of α. The solution to Equation (31), i.e., w_3, provides a bias value equal to b. Equation (32) has as solution w_4, that represents the unknown and bounded constant deviation $c = \Delta E$ of the nominal DC input voltage. It is a common practice in the output regulation theory to assume that perturbations are generated by an autonomous exosystem. Since ΔE in (27) is unknown, Equation (32) is not implemented, thus it is only used for analysis purposes. Finally, with respect to (3), the following vectors that will be used in the following subsections are defined $w = (w_1, w_2, w_3, w_4)^T$ and $s(w) = (-\alpha w_2, \alpha w_1, 0, 0)^T$.

3.2. Manifold Computation

Let us now introduce the steady state for x_1 and x_2 as $\pi_1(w) : \mathbb{W}_0 \to \mathbb{R}$ and $\pi_2(w) : \mathbb{W}_0 \to \mathbb{R}$ (where $\mathbb{W}_0$ is an open neighborhood of $w = 0$), respectively, with $\pi_1(0) = 0$ and $\pi_2(0) = 0$. These smooth mappings are such that the pair $(\pi_1(w), \pi_2(w))$ is the unique solution of the following nominal partial differential equations (PDEs) (FBI equations):

$$\frac{\partial \pi_1(w)}{\partial w} s(w) = -\frac{c(w)\pi_2(w)}{L} + \frac{E_\circ}{L} \tag{34}$$

$$\frac{\partial \pi_2(w)}{\partial w} s(w) = \frac{c(w)\pi_1(w)}{C} - \frac{\pi_2(w)}{RC} \tag{35}$$

$$0 = \pi_2(w) - q(w) \tag{36}$$

where $c(w)$ represents the steady state for the input variable v. Equations (34)–(36) are obtained when substituting $x_1 = \pi_1(w)$, $x_2 = \pi_2(w)$, $v = c(w)$ in Equations (24), (25) and (28). It is clear that a desired steady state for e is zero.

From (33) and (36) one can determine the solution for Equation (35) as $\pi_2(w) = w_1 + w_3$, i.e., a sinusoidal biased signal. Then, one can calculate $c(w)$ from (35) as follows:

$$c(w) = \frac{C}{\pi_1(w)} \frac{\partial \pi_2(w)}{\partial w} s(w) + \frac{\pi_2(w)}{R\pi_1(w)} \tag{37}$$

and by replacing (37) in (34) yields

$$\frac{\partial \pi_1(\omega)}{\partial \omega} s(w) = -\frac{C\pi_2(\omega)}{L\pi_1(\omega)}\frac{\partial \pi_2(\omega)}{\partial \omega}s(w) - \frac{\pi_2^2(\omega)}{LR\pi_1(\omega)} + \frac{E_o}{L}. \tag{38}$$

Finding a solution to this PDE results in a difficult task that can be solved by proposing an approximated solution as in [37–39]. Thus, one proposes the following polynomial as an approximated solution for (38)

$$\pi_1(\omega) = a_0 + a_1\omega_1^3 + a_2\omega_1^2 + a_3\omega_2\omega_1^2 + a_4\omega_3\omega_1^2 + a_5\omega_1\omega_2^2 + a_6\omega_1\omega_2 + a_7\omega_3\omega_1\omega_2 + a_8\omega_1 + a_9\omega_3\omega_1$$
$$+ a_{10}\omega_3^2\omega_1 + a_{11}\omega_2^3 + a_{12}\omega_2^2 + a_{13}\omega_3\omega_2^2 + a_{14}\omega_2 + a_{15}\omega_3\omega_2 + a_{16}\omega_3^2\omega_2 + a_{17}\omega_3 + a_{18}\omega_3^2$$
$$+ a_{19}\omega_3^3 + \mathcal{O}(\|\omega\|_1). \tag{39}$$

Multiplying Equation (38) by $\pi_1(\omega)$ and then replacing (39) in the resulting equation, one can find the values a_i ($i = \{0, \ldots 19\}$) if the coefficients of the same monomials appearing in both sides of such equation are equalized. In that case, the unique coefficients with values different from zero are:

$$a_2 = \frac{1}{RE_o}, \ a_6 = -\frac{\alpha C}{E_o}, \ a_9 = \frac{2}{RE_o}, \ a_{15} = -\frac{\alpha C}{E_o}, \ a_{18} = \frac{1}{RE_o}. \tag{40}$$

Please note that with the help of (40), terms different from zero in (39) are consistent with the ampere unit.

3.3. Discontinuous Output Regulation Design for a Boost Power Converter

The steady state error is defined as:

$$z = x - \pi(\omega) = (z_1, z_2)^1, \tag{41}$$

where $x = (x_1, x_2)^T$, $\pi(\omega) = (\pi_1(\omega), \pi_2(\omega))^T$. Then, the dynamic equation for (41) with tracking error (28) can be obtaining by using (24) and (25) as follows:

$$\dot{z}_1 = -\frac{v(z_2 + \pi_2(\omega))}{L} + \frac{E_o}{L} + \frac{\omega_4}{L} - \frac{\partial \pi_1(\omega)}{\partial \omega}s(\omega) \tag{42}$$

$$\dot{z}_2 = \frac{v(z_1 + \pi_1(\omega))}{C} - \frac{z_2 + \pi_2(\omega)}{RC} - \frac{\partial \pi_2(\omega)}{\partial \omega}s(\omega) \tag{43}$$

$$e = z_2 + \pi_2(\omega) - q(\omega). \tag{44}$$

With respect to the output capacitor voltage variable, system (24) and (25) has relative degree one and is nonminimum-phase, i.e., the inductor current dynamics is unstable. Hence, in order to satisfactorily solve the stated control problem, the sliding function is proposed of the following form:

$$\sigma = z_2 + c_1 z_1 + c_2 \int z_1 dt \tag{45}$$

with c_1 and c_2 as constant design parameters that will be determined in the following lines. The linear combination of z_1 and z_2 in the sliding function can stabilize nonminimum-phase systems with unitary relative degree as in [26], and the integral term can deal with constant perturbations in the sliding mode dynamics. The dynamics of the sliding function results as follows:

$$\dot{\sigma} = \eta(z, \omega) + \delta(z, \omega)v + c_1\frac{\omega_4}{L} \tag{46}$$

where

$$\eta(z,w) = c_1 \frac{E_\circ}{L} - c_1 \frac{\partial \pi_1(\omega)}{\partial \omega} s(\omega) - \frac{z_2 + \pi_2(\omega)}{RC} - \frac{\partial \pi_2(\omega)}{\partial \omega} s(\omega) + c_2 z_1 \tag{47}$$

$$\delta(z,\omega) = \frac{z_1 + \pi_1(\omega)}{C} - c_1 \frac{z_2 + \pi_2(\omega)}{L}. \tag{48}$$

All terms in (47) and (48) are assumed to be known or measured. Please note that ω_4 in Equation (46) is an unknown constant variable due to variations in E. Now, the sliding control is designed for stabilizing the dynamics of the sliding mode function (46) as follows:

$$v = -\frac{M}{\delta(z,\omega)} sign(\sigma) - v_{eq,n}(z_1, z_2, w), \tag{49}$$

where $v_{eq,n}$ as the nominal equivalent control calculated from the nominal dynamics of the sliding function $\dot\sigma = \eta(z,\omega) + \delta(z,\omega)v_{eq,n} = 0$ as follows:

$$v_{eq,n}(z_1, z_2, w) = -\frac{\eta(z,\omega)}{\delta(z,\omega)}. \tag{50}$$

The equivalent control term renders invariant the sliding manifold ($\sigma = 0$). Although the sign function can also render invariant the sliding manifold, its functionality is limited by the switching frequency of transistors. To prove convergence to the sliding manifold $\sigma = 0$ of system (42), (43) closed–loop by (49), let us consider the following Lyapunov candidate function:

$$V = \frac{1}{2}\sigma^2. \tag{51}$$

Taking the derivative of (51) along the trajectories of the closed-loop system (42), (43), (49) results in $\dot V \leq -(M - |c_1 \omega_4|/L)|\sigma|$, and if $M > |c_1 d_1|/L$, then condition (SMS_{SF}) is met.

Remark 1. *It is a common practice in the sliding mode control design to incorporate the equivalent control as in (49) for improving the reaching phase of the projection motion. Moreover, due to the smoothness of the equivalent control term, the real time implementation of the control action (49) will require of a pulse width modulation (PWM) stage.*

3.4. Sliding Mode Dynamics Stability Analysis

After the sliding mode occurs, i.e., $\sigma = 0$, it is clear that from Equation (45), the relation

$$z_2 = -c_1 z_1 - c_2 \int z_1 dt \tag{52}$$

holds, where z_2 depends on z_1. Hence the sliding mode dynamics reduces to that of z_1, which it is well known to be unstable. For the stabilization of this dynamics, the nominal equivalent control (50) is substituted in the dynamics of z_1 in (42), then, a linear approximation results in the following expression:

$$\dot z_1 = a_{11}z_1 + a_{12}z_2 + \frac{\omega_4}{L} + R_1 w + \varphi_1(z,\omega) \tag{53}$$

with $a_{1i} = (\partial f_1/\partial z_i)_{(0,0)}$, $R_1 = a_{11}\Pi_1 + a_{12}\Pi_2 - \Pi_1 S$, $\Pi_i = (\partial \pi_i/\partial \omega)_{(0)}$ for $i = \{1,2\}$, moreover, f_1 is the right side of (42) evaluated with $v = v_{eq,n}$, and $\varphi_1(z,\omega)$ as a function of H.O.T. that vanish at the origin with their first derivative. Substituting the relation (52) under the sliding regime in the linear approximation (53) and in the output tracking error (44) yields

$$\dot{z}_1 = (a_{11} - c_1 a_{12})z_1 - c_2 a_{12}\zeta_1 + \frac{\omega_4}{L} + R_1\omega + \varphi_1(z, \omega) \tag{54}$$

$$\dot{\zeta}_1 = z_1 \tag{55}$$

$$e = -c_1 z_1 - c_2 \zeta_1 + \pi_2(\omega) - q(\omega) \tag{56}$$

with $\zeta_1 = \int z_1 dt$ as the integral action and solution to Equation (55). If condition (38) holds, then

$$R_1\omega + \varphi_1(z, \omega) = -\frac{C\pi_2(\omega)}{L\pi_1(\omega)}\frac{\partial \pi_2(\omega)}{\partial \omega}s(w) - \frac{\pi_2^2(\omega)}{LR\pi_1(\omega)} + \frac{E_\circ}{L} - \frac{\partial \pi_1(\omega)}{\partial \omega}s(\omega) = 0. \tag{57}$$

By using (57) in (54) it reduces to the following expression:

$$\dot{z}_1 = (a_{11} - c_1 a_{12})z_1 - c_2 a_{12}\zeta_1 + \frac{\omega_4}{L}. \tag{58}$$

Please note that the constant parameter c_1 is responsible for the stabilization of the sliding mode dynamics that in this case coincides with the tracking error of the inductor current, meanwhile, the integral term can deal with unknown constant perturbations, hence, a suitable design of the sliding function (45) is a fundamental task. It can be appreciated that when the sliding mode occurs, i.e., $\sigma = 0$, z_2 in (52) acts as a pseudo-control input for the dynamics of z_1, introducing a Proportional+Integral (PI) control action. Hence, by choosing $c_1 = (a_{11} - p_1 - p_2)/a_{12}$, $c_2 = p_1 p_2/a_{12}$, with $p_1 < 0$ and $p_2 < 0$ as desired poles locations for subsystem (55)–(58) then, its corresponding steady state is $(z_{1,ss}, \zeta_{1,ss})^T = (0, \omega_4/(Lc_2 a_{12}))^T$. Please note that the steady state of the integral term ζ_1 is exactly the one required for canceling out the perturbation term ω_4/L, thus the requirement (S_{SF}) is satisfied. By continuity, the output tracking error (56) converges to $-\omega_4/(La_{12})$ and condition (R_{SF}) is satisfied in a practical sense.

4. Simulations

The proposed solution is simulated and compared with the work [8], in which an indirect method based on the sliding mode control technique is used, and with [40], where an indirect method based on the inverse optimal control technique is applied. The parameters of the boost converter are as follows: $L = 800\,\mu\text{H}$, $C = 40\,\mu\text{F}$, $R = 30\,\Omega$, and $E = 118$ V, and the initial value of the exosystem $c = \Delta E = 24$ V. The output voltage capacitor is forced to track a sinusoidal signal with a peak value of 305 V and a bias value of 235 V, hence the initial conditions for the exosystem (29)–(32) are chosen as $a = 49.49$ and $b = 235$. For tracking a sinusoidal signal with a frequency value of 60 Hz, then, $\alpha = 377$.

Figure 2 shows the comparison results for the output voltage. It can be noted that the disadvantage of using indirect control methods, as the tracking of a reference signal for the output capacitor voltage is not accurately done (Figure 2b,c). However, in the case of the proposed direct control method (Figure 2a), the tracking is more accurate.

Since indirect methods are directly controlling the inductor current, the tracking of a reference signal for the inductor current should be more exact as can be appreciated in Figure 3b,c. However, it is expected that in direct control methods the tracking of the inductor current to be less accurate with respect to the direct method, but thanks to the proposed discontinuous output regulator, the tracking of a reference signal for the inductor current is accurate.

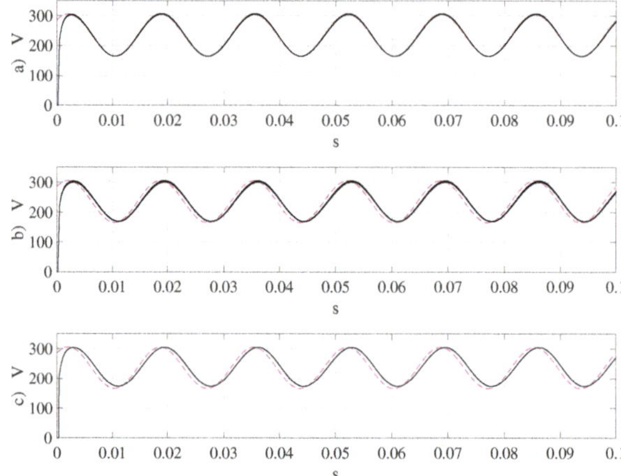

Figure 2. Output tracking of the capacitor voltage. Reference signal in dashed line, and capacitor voltage in solid line. (**a**) Proposed controller. (**b**) Controller in [8]. (**c**) Controller in [40].

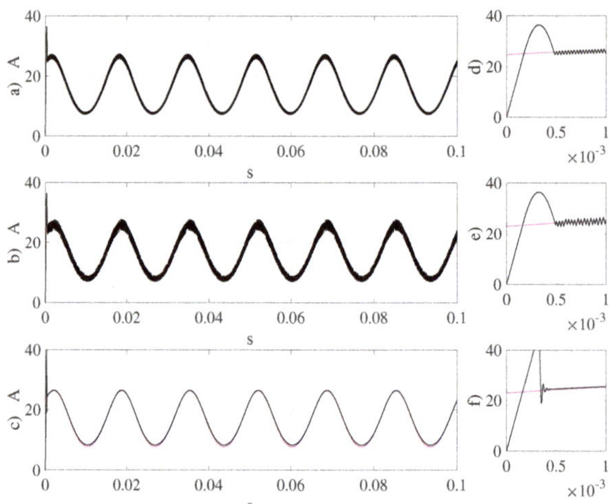

Figure 3. Output tracking of the inductor current. Reference signal in dashed line, and inductor current in solid line. (**a**) Proposed controller. (**b**) Controller in [8]. (**c**) Controller in [40].

5. Real Time Experimentation

The parameters of the boost converter are the same to those in the Simulation section. The experimental setup is similar to that reported in [41,42], and consists of a VARIAC (three-phase variable autotransformer) that is fed from a three-phase voltage source. By rotating the knob of the VARIAC, the amplitude of the three-phase voltage source is regulated. These voltages are fed to the power module (Semikron) that incorporates a three-phase rectifier and a single switching transistor (DC chopper) that is connected to the boost power converter elements. The Semikron power module also incorporates a three-phase inverter, but in this application is not used. The control algorithm and PWM generation are programmed in Simulink and implemented with a digital signal processing (DSP) board (dSPACE 1104). This board comes along with a library that easily incorporates with Simulink. Analog-to-digital converters included in the DSP board acquire the signals from the inductor current and capacitor voltage by means of hall-type sensors, as the HX 10-P and LV 25-P, respectively; both manufactured by LEM. Once the DSP executes the control algorithm in each sampling step, it generates one digital signal for switching the insulated-gate bipolar transistor (IGBT). This digital signal is TTL level and is converted to a CMOS level of 15 V. This voltage level is the required one for switching on the IGBT. A block diagram of this setup is shown in Figure 4.

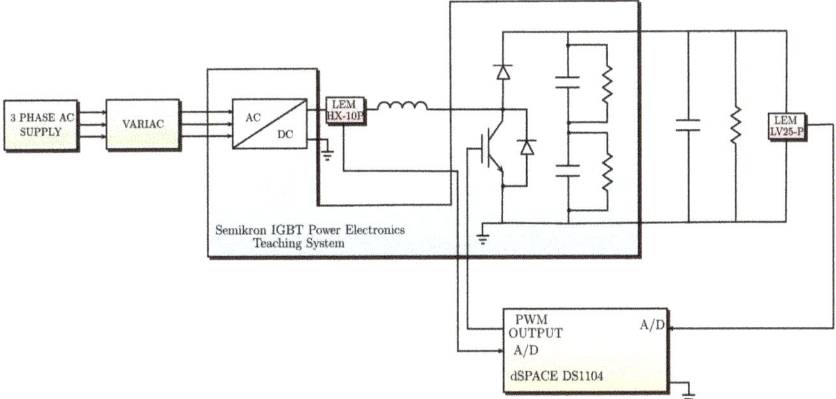

Figure 4. Block diagram of the experimental setup.

The sampling period in the DSP board has been fixed to 60 µs. The output voltage capacitor is forced to track the same sinusoidal signal described in the Simulations section. Two values were chosen for α, 377 and 314 that correspond to frequencies values of 60 Hz and 50 Hz respectively. The real time results for the tracking of a biased sinusoidal signal with a frequency value of 60 Hz are shown in Figures 5 and 6, and for a frequency value of 50 Hz are shown in Figures 7 and 8.

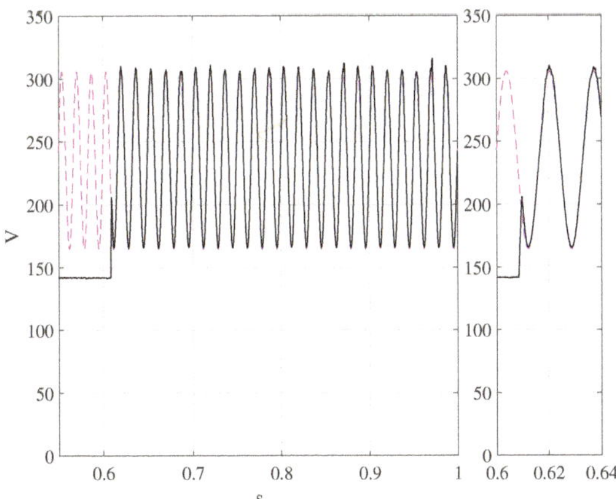

Figure 5. Output tracking of the capacitor voltage with $\alpha = 377$ (60 Hz). Reference signal in dashed line, and capacitor voltage in solid line.

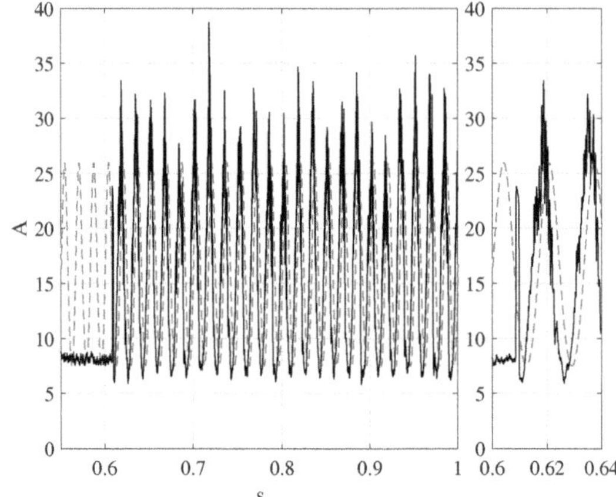

Figure 6. Output tracking of the inductor current with $\alpha = 377$ (60 Hz). Reference signal in dashed line, and inductor current in solid line.

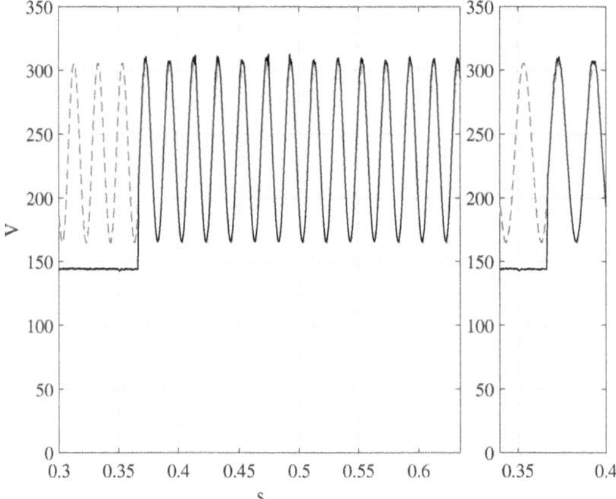

Figure 7. Output tracking of the capacitor voltage with $\alpha = 314$ (50 Hz). Reference signal in dashed line, and capacitor voltage in solid line.

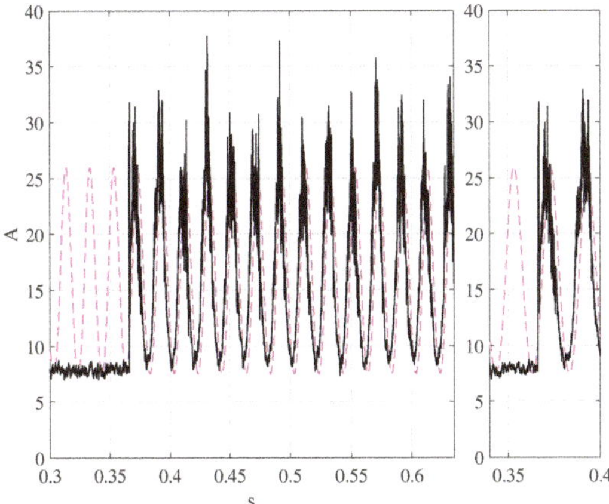

Figure 8. Output tracking of the inductor current with $\alpha = 314$ (50 Hz). Reference signal in dashed line, and inductor current in solid line.

In general, a similar behaviour for the two frequencies values can be appreciated, i.e., a good output tracking performance for the voltage capacitor where the transient response is fast without overshoot in all cases. The sliding mode dynamics performance that corresponds to the inductor current, which is commonly known to be unstable when directly controlling the capacitor voltage [17], its linear stabilization was possible with an adequate sliding function design. It is worth mentioning that the reference signal for the inductor current (39) is not updated with parameter variations. Hence, for compensating the deviations of the inductor current, the parameters in the reference signal (39) must be updated with an estimation scheme.

Finally, Figures 9 and 10 show the magnitude of the frequency spectrum for the output voltage for unbiased sinusoidal signals with a frequency value of 60 Hz and 50 Hz respectively. It can be appreciated that the peaks in both figures correspond to proper frequency values. The total harmonic distortion (THD) for the unbiased sinusoidal signal of frequency value of 60 Hz is 4.15%, and in the case of 50 Hz is 4.72%, which are admissible values inside the 5% standard limit.

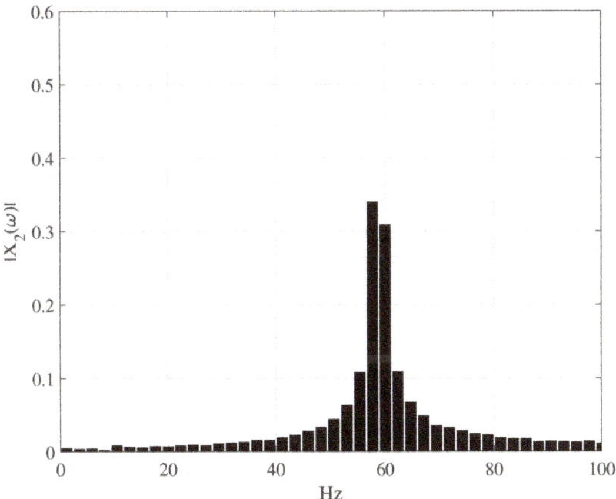

Figure 9. Output voltage harmonic content with $\alpha = 377$ (60 Hz).

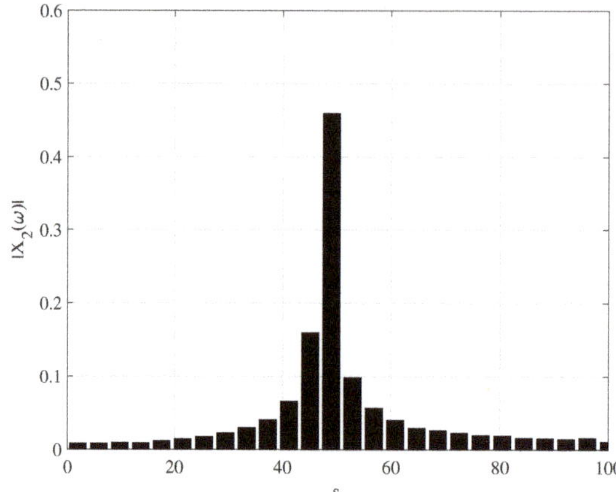

Figure 10. Output voltage harmonic content with $\alpha = 314$ (50 Hz).

6. Conclusions

The boost inverter proposed in [17] has motivated several researchers in the control engineering field to work at least with the fundamental problem of controlling a single boost power converter for the tracking of a DC biased sinusoidal signal. This is a challenging task due to the fact that the boost power converter is a nonminimum-phase system, and that the inductor current reference signal is difficult to compute. Despite the tremendous effort by researchers, this problem have not been addressed by output regulation theory. This theory effectively solves the problem of trajectory tracking and disturbance rejection. An enhanced version of this theory is obtained when combined with sliding modes, known as discontinuous output regulation. Hence, the discontinuous output regulation when applied to the tracking of a DC biased sinusoidal signal for a boost power converter yields to favorable results. With the proper design of the sliding function, which includes a linear combination of tracking errors and an integral term, the first part helps to effectively stabilize the inductor current dynamics, while the second part cancels out unknown constant perturbations. Hence, when the sliding mode occurs, i.e., $\sigma = 0$, z_2 can be considered as a pseudo-control that introduces a PI action to the unstable dynamics that corresponds to the tracking error of the inductor current, hence, with the current selection of the sliding function, the stabilization of the nonminimum-phasesystem was possible. To make possible $\sigma = 0$, the addition of the equivalent control term to the sign function renders invariant the sliding manifold ($\sigma = 0$). Although the sign function can also render invariant the sliding manifold, its functionality is limited by the switching frequency of transistors. On the other hand, the reference signal for the inductor current was proposed as a polynomial of order 3 that solves the corresponding solution of one of the FBI equations. Experimental results illustrate the good performance of the boost power converter when closed-loop with the discontinuous regulator strategy, where the input voltage presented an unknown increment. In general, the performance of the output capacitor voltage was satisfactory in terms of THD, whose values are inside of the 5% standard limit, thus highlighting the merits of this control technique. Some interesting issues still remain as the updating of $\pi_1(\omega)$ for robustifying the sliding mode dynamics in the presence parameter variations.

Author Contributions: Conceptualization, J.R., S.O.-C. and F.C.; Methodology, J.R. and F.C.; Software, J.R. and S.O.-C.; Validation, J.R., S.O.-C. and F.C.; Formal Analysis, J.R. and F.O.; Investigation, J.R.; Resources, F.C.; Data Curation, S.O.-C.; Writing—Original Draft Preparation, J.R.; Writing—Review & Editing, J.R.

Funding: This research received no external funding.

Conflicts of Interest: The authors declare no conflict of interest.

References

1. Guo, X.; Yuan, J.; Tang, Y.; You, X. Hardware in the Loop Real-Time Simulation for the Associated Discrete Circuit Modeling Optimization Method of Power Converters. *Energies* **2018**, *11*, 3237.
2. Zapata, J.W.; Kouro, S.; Carrasco, G.; Renaudineau, H. Step-Up Partial Power DC-DC Converters for Two-Stage PV Systems with Interleaved Current Performance. *Energies* **2018**, *11*, 357.
3. Nizami, T.K.; Mahanta, C. An intelligent adaptive control of DC–DC buck converters. *J. Frankl. Inst.* **2016**, *353*, 2588–2613.
4. Ortega, R. *Passivity-Based Control of Euler-Lagrange Systems: Mechanical, Electrical, and Electromechanical Applications*; Communications and Control Engineering; Springer Science & Business Media: London, UK, 1998.
5. Ciezki, J.; Ashton, R. The design of stabilizing controls for shipboard DC-to-DC buck choppers using feedback linearization techniques. In Proceedings of the 29th Annual IEEE Power Electronics Specialists Conference, Fukuoka, Japan, 22 May 1998; Volume 1, pp. 335–341.
6. Ayachit, A.; Reatti, A.; Kazimierczuk, M.K. Small-signal modeling of the PWM boost DC-DC converter at boundary-conduction mode by circuit averaging technique. In Proceedings of the 2015 IEEE International Symposium on Circuits and Systems (ISCAS), Lisbon, Portugal, 24–27 May 2015; pp. 229–232.
7. Kassakian, J.; Schlecht, M.; Verghese, G. *Principles of Power Electronics*; Addison-Wesley Series in Electrical Engineering; Addison-Wesley: Boston, MA, USA, 1991.
8. Sira-Ramírez, H. DC-to-AC power conversion on a 'boost'converter. *Int. J. Robust Nonlinear Control* **2001**, *11*, 589–600.
9. Utkin, V. Sliding mode control of DC/DC converters. *J. Frankl. Inst.* **2013**, *350*, 2146–2165. [CrossRef]
10. Yasin, A.R.; Ashraf, M.; Bhatti, A.I. Fixed Frequency Sliding Mode Control of Power Converters for Improved Dynamic Response in DC Micro-Grids. *Energies* **2018**, *11*, 2799.
11. Choi, B.; Lim, W.; Choi, S.; Sun, J. Comparative Performance Evaluation of Current-Mode Control Schemes Adapted to Asymmetrically Driven Bridge-Type Pulsewidth Modulated DC-to-DC Converters. *IEEE Trans. Ind. Electron.* **2008**, *55*, 2033–2042.
12. Marie-Francoise, J.N.; Gualous, H.; Berthon, A. DC to DC converter with neural network control for on-board electrical energy management. In Proceedings of the 4th International Power Electronics and Motion Control Conference (IPEMC 2004), Xi'an, China, 14–16 August 2004; Volume 2, pp. 521–525.
13. Taeed, F.; Salam, Z.; Ayob, S. Implementation of Single Input Fuzzy Logic Controller for Boost DC to DC power converter. In Proceedings of the 2010 IEEE International Conference on Power and Energy (PECon), Kuala Lumpur, Malaysia, 29 November–1 December 2010; pp. 797–802.
14. Linares Flores, J.; Avalos, J.; Espinosa, C. Passivity-Based Controller and Online Algebraic Estimation of the Load Parameter of the DC-to-DC power converter Cuk Type. *IEEE Lat. Am. Trans.* **2011**, *9*, 784–791.
15. Luchetta, A.; Manetti, S.; Piccirilli, M.C.; Reatti, A.; Kazimierczuk, M.K. Effects of parasitic components on diode duty cycle and small-signal model of PWM DC-DC buck converter in DCM. In Proceedings of the 2015 IEEE 15th International Conference on Environment and Electrical Engineering (EEEIC), Rome, Italy, 10–13 June 2015; pp. 772–777.
16. Luchetta, A.; Manetti, S.; Piccirilli, M.C.; Reatti, A.; Kazimierczuk, M.K. Derivation of network functions for PWM DC-DC Buck converter in DCM including effects of parasitic components on diode duty-cycle. In Proceedings of the 2015 IEEE 15th International Conference on Environment and Electrical Engineering (EEEIC), Rome, Italy, 10–13 June 2015; pp. 778–783.
17. Caceres, R.O.; Barbi, I. A boost DC-AC converter: analysis, design, and experimentation. *IEEE Trans. Power Electron.* **1999**, *14*, 134–141. [CrossRef]
18. Flores-Bahamonde, F.; Valderrama-Blavi, H.; Bosque-Moncusi, J.M.; García, G.; Martínez-Salamero, L. Using the sliding-mode control approach for analysis and design of the boost inverter. *IET Power Electron.* **2016**, *9*, 1625–1634.
19. Shtessel, Y.B.; Zinober, A.S.; Shkolnikov, I.A. Sliding mode control of boost and buck-boost power converters using method of stable system centre. *Automatica* **2003**, *39*, 1061–1067. [CrossRef]
20. Shtessel, Y.B.; Zinober, A.S.I.; Shkolnikov, I.A. Sliding mode control of boost and buck-boost power converters using the dynamic sliding manifold. *Int. J. Robust Nonlinear Control* **2003**, *13*, 1285–1298.

21. Fossas, E.; Olm, J. Galerkin method and approximate tracking in a non-minimum phase bilinear system. *Discret. Contin. Dyn. Syst. Ser. B* **2007**, *7*, 53–76.

22. Olm, J.; Ros, X. Approximate tracking of periodic references in a class of bilinear systems via stable inversion. *Discret. Contin. Dyn. Syst. Ser. B* **2011**, *15*, 197–215.

23. Almawlawe, M.; Mitić, D.; Antić, D.; Icić, Z. An Approach to Microcontroller-Based Realization of Boost Converter with Quasi-Sliding Mode Control. *J. Circ. Syst. Comput.* **2017**, *26*, 1750106. [CrossRef]

24. Pandey, S.K.; Patil, S.L.; Phadke, S.B. Regulation of Nonminimum Phase DC–DC Converters Using Integral Sliding Mode Control Combined With a Disturbance Observer. *IEEE Trans. Circ. Syst. II Express Briefs* **2018**, *65*, 1649–1653. [CrossRef]

25. Repecho, V.; Biel, D.; Olm, J.M.; Fossas, E. Robust sliding mode control of a DC/DC Boost converter with switching frequency regulation. *J. Frankl. Inst.* **2018**, *355*, 5367–5383.

26. Bonivento, C.; Marconi, L.; Zanasi, R. Output regulation of nonlinear systems by sliding mode. *Automatica* **2001**, *37*, 535–542. [CrossRef]

27. Isidori, A.; Byrnes, C. Output regulation of nonlinear systems. *IEEE Trans. Autom. Control* **1990**, *35*, 131–140.

28. Francis, B.A. The Linear Multivariable Regulator Problem. *SIAM J. Control Optim.* **1977**, *15*, 486–505. [CrossRef]

29. Loukianov, A.G.; Domínguez, J.R.; Castillo-Toledo, B. Robust sliding mode regulation of nonlinear systems. *Automatica* **2018**, *89*, 241–246. [CrossRef]

30. Ma, Y.; Zhao, J. Cooperative output regulation for nonlinear multi-agent systems described by T-S fuzzy models under jointly connected switching topology. *Neurocomputing* **2018**, *332*, 351–359. [CrossRef]

31. Liu, C.; Ping, Z.; Hu, H. Approximate discrete-time nonlinear output regulation of linear motor inverted pendulum and experimental study. In Proceedings of the 33rd Youth Academic Annual Conference of Chinese Association of Automation (YAC), Nanjing, China, 18–20 May 2018; pp. 1049–1054.

32. Utkin, V. Discussion Aspects of High-Order Sliding Mode Control. *IEEE Trans. Autom. Control* **2016**, *61*, 829–833. [CrossRef]

33. Elmali, H.; Olgac, N. Robust output tracking control of nonlinear MIMO systems via sliding mode technique. *Automatica* **1992**, *28*, 145–151. [CrossRef]

34. Loukianov, A.; Rivera, J.; Chavira, F.; Ortega, S. Discontinuous output regulation for a DC-to-AC boost converter. In Proceedings of the 9th International Conference on Electrical Engineering, Computing Science and Automatic Control (CCE), México City, México, 26–28 September 2012; pp. 1–6.

35. Loukianov, A.; Utkin, V. Methods for reducing dynamic systems to regular form. *Autom. Remote Control* **1981**, *42*, 413–420.

36. Utkin, V.I.; Loukianov, A.G.; Castillo-Toledo, B.; Rivera, J. Sliding mode regulator design. In *Variable Structure Systems: From Principles to Implementation*; IET Control Engineering Series; Sabanovic, A., Fridman, L.M., Spurgeon, S., Eds.; Institution of Engineering and Technology: London, UK, 2004; Volume 66, pp. 19–44.

37. Ramos, L.E.; Castillo-Toledo, B.; Alvarez, J. Nonlinear regulation of an underactuated system. In Proceedings of the International Conference on Robotics and Automation, Albuquerque, NM, USA, 25 April 1997.

38. Rivera, J.; Loukianov, A.; Castillo-Toledo, B. Discontinuous output regulation of the Pendubot. In Proceedings of the 17th World Congress, the International Federation of Automatic Control, Seoul, Korea, 6–11 July 2008.

39. Tarn, T.J.; Sanposh, P.; Cheng, D.; Zhang, M. Output Regulation for Nonlinear Systems: Some Recent Theoretical and Experimental Results. *IEEE Trans. Control Syst. Technol.* **2005**, *13*, 605–610. [CrossRef]

40. Ornelas-Tellez, F.; Rico, J.; Zunig, G.; Casarrubias, G. Inverse Optimal Trajectory Tracking for Discrete–Time Nonlinear Systems: Application to the Boost Converter. In Proceedings of the 2012 ROPEC INTERNACIONAL, Colima, México, 7–9 November 2012; pp. 31–36.

41. Jorge, R.; Florentino, C.; Alexander, L.; Susana, O.; Juan, J.R. Discrete-time modeling and control of a boost converter by means of a variational integrator and sliding modes, *J. Frankl. Inst.* **2014**, *351*, 315–339.

42. Jorge, R.; Florentino, C.; Alexander, L. On the Discrete-Time Modeling of a DC-to-DC Power Converter and Control Design with Discrete-Time Sliding Modes. *Math. Probl. Eng.* **2013**, *2013*, 1–17.

Article

Control of a Multiphase Buck Converter, Based on Sliding Mode and Disturbance Estimation, Capable of Linear Large Signal Operation

Rok Pajer [1,2,*], Amor Chowdhury [1,2] and Miran Rodič [2]

[1] Margento R&D, Turnerjeva ulica 17, SI-2000 Maribor, Slovenia
[2] Faculty of Electrical Engineering and Computer Science, University of Maribor, Koroška cesta 46, SI-2000 Maribor, Slovenia
* Correspondence: rok.pajer@margento.com

Received: 20 June 2019; Accepted: 16 July 2019; Published: 19 July 2019

Abstract: Power-hardware-in-the-loop systems enable testing of power converters for electric vehicles (EV) without the use of real physical components. Battery emulation is one example of such a system, demanding the use of bidirectional power flow, a wide output voltage range and high current swings. A multiphase synchronous DC-DC converter is appropriate to handle all of these requirements. The control of the multiphase converter needs to make sure that the current is shared equally between phases. It is preferred that the closed-loop dynamic model is linear in a wide range of output currents and voltages, where parameter variations, control signal limits, dead time effects, and so on, are compensated for. In the case presented in this paper, a cascade control structure was used with inner sliding mode control for phase currents. For the outer voltage loop, a proportional controller with output current feedforward compensation was used. Disturbance observers were used in current loops and in the voltage loop to compensate mismatches between the model and the real circuit. The tuning rules are proposed for all loops and observers, to simplify the design and assure operation without saturation of control signals, that is, duty cycle and inductor current reference. By using the proposed control algorithms and tuning rules, a linear reduced order system model was devised, which is valid for the entire operational range of the converter. The operation was verified on a prototype 4-phase synchronous DC-DC converter.

Keywords: DC-DC converter; multiphase converter; sliding mode control; disturbance observer; electric vehicles; power-hardware-in-the-loop

---

1. Introduction

The global transport sector is currently moving from fossil fuel vehicles to electrically driven vehicles (EV). An electric motor requires a power controller to control speed, torque and power. In addition to powering the motor, the controller can support the regeneration of the energy in the battery and, thus, establishes a two-way flow of energy. In order to increase efficiency and reduce size, controllers will be under constant development in the future. When testing a controller for EV and hybrid electric vehicles (HEV), real operating conditions are needed, which involves a battery. This battery should be preconditioned, which takes time. The high capacity batteries of today are lithium-based and very vulnerable to abuse. In case of failure, for example, high temperature, overcurrent, or mechanical intervention, they can catch fire, or, in the worst cases, even explode. The life expectancy of a battery depends on a number of cycles; as it ages, the characteristics change and the repeatability of tests is worse. Batteries are in the form of big packs, so the testing facility also needs enough room. A battery emulator (BE) is a solution to the problems mentioned. A real battery is replaced by a DC power supply, which has programmable characteristics. That way, different types of batteries can be emulated, simply by changing the battery model structure and parameters. Preconditioning is not needed, as in the case of real batteries, because the state of charge (SOC) is set in a program, and the emulator sets the corresponding output voltage. Repeatability of tests is better, because a BE always has the same characteristics for the same parameters.

The DC power supply for battery emulation should supply a high enough current to the motor controller, with a wide output voltage range that mimics a real battery. Power converters of the switching type are preferred as the power demand of the motor controller is high. Switching the power supply brings more noise into the circuit and the wanted characteristic of the dynamic is hard to achieve, especially in the case of big power swings, as is the case of an EV. The power flow is preferred to be bidirectional, as in HEV, where, for example, the battery is a power source as well as a power drain. The control algorithm of such a converter would operate it to behave according to a real battery model [1]. Because of the controller's finite bandwidth, external disturbances, sensor noise, and an unmatched converter model, this can be achieved in a limited sense. The power supply for a BE should operate at a wide range of voltages (from empty to full battery) and with large current swings (transitions from no load to full load). Power converters, which are typically designed for the nominal operating point, behave differently at other operating points. When exposed to large signal change, they incorporate some nonlinear characteristics, like soft start and current limiting, which limit their performance, and, of course, the dynamic characteristics are different to those in small signal operation. These effects were described and analyzed for the case of a buck converter in a previous study [2]. Power converters can also exhibit unstable behavior in such cases. Nonlinearities in the forms of converter nonlinear characteristic, parameter nonlinear dependence (for example, inductance is dependent on current), control signal saturation, and dead time, effects in the transistor legs [3], impact the dynamical response and stability, so they can be significantly different than in the designed nominal case.. A controller for BE should compensate for these nonlinearities, so that the dynamical characteristic is operating point independent; if it is linear, the control design for battery model tracking can be simpler. The motor controller acts as a Constant Power Load (CPL), which is described as negative incremental resistance in the literature [4]. The behavior of a converter with such a load could become unstable, which the controller has to account for [5].

The aim of this work was to design a power converter control algorithm suitable for use in BEs. The power converter used was a multiphase buck converter (Figure 1), which is used widely for high power applications in different configurations, especially in the EV and HEV sector [6–10]. Because all the active elements are MOSFETs (Metal Oxide Semiconductor Field Effect Transistor—$Q_{1...8}$), topology is synchronous and power flow is bidirectional. Other benefits of such a converter are current sharing between phases, and low current ripple on output, and it can be used in buck and boost configurations. As the phases of such a converter are never completely identical, currents will not be completely balanced without a special balancing algorithm, which is one of the main problems of such a topology. For any balancing algorithm, phase currents need to be measured directly, or estimated from voltage drop on the equivalent series resistance (ESR) of the input [11] or output [12] capacitors. Direct measurement is preferred, as the latter puts a limit on the duty cycle. It is also important to sample the currents at the right moments, which was investigated in a previous study [13]. There is a lot of ongoing research on balancing algorithms, and solutions can be divided into two major types:

- Current balancing by varying duty ratios on phases.
- A current control loop on each phase with the same reference current for all phases.

For the first type, balancing is implemented by comparing a corresponding phase current to the average of currents, or to the master phase current, which is fixed or selected dynamically [14]. The method may work well, but the dynamic operation is not very predictable, which can be a problem if one wants to get a system dynamic description. Peak current mode control fits into the second type of balancing algorithms; it is attractive and used widely because of the high bandwidth [15]. Another benefit is that the effect of input voltage on the output is cancelled in one cycle and the system order is reduced. However, the model presented in a previous study [16] is dependent on the rising slope of the inductor current, which means it is dependent on the operating point—input and output voltage. This nonlinearity is not desirable for emulation purposes, where the converter's output voltage range is wide. There are some other variants, like current mode control based on integral values [10,17] and average current mode control [18], but their model descriptions are also nonlinear. To give an operating point independent dynamic operation, feedforward blocks can be combined with classical Proportional-Integral (PI) controllers, as in Karimi et al. [7]. However, the integral part in the controller adds undesirable effects [19] to the system; the most problematic is control signal saturation, leading again to a nonlinear model. This implies large overshoots, slow settling time and control signal saturation, which is undesirable for emulation purposes, as the dynamic model becomes nonlinear, and it is not possible to find its inverse. One method to reduce the impact has been proposed in the form of constrained Proportional-Integral-Derivative (PID) algorithms [20].

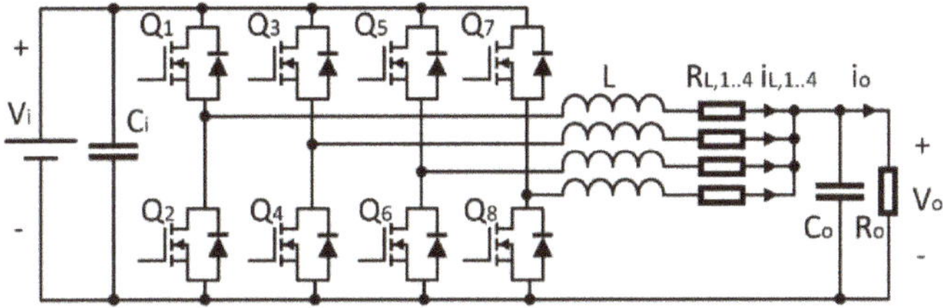

Figure 1. Multiphase synchronous DC-DC converter.

Nonlinear control methods need to be used to gain a globally linear control model. Sliding mode control (SMC) is one such method [21], and it is being used in switching converters because of its robustness, stability, transient performance and simple implementation. In SMC, states of a system will reach the sliding surface, $\sigma = 0$, from any initial condition in finite time and stay on it. System stability is, therefore, proven globally, and system dynamics can be prescribed for large signal changes. SMC has already been applied to multiphase converters, mainly with one master sliding surface for voltage control and slave sliding surfaces for phase current control. In an earlier work [22] these sliding surfaces make sure control signals of phases follow each other. The drawback, however, is lack of current control, so current sharing needs to be implemented separately. A similar procedure was used in another work [23], where switching frequency also needed to be controlled. Constant switching frequency SMC [5] is preferable, because of easier filter design and component efficiency optimization. Another way of controlling currents is by incorporating them into sliding surfaces. In previous studies [24,25] current sharing has been achieved by incorporating the average of currents into the sliding surfaces of each phase. Because surfaces are coupled, the dynamics of such a system are hard to describe in a reduced order model, which is preferable for emulation purposes. An interesting point of the principle is the voltage control part that is included in all surfaces, so the configuration can be thought of as multi-master. A similar structure was used in an earlier study [26]. When a phase malfunctions, voltage remains under control of the sliding surfaces of other phases. However, there are 3 parameters which can be difficult to tune. A better approach to controlling the power factor in a multiphase configuration was used in another study [27], where one reference current was applied to the sliding surfaces of all phases. This way, phases were decoupled from each other, and current control could be designed separately. This resulted in a simpler model, and is appropriate for emulation algorithms. An interesting use of sliding mode to control the current part of a single-phase buck converter is described in another work [28], where the inner current loop is controlled by the sliding surface, and the outer loop is an ordinary proportional-integral (PI) controller. It could be applied to multiphase converters, where they would share the same reference current. The drawback of the proposed algorithm is that it is based on one-step reaching of the sliding surface, which, in turn, means higher control effort and saturated control signals in the reaching phase, resulting in a nonlinear model. To cope with this problem, different variants of SMC have been proposed, based on a reaching law, such as Gao's approach [29] and the linear reaching law [30]. By prescribing system dynamics in the reaching phase, control parameters can be designed in a way that the control signal is never saturated (the duty cycle is always within $[0, 1]$) and the closed-loop dynamics are globally linear, also for large signal operations. This is different from conventional designs, where control is designed in the sliding surface, but, while reaching it, the control signal can be in saturation, and the system dynamics are different. Another solution to the problem can be found in another class of control algorithms, model predictive control (MPC) [8,31], which are also used for battery emulation purposes. The algorithm searches for control solutions inside bounds to prevent saturation. The disadvantage of this approach is computational complexity, making it inappropriate for use in digital signal processor (DSP)-controlled power converters. Moreover, in some cases, this method fails to find a solution [8], and the control signal is still saturated.

Nearly all of the mentioned sliding mode control algorithms for multiphase converters are based on an ideal model, where model uncertainties and unmatched phases are not accounted for. Unmatched characteristics can be compensated for by incorporating the integrative state in the sliding surface, similarly to the aforementioned linear PI controllers. This was applied to power converters in previous works [32,33], for discrete SMC, and this is called integral sliding mode (ISM) [34], where the integral part will make sure that the system reaches the sliding surface despite the uncertainties. However, as already mentioned for linear systems control theory, the integral part has its drawbacks [19], which are also manifested in ISM as integral dynamics in the form of large overshoots; slow settling time is also added to the system when the model matches the real system perfectly. Disturbance compensation, on the other hand, makes no intervention in the system when the model is matched. The observed disturbance value is bounded, unlike the integral, which can reach infinite values and requires anti-windup schemes. In Zongxiang et al. [35], the disturbance observer was used, along with the ISM, for motor control. However, the integral part was not eliminated completely, only its impact on the system dynamics was reduced. In Marcos-Pastor et al. [36], the integral part was completely replaced by the disturbance estimate for the magnetic levitation system, and its superior performance was shown. Some improvements of the method were presented in another study [37], where disturbance dynamics were also accounted for in the control algorithm. While SMC with a disturbance observer is used in various fields, there are only a few reported examples of its usage for controlling DC/DC converters. In a previous study [38], load current was an estimated disturbance, and the converter was controlled in a cascade of super-twisting-algorithms (STAs), which are based on SMC. In another study [39], the load resistance variation disturbance was estimated and used in terminal-sliding-mode (TSM) control of a buck converter. In another earlier study [40], disturbances of input voltage and load resistance are estimated by a linear observer, and used in SMC for a half bridge isolated converter. In another study [41], a disturbance estimate was used in the sliding surface to control a buck converter. However, for mismatched disturbances, integral state was still used, leading to the abovementioned problems.

The main contribution of this paper is a control for a multiphase buck converter that fits into the class of SMC algorithms. The aim was to control the converter appropriately for later use in battery emulation algorithms for EV. The control is based on cascade control with N identical inner sliding surfaces for inductor currents and an outer voltage control loop with output current feedforward. With the same current reference on all sliding surfaces, equal current sharing was accomplished. To steer states towards the sliding surface, the linear reaching law was used [30], which is based on Gao's approach [29], but without the chattering effect. It will be shown, by using this approach, that control parameters can be designed in such a way that the control signal is never saturated (the duty cycle is always within $[0, 1]$). By accounting for control limits in the design phase, the SMC design of this work resulted in a computationally simple control law, which can easily be applied to DSP processors, compared to MPC [8,31], which requires more computationally powerful processors. To compensate for model uncertainties and mismatch between phases, a disturbance observer of the Luenberger type, similar to the one in previous studies [36,41,42], was used, which made it possible to achieve zero steady state current error. The observer was of the linear type, which is simpler, and does not generate an undesired chattering effect like the sliding type observer used in a previous study [39]. Observer stability was proven with the linear system theory, and was easily implementable on digital controllers. Estimated disturbance was bounded, so the calculated control signal never went into saturation, as can be the case when using integral states [19,34]. The outer voltage loop was of proportional type, with feedforward of the output current that was measured. To compensate for the model's mismatch in the voltage part, an observer was used of the same structure as in the case of the current part. In contrast to previous studies [38], the observer acted on a closed loop equation rather than on a system model. In contrast to another study [40], where the observer and controller were based on continuous domain, in this work they are discrete and directly applicable to DSP implementation. In discrete form, implementation is even simpler, because of the elimination of derivative calculations. A similar disturbance observer was used in another study [41], where a disturbance estimate was used in the sliding surface to control a buck converter, but for mismatched disturbances, the integral state was used, with all its drawbacks. In the present work, it will be shown that the integral state is not necessary; mismatched disturbances can also be cancelled effectively with a disturbance observer and appropriate tuning of current and voltage loop bandwidths. The rules are given to tune all parameters of the control system. The designed algorithms were tested on a prototype converter.

2. Multiphase Converter Modeling

A multiphase converter from Figure 1 can be modeled by the system Equation (1) with N identical phase models, where $n \in \{1, 2, \ldots, N\}$. Phase inductor currents $i_{L,n}$ and output voltage v_O are used as state variables. The inputs of the model u_n represent the duty cycles of the pulse width modulators (PWM). Other variables are input voltage v_i and output current i_O. The parameters for the current part are inductance L and inductor ohmic resistance R_L, which can also include MOSFETs' on-state resistances R_{ds-ON} and current shunt resistances. C_o in the voltage part of the model represents the output capacitance.

$$\frac{di_{L,n}}{dt} = -\frac{R_L}{L} i_{L,n} - \frac{1}{L} v_0 + \frac{1}{L} v_i u_n$$
$$\frac{dv_o}{dt} = \frac{1}{C_o} \sum_{n=1}^{N} i_{L,n} - \frac{1}{C_o} i_o \tag{1}$$

To design an SMC law, it needs to be transferred to the discrete domain. This can be evaluated by first order approximation of the z-transform [43], which gives a simple model (2), with sample time T, where k represents the time step in sample intervals.

$$i_{L,n}(k+1) = \left(1 - \frac{R_L T}{L}\right) i_{L,n}(k) - \frac{T}{L} v_0(k) + \frac{T}{L} v_i(k) u_n(k)$$
$$v_o(k+1) = \frac{T}{C_o} \sum_{n=1}^{N} i_{L,n}(k) + v_o(k) - \frac{T}{C_o} i_o(k) \tag{2}$$

In a real system there are some parameter uncertainties in the forms of inductor ohmic resistance deviation ΔR_L inductance drops and tolerances ΔL, and capacitance deviation ΔC_o. A real effective duty ratio is different from the ideal because of the dead times [3] and rise times of MOSFETs. There are some unknown disturbances in the form of sensor offsets and noise. This can all be lumped into disturbances for current part $d_{i,n}$ and voltage part d_v and added to the model (2), obtaining a model with disturbances (3).

$$i_{L,n}(k+1) = \left(1 - \frac{R_L T}{L}\right) i_{L,n}(k) - \frac{T}{L} v_0(k) + \frac{T}{L} v_i(k) u_n(k) + d_{i,n}(k)$$
$$v_o(k+1) = \frac{T}{C_o} \sum_{n=1}^{N} i_{L,n}(k) + v_o(k) - \frac{T}{C_o} i_o(k) + d_{i,n}(k) \tag{3}$$

3. Phase Current Control

Model (3) is of the $N+1$ order, where each phase current $i_{L,n}(k)$ is different, which can be a complex control problem from the parameter selection point of view, as well as closed loop dynamics could also be complex. With appropriate control algorithms for phase currents, they can be made equal, and the voltage model can be simplified to (4). This can be true if phase current control is faster than voltage control.

$$v_o(k+1) = \frac{NT}{C_o} i_L(k) + v_o(k) - \frac{T}{C_o} i_o(k) + d_v(k) \tag{4}$$

In Equation (3) it can be observed that the current model has nonlinearity on input, because of the multiplication of input voltage $v_i(k)$ with the duty ratio $u_n(k)$. The duty ratio $u_n(k)$ in a real system is limited, depending on the hardware; current sensors, MOSFET drivers, and so on. The converter presented here, neglecting dead times, enables operation from 0–1 duty cycle, so the assumed range of $u_n(k)$ will be on the interval $[0,1]$. Control signals outside of this range would result in duty cycle limiting and saturation, which implies a nonlinear model. It would be hard to find an inverse of such a model, which is desired for emulation algorithms. Another problem is disturbances $d_{i,n}(k)$, which are different and unknown. Despite disturbances, phase currents should have equal values in the stationary state, which implies accurate model description (4), distributed heating in the circuit, and efficient use of components.

3.1. Sliding Mode Current Control

Sliding mode control is a nonlinear control method, which is robust to model uncertainties [44] and performs well with large signal changes. The basis of this method is a prescribed sliding surface σ, made from controlled variables. The system variables are steered to reach the sliding surface and stay on it. The conventional method of implementing the SMC control is by using a relay of the signum function given along the sliding surface. For switching power converters, this results in unpredictable frequency, so the PWM-based continuous SMC [45] is more appropriate. The converter can be operated with fixed switching frequency and control signal governing system to follow ideal sliding dynamics—quasi SMC.

The control used in this paper is based on N identical sliding surfaces (5), where n is the phase index. All surfaces use the same reference current $i_{L,r}$, and compare it to the phase current $i_{L,n}$. Sliding surface σ_n is like in [28], where the reference current is compared to the phase current.

$$\sigma_n(k) = i_{L,r}(k) - i_{L,n}(k) \tag{5}$$

An equivalent control in a previous study [28] was evaluated by setting $\sigma(k+1) = 0$, the so-called "one-step reaching", but this is undesirable, as it results in large control signals, leading to limiting, and making the dynamics become nonlinear and noninvertible. Different reaching laws [30,46,47] can be applied to prevent overlarge control effort and to control the reaching phase of SMC. Among them, the linear reaching law (6) is simple and linear, meaning the closed loop system description will be linear in a global sense. The linear reaching law (6) is based on Gao's reaching law [29] without the switching part to eliminate chattering. Sample time independent convergence rate parameter q determines convergence speed in the reaching phase. For the simplification of equations, it will be replaced by sample time-dependent equivalent Q in the following derivations.

$$\sigma_n(k+1) = (1 - qT)\sigma_n(k), \text{ where } qT = Q \in (0,1) \tag{6}$$

There are, however, some improvements in the approach, in the form of variable convergence parameters in [48,49], but they result in nonlinearities in the closed loop description, so they are not applicable here. By combining the sliding surface (5) with the reaching law (6), Equation (7) can be obtained.

$$i_{L,r}(k+1) - i_{L,n}(k+1) = (1-Q)(i_{L,r}(k) - i_{L,n}(k)) \tag{7}$$

Then, by introducing (3) into (7) and solving for u_n, the equivalent control law (8) can be derived.

$$u_n(k) = \frac{L}{Tv_i(k)}\left(i_{L,r}(k+1) + (Q-1)i_{L,r}(k) + \left(-Q + \frac{R_L T}{L}\right)i_{L,n}(k) + \frac{T}{L}v_o(k) - d_{i,n}(k)\right) \tag{8}$$

The control law derived (8) is not implementable, because disturbance $d_{i,n}(k)$ is not known. By removing it from the control law, that is, $d_{i,n}(k) = 0$, reaching dynamics (9) can be obtained. It deviates from (7) by $d_{i,n}(k)$, so additional measures need to be implemented for cancelling its impact.

$$i_{L,r}(k+1) - i_{L,n}(k+1) = (1-Q)(i_{L,r}(k) - i_{L,n}(k)) - d_{i,n}(k) \tag{9}$$

3.2. Design of a Disturbance Observer for Current Control

According to (9), unknown disturbance alters the reaching law, and moreover, as $k \to \infty$, the system does not reach the sliding surface, it only converges to the disturbance's steady state value $d_{i,n}(\infty)$. The first solution to the problem can be modification of the sliding surface, with the addition of an integral term, called the integral sliding mode (ISM) [34]. However, the integral has some drawbacks, observed from the linear systems control theory [19], which also manifest themselves in ISM [37]. Integral dynamics are also added to the system when the model matches the real system perfectly. This implies large overshoots, slow settling time, and so on. As the maximum value of the integral is hard to predict, it can result in control signal saturation. The second solution to deal with disturbance is to prescribe model uncertainties, and include their assumptions in the sliding surface, as in [48,50]. The control signal can, therefore, be predicted in the development phase of control to be always within the interval $[0,1]$, so there is no need for limitation. However, without the use of switching functions in the sliding surface, it is hard to achieve zero steady state error, where system states are only in the vicinity of the sliding surface. Disturbance estimation and rejection techniques [19,36,51] solve the drawbacks of both solutions mentioned. As disturbance estimate has a finite and predictable value, the control signal can be designed to be always inside the bounds. The disturbance observer can estimate steady state disturbances, so the system can reach the sliding surface, not just the vicinity of it. Disturbance estimation is especially attractive in discrete time control [42], because of the easier implementation of delayed states, compared with computing state derivatives, as in the case of continuous time control [51]. With the use of estimated disturbance $\hat{d}_{i,n}$ instead of real $d_{i,n}$, the control law (8) can be changed. It can also be assumed that the reference current is slow time varying, $i_{L,r}(k+1) \approx i_{L,r}(k)$, which all results in an implementable control law (10).

$$u_n(k) = \frac{L}{Tv_i(k)}\left(Qi_{L,r}(k) + \left(-Q + \frac{R_L T}{L}\right)i_{L,n}(k) + \frac{T}{L}v_o(k) - \hat{d}_{i,n}(k)\right) \tag{10}$$

By putting (10) into system Equation (3), the closed loop response of the system can be evaluated (11).

$$i_{L,n}(k+1) = (1-Q)i_{L,n}(k) + Qi_{L,r}(k) + d_{i,n}(k) - \hat{d}_{i,n}(k) \tag{11}$$

It can be observed that the disturbance estimate cancels the disturbance; hence, the system reaches the sliding surface in steady state if the disturbance is matched. If all states of the system are measurable, disturbance can be estimated from the system model. It is convenient to estimate disturbance from closed loop dynamics (11), as a simple linear Luenberger observer type can be used with observer gain l_i (12).

$$\hat{d}_{i,n}(k+1) = \hat{d}_{i,n}(k) + l_i\left(i_{L,n}(k) - \hat{i}_{L,n}(k)\right) \tag{12}$$

Equation (13) is proposed to calculate the estimated current $\hat{i}_{L,n}(k)$. It has a similar structure to (11), with the disturbance parts omitted. In the following analysis it will be proven that it can form a stable observer in conjunction with Equations (11) and (12).

$$\hat{i}_{L,n}(k+1) = (1-Q)\hat{i}_{L,n}(k) + Qi_{L,r}(k) \tag{13}$$

First, the current estimation error $e_i(k)$ and disturbance estimation error $e_d(k)$ variables are defined as (14).

$$\begin{aligned} e_i(k) &= i_{L,n}(k) - \hat{i}_{L,n}(k) \\ e_d(k) &= d_{i,n}(k) - \hat{d}_{i,n}(k) \end{aligned} \tag{14}$$

By rewriting (14) for $k+1$, estimation errors are gained for the next time step (15).

$$\begin{aligned} e_i(k+1) &= i_{L,n}(k+1) - \hat{i}_{L,n}(k+1) \\ e_d(k+1) &= d_{i,n}(k+1) - \hat{d}_{i,n}(k+1) \end{aligned} \tag{15}$$

Then, current estimation error for $k+1$ step (16) is derived by putting (11) and (13) into (15) and simplifying by (14).

$$e_i(k+1) = e_d(k) \tag{16}$$

The disturbance estimation error for $k+1$ step can follow the same derivation as used in [42], only the disturbance estimate (12) is different here. The real disturbance dynamics are unknown, but can be assumed by (17), where the incremental disturbance change is defined as $\Delta d_{i,n}(k+1) = d_{i,n}(k+1) - d_{i,n}(k)$.

$$d_{i,n}(k+1) = d_{i,n}(k) + \Delta d_{i,n}(k+1) \tag{17}$$

Putting (12) and (17) into (15) and simplifying by (14), the disturbance estimation error for $k+1$ step can be evaluated in (18).

$$e_d(k+1) = e_d(k) + \Delta d_{i,n}(k+1) - l_i e_i(k) \tag{18}$$

The final control diagram with SMC and the disturbance observer for phase current control is shown in Figure 2.

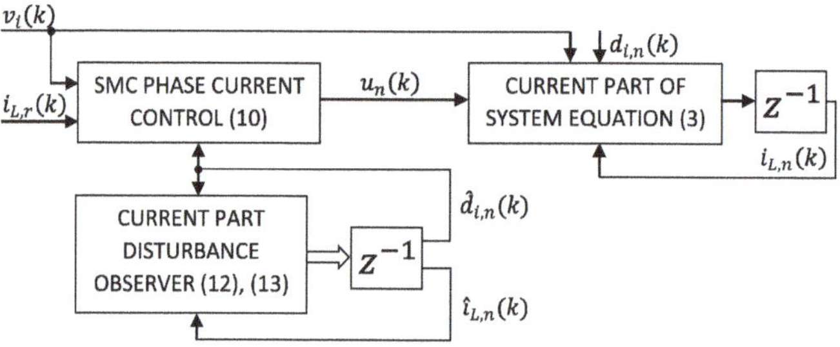

Figure 2. Diagram of sliding mode control (SMC) and disturbance observer for phase current control.

3.3. Proposed Rules to Tune the Sliding Mode Current Controller and the Disturbance Observer

For tuning controller parameter Q and observer parameter l_i, first the tuning rules will be proposed, based on pole location analysis. Then, the second rule will be proposed, where operation inside control limits and operation without saturation effects is guaranteed. Equations (11), (16) and (18) can be compacted in the matrix form (19) with system matrix A_i and system vectors x_i, b_i, g_i.

$$\underbrace{\begin{bmatrix} i_{L,n}(k+1) \\ e_d(k+1) \\ e_i(k+1) \end{bmatrix}}_{x_i(k+1)} = \underbrace{\begin{bmatrix} 1-Q & 1 & 0 \\ 0 & 1 & -l_i \\ 0 & 1 & 0 \end{bmatrix}}_{A_i} \underbrace{\begin{bmatrix} i_{L,n}(k) \\ e_d(k) \\ e_i(k) \end{bmatrix}}_{x_i(k)} + \underbrace{\begin{bmatrix} Q \\ 0 \\ 0 \end{bmatrix}}_{b_i} i_{L,r}(k) + \underbrace{\begin{bmatrix} 0 \\ \Delta d_{i,n}(k+1) \\ 0 \end{bmatrix}}_{g_i} \tag{19}$$

The system is linear, and stability can be checked by calculating eigenvalues λ_i of the system matrix A_i by (20), where I is the identity matrix.

$$\det(\lambda_i I - A_i) = 0 \tag{20}$$

By putting (19) into (20), eigenvalues λ_1, $\lambda_{2,3}$, that represent system poles, are gained in (21). To be stable, all should be inside the unit circle. Without common parameters, sliding mode controller pole λ_1 and observer poles $\lambda_{2,3}$ do not depend on each other, so the poles can be treated separately from the stability point of view.

$$\begin{aligned} \lambda_1 &= 1 - Q \\ \lambda_{2,3} &= \tfrac{1}{2} \pm \tfrac{\sqrt{1-4L_i}}{2} \end{aligned} \tag{21}$$

Figure 3 shows how the poles move in the z-plane when Q is increased from 0 to infinity, and l_i is increased from 0 to infinity. Observer poles $\lambda_{2,3}$ do not affect controller stability, but in (19) it can be seen that the disturbance estimation error $e_d(k)$ affects controller response. The poles of controller and observer therefore need to be placed as far apart as possible, with the controller pole much closer to the unit circle. Thus, the dynamics of the controller are predominant. As poles $\lambda_{2,3}$ start from different locations, the optimal case would be to place them in $\lambda_{2,3} = \tfrac{1}{2}$. That way, the location of both poles would be far away from the unit circle, that is, they would be less dominant, with high natural frequency. Putting $\lambda_{2,3} = \tfrac{1}{2}$ into (21) gives the needed observer gain l_i (22). That way, observer pole placement is solved analytically, and the observer parameter is independent of system parameters.

$$l_i = \frac{1}{4} \tag{22}$$

To tune the controller parameter Q, two major requirements should be fulfilled. From Figure 3, the first requirement is that the controller pole λ_1 is dominant, that is, separated enough from observer poles $\lambda_{2,3}$ and closer to the unit circle. The second requirement is for the control signal, that is, the duty ratio, to be always inside bounds, $u_n(k) \in [0,1]$. This way, it would be possible to gain reduced order dynamic description without saturation effects—nonlinearities. To make λ_1 dominant, it should be closer to the unit circle. To get some more understanding of poles, their natural frequencies ω_1, $\omega_{2,3}$ and damping ratio ζ in continuous space (s-domain) can be gained by equating eigenvalues with the equation for poles of the first order system (23) and second order system (24). By setting ζ to 1, the observer's poles are put to the same location on the real axis ($\lambda_{2,3} = \tfrac{1}{2}$) and pole location Equation (24) is simplified to $\lambda_{2,3} = e^{-\omega_{2,3}T}$, which is of the same structure as (23).

$$s = -\omega_1 \overset{z=e^{sT}}{\rightarrow} \lambda_1 = e^{-\omega_1 T} \tag{23}$$

$$s = -\zeta\omega_{2,3} \pm j\omega_{2,3}\sqrt{1-\zeta^2} \overset{z=e^{sT}}{\rightarrow} \lambda_{2,3} = e^{-\zeta\omega_{2,3}T \pm j\omega_{2,3}T\sqrt{1-\zeta^2}} \overset{\zeta=1}{\rightarrow} \lambda_{2,3} = e^{-\omega_{2,3}T} \tag{24}$$

That way, pole dominance can be assured only by setting poles' natural frequencies (ω_1 and $\omega_{2,3}$) appropriately. For λ_1 to be dominant, its natural frequency ω_1 should be lower than the pole's $\lambda_{2,3}$ natural frequency $\omega_{2,3}$. Generally, a ratio of 5 or more is sufficient, so (25) can be used as a basic rule for further tuning.

$$\omega_{2,3} \geq 5\omega_1 \tag{25}$$

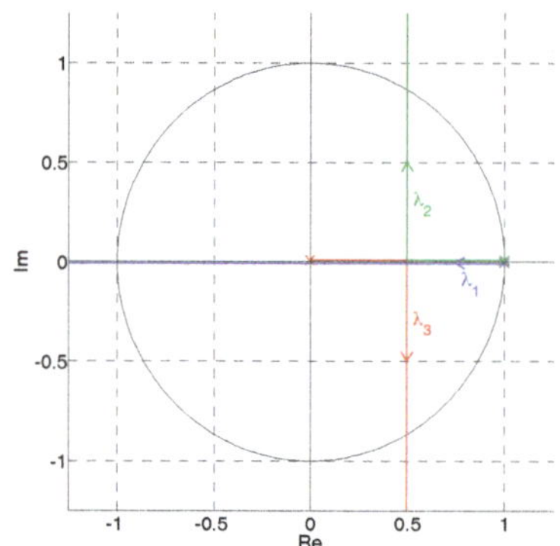

Figure 3. Root-locus chart for the SCM pole λ_1 and observer poles $\lambda_{2,3}$.

By solving ω_1 from (23) and $\omega_{2,3}$ from (24), using (21), (22) and putting all into (25), Q can be derived (26).

$$Q \leq 1 - \sqrt[5]{\lambda_{2,3}} = 0.13 \tag{26}$$

As value Q depends on $\lambda_{2,3}$, which has been determined before, parameter tuning selection has been greatly simplified. However, this is the first requirement for tuning the SCM, which assures its dominance over the observer.

The second requirement, $u_n(k) \in [0,1]$, is needed to guarantee that the control signal will never be saturated and that globally linear system dynamics can be gained. This is different than in the general procedure, where $u_n(k)$ can be calculated outside the range and then limited. The philosophy has been used in the design of multiphase converters for transient performance, where converters' bandwidths are designed so that voltage spikes are symmetrical [52,53]. In the work presented here, the controller was designed with bandwidth based on a physical system, but it should be noted that in the sources mentioned a different approach is also possible; to design a physical system—inductances based on the required control bandwidth. The idea from previous works [52,53] is that the control signal can be guaranteed globally inside bounds ($u_n(k) \in [0,1]$) when the inductor current slope $\frac{di_L}{dt}$, demanded in the control algorithm, is never higher than the physical system can handle, that is, it is dependent on inductance. For discrete time system description, this can be translated into incremental current change $\Delta i_{L,n}(k+1)$, expressed in (27).

$$\Delta i_{L,n}(k+1) = i_{L,n}(k+1) - i_{L,n}(k) \tag{27}$$

Putting closed loop current dynamics ($i_{L,n}(k+1)$) from (19) into (27) and assuming $e_d(k) \approx 0$ (this is valid, as the observer's bandwidth is higher than the controller's (25)), the worst case of value of $\Delta i_{L,n}(k+1)$, that is, $\Delta I_{L,C}$, is gained in (28). By selecting the appropriate border value of $i_{L,r}$, that is, $I_{L,r}$, and border value of i_L, that is, I_L, the equation can give the worst case demanded incremental current change $\Delta I_{L,C}$ from the controller.

$$\Delta I_{L,C} = Q I_{L,r} - Q I_L \tag{28}$$

Putting open loop current dynamics ($i_{L,n}(k+1)$), from (2) into (27), the worst case of value of $\Delta i_{L,n}(k+1)$, that is, $\Delta I_{L,P}$ is gained in (29). By selecting appropriate border values of system variables (I_L for i_L, V_o for v_o, V_i for v_i and U for u_n), the equation can give the worst case possible incremental current change $\Delta I_{L,P}$ of the physical system.

$$\Delta I_{L,P} = -\frac{R_L T}{L} I_L - \frac{T}{L} V_o + \frac{T}{L} V_i U \tag{29}$$

As $\Delta I_{L,C}$ from (28) can be positive (for $I_{L,r} \geq I_L$) or negative (for $I_{L,r} < I_L$), two different requirements are given in inequalities (30).

$$a)\ \Delta I_{L,C} \leq \Delta I_{L,P}\ when\ \ I_{L,r} \geq I_L,$$
$$b)\ \Delta I_{L,C} \geq \Delta I_{L,P}\ when\ \ I_{L,r} < I_L \tag{30}$$

By using (28), (29) in (30), two requirements for Q value can be gained in (31).

$$a)\ Q \leq \frac{1}{I_{L,r,max}-I_{L,min}}\left(-\frac{R_L T}{L}I_{L,min} - \frac{T}{L}V_{o,max} + \frac{T}{L}V_{i,min}U_{max}\right)$$
$$b)\ Q \leq \frac{1}{I_{L,r,min}-I_{L,max}}\left(-\frac{R_L T}{L}I_{L,max} - \frac{T}{L}V_{o,min} + \frac{T}{L}V_{i,max}U_{min}\right) \tag{31}$$

Equations (26) and (31) therefore form two restrictions on the selection of control parameter Q. Both should be satisfied when designing control.

3.4. Current Control Model Reduction

The dynamics of the current controller and disturbance observer (19) are of the third order. By combining it with the voltage control part, the system would become even more complex. As in the former controller and observer tuning, it has been assured that when the controller dynamics are dominant over the observer, it is possible to reduce model order. The methods to achieve model reduction are given in [54]. Among them, singular perturbation approximation (SPA) has been selected as the most appropriate, as it preserves steady-state gain, and there is no need to apply special system transformations to achieve it. The least dominant states with fast dynamics are considered to be in a steady state ($\Delta x(k+1) = x(k+1) - x(k) = 0$). For model (19), SPA can be applied to observer states $e_d(k)$ and $e_i(k)$, with the use of Equation (32)

$$e_d(k+1) = e_d(k)$$
$$e_i(k+1) - e_i(k) \tag{32}$$

By putting (16) and (18) into (32) and neglecting disturbance dynamics ($\Delta d_{i,n}(k+1) = 0$, disturbance is assumed constant), (33) can be gained.

$$e_d(k) = e_i(k) = 0 \tag{33}$$

By using (33) in model (19), the final reduced current system of the first order is gained (34), which is appropriate for use in output voltage control design.

$$i_{L,n}(k+1) = (1-Q)i_{L,n}(k) + Qi_{L,r}(k) \tag{34}$$

4. Output Voltage Control

By using the phase current control designed, the current part of the system (1) has been simplified to (34) and it is valid to use (4) for the voltage part. If it is certain that phase currents $i_{L,n}$ follow the same reference current $i_{L,r}$, only a model of one phase is enough for voltage control design, that is, index n is omitted. The model for voltage control design is then (35), where $i_{L,r}$ is input and output voltage and v_o is the variable to be controlled. Output current i_o is measurable disturbance, which can be justified for emulation applications, where battery voltage is emulated depending on i_o, so it should be sensed. Disturbance of the voltage model d_v is in a different channel as input $i_{L,r}$, so it is regarded as mismatched disturbance, which is generally difficult to cancel.

$$i_L(k+1) = (1-Q)i_L(k) + Qi_{L,r}(k)$$
$$v_o(k+1) = \frac{NT}{C_o}i_L(k) + v_o(k) - \frac{T}{C_o}i_o(k) + d_v(k) \tag{35}$$

As current part dynamics will be faster than the voltage, the system can be reduced using SPA. Similar to in Section 3.4., SPA is used for inductor current using (36).

$$i_L(k+1) = i_L(k) \tag{36}$$

By using (36) in (35), $i_{L,r}(k) = i_L(k)$ can be derived, and system (35) is reduced to one voltage Equation (37).

$$v_o(k+1) = \frac{NT}{C_o}i_{L,r}(k) + v_o(k) - \frac{T}{C_o}i_o(k) + d_v(k) \tag{37}$$

It should be noted here that input $i_{L,r}$ now acts in the same channel as disturbance d_v, which can now be canceled directly by the control algorithm.

Proposition 1. *Based on the reduced voltage part model (37), a control law is proposed (38) which would compensate disturbances i_o and d_v and make voltage $v_{o,r}(k)$ converge to its reference $v_{o,r}(k)$ with zero steady state error. To compensate $d_v(k)$, its estimate $\hat{d}_v(k)$ can be used. Convergence speed can be adjusted by K_p.*

$$i_{L,r}(k) = \frac{C_o}{NT}\left(K_p(v_{o,r}(k) - v_o(k)) + \frac{T}{C_o}i_o(k) - \hat{d}_v(k)\right) \tag{38}$$

Proof. The closed loop Equation (39) can be gained by putting (38) into (37). From the equation, it can be seen that, by assuring $\hat{d}_v(k)$ matches $d_v(k)$ in steady state, the disturbance effect is canceled. Disturbance i_o has disappeared from the dynamics' description and the dynamics are linear, with K_p being the only tunable parameter.

$$v_o(k+1) = \left(1 - K_p\right)v_o(k) + K_p v_{o,r}(k) + d_v(k) - \hat{d}_v(k) \tag{39}$$

□

4.1. Design of a Disturbance Observer for Voltage Control

For the proposed control law (38) to be implementable, a voltage disturbance observer needs to be designed that would give disturbance estimate $\hat{d}_v(k)$. As the closed loop voltage Equation (39) is linear and similar to (11), a Luenberger observer, like the one from Section 3.2, could be used for disturbance estimation. Equation (40) presents the disturbance estimation with observer gain l_v, where the output voltage estimate $\hat{v}_o$ is needed.

$$\hat{d}_v(k+1) = \hat{d}_v(k) + l_v(v_o(k) - \hat{v}_o(k)) \tag{40}$$

The estimated voltage $\hat{v}_o$ is a prediction gained from the algorithm's previous step $(k-1)$ and calculated by (41), which is similar to (13), so observer derivation and stability rules are the same as in Section 3.2.

$$\hat{v}_o(k+1) = \left(1 - K_p\right)\hat{v}_o(k) + K_p v_{o,r}(k) \tag{41}$$

First, voltage estimation error $e_v(k)$ and disturbance estimation error $e_{dv}(k)$ variables are defined in (42).

$$\begin{aligned} e_v(k) &= v_o(k) - \hat{v}_o(k) \\ e_{dv}(k) &= d_v(k) - \hat{d}_v(k) \end{aligned} \tag{42}$$

By rewriting (42) for step $k+1$, estimation errors (43) are gained for the next time step.

$$\begin{aligned} e_v(k+1) &= v_o(k+1) - \hat{v}_o(k+1) \\ e_{dv}(k+1) &= d_v(k+1) - \hat{d}_v(k+1) \end{aligned} \tag{43}$$

Then, by putting (39) and (41) into (43) and simplifying by (42), the voltage estimation error for $k+1$ step can be derived in (44).

$$e_v(k+1) = e_{dv}(k) \tag{44}$$

Real disturbance dynamics are unknown, but can be replaced with incremental disturbance change, that is, $\Delta d_v(k+1) = d_v(k+1) - d_v(k)$, to gain (45) (similar to (17)).

$$d_v(k+1) = d_v(k) + \Delta d_v(k+1) \tag{45}$$

Putting (40) and (45) into (43), and simplifying by (42), the disturbance estimation error for $k+1$ step can be evaluated in (46) (similar to (18)).

$$e_{dv}(k+1) = e_{dv}(k) + \Delta d_v(k+1) - L_v e_v(k) \tag{46}$$

As has already been justified in 0 for the same observer type, $l_v = \frac{1}{4}$ was selected for optimal placement of observer poles.

The final control diagram with voltage control and voltage part disturbance observer, which could also be referred to as outer control, is shown in Figure 4. It can be seen that the same reference current from voltage control is applied to N identical current phase control blocks of structure, shown in Figure 2.

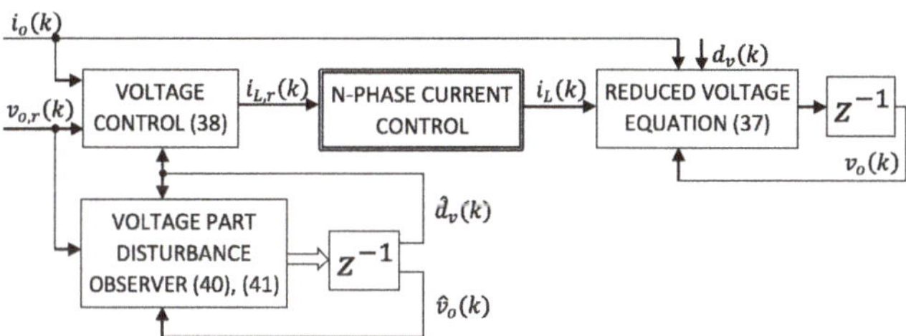

Figure 4. A diagram of voltage control, disturbance observer and N-inner current control loops. Proposed rules to tune the output voltage controller.

The tuning rules that will be proposed for the voltage controller are similar to the rules from Section 3.3. There are two major requirements to follow: making the voltage control loop dominant over the current control loop and guaranteeing that operation is always inside the prescribed limits on $i_{L,r}$. By assuring voltage part dominance, the system can be reduced to a simple and linear description, for ease of use later in the emulation algorithms. By guaranteeing $i_{L,r}$ is always inside limits, saturation effects can be omitted, and globally linear invertible dynamics can be attained.

Equation (35) can be augmented with voltage disturbance observer states (44) and (46), and $i_{L,r}$ can be replaced from the voltage control law (38). All states' dynamics are then compacted in matrix form (47) with system matrix A_v and system vectors x_v, b_v, g_v, and h_v.

$$
\underbrace{\begin{bmatrix} i_L(k+1) \\ v_o(k+1) \\ e_v(k+1) \\ e_{dv}(k+1) \end{bmatrix}}_{x_v(k+1)} = \underbrace{\begin{bmatrix} 1-Q & -\frac{QC_oK_p}{NT} & 0 & \frac{QC_o}{NT} \\ \frac{NT}{C_o} & 1 & 0 & 0 \\ 0 & 0 & 0 & 1 \\ 0 & 0 & -l_v & 1 \end{bmatrix}}_{A_v} \underbrace{\begin{bmatrix} i_L(k) \\ v_o(k) \\ e_v(k) \\ e_{dv}(k) \end{bmatrix}}_{x_v(k)} + \underbrace{\begin{bmatrix} \frac{QC_oK_p}{NT} \\ 0 \\ 0 \\ 0 \end{bmatrix}}_{b_v} v_{o,r} + \underbrace{\begin{bmatrix} \frac{Q}{N} \\ -\frac{T}{C_o} \\ 0 \\ 0 \end{bmatrix}}_{g_v} i_o + \underbrace{\begin{bmatrix} \frac{QC_o}{NT} \\ -1 \\ 0 \\ 0 \end{bmatrix}}_{h_v} d_v \tag{47}
$$

The system is linear, and its stability can be checked by calculating eigenvalues λ_v of the system matrix A_v (48).

$$\det(\lambda_v I - A_v) = 0 \tag{48}$$

By putting (47) into (48), eigenvalues λ_{v1}, λ_{v2} from the controller and eigenvalues λ_{v3}, λ_{v4} from the observer are gained in (49). For the system to be stable, all eigenvalues should be inside the unit circle. Because of uncommon parameters, controller poles λ_{v1}, λ_{v2} are not dependent on observer poles λ_{v3}, λ_{v4}, and vice versa, so poles can be treated separately from the stability point of view.

$$\lambda_{v1} = 1 - \frac{Q}{2} + \frac{\sqrt{-Q\left(4K_p-Q\right)}}{2}, \ \lambda_{v2} = 1 - \frac{Q}{2} - \frac{\sqrt{-Q\left(4K_p-Q\right)}}{2}$$
$$\lambda_{v3} = \frac{1}{2} + \frac{\sqrt{1-4l_v}}{2}, \ \lambda_{v4} = \frac{1}{2} - \frac{\sqrt{1-4l_v}}{2}$$

$$(49)$$

Figure 5 shows the root locus chart of poles with varying parameter K_p from 0 to infinity. Observer poles were expected to be closer to the unit circle origin, so their impact on the system dynamics can be neglected. By setting $K_p = 0$ in (49) and comparing it to Equation (21), it can be observed that pole λ_{v1} originates from the voltage part and pole λ_{v2} originates from the current part. As K_p is increased, both poles approach each other. For cascade control it is desired that the inner current control loop has higher bandwidth than the outer voltage loop. In other words, the voltage pole must be dominant over the current pole. This is also desired from the emulation perspective, as, with the voltage pole being dominant, reduction of system order is possible. The same rule for pole dominance as in Section 3.3 can also be applied here. From Figure 5 it can be deduced that poles λ_{v1} and λ_{v2} must remain on the real axis, and must be far enough apart so that λ_{v1} stays closer to the unit circle, therefore being dominant. For poles to be real, the range for the control parameter is $K_p \leq \frac{Q}{4}$, which can be deduced from (49).

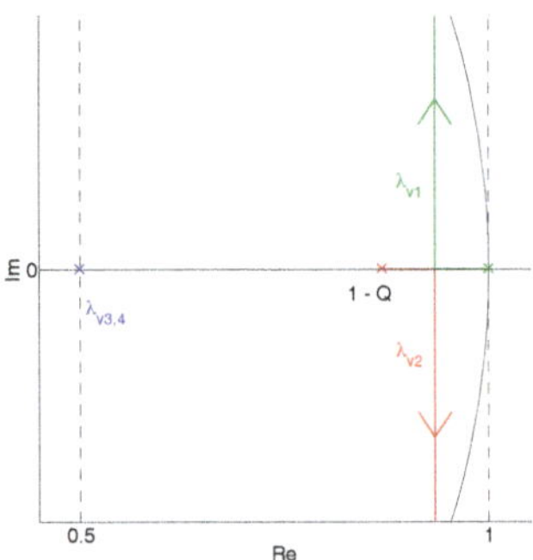

Figure 5. Root-locus chart for voltage control poles $\lambda_{v1,2}$ and disturbance observer poles $\lambda_{v3,4}$

The natural frequency ω_{v2} of pole λ_{v2} should be more than five times the natural frequency ω_{v1} of pole λ_{v1}. This is possible only if both poles are placed on the real axis, and the condition is (50) (similar as for the current part (25)).

$$\omega_{v2} > 5\omega_{v1}$$

$$(50)$$

Inequality (51) is derived by using the same equations as in (24), that is, $\omega_{v1} = \frac{-\ln(\lambda_{v1})}{T}$ and $\omega_{v2} = \frac{-\ln(\lambda_{v2})}{T}$, which are put into (50). By replacing Q with its value, a solution for K_p can be found with numeric methods.

$$\left(1 - \frac{Q}{2} + \frac{\sqrt{-Q\left(4K_p - Q\right)}}{2}\right)^5 > \left(1 - \frac{Q}{2} - \frac{\sqrt{-Q\left(4K_p - Q\right)}}{2}\right)$$

$$(51)$$

The second requirement for tuning K_p is that $i_{L,r}$ is always inside the limits, thus obtaining globally linear behavior without saturation effects. It was also pointed out in a previous study [5] that, due to

$i_{L,r}$ saturation, the voltage loop becomes open and possibly unstable. Like in the case of the current control part (27), the incremental voltage change $\Delta v_o(k+1)$ can be expressed by (52).

$$\Delta v_o(k+1) = v_o(k+1) - v_o(k) \tag{52}$$

Putting closed loop voltage Equation (39) into (52), using (42), and neglecting disturbance observer estimate error $e_{dv}(k) \approx 0$ (this is valid, as the observer has a sufficiently higher natural frequency), the worst case of value of $\Delta v_o(k+1)$, that is, $\Delta V_{o,C}$, is gained in (53). By selecting the appropriate border value of $v_{o,r}$, that is, $V_{o,r}$, and border value of v_o, that is, V_o, the equation can give the worst case demanded incremental voltage change $\Delta V_{o,C}$ from the controller.

$$\Delta V_{o,C} = K_p V_{o,r} - K_p V_o \tag{53}$$

Putting a simplified converter model (37) into (52) and neglecting mismatch $d_v(k)$, the worst case possible $\Delta v_o(k+1)$, i.e., $\Delta V_{o,P}$, can be gained in (54). By selecting the appropriate border value of $i_{L,r}$, that is, $I_{L,r}$, and border value of i_o, that is, I_o, the equation can give the worst case achievable incremental voltage change $\Delta V_{o,P}$. $I_{L,r}$ as well as I_o can be positive when current flows from the converter, emulating a power source, or negative when current flows into the converter, emulating a power drain. It should be noted here that the converter needs to be designed with sufficient margin on the inductor current, i.e., $I_{L,r} > \frac{I_o}{N} > 0$ for positive values of I_o or $I_{L,r} < \frac{I_o}{N} < 0$ for negative values of I_o.

$$\Delta V_{o,P} = \frac{T}{C_o}(NI_{L,r}(k) - I_o(k)) \tag{54}$$

As $\Delta V_{o,C}$ from (53) can be positive (when $V_{o,r} \geq V_o$), or negative (when $V_{o,r} < V_o$), two different cases of requirements are given in (55).

$$\begin{aligned} a)\; \Delta V_{o,C} \leq \Delta V_{o,P} \; when \; V_{o,r} \geq V_o, \\ b)\; \Delta V_{o,C} \geq \Delta V_{o,P} \; when \; V_{o,r} < V_o \end{aligned} \tag{55}$$

By using (53) and (54) in (55), two cases of requirements of K_p are gained in (56). Equations for both are similar, only the minimal (min) and maximal (max) worst-case values are different, as they were selected to give the most restrictive worst case.

$$\begin{aligned} a)\; K_p \leq \frac{1}{V_{o,r,max}-V_{o,min}} \frac{T}{C_o}(NI_{L,r,max}(k) - I_{o,max}(k)) \\ b)\; K_p \leq \frac{1}{V_{o,r,min}-V_{o,max}} \frac{T}{C_o}(NI_{L,r,min}(k) - I_{o,min}(k)) \end{aligned} \tag{56}$$

Equations (51) and (56) are, therefore, the main rules to select parameter K_p.

4.2. Voltage Control Model Reduction

The dynamics of the closed voltage loop, current loop and disturbance observer (47) are of the fourth order, which is complex for future use in emulation algorithms. However, in the preceding derivation, it has been assured that the voltage model is dominant over current and observer models. Model reduction is, therefore, possible like in the case of the current controller (Chapter 3.4). SPA can be applied to observer states $e_v(k)$, $e_{dv}(k)$ and inductor current $i_L(k)$ by using (57).

$$\begin{aligned} e_{dv}(k+1) &= e_{dv}(k) \\ e_v(k+1) &= e_v(k) \\ i_L(k+1) &= i_L(k) \end{aligned} \tag{57}$$

By using (57) in (47) approximated system states are gained in (58).

$$e_v(k) = e_{dv}(k) = 0$$
$$i_{L,r}(k) = \frac{K_p C_o}{NT}(v_{o,r}(k) - v_o(k)) + \frac{1}{N}i_o(k) - \frac{C_o}{NT}d_v(k)$$

$$(58)$$

Putting (58) into the voltage part of (47) ($v_o(k+1)$), reduced order voltage Equation (59) is obtained.

$$v_o(k+1) = \left(1 - K_p\right)v_o(k) + K_p v_{o,r}(k)$$

$$(59)$$

It is a simple reduced model of the first order, which gives an accurate enough description of the system dynamics. By designing control parameters with the use of the rules presented, they will be globally linear and valid for large signal changes.

5. Results

A prototype 4-phase synchronous buck converter was built to prove the concept of the proposed control algorithm. Figure 6 shows the prototype converter and the experimental setup with a power supply unit (PSU) and load (R_o). The input voltage (v_i) range was from 10 V to 14.4 V, and output voltage (v_o) was from 2 V to 8.5 V. The converter was designed for bidirectional operation, where the output current ranges from −2.5 A to 2.5 A. Load is adjustable by switching different combinations of 12 Ω resistors on-off. Four dual MOSFETs of type FDMS9620 were used for the power part. A3946 was used as a gate driver in all phases, which enabled 0–100 % duty ratio operation. The inductors were MSS1583, with saturation current of 2.0 A (for 10 % inductance drop). In this work, the inductor current range used was from −1.0 A to 1.0 A, so it stayed in the linear region. Input and output voltages (v_i, v_o) were sensed; for currents, the current shunts were on all 4 phases for sensing $i_{L1...4}$, and on output for sensing i_o. The basic power part component values are presented in Table 1. A Texas Instruments DSP, TMS320F28377S, was used to implement the control algorithm. The control algorithm and PWM executed with 20 kHz, where the same algorithm was applied for each phase current control, and one master algorithm was applied for voltage control. PWM pulses were synchronized and delayed by 90° from each other. Center aligned PWM sampling was used for phase currents, so average current values were obtained for a defined period.

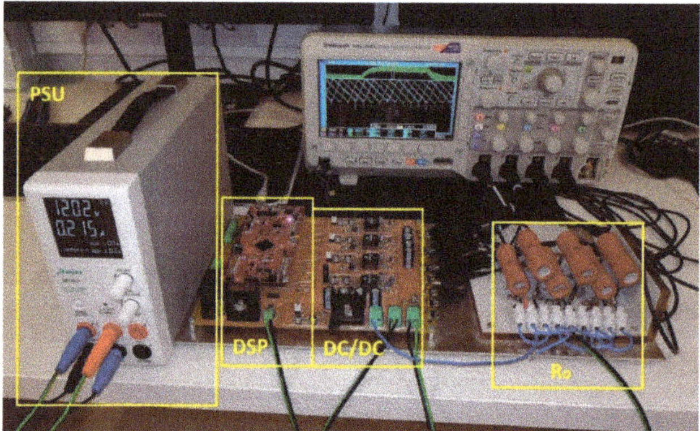

Figure 6. Test setup for prototype converter with ohmic load.

Table 1. Prototype converter parameters.

Parameter	Symbol	Value
Inductor inductances	$L_{1\ldots4}$	330 µH
Inductor resistances	$R_{L,1\ldots4}$	300 mΩ
Input, output capacitance	C_i, C_o	1880 µF
Input, output ESR	ESR_{Ci}, ESR_{Co}	20 mΩ
High MOSFETs ($Q_{1,3,5,7}$) on-state resistance	$R_{ds-ON,U}$	21.5 mΩ
Low MOSFETs ($Q_{2,3,6,8}$) on-state resistance	$R_{ds-ON,L}$	13 mΩ
Current sense shunt resistances	R_{cs}	3 Ω

5.1. Sliding Mode Current Control with a Disturbance Observer

By using SMC, system dependence on nominal circuit parameters was canceled; any parameter deviations and sensor errors are included in disturbance estimate $\hat{d}_i(k)$ from the observer. Using the proposed tuning rules, system stability can be gained in a global sense and operating point independent dynamics. Table 2 contains selected system marginal variables, which were used in (31) together with the parameters from Table 1, to gain two conditions ($Q < 0.14$ and $Q < 0.18$) for the algorithm to operate always inside the saturation limits. The third condition for parameter Q ($Q \leq 0.13$) is given in (26). Based on that, control parameter $Q = 0.13$ was selected, which gives the fastest system response. Observer gain l_i was selected in (22) and is independent of system parameters. By using Q, l_i, and system parameters, the control law (10) and the disturbance observer (12), (13) were implemented on DSP in the control algorithm of each phase. All tests were performed at a constant input voltage $V_i = 12\ V$.

The operation of the converter was first checked for current control only. The load was ohmic resistance $R_o = 3\ \Omega$, and the current reference $i_{L,r}$ was varied from 0 to 1 A. The period of reference value was set to 10 ms; this way, the output voltage was in the operational range (as the voltage control loop was opened, the voltage could rise up to its limits). In order to present the benefits of using SMC with Disturbance Observer (DO), operation of SMC was first investigated based on a nominal model. This can be accomplished by using control law (10) without disturbance estimate, that is, $\hat{d}_{i,n}(k) = 0$. The responses of phase currents are shown in Figure 7a. Figure 7b shows the values of currents, sampled by DSP, which were used in the control algorithm. It can be observed that reference tracking was poor and currents were not equal. This is because of model imperfections which were not compensated for.

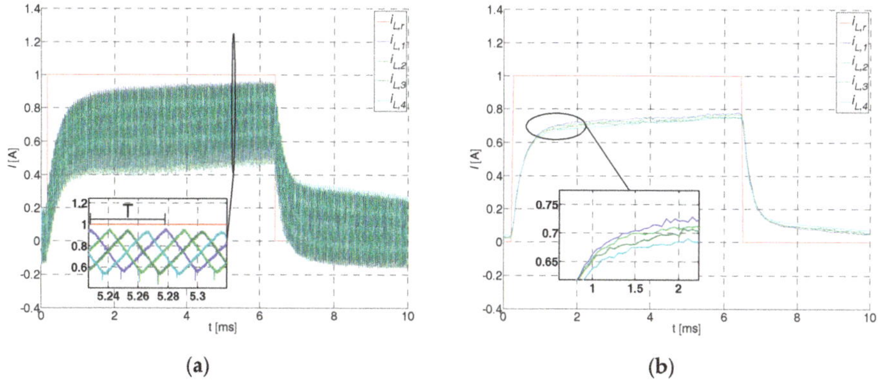

Figure 7. Phase currents' $i_{L,1\ldots4}$ responses to step change of current reference $i_{L,r}$ from 0 to 1 A when SMC is used, based on a nominal model without uncertainties; (**a**) real current values and (**b**) current values, sampled by digital signal processor (DSP).

Table 2. Current subsystem marginal values.

Variable	Value
$I_{L,r,min}$, $I_{L,min}$	−1 A
$I_{L,r,max}$, $I_{L,max}$	1 A
$V_{i,min}$	10 V
$V_{i,max}$	14.4 V
$V_{o,min}$	2 V
$V_{o,max}$	8.5 V
U_{min}	0
U_{max}	1

Then, the control performance was checked for the well-known integral sliding mode (ISM). This can be accomplished by using the additional integrative state in (10), and setting $\hat{d}_{i,n}(k) = 0$. Figure 8a shows current responses; compared to the previous case, the currents reached the current reference, that is, model imperfections were compensated for. Figure 8b shows current values, sampled by DSP, and it can be seen that, while reaching the current reference, currents were unbalanced. During that phase, the integral part of ISM was in operation, its dynamics were added to the system's and it can be observed from the response.

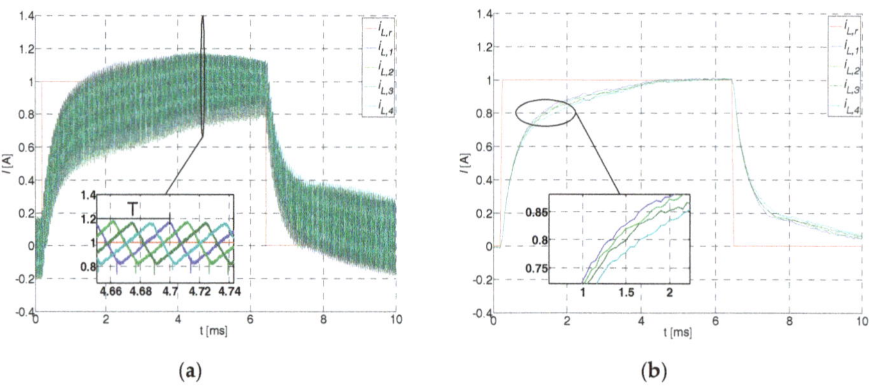

(a)   (b)

Figure 8. Phase currents' $i_{L,1...4}$ responses to step change of current reference $i_{L,r}$ from 0 to 1 A when integral sliding mode (ISM) is used to account model uncertainties; (**a**) real current values and (**b**) values, sampled by DSP.

Then, the operation of the SMC algorithm was verified with the disturbance observer, according to (10). Figure 9a shows the current responses, which reached reference faster than in previous examples (Figures 7 and 8). Figure 9b shows the sampled currents, where good tracking of reference can be observed. Compared to previous cases, currents were balanced, even while reaching the reference; this is convenient, as their dynamical description is the same, as well as from the electronic circuit perspective, because better current distribution and equal heating of phases was accomplished.

Next, response to load change was observed, with $i_{L,r} = 0.5\ A$ and R_o changed from 3 Ω to 2 Ω.

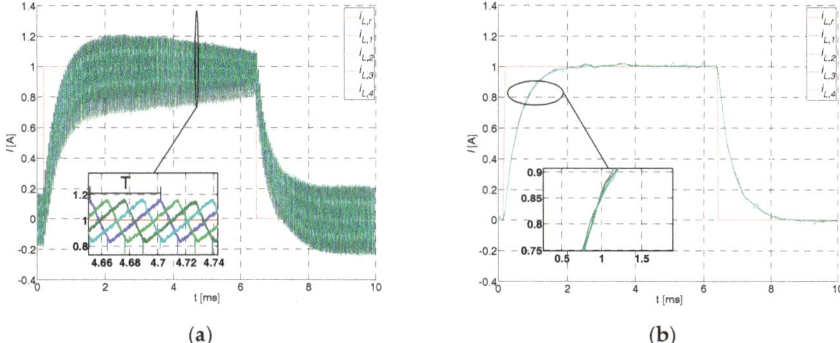

(a)

(b)

Figure 9. Phase currents' $i_{L,1...4}$ to step change of current reference $i_{L,r}$ from 0 to 1 A when a disturbance observer is used to account for model uncertainties; (**a**) real current values and (**b**) values, sampled by DSP.

Figure 10 shows the results, where the output current step and output voltage can be seen, the latter being inside the operational range of the converter. Output load change does not affect controller performance significantly for non-observer or observer cases, although, by using the observer, current value is recovered to value before the load change.

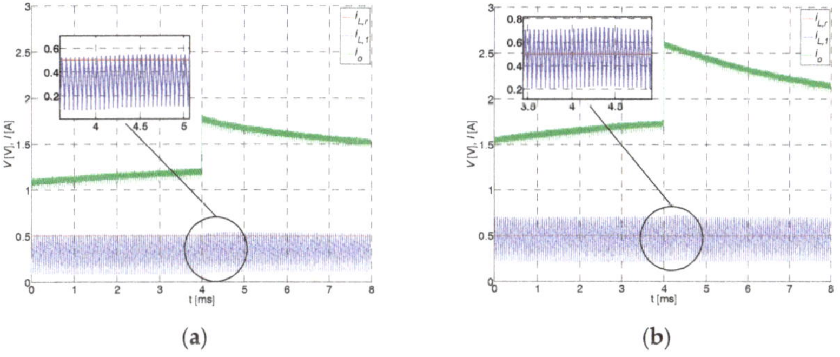

(a)

(b)

Figure 10. Phase 1 current $i_{L,1}$ response with reference current $i_{L,r} = 0.5$ A and load change from 3 Ω to 2 Ω without observer (**a**) and with observer (**b**).

5.2. Voltage Control and Disturbance Observer

As in the current control, system dependence on nominal parameters was canceled by the proposed control law; any parameter deviations and sensor errors were included in the disturbance estimate $\hat{d}_v(k)$. By following the proposed tuning rules, a globally stable system could be gained with operating point independent dynamics. Table 3 contains the marginal values of system variables, which were used in (56), together with the parameters from Table 1, in order to obtain the value of parameter K_p, which will assure us that the algorithm always operates inside the inductor current marginal values, that is, $K_p \leq 0.00614$ for both cases. Equation (51) gives the second condition for the choice of K_p, with the use of Q applied in the current control design. The value used was $K_p = 0.006$. By using this value and system parameters, control law (38) with disturbance observers (40) and (41), could be implemented on DSP. The result from (38) was then used as a reference for all phase current control loops. It can easily be compared to a P controller with output current feedforward control, which was accomplished by setting $\hat{d}_v(k) = 0$ in (38). By using integral of voltage error instead of $\hat{d}_v(k)$, the well-known PI controller could be implemented with output current feedforward.

Table 3. Voltage subsystem marginal values.

Variable	Value
$V_{o,min}$, $V_{o,r,min}$	2 V
$V_{o,max}$, $V_{o,r,max}$	8.5 V
$I_{L,r,min}$	−1 A
$I_{L,r,max}$	1 A
$I_{L,r,max}$, $I_{L,max}$	1 A
$I_{o,min}$	−2.5 A
$I_{o,max}$	2.5 A

The operation of all controllers was verified by reference voltage steps from 3 V to 4 V, with the $R_o = 2\ \Omega$. Figure 11a shows the operation for output current feedforward, combined with a P-type controller. A stationary error can be observed, which was the result of model imperfection, sensor errors, and so on. Figure 11b shows the improved version (PI controller with output current feedforward). As can be seen, stationary state error has been removed, although, some additional system dynamics can be observed in the form of overshoot. Figure 11c shows the operation of output current feedforward combined with the disturbance observer, which was derived in the current work. The results controller removes the stationary error as with the controller in Figure 11b, but, without additional dynamics, there was no overshoot. This could be accomplished as the observer was active only for model imperfections, compared with the integral part of PI, which was active all the time, resulting in overshoots, control signal saturation and long settling times. The observer dynamics are magnified in Figure 11c; by using the proposed tuning rules it was made negligible.

Next, the operation of all three controllers was checked in response to load step. Voltage reference was set to 4 V and the load R_o was changed from 6 Ω to 3 Ω and back. The results in Figure 12a show that there was some stationary error in the operation of the P type controller with i_o feedforward, which was the result of an unmatched model. Figure 12b shows improvement when using a PI controller with i_o feedforward, as the stationary error was reduced. Figure 12c shows the operation of a P controller with i_o feedforward and disturbance observer. Stationary state error was reduced, as in the case of PI, but settling is much faster. There were some oscillations in inductor current and output voltage. They can be explained by Figure 5, where they appear as observer poles which have been placed in the marginal point before becoming complex values. The model of the converter is not perfect; some simplifications were applied in (36) and (37), where one sample time delay was neglected, and this could explain the oscillations. Oscillations could be eliminated by manually tuning parameter l_v to a lower value, but it is not necessary, as their amplitude is low. The contribution of this work is to show that observer poles can be placed analytically, and that they are not dominant in voltage response.

Operation in the whole operational range of output voltage was checked by different voltage reference steps between 2 V, 4 V, 6 V and 8 V, with $R_o = 4\ \Omega$. Figure 13 shows the results, where voltage responses can be regarded as the same for all operating points. It can be assumed from phase 1 current i_{L1} that it is inside the designed bounds presented in Table 3. The oscillations in the current are not problematic, as its value remains well below $I_{L,max}$ (this can be assumed since for $V_o = 8$ V, and $R_o = 4\ \Omega$ the output current is 2 A, which is close to the maximum rated load). As already mentioned before, they can be eliminated by tuning l_v manually if necessary.

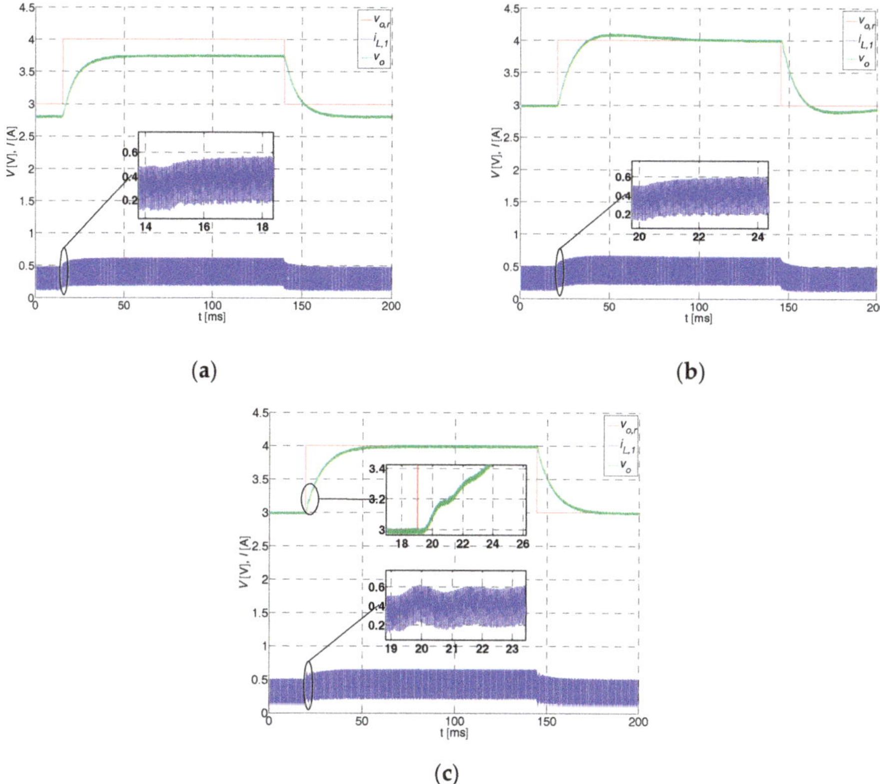

(a)

(b)

(c)

Figure 11. Output voltage response to step voltage reference change when using (**a**) P type control with i_o feedforward, (**b**) PI control with i_o feedforward, and (**c**) P control with i_o feedforward and disturbance observer. The load was 2 Ω.

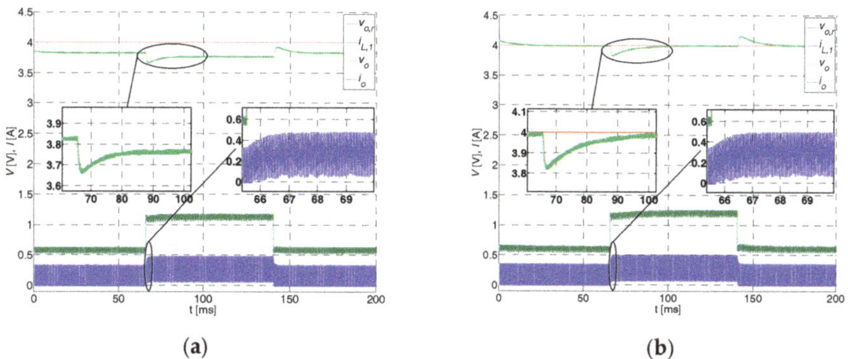

(a)

(b)

Figure 12. *Cont.*

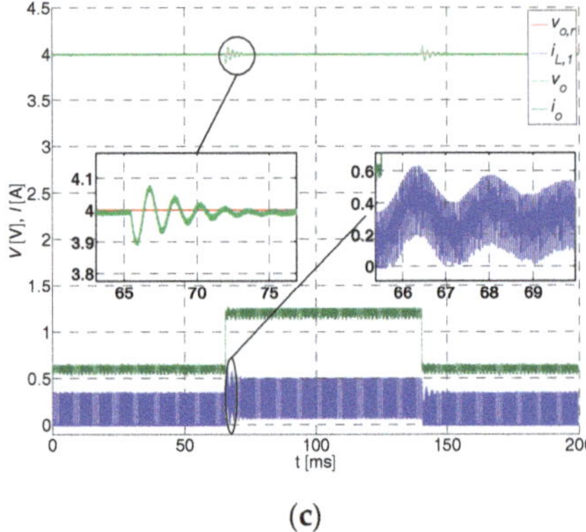

(**c**)

Figure 12. Output voltage v_o and phase 1 current $i_{L,1}$ response to load change from 6 Ω to 3 Ω and back when using (**a**) P type control with i_o feedforward, (**b**) PI control with i_o feedforward, (**c**) P control with i_o feedforward and disturbance observer.

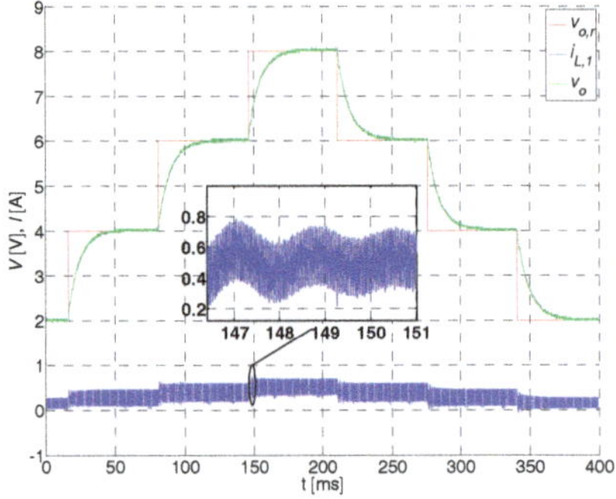

Figure 13. Output voltage and phase 1 current response to 2 V reference voltage steps in the whole operating range.

6. Discussion

A control algorithm has been proposed for control of a multiphase buck converter. Inner control sliding surfaces take care of current control for each phase, and make the current loops' model linear and operating points independent. The saturation of duty ratio is prevented with the proposed tuning rules. Model imperfections in the form of discretization errors, dead time effects, and parameter uncertainties, have been compensated for by the introduction of disturbance observers. Compared to ISM [34], a disturbance observer is active only when model imperfections exist, and it alters the system dynamics minimally. When model imperfections are present, observer dynamics are non-dominant if the proposed tuning rules are followed. Parameters are determined analytically. Compared to ISM, a disturbance observer does not suffer from wind-up effects, as the estimated disturbance value is bounded. Its maximum value could be calculated based on predicted model imperfections with some additional analysis, but this has been omitted in the present work to simplify the design. Another benefit of a disturbance observer over the ISM is phase currents are equal while reaching current reference, resulting in equal dynamics and better heating distribution in a circuit. Control signal limiting is not needed with the use of the proposed tuning rules, resulting in a globally linear model. For the voltage part control, a disturbance observer of the same structure has been used, and the same rules can be used for tuning, which simplifies the design. Although disturbance in the voltage part is mismatched, it has been shown that it can be compensated for sufficiently by current loops tuned to a higher bandwidth than the voltage loop. By evaluating the estimated disturbance value, it would be possible also to evaluate model accuracy. If necessary, accuracy can be improved with model expansion by introduction of additional parameters, like output capacitance, equivalent series resistance (ESR), dead times, or use of discretization of higher orders, and so on.

It was confirmed that the proposed algorithm makes the converter characteristic linear and operating points independent. Voltage responses were identical in the whole output voltage range, which is convenient in later use for emulation of batteries in a wide range of voltages. Compared to the PI voltage control with the output current feedforward, there were no additional dynamics, which would manifest in form of overshoots. Model reduction based on SPA was applied to gain a simple closed loop description of the converter without nonlinear effects, which can be used later by the battery emulation algorithm. It was confirmed that the control algorithm can be implemented on a DSP, unlike MPC [8,31], which is computationally more complex, needing more powerful processors. Controller delay because of sample and hold operation was neglected in this work, as the switching frequency was sufficiently higher than the controller bandwidth. If needed, its effects can be analyzed and compensated for in future work.

Author Contributions: Conceptualization, R.P. and M.R.; methodology, R.P. and M.R.; software, R.P.; validation, R.P., A.C. and M.R.; formal analysis, R.P.; investigation, R.P.; resources, A.C. and M.R.; data curation, R.P.; writing—original draft preparation, R.P.; writing—review and editing, R.P, A.C. and M.R.; visualization, R.P.; supervision, M.R.; project administration, R.P. and M.R.; funding acquisition, A.C. and M.R.

Funding: This research received no external funding.

Acknowledgments: R.P. would like to thank the Faculty of Electrical Engineering and Computer Science, University of Maribor, Republic of Slovenia and Margento R&D for funding of the PhD study this work is part of.

Conflicts of Interest: The authors declare no conflict of interest.

References

1. Mousavi, G.S.M.; Nikdel, M. Various battery models for various simulation studies and applications. *Renew. Sustain. Energy Rev.* **2014**, *32*, 477–485. [CrossRef]
2. Rivetta, C.; Williamson, G.A. Large-signal analysis and control of buck converters loaded by DC-DC converters. In Proceedings of the 2004 IEEE 35th Annual Power Electronics Specialists Conference (IEEE Cat. No.04CH37551), Aachen, Germany, 20–25 June 2004; Volume 3675, pp. 3675–3680.
3. Qiu, W.; Mercer, S.; Liang, Z.; Miller, G. Driver deadtime control and its impact on system stability of synchronous buck voltage regulator. *IEEE Trans. Power Electron.* **2008**, *23*, 163–171. [CrossRef]
4. Emadi, A.; Khaligh, A.; Rivetta, C.H.; Williamson, G.A. Constant power loads and negative impedance instability in automotive systems: definition, modeling, stability, and control of power electronic converters and motor drives. *IEEE Trans. Veh. Technol.* **2006**, *55*, 1112–1125. [CrossRef]
5. El Aroudi, A.; Martínez-Treviño, B.A.; Vidal-Idiarte, E.; Cid-Pastor, A. Fixed switching frequency digital sliding-mode control of DC-DC power supplies loaded by constant power loads with inrush current limitation capability. *Energies* **2019**, *12*, 1055. [CrossRef]
6. König, O.; Hametner, C.; Prochart, G.; Jakubek, S. Battery Emulation for Power-HIL Using Local Model Networks and Robust Impedance Control. *IEEE Trans. Ind. Electron.* **2014**, *61*, 943–955. [CrossRef]
7. Karimi, R.; Kaczorowski, D.; Zlotnik, A.; Mertens, A. Loss optimizing control of a multiphase interleaving DC-DC converter for use in a hybrid electric vehicle drivetrain. In Proceedings of the 2016 IEEE Energy Conversion Congress and Exposition (ECCE), Milwaukee, WI, USA, 18–22 September 2016; pp. 1–8.
8. König, O.; Gregorčič, G.; Jakubek, S. Model predictive control of a DC–DC converter for battery emulation. *Control Eng. Pract.* **2013**, *21*, 428–440. [CrossRef]
9. König, O.; Jakubek, S.; Prochart, G. Battery impedance emulation for hybrid and electric powertrain testing. In Proceedings of the 2012 IEEE Vehicle Power and Propulsion Conference, Seoul, Korea, 9–12 October 2012; pp. 627–632.
10. Rodič, M.; Milanovič, M.; Truntič, M. Digital control of an interleaving operated buck-boost synchronous converter used in a low-cost testing system for an automotive powertrain. *Energies* **2018**, *11*, 2290. [CrossRef]

11. Eirea, G.; Sanders, S.R. Phase current unbalance estimation in multiphase buck converters. *IEEE Trans. Power Electron.* **2008**, *23*, 137–143. [CrossRef]

12. Mariethoz, S.; Beccuti, A.G.; Morari, M. Model predictive control of multiphase interleaved DC-DC converters with sensorless current limitation and power balance. In Proceedings of the 2008 IEEE Power Electronics Specialists Conference, Rhodes, Greece, 15–19 June 2008; pp. 1069–1074.

13. Schumacher, D.; Magne, P.; Preindl, M.; Bilgin, B.; Emadi, A. Closed loop control of a six phase interleaved bidirectional dc-dc boost converter for an EV/HEV application. In Proceedings of the 2016 IEEE Transportation Electrification Conference and Expo (ITEC), Dearborn, MI, USA, 27–29 June 2016; pp. 1–7.

14. Ou, S.Y.; Liu, L.Y. Design and implementation of a four-phase converter with digital current sharing control for battery charger. In Proceedings of the 2015 IEEE International Telecommunications Energy Conference (INTELEC), Osaka, Japan, 18–22 October 2015; pp. 1–6.

15. Yang, Q.; Juanjuan, S.; Ming, X.; Kisun, L.; Lee, F.C. High-bandwidth designs for voltage regulators with peak-current control. In Proceedings of the Twenty-First Annual IEEE Applied Power Electronics Conference and Exposition, APEC '06, Dallas, TX, USA, 19–23 March 2006; p. 7.

16. Ridley, R.B. *A New Small-Signal Model for Current-Mode Control*; Virginia Polytechnic Institute and State University: Blacksburg, VA, USA, 1999.

17. Truntič, M.; Milanovič, M. Voltage and current-mode control for a buck-converter based on measured integral values of voltage and current implemented in FPGA. *IEEE Trans. Power Electron.* **2014**, *29*, 6686–6699. [CrossRef]

18. Cooke, P. Modeling average current mode control. In Proceedings of the APEC 2000—Fifteenth Annual IEEE Applied Power Electronics Conference and Exposition (Cat. No.00CH37058), New Orleans, LA, USA, 6–10 February 2000; Volume 251, pp. 256–262.

19. Han, J. From PID to active disturbance rejection control. *IEEE Trans. Ind. Electron.* **2009**, *56*, 900–906. [CrossRef]

20. Szeifert, F.; Nagy, L.; Chován, T.; Abonyi, J. Constrained PI(D) algorithms (C-PID). *Hungarian J. Ind. Chem.* **2005**, *33*, 81–88.

21. Halihal, Y. *State-Space Oriented Advanced Analysis And Control Mehtods of Switching Converters*; Ben-Gurion University of the Negev: Beersheba, Israel, 2015.

22. Repecho, V.; Biel, D.; Ramos-Lara, R.; Vega, P.G. Fixed-switching frequency interleaved sliding mode eight-phase synchronous buck converter. *IEEE Trans. Power Electron.* **2018**, *33*, 676–688. [CrossRef]

23. Repecho, V.; Biel, D.; Ramos, R.; Garcia, P. Sliding mode control of a m-phase DC/DC buck converter with chattering reduction and switching frequency regulation. In Proceedings of the 2016 14th International Workshop on Variable Structure Systems (VSS), Nanjing, China, 1–4 June 2016; pp. 290–295.

24. Lopez, M.; Vicuna, L.G.D.; Castilla, M.; Lopez, O.; Majo, J. Interleaving of parallel DC-DC converters using sliding mode control. In Proceedings of the IECON '98 24th Annual Conference of the IEEE Industrial Electronics Society (Cat. No.98CH36200), Aachen, Germany, 31 August–4 September 1998; Volume 1052, pp. 1055–1059.

25. Mazumder, S.K. *Nonlinear Analysis and Control of Standalone, Parallel DC-DC, and Parallel Multi-Phase PWM Converters*; Faculty of the Virginia Polytechnic Institute and State University: Blacksburg, VA, USA, 2001.

26. Zongxiang, C.; Ya-nan, G.; Mingxing, C.; Lusheng, G. Study on PI sliding mode controller for paralleled dc-dc converter. In Proceedings of the 2016 IEEE 8th International Power Electronics and Motion Control Conference (IPEMC-ECCE Asia), Hefei, China, 22–26 May 2016; pp. 3079–3083.

27. Marcos-Pastor, A.; Vidal-Idiarte, E.; Cid-Pastor, A.; Martinez-Salamero, L. Interleaved digital power factor correction based on the sliding-mode approach. *IEEE Trans. Power Electron.* **2016**, *31*, 4641–4653. [CrossRef]

28. Vidal-Idiarte, E.; Marcos-Pastor, A.; Giral, R.; Calvente, J.; Martinez-Salamero, L. Direct digital design of a sliding mode-based control of a PWM synchronous buck converter. *IET Power Electron.* **2017**, *10*, 1714–1720. [CrossRef]

29. Weibing, G.; Yufu, W.; Homaifa, A. Discrete-time variable structure control systems. *IEEE Trans. Ind. Electron.* **1995**, *42*, 117–122. [CrossRef]

30. Monses, G. *Discrete-Time Sliding Mode Control*; Technische Universiteit Delft: Delft, The Netherlands, 2002.

31. König, O.; Jakubek, S.; Prochart, G. Model predictive control of a battery emulator for testing of hybrid and electric powertrains. In Proceedings of the 2011 IEEE Vehicle Power and Propulsion Conference, Chicago, IL, USA, 6–9 September 2011; pp. 1–6.

32. Orosco, R.; Vazquez, N. Discrete sliding mode control for DC/DC converters. In Proceedings of the 7th IEEE International Power Electronics Congress, Technical Proceedings, CIEP 2000 (Cat. No.00TH8529), Acapulco, Mexico, 15–19 October 2000; pp. 231–236.

33. Cao, J.; Chen, Q.; Zhang, L.; Quan, S. Sliding mode control of bidirectional DC/DC converter. In Proceedings of the 2018 33rd Youth Academic Annual Conference of Chinese Association of Automation (YAC), Atlanta, GA, USA, 18–20 May 2018; pp. 717–721.

34. Abidi, K.; Jian-Xin, X. A Discrete-Time Integral Sliding Mode Control Approach for Output Tracking with State Estimation. *IFAC Proc. Vol.* **2008**, *41*, 14199–14204. [CrossRef]

35. Zheng, C.; Zhang, J. Finite-time nonlinear disturbance observer based discretized integral sliding mode control for PMSM drives. *J. Power Electron.* **2018**, *18*, 1075–1085. [CrossRef]

36. Yang, J.; Li, S.; Yu, X. Sliding-mode control for systems with mismatched uncertainties via a disturbance observer. *IEEE Trans. Ind. Electron.* **2013**, *60*, 160–169. [CrossRef]

37. Yang, J.; Su, J.; Li, S.; Yu, X. High-order mismatched disturbance compensation for motion control systems via a continuous dynamic sliding-mode approach. *IEEE Trans. Ind. Inf.* **2014**, *10*, 604–614. [CrossRef]

38. Yin, Y.; Liu, J.; Vazquez, S.; Wu, L.; Franquelo, L.G. Disturbance observer based second order sliding mode control for DC-DC buck converters. In Proceedings of the IECON 2017—43rd Annual Conference of the IEEE Industrial Electronics Society, Beijing, China, 29 October–1 November 2017; pp. 7117–7122.

39. Wang, J.; Li, S.; Yang, J.; Wu, B.; Li, Q. Finite-time disturbance observer based non-singular terminal sliding-mode control for pulse width modulation based DC–DC buck converters with mismatched load disturbances. *IET Power Electron.* **2016**, *9*, 1995–2002. [CrossRef]

40. Sanjeev Kumar, P.; Patil, S.L.; Chaudhari, B.N. Disturbance observer based sliding mode control for DC-DC power converters. In Proceedings of the IECON 2016—42nd Annual Conference of the IEEE Industrial Electronics Society, Florence, Italy, 23–26 October 2016; pp. 3366–3371.

41. Pandey, S.K.; Patil, S.L.; Chaskar, U.M.; Phadke, S.B. Direct Duty Ratio Control of Buck DC-DC Converters Using Disturbance Observer Based Integral Sliding Mode Control. In Proceedings of the IECON 2018—44th Annual Conference of the IEEE Industrial Electronics Society, Washington, DC, USA, 21–23 October 2018; pp. 5507–5512.

42. Kim, K.-S.; Rew, K.-H. Reduced order disturbance observer for discrete-time linear systems. *Automatica* **2013**, *49*, 968–975. [CrossRef]

43. Maksimovic, D.; Zane, R. Small-signal Discrete-time Modeling of Digitally Controlled DC-DC Converters. In Proceedings of the 2006 IEEE Workshops on Computers in Power Electronics, Troy, NY, USA, 16–19 July 2006; pp. 231–235.

44. Feng, C.; Lin, Y. Robust control design based-on integral sliding-mode for systems with norm-bounded uncertainties. In Proceedings of the 2011 11th International Conference on Control, Automation and Systems, Gyeonggi-do, Korea, 26–29 October 2011; pp. 1178–1183.

45. Siew-Chong, T.; Yuk-Ming, L.; Chi Kong, T. *Sliding Mode Control of Switching Power Converters*; Taylor & Francis Group, LLC: New York, NY, USA, 2012.

46. Xiong, L.; Li, P.; Li, H.; Wang, J. Sliding mode control of DFIG wind turbines with a fast exponential reaching law. *Energies* **2017**, *10*, 1788. [CrossRef]

47. Yang, Z.; Wan, L.; Sun, X.; Li, F.; Chen, L. sliding mode variable structure control of a bearingless induction motor based on a novel reaching law. *Energies* **2016**, *9*, 452. [CrossRef]

48. Bartoszewicz, A.; Leśniewski, P. Refined reaching laws for sliding mode control of discrete time systems. In Proceedings of the 2013 18th International Conference on Methods & Models in Automation & Robotics (MMAR), Miedzyzdroje, Poland, 26–29 August 2013; pp. 824–829.

49. Lee, C.W.; Chung, C.C. An approach to discrete-time sliding mode control with variable convergence rate to sliding surface. In Proceedings of the 2008 SICE Annual Conference, Tokyo, Japan, 20–22 August 2008; pp. 2358–2363.

50. Bartoszewicz, A.; Leśniewski, P. New Switching and Nonswitching Type Reaching Laws for SMC of Discrete Time Systems. *IEEE Trans. Control Syst. Technol.* **2016**, *24*, 670–677. [CrossRef]

51. Wen-Hua, C. Disturbance observer based control for nonlinear systems. *IEEE/ASME Trans. Mechatron.* **2004**, *9*, 706–710. [CrossRef]

52. Kaiwei, Y.; Yuancheng, R.; Lee, F.C. Critical bandwidth for the load transient response of voltage regulator modules. *IEEE Trans. Power Electron.* **2004**, *19*, 1454–1461. [CrossRef]

53. Pit-Leong, W.; Lee, F.C.; Peng, X.; Kaiwei, Y. Critical inductance in voltage regulator modules. *IEEE Trans. Power Electron.* **2002**, *17*, 485–492. [CrossRef]

54. Rydel, M.; Stanisławski, R.; Latawiec, K.J.; Gałek, M. Model order reduction of commensurate linear discrete-time fractional-order systems. *IFAC-PapersOnLine* **2018**, *51*, 536–541. [CrossRef]

Article

Fixed Switching Frequency Digital Sliding-Mode Control of DC-DC Power Supplies Loaded by Constant Power Loads with Inrush Current Limitation Capability

Abdelali El Aroudi *, **Blanca Areli Martínez-Treviño** and **Enric Vidal-Idiarte** and **Angel Cid-Pastor**

Departament d'Enginyeria Electrònica, Elèctrica i Automàtica, Universitat Rovira i Virgili, Av. Paisos Catalans, No. 26, 43007 Tarragona, Spain; blancaareli.martinez@urv.cat (B.A.M.-T.); enric.vidal@urv.cat (E.V.-I.); angel.cid@urv.cat (A.C.-P.)
* Correspondence: abdelali.elaroudi@urv.cat; Tel.: +34977558522

Received: 13 February 2019; Accepted: 12 March 2019; Published: 19 March 2019

Abstract: This paper proposes a digital sliding-mode controller for a DC-DC boost converter under constant power-loading conditions. The controller has been designed in two steps: the first step is to reach the sliding-mode regime while ensuring inrush current limiting; and the second one is to move the system to the desired operating point. By imposing sliding-mode regime, the equivalent control and the discrete-time large-signal dynamic model of this system are derived. The analysis shows that unlike with a resistive load, the boost converter under a fixed-frequency digital sliding-mode current control with external voltage loop open and loaded by a constant power load, is unstable. Furthermore, as with a resistive load, the system presents a right-half plane zero in the control-to-output transfer function. After that, an outer controller is designed in the z-domain for system stabilization and output voltage regulation. The results show that the system exhibits good performance in startup in terms of inrush current limiting and in transient response due to load and input voltage disturbances. Numerical simulations from a detailed switched model are in good agreement with the theoretical predictions. An experimental prototype is implemented to verify the mathematical analysis and the numerical simulation, which results in a perfect agreement in small-signal and steady-state behavior but also in a small discrepancy in the current limitation due a small propagation delay. Some efficient solutions have been proposed to mitigate the inrush current in the experimental results.

Keywords: DC-DC converters; boost converter; constant power load (CPL); fixed switching frequency; sliding-mode control; inrush current mitigation

---

1. Introduction

Many power systems call for a DC-DC multiconverter approach to provide various power and voltage forms [1–3]. Cascade connection of DC-DC converters arises in many industrial applications such as in modern electric vehicles (EV) [4,5], sea and undersea vehicles [6], and DC microgrids [7–10]. When the downstream converter in a two-stage cascade connection is tightly controlled to maintain an output voltage fixed on the load, it behaves as a constant power load (CPL) [1,11]. Other loads such as motor drives or electronic loads with tightly regulated controllers behave also as a CPL [11].

Figure 1 shows typical configurations of cascade connection of switching converter where CPL behavior may appear. CPLs exhibit a negative impedance behavior leading to a high risk of instability in this type of interconnection [1,11,12]. The control design of upstream DC-DC converters supplying constant power to the downstream converter becomes challenging due to the nonlinearity of the CPL.

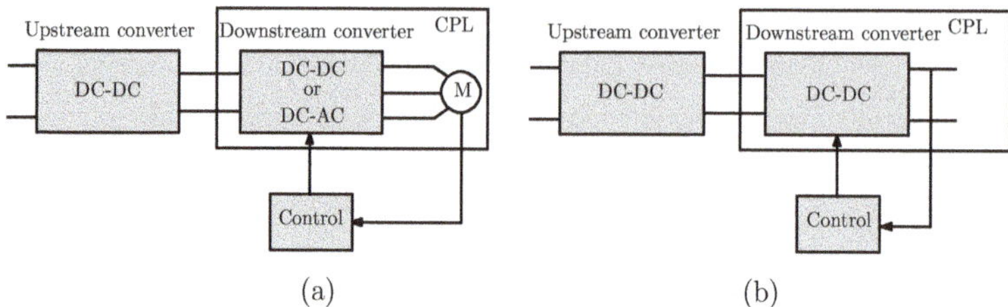

Figure 1. Power supply systems using a multiconverter approach where the CPL behavior may appear. (**a**) an Upstream converter loaded by a downstream converter regulating a DC-DC or DC-AC motor drive; (**b**) an upstream converter supplying a tightly regulated downstream converter.

Several methods have been proposed to cope with the mentioned CPL instability. Passive damping added to one of the filter elements in the cascade connection of a voltage source, an LC filter, and a CPL is used in [13] to stabilize the system without requiring the modification of the source or the load control. An active damping method based on the insertion of a virtual resistor to compensate for the negative incremental impedance of the CPL is successfully employed in [14]. Feedback linearization is reported in [15] in the context of a medium voltage DC bus for power distribution on ships to compensate the nonlinearity introduced by the CPL. Active compensation has been also explored in the case of a source converter of boost type by using current-control mode to introduce damping into the system dynamics [16]. Also, the existence of a stable behavior in the cascade connection has been proved for elementary hard-switching converters acting as source converters and operating in open-loop and in discontinuous conduction mode [17]. In [18], a robust control approach has been considered for the elementary power electronics switching converters with a CPL. In [19] robust controller based on linear programming is proposed to regulate the output of buck converters loaded by another buck converter acting as a CPL. A comprehensive review of the compensation techniques for switching converters with CPL can be found in [20]. In the solutions based on linear controllers, the starting point is an unstable transfer function relating either the control-to-output voltage or the control to inductor current. The transfer function is unstable because it is derived by simple substitution of the resistive load corresponding to a conventional supplying case by the negative incremental resistance of the CPL. This substitution results in a negative value of the damping factor or, equivalently, in the existence of right half plane poles. The open-loop unstable transfer function of the power converter with CPL describes the dynamic behavior in the vicinity of the steady-state operating point. However, there is no steady-state in open-loop due to the unstable nature of the converter. For that reason, the steady-state values of the state variables required in the transfer function are the ones imposed by the closed-loop behavior of the system provided that an appropriate controller stabilizes the converter. However, in some cases, the hypothesis of stable closed-loop steady state is not achieved despite introducing some control loops. For example, the introduction of an analog inner current loop for the average value regulation of the inductor current stabilizes a boost converter with CPL but it fails in a buck converter with the same type of load as demonstrated in [3].

Conventional linear control methods when applied to switching converters with CPL have limited stability region in the vicinity of the open-loop operating point which does not even exist in most cases of switching converters with loaded by CPL. Sliding-mode control (SMC) is a large-signal time-domain analytical technique for controlling the dynamic behavior of switching systems [21,22] that has been applied in the power electronics field in the early 1980s [23]. The first step in designing a SMC is to select a switching manifold in the spate space to which the system trajectories must be conducted. For this, the error between a suitable output signal and its desired reference is forced to be zero by an appropriate switching action. It could seem that any linear controller, properly designed, satisfies this control target naturally and that there is no need for a nonlinear SMC. However, SMC techniques result in reduced-order dynamics of the controlled system on the switching manifold in such a way that the error is zero not only in steady state as with linear controllers but also during transient provided that sliding-mode conditions are satisfied. In earlier works in this field, the typical choice for the controlling function σ is a linear combination of the error of the variable to be controlled and its $r-$th time derivative [23,24]. The order r of the derivation must be selected in such a way that the relative degree between the function σ and the discontinuous square wave signal u is equal to 1 for the sliding-mode conditions to be fulfilled [25]. In switching converters, it turns out to be that there is always an inductor current fulfilling this condition when it is used to construct the switching function σ without using the derivative of the error signal. SMC technique has therefore been later evolved to the use of an inductor current to be the variable to be controlled either for current limiting [22,26] or for different control purposes such as current balancing in parallel connected interleaved converters [27], impedance matching in PV systems [28,29], power factor correction in AC-DC rectifiers [30] among others. This type of control, when applied to switching converters, normally leads to unacceptable high switching frequencies due to chattering phenomenon. This has been solved by using hysteresis comparators resulting in a switching function σ to have a triangular shape with a variable switching frequency that can be adjusted by tuning the hysteresis width, but it will still be dependent on the operating point of the converter. This has motivated many studies aiming at solving this problem and getting limited and constant switching frequencies. In [25] it was shown that the modulation technique in switching converter under SMC is not necessarily of a variable frequency type such that using a hysteretic comparator and that their dynamics when they are under fixed-frequency strategies such as peak and valley current-mode control, can still be interpreted using SMC theory demonstrating that peak and valley current-mode control in switching converters are a kind of SMC regardless of the modulator used. Most of the existing works on using SMC in switching converters consider a linear resistive load. However, there are many cases in which the load is nonlinear. Some recent works consider loads containing nonlinear CPL. For instance, in [31], the authors use a variable frequency continuous-time SMC approach to regulate a boost converter feeding a CPL connected in parallel with a resistive load. An SMC-based fixed frequency pulse width modulation (PWM) approach is applied in [32] to boost converter supplying a pure CPL. In that work, the control law is derived by using a nonlinear switching surface. With the aim to improve the output voltage regulation, a linear term proportional to the voltage error was included in the same switching condition used in [10]. The control function used in that work contains a sign function inducing undesirable multiple pulsing or chattering due to an additional discontinuity in the system equations apart from the one induced naturally by the comparator. Moreover, a consistent performance evaluation in the whole operating range including the system response during startup and under parameter changes was not presented.

Most of the controllers used for switching converters were of analog nature although the final implementation is performed using digital platforms. Analog controllers are currently being substituted by digital controllers since the speed of computer hardware has increased exponentially in many industrial application fields. This increase in the processing speed has made it possible to sample and process control signals at very high frequencies. Digital control offers many advantages over analog control that explains the wide popularity in recent literature. However, most of existing digital controllers are based on emulation of analog controllers. Digital SMC (DSMC) is a direct approach offering the advantages of analog SMC combined with a fixed switching frequency operation.

The application of DSMC to the switching converters regulation has been always conditioned by the high switching frequency of the converters and the quasi-sliding effects caused by the sampling frequency [33,34] constraining the application to slow system variables [35] or to reduced switching frequency cases some of them eventually requiring a sophisticated digital hardware environment [32]. However, the application of predictive strategies has allowed the use of DSMC in the regulation of fast system variables as the inductor current operating at high switching frequencies [36]. Fixed frequency

DSMC technique has only very recently appeared in the power electronics field. The application of DSMC theory [37] has used first the discrete-time representation of the converters dynamics and subsequently has been used as a natural technique to analyze and to digitally implement SMC-based controllers with fixed-frequency PWM, which were validated in classical two-loop control strategies such as in [38,39] and in the synthesis of a discrete-time loss-free resistors for AC-DC PFC applications [40].

Fixed-frequency digital control of DC-DC converters with CPL has not been addressed as far as the authors are aware. Direct application of the results in [38], is not possible because the discrete-time model cannot be obtained in closed form without approximations. Here, a direct digital control design is provided by first deriving an approximate discrete-time model which is demonstrated to faithfully predict the dynamics of the exact switched system. A digital control design based on DSMC theory is proposed. For the verification of the proposed approach, the discrete-time sliding-mode controller of a boost converters loaded by a CPL is implemented using a digital signal processor (DSP).

The rest of this paper is organized as follows: In Section 2, the destabilizing negative impedance effect of the CPL are revisited. In Section 3, a discrete-time model of the boost converter loaded by a CPL, suitable for digital controller design, is derived. The digital control law that is based on current-mode control with constant current reference using discrete-time sliding-mode approach is derived in Section 4 demonstrating that the resulting system is unstable. Then, a two-loop control strategy is adopted in Section 5 where an outer voltage loop is added. Stability analysis of the closed-loop system is presented in Section 6. Thereafter, Section 7 presents simulation results illustrating the performance of the proposed control approach and illustrating the importance of working under sliding-mode regime for inrush current limitation. Experimental results are given in Section 8. The paper is summarized in the last section where concluding remarks are drawn.

2. The Destabilizing Effect Associated with the Negative Impedance Due to the CPL Behavior

CPLs do not exist in nature but their behavior arises in switching converters feeding either other downstream converters or motor drives. A block diagram of a CPL is shown in Figure 2a and its voltage-current characteristic is shown in Figure 2b. Often, the CPL model considered is a static nonlinear current sink whose power is constant. This is because an increase in the CPL voltage results in a decrease in its current and the hence the product of both variables is kept constant [1,9,11,12]. The equation describing the current through a CPL in terms of the voltage across it is given by the following expression [14]:

$$i_o = \frac{P}{v_o}. \tag{1}$$

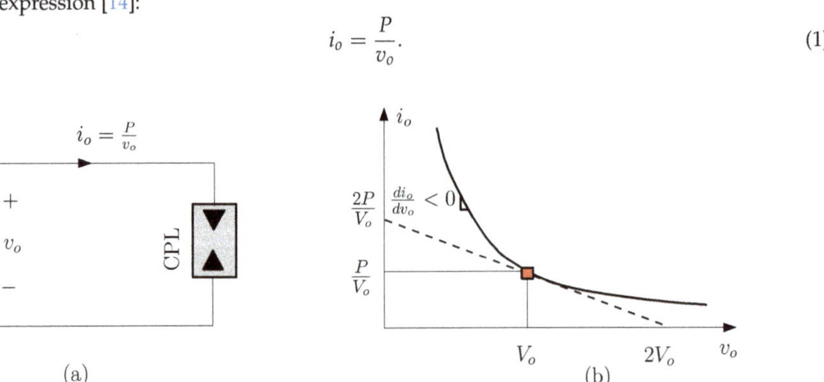

Figure 2. Instantaneous constant power load consisting of a nonlinear current sink. (**a**) Schematic circuit diagram of a CPL; (**b**) voltage-current characteristic showing a negative conductance/resistance.

In linear terms, the local resistance of the CPL can be obtained by linearizing the previous equation in the vicinity of an operating point $(V_o, P/V_o)$ established by the load converter or the motor drive system. Close to this operating point, a CPL can be described by the following linearized model [14]

$$i_o \approx \frac{P}{V_o} + \frac{v_o - V_o}{R_o}.$$ (2)

where $R_o = \left(\frac{\partial i_o}{\partial v_o}\right)^{-1}$. Therefore, the model of the CPL in the vicinity of the operating point can be represented by a straight line that is tangent to the nonlinear hyperbolic curve at the operating point with a negative slope equal to R_o^{-1} in the voltage-current space. The equation for this line is given by (2). This model represents a current source I_0 in parallel with the negative resistance R_o and these are given by the following equations:

$$R_o = -\frac{V_o^2}{P}, \quad I_0 = \frac{2P}{V_o}$$ (3)

As a result, although the local resistance at a certain point is positive, the corresponding incremental resistance R_o is negative and this is known to produce instabilities to the system to which the CPL is connected [1,7,11]. As an example, consider the first-order nonlinear network depicted in Figure 3 which could represent an approximate circuit model for a boost converter under current-mode control and loaded by a CPL. The current at the input port is imposed to be the current limit $I_{\lim}$ and this port can be considered as a constant power source (CPS) whose power is totally delivered to the nonlinear network. This power source models how this ideal power transfer takes place from the input port to the output port.

From KCL, the equation describing the dynamic behavior of the network of Figure 3 can be written in the following form:

$$C\frac{dv_o}{dt} = \frac{\delta P}{v_o},$$ (4)

where $\delta P = P_g - P$, P being the power of the CPL and $P_g = V_g I_r$ the power delivered from the input voltage generator.

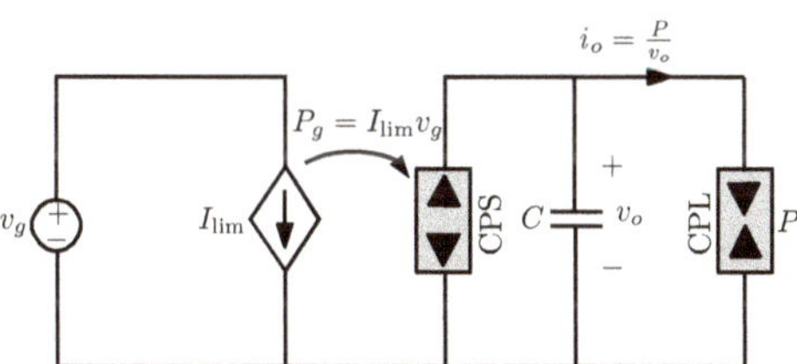

Figure 3. An electrical network loaded by a CPL.

For this particular system, the response can be derived mathematically without linearization. Indeed, by making the change of variable $x(t) = v_o^2(t)$, the previous equation can be expressed as follows:

$$\frac{dx}{dt} = 2\frac{\delta P}{C},$$ (5)

whose time-domain solution is given by the following expression:

$$x(t) = x(0) + 2\frac{\delta P}{C}t,$$ (6)

or equivalently in terms of the original state variable v_o,

$$v_o(t) = \sqrt{v_o^2(0) + 2\frac{\delta P}{C}t} \tag{7}$$

Three different cases arise depending on δP. These are:

- $\delta P > 0$, the response is unbounded, and therefore the system is unstable.
- $\delta P = 0$, the response is bounded but present an infinite number of equilibria depending on the initial condition $v_o(0)$. Indeed, in this case, one has $v_o(t) = v_o(0)\ \forall t$.
- $\delta P < 0$, the response collapses at a certain time instant t_c given by

$$t_c = -\frac{C}{2\delta P}x(0) \tag{8}$$

At this time instant, the voltage v_o across the CPL becomes zero and its current becomes infinite. For $t > t_c$, no real solution exists for the network equation. Figure 4 shows the responses corresponding to the previous three different cases.

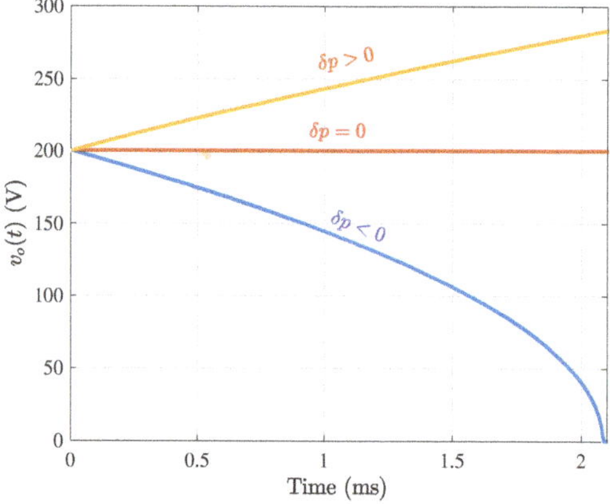

Figure 4. Response of the nonlinear electrical network of Figure 3. $C = 20.8\ \mu\text{F}$. $v_o[0] = v_g = 200\ \text{V}$. The current I_{lim} has been varied in the set $\{4, 5, 6\}$ A to consider the cases $\delta P = -200\ \text{W} < 0$ ($I_{\text{lim}} = 4$ A), $\delta P = 0$ ($I_{\text{lim}} = 5$ A) and $\delta P = 200\ \text{W} > 0$ ($I_{\text{lim}} = 6$ A). The collapse time in the case of $\delta P = -200\ \text{W} < 0$ is $t_c \approx 2.1$ ms in perfect agreement with (8).

3. Discrete-Time Modeling of a Boost Converter Loaded by a CPL

3.1. System Description

The results presented in this section and the sections coming later correspond to the boost converter depicted in Figure 5. However, the same approach can be applied to other converter topologies. The aim of the digital controller is to provide the suitable duty cycle for ensuring output voltage regulation and inducing sliding-mode regime in discrete-time. For that, the variables needed for the synthesis of the controller, are first converted into digital signals using analog-digital converters (ADC) at the rate of the sampling frequency and then processed by the controller. Selecting a proper sampling rate is important. Multi-sampling is a recently used approach in switching converters resulting in a sampling frequency larger than the switching frequency. This possibly will lead to unnecessarily overloading the digital processor. On the other hand, there are also some approaches using a sampling frequency shorter than the switching frequency. With this approach the controller will possibly miss dynamics of the power stage downgrading the performance of the closed-loop system. In switching converters applications, the duty cycle is updated once per switching period, for acceptable performance, it is quite appropriate to select the sampling frequency equal to the switching frequency which lead to a good compromise between accuracy and computing efficiency.

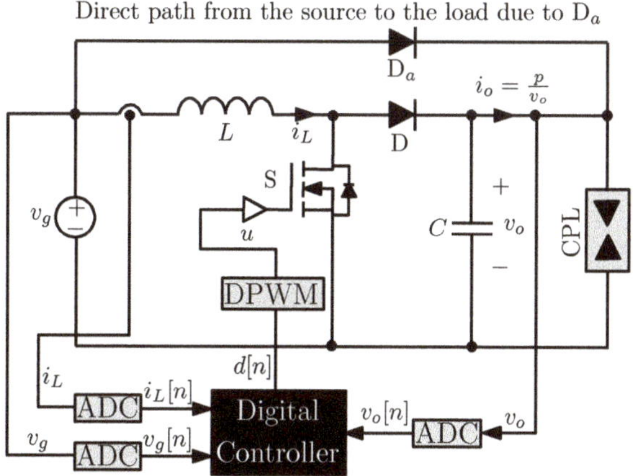

Figure 5. Schematic circuit diagram of a boost converter with CPL.

The regulation of the output voltage v_o is required since disturbances in the output power p and the input voltage v_g can take place. What is more, with CPL, closing the output voltage loop is also necessary for stabilizing the system since with voltage loop open it is unstable as will be shown later. The regulation can be accomplished by an outer voltage loop making the current reference $i_{\text{ref}}[n]$ to be the output of this loop. Figure 6 shows a double-loop control scheme, which is used for output voltage regulation while performing current limiting. The voltage controller consists of two stages. Namely, a digital PI block to process the error $\varepsilon[n] := V_{\text{ref}} - v_o[n]$ and a limiter to avoid that the current reference overpasses an admissible level. The current controller is based on a DSMC strategy to be described below. This controller together with the digital PWM (DPWM) directly provides the duty cycle of the driving signal u of the converter MOSFET. The auxiliary diode D_a in the power stage of Figure 5 is usually not present in the conventional DC-DC boost converter topology. It has been added to create a unidirectional path from the source to the load hence guaranteeing the condition $v_o = v_g$ at the starting time and this helps to minimize the effect of the inrush current in the inductor as will be detailed later. It is worth noting that this condition is also required for the model of the load converter as a CPL to be valid. Indeed, the ideal static behavior of a CPL corresponds to the actual power absorbed by a converter acting as a load in a cascade connection with the source converter. However, while the tightly regulated load converter can be considered locally as a CPL for the source converter, this model is not valid for all the operating range of the interconnected system. In fact, the voltage drop across an ideal current sink (the CPL) is not defined unless there is some voltage applied to it. In [7], a small voltage limit is used as a threshold value for the load voltage to decide if an ideal static CPL behavior or an open circuit behavior must be used when performing numerical simulations. In cascaded converters, this voltage is the output of the source converter which cannot be tightly regulated in the case of a boost converter unless it is larger than the input voltage of the same converter. The requirement of output voltage at least equal to the input voltage in this converter can be easily met in the presence of the auxiliary diode D_a in the boost converter power stage as shown in Figure 5. This diode guarantees that $v_o \geq v_g \; \forall t$ in a normal operation of the boost converter.

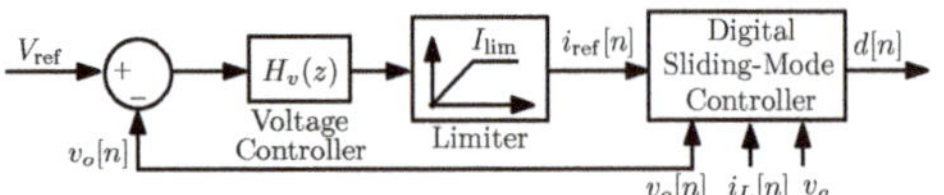

Figure 6. Control scheme with a two-loop voltage regulation.

3.2. Discrete-Time Mathematical Modeling

The development of a digital controller using DSMC concepts is performed in a discrete-time state-space formulation. Hence, for DSMC design, a discrete-time model of the power stage is first needed. Due to the presence of the nonlinear CPL, such a model cannot be exactly obtained in closed form. This is because, in contrast to switching converters with linear loads, the differential equations of the system for each switch position are nonlinear and cannot be solved in closed form. To overcome this handicap, we deal with the problem approximately by discretizing the averaged model which can be written as follows

$$\frac{di_L}{dt} = -\frac{v_o}{L}(1-d) + \frac{v_g}{L} \tag{9}$$

$$\frac{dv_o}{dt} = -\frac{p}{Cv_o} + \frac{i_L}{C}(1-d) \tag{10}$$

where d is the duty cycle, all parameters and variables appearing in (9)–(10) can be identified in Figure 5. The key issue in the discrete model is the nonlinear differential Equation (10) associated with the dynamics of the capacitor in parallel with the CPL. The discrete-time model corresponding to this equation cannot be obtained in closed form. Different approaches can be used for obtaining an approximate discrete-time model. These are the Euler forward, Euler backward, and the Tustin (trapezoidal) methods [41]. For sufficiently small switching/sampling period, all these approximations yield similar results. For the sake of simplicity, let us choose the Euler forward approach for obtaining the discrete-time model. A discrete-time model can be obtained by approximating the continuous-time derivatives by their equivalent rate of change, hence assuming in the averaged model (9)–(10) that $di_L/dt \approx (i_L[n+1] - i_L[n])/T$ and that $dv_o/dt \approx (v_o[n+1] - v_o[n])/T$, where T is the switching/sampling period. The previous forward Euler approximation leads to the following discrete-time model of the system:

$$i_L[n+1] \;=\; i_L[n] + \frac{T}{L}(v_g - v_o[n]) + \frac{T}{L}v_o[n]d[n] \tag{11}$$

$$v_o[n+1] \;=\; v_o[n] - \frac{T}{C}i_L[n]d[n] + \frac{T}{C}\left(i_L[n] - \frac{p}{v_o[n]}\right) \tag{12}$$

Notice that with a constant duty cycle value $d[n] = D$ (open-loop operation), the coordinates of the equilibrium point are $V_o = V_g/(1-D)$ and $I_L = P/V_g$, where P and V_g are the nominal values of the power p and the input voltage v_g. It should be noticed that the steady-state inductor current, in contrast to the case of resistive load, inductor current coordinate of the equilibrium point does not depend on the operating duty cycle and is only imposed by the power P of the CPL and the input voltage V_g. Moreover, the equilibrium point is unstable for all values of $D \in (0,1)$.

When sampling the state variables at the beginning of the switching cycle, depending on the modulation strategy, the samples in (11)–(12) can correspond to the peak values (leading-edge modulation), the valley values (trailing-edge modulation) or the average values (double-edge modulation). Here, a double-edge modulation will be used and the samples at the starting of each switching period will coincide with the averages.

3.3. Open-Loop Model Validation

We will show below that the approximate discrete-time model (11)–(12) is enough accurate for control design. The results from this model are compared with those from the circuit-level switched model implemented in PSIM© software. Figure 7 shows the samples of the capacitor voltage and the inductor current obtained from (11)–(12) and the waveforms of the same variables obtained from the switched model. As can be observed, there is a good agreement between the results and therefore (11)–(12) can be faithfully used for digital control purposes. Please note that the dynamics of the inductor current is accurately predicted, and that the approximation only induces a relatively small loss of accuracy in predicting the samples of the output capacitor voltage. The small deviation can be perfectly compensated by the controller imposing the closed-loop poles at a desired position.

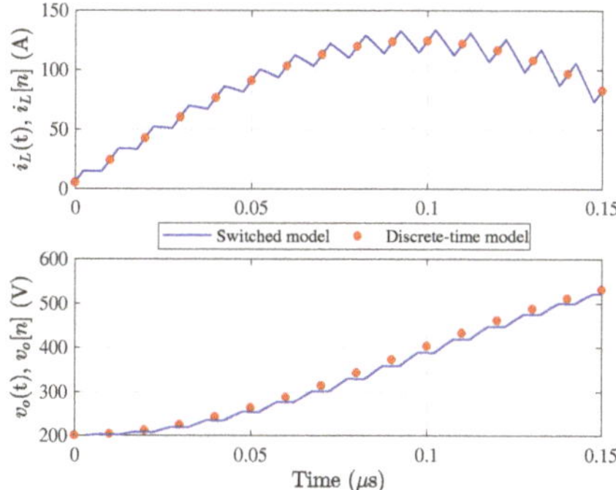

Figure 7. Comparison between the evolution of the state variables from the approximate discrete-time model and from the switched model implemented in PSIM© software. $V_g = 200$ V, $T = 10$ µs, $L = 326$ µH and $d[n] = D = 0.5$.

4. Digital Sliding-Mode Inner Loop Control Design

4.1. Large-Signal Model with Voltage Loop Open

Let $\mathbf{x} = (i_L, \quad v_o)^\mathsf{T}$ be the vector of the state variables of the power stage circuit. With the aim to control the samples of the inductor current $i_L[n]$ to their desired reference $i_{\text{ref}}[n]$, the following discrete-time sliding surface is used

$$\Sigma = \{\mathbf{x}|\sigma[n] := i_{\text{ref}}[n] - i_L[n] = 0\} \tag{13}$$

When the voltage loop is open, the current reference $i_{\text{ref}}[n]$ is given in a fixed pattern i.e., without any feedback loop. Although this is a not a normal operation of the converter, the situation arises during startup while limiting the inrush current and the current reference remains constant at a certain constant limit I_{lim} during this phase.

4.2. Equivalent Control

In a fixed-frequency DSMC of switching converters, the control variable $d[n]$ during a certain switching period is selected in such a way that the controlled variable is imposed to catch its reference one period later. Therefore, the duty cycle $d[n]$ is obtained by imposing the discrete-time sliding-mode condition $\sigma[n+1] = 0$ in (13) and solving for $d[n]$. In doing so, the following expression for the duty cycle (equivalent control) is obtained

$$d_e[n] = \frac{L(i_{\text{ref}}[n+1] - i_L[n])}{Tv_o[n]} + \frac{v_o[n] - v_g[n]}{v_o[n]} \tag{14}$$

The value of the duty cycle is constrained within the interval (0,1) and the effective expression of the duty cycle becomes

$$d[n] = \text{sat}(d_e[n]) \tag{15}$$

where $\text{sat}(\cdot)$ stands for the saturation function defined by:

$$\text{sat}(x) = \frac{1}{2}\left(1 + |x| - |x - 1|\right) \tag{16}$$

The saturation will not take place whenever $0 < d_e[n] < 1$ requiring the following condition to be satisfied:

$$i_{\text{ref}}[n+1] - \frac{Tv_g[n]}{L} < i_L[n] < i_{\text{ref}}[n+1] + \frac{T(v_o[n] - v_g[n])}{L} \tag{17}$$

At the initial time ($n = 0$) without the presence of the diode D_a, the previous constraints have no solutions and the system may have serious problems to startup. With initial conditions $i_L(0) = 0$ and $v_o[0] = v_g[0]$ (presence of diode D_a), the previous condition becomes

$$i_{\text{ref}}[1] < \frac{Tv_g[0]}{L} \tag{18}$$

If the previous constraint is not fulfilled, the system can startup easily, but the duty cycle will be saturated during a few switching cycles.

4.3. Comparison with State-of-Art Digital Predictive Control

Before continuing our study on the DSMC of the boost converter loaded by a CPL, we will present a short comparison with the Digital Predictive Control (DPC) published in [42]. To make the comparison clear and easy to follow, let us consider that the current reference i_{ref}, the output voltage v_o and the input voltage reference v_g are constant ($v_o[n] = V_o$ and $v_g[n] = V_g$ $\forall n \in \mathbb{N}$). Please note that these are the same operating conditions used in [42] when deriving the control law. The fixed-frequency DSMC will be later applied separately using the full-order model of a boost converter loaded by a CPL. When applied to the DC-DC boost converter, the duty cycle in the case of DPC is given by the following expression:

$$d[n] = 2D - d[n-1] + \frac{L(i_{\text{ref}} - i_L[n-1])}{TV_o} \tag{19}$$

Under the same conditions, from (14), the expression for the duty cycle (equivalent control in (14)) for the case of the DSMC becomes

$$d[n] = D + \frac{L(i_{\text{ref}} - i_L[n])}{TV_o} \tag{20}$$

where in both cases $D = (V_o - V_g)/V_o$ is the steady-state value of the duty cycle $d[n]$. There is a fundamental difference between the two control laws. While it can be observed that the duty cycle $d[n]$ synthesized by the fixed-frequency DSMC approach is generated according to a *static* control law in which, at a certain switching cycle, its value is given directly in terms of the samples of the state variables and parameters at the same cycle, it is not the case of (19) which describes a *dynamic* control law for synthesizing the duty cycle $d[n]$, hence increasing the total order of the system by one. The equivalent control law based on the DSMC is *static* and does not change the order of the system. In the DSMC case, the current reference i_{ref} will be caught in one cycle by the inductor current $i_L[n]$ assuming the output voltage constant while with the DPC two switching cycles are needed for the reference to be reached by the inductor current under the same operating conditions. Indeed, by linearizing the closed-loop system corresponding to DSMC and DPC, both approaches correspond to a *dead-beat* response. Notice that in linear digital control theory, the *dead-beat* control consists of finding the control law that when applied to a system, it brings the output to the steady state in several sampling cycles equal to the order of the system. To achieve this target, the control law places all the poles of the closed-loop transfer function at the origin of the z-plane.

Let I_L and D be the nominal steady-state values of $i_L[n]$ and $d[n]$ respectively. Let $\hat{i}_L[n] = i_L[n] - I_L$, $\hat{d}[n] = d[n] - D$. The linearized model of the closed-loop boost converter under the DSMC is given by:

$$\hat{i}_L[n+1] = 0 \tag{21}$$

This is a 1-dimensional system with a pole at 0 and the current reference will be reached in one switching period. The linearized model of the closed-loop boost converter under the DPC is given by:

$$\hat{i}_L[n+1] \;=\; \hat{i}_L[n] + \frac{T}{L}V_o\hat{d}[n] \tag{22}$$

$$\hat{d}[n+1] \;=\; -\hat{d}[n] - \frac{L}{TV_o}\hat{i}_L[n] \tag{23}$$

The Jacobian matrix corresponding to the previous model can be expressed as follows

$$\mathbf{J}_p = \begin{pmatrix} 1 & \dfrac{TV_o}{L} \\ -\dfrac{L}{TV_o} & -1 \end{pmatrix} \tag{24}$$

The eigenvalues of $\mathbf{J}_p$ are the closed-loop poles of the boost converter under the DPC and are both located at 0. Therefore, with this control, the current reference will be reached in two switching periods. The numerical simulations depicted in Figure 8 shows the inductor current responses of the system under the DPC and the DSMC in front of positive and negative step changes in the current reference i_{ref} between 5 A to 10 A. For ease of comparison, only the dynamics of the inductor current is conserved while the output voltage was fixed to $V_o = 380$ V with $L =326$ µH and $f_s = 100$ kHz and $V_g = 200$ V. These responses confirm the previous theoretical remarks about the *dead-beat* nature of the responses of the two control strategies.

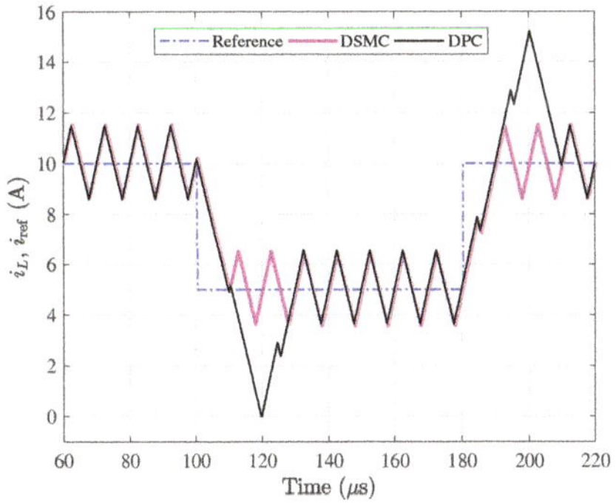

Figure 8. Comparison between the inductor current response corresponding to the predictive control and the DSMC. Under the DSMC, the inductor current reaches its reference in one switching/ sampling cycle. Under the predictive control, the inductor current reaches its reference in two switching/ sampling cycles.

4.4. DSMC Design

To guarantee convergence of the trajectories of the system to the sliding surface Σ, the following reaching conditions must hold

$$\sigma[n+1] \leq 0 \text{ if } \Delta\sigma[n] \geq 0 \tag{25}$$

$$\sigma[n+1] \geq 0 \text{ if } \Delta\sigma[n] \leq 0 \tag{26}$$

where $\Delta\sigma[n] := \sigma[n+1] - \sigma[n]$ is the increment in function variable $\sigma[n]$ during one switching cycle which can be obtained as follows

$$\Delta\sigma[n] = i_{\text{ref}}[n+1] - i_{\text{ref}}[n] + \frac{T}{L}((1-d[n])v_o[n] - v_g[n]) \tag{27}$$

Accordingly, (25)–(26) become as follows

$$\frac{v_g[n] - (1-d[n])v_o[n]}{L} \leq \frac{i_{\text{ref}}[n+1] - i_{\text{ref}}[n]}{T} \leq \frac{v_g[n]}{L} \tag{28}$$

The above inequalities mean that the reference current rate change must be bounded between the negative and the positive slopes of the inductor current during the charging and the discharging time intervals, respectively. In steady-state operation, these conditions are easily met. However, they can be violated during startup or during transient due to abrupt changes. The loss of sliding-mode operation could lead performance degradation manifested by either large overshoots or slow response.

When $i_{\text{ref}}[n+1] = i_{\text{ref}}[n]$ (either constant or T-periodic), (28) becomes $0 \leq v_g \leq v_o$. In this case, starting from zero initial conditions and without the presence of the auxiliary diode D_a (See Figure 5), sliding condition will not be fulfilled at startup. With the presence of the auxiliary diode D_a, the condition $v_o = v_g$ is guaranteed from the beginning, the system starts in sliding-mode and immediately the system trajectory is constrained in the discrete sliding-mode domain defined by the constraint $i_{\text{ref}}[n] - i_L[n] = 0$. The worse cases take place when the duty cycle is saturated. For $d[n] = 0$, $\Delta\sigma[n] = \frac{T}{L}(v_o[n] - v_g[n]) \geq 0$ (if $v_o[n] \geq v_g[n]$) and $\sigma[n+1] \leq 0$. For $d[n] = 1$, $\Delta\sigma[n] = -\frac{T}{L}v_g[n] \leq 0$ and $\sigma[n+1] \geq 0$, which ensures the convergence to the switching surface. It will be shown later that at startup, the current reference could be saturated, the voltage loop becomes open and the resulting system is unstable making the voltage v_o to increase above v_g hence guaranteeing the sliding-mode condition. When the output voltage reaches the vicinity of its desired value, saturation disappears and the PI regulator starts regulating the output voltage to its desired value according to the imposed performance by the outer loop controller design.

4.5. Control-Oriented Full-Order Discrete-Time Small-Signal Model

As stated before, in linear terms, the current loop presents a *dead-beat* response under the fixed-frequency DSMC and the DPC. However, the interaction with other state variables can only be revealed by considering the full-order model of the system. The focus on this paper is on the DSMC. Similar procedures can be followed for studying the DPC.

By substituting the expression of the equivalent control (14) in (12) and imposing the discrete-time sliding-mode constraint $i_L[n] = i_{\text{ref}}[n]$ imposed by (13), one obtains the following equation describing the output capacitor voltage v_o in the discrete-time domain

$$v_o[n+1] = f_v(v_o[n], i_{\text{ref}}[n], i_{\text{ref}}[n+1], v_g[n], p[n]) \tag{29}$$

where the function f_v is given by the following expression

$$f_v(v_o[n], i_{\text{ref}}[n], i_{\text{ref}}[n+1], v_g[n], p[n]) = v_o[n] + i_{\text{ref}}[n]\frac{L(i_{\text{ref}}[n] - i_{\text{ref}}[n+1]) + Tv_g[n]}{Cv_o[n]} - T\frac{p[n]}{Cv_o[n]} \tag{30}$$

The equilibrium point of the discrete-time dynamical system described by (29)–(30) can be obtained by imposing $v_o[n+1] = v_o[n]$ and $i_{\text{ref}}[n+1] = i_{\text{ref}}[n]$ in the same equations. Imposing these constraints implies that $v_o[n]$ takes an infinite value unless the current reference $i_{\text{ref}}[n]$ is chosen to be exactly equal to $p[n]/v_g[n]$. Indeed, this is the only inductor current value that correspond to a balance between the input power delivered by the voltage source and the output power imposed by the CPL. Therefore, during startup and while the system is under inrush current limiting phase, it is feeding a CPL with a constant current different from the one that balances input and output powers in the system and this explains the output voltage divergence in a similar way to the first case of the nonlinear network of Section 2. It is worth noting that for inrush current limitation during startup, the converter will unavoidably work under this condition. The output voltage will collapse if the current reference is smaller than P/V_g. This fact appears in a clear contrast with the case of resistive load for which the voltage reaches a finite value in steady state when the system is under pure current-mode control [38]. The case studied here is similar to the analog control based on the average inductor current regulation in a buck converter loaded by a CPL, which is still unstable after the introduction of the current control loop [3]. In both cases, the introduction of an outer voltage loop will contribute to the global stabilization of the system apart from ensuring output voltage regulation. It should also be noted that when an outer loop is added to stabilize the output voltage while establishing the current reference, this current will be imposed to be $I_{ref} = P/V_g$ in steady state regardless the desired reference value V_{ref} of the output voltage v_o.

Let V_o and I_{ref} be the nominal steady-state values of $v_o[n]$ and $i_{ref}[n]$ respectively. Let $\hat{v}[n] = v_o[n] - V_o$, $\hat{p}[n] = p[n] - P$, $\hat{v}_g[n] = v_g[n] - V_g$, $\hat{i}_{ref}[n] = i_{ref}[n] - I_{ref}$ the small deviations of the output voltage $v_o[n]$, the input voltage $v_g[n]$, the current reference $i_{ref}[n]$ and the power $p[n]$ with respect to their steady-state values V_{ref}, V_g, I_{ref} and P respectively. Therefore, the small-signal model of the system under current-mode control can be written as follows:

$$\hat{v}_o[n+1] = \frac{\partial f_v}{\partial v_o[n]}\hat{v}_o[n] + \frac{\partial f_v}{\partial i_{ref}[n]}\hat{i}_{ref}[n] + \frac{\partial f_v}{\partial i_{ref}[n+1]}\hat{i}_{ref}[n+1] + \frac{\partial f_v}{\partial v_g[n]}\hat{v}_g[n] + \frac{\partial f_v}{\partial p[n]}\hat{p}[n] \qquad (31)$$

The different partial derivatives appearing in (31) are

$$\frac{\partial f_v}{\partial v_o[n]} = 1 + \frac{T}{Cv_o^2[n]}(P - I_{ref}V_g), \qquad \frac{\partial f_v}{\partial i_{ref}[n]} = \frac{TV_g - LI_{ref}}{Cv_o[n]}$$

$$\frac{\partial f_v}{\partial i_{ref}[n+1]} = \frac{-LI_{ref}}{Cv_o[n]}, \qquad \frac{\partial f_v}{\partial v_g[n]} = \frac{TI_{ref}}{Cv_o[n]}, \qquad \frac{\partial f_v}{\partial p[n]} = -\frac{T}{Cv_o}$$

With abuse of notation, let $\hat{v}_o(z)$, $\hat{v}_g(z)$, $\hat{p}(z)$ and $\hat{i}_{ref}$ be the z-transforms of $v_o[n]$, $v_g[n]$, $p[n]$ and i_{ref} respectively.

Taking the $z-$ transform of (31), the i_{ref}-to-v_o, the v_g-to-v_o and the p-to-v_o small-signal transfer functions of the digital sliding current-mode-controlled boost converter with voltage loop open and supplying a constant power can be expressed as follows

$$H_i(z) = \frac{\hat{v}_o(z)}{\hat{I}_{ref}(z)} = -R_i\frac{z - z_c}{z - z_p}, \qquad (32)$$

$$H_g(z) = \frac{\hat{v}_o(z)}{\hat{V}_g(z)} = \frac{I_{ref}T}{CV_o}\frac{1}{z - z_p}, \qquad (33)$$

$$H_p(z) = \frac{\hat{v}_o(z)}{\hat{P}(z)} = -\frac{T}{CV_o}\frac{1}{z - z_p} \qquad (34)$$

where R_i, z_c and z_p are given by

$$R_i = \frac{LI_{ref}}{CV_o}, \qquad z_c = 1 + \frac{TV_g}{I_{ref}L}, \qquad z_p = 1 + \frac{T}{CV_o^2}(I_{ref}V_g - P) \qquad (35)$$

The previous transfer functions represent the discrete-time small-signal model of the boost converter under an inner current control loop based on a DSMC strategy and can be used to design an outer digital voltage control loop in the $z-$domain. Please note that the zero z_c of the i_ref-to-v_0 transfer function $H_i(z)$ in (32) is outside the unit circle, which explains the well-known non-minimum phase characteristics of the control-to-output voltage transfer function in a boost converter. Also note that during the startup, the current reference i_ref will be limited by an upper bound I_lim which must be selected larger than the desired steady-state P/V_g, hence, the pole z_p of the previous transfer functions is outside the unit circle, which corresponds to an unstable system. Therefore, any designed controller must stabilize the system while regulating the output voltage and exhibiting desired performance in terms of disturbance rejection and transient response.

5. Digital Control for Output Voltage Regulation

To ensure an output voltage regulation, an outer and slower control loop in cascade with the inner DSMC current loop must be added. This loop is designed in the $z-$domain based on the i_ref-to-v_0 transfer function $H_i(z)$ in (32) representing the small-signal model around a desired operating point. This second control loop will regulate the output voltage to a desired value V_ref. The steady-state current reference I_ref will be equal to P/V_g regardless the value of V_ref as mentioned before. To stabilize the system, a two-loop control strategy will be used, hence the outer voltage loop provides the reference for the inner current loop. Let $\varepsilon[n] = V_\text{ref} - v_0[n]$ be the output voltage error. The current reference is updated from the output of a digital PI compensator as follows

$$i_r[n] = K_p\varepsilon[n] + q[n] \tag{36}$$

where $q[n] = K_i \sum_{k=0}^{n} \varepsilon[k]$ is the discrete-time accumulative sum of the error voltage weighted by the integral gain K_i being K_p the proportional gain. In order to avoid high inrush current in startup, the current reference must be limited and the final expression for the current reference becomes

$$i_\text{ref}[n] = \begin{cases} i_r[n] & \text{if} & i_r[n] < I_\text{lim} \\ I_\text{lim} & \text{if} & i_r[n] \geq I_\text{lim} \end{cases} \tag{37}$$

Different approaches can be often used for integral control emulation. For sufficiently small switching/sampling period and all the approaches yield a controller which produces a closed-loop behavior similar to the one provided by a continuous-time controller. For the sake of simplicity, let us choose the Euler forward approach for emulating the integrator. This approximation yields the following recurrence equation for the discrete-time integral variable:

$$\zeta[n+1] = \zeta[n] + K_i\varepsilon[n] \tag{38}$$

where K_i is the integral gain. To avoid windup phenomenon, the integral variable $\zeta[n]$ is also limited to an upper bound Z_lim and the expression of the variable $q[n]$ becomes as follows

$$q[n] = \begin{cases} \zeta[n] & \text{if } \zeta[n] < Z_\text{lim} \\ Z_\text{lim} & \text{if } \zeta[n] \geq Z_\text{lim} \end{cases} \tag{39}$$

The presence of an advanced sample of the current reference $i_\text{ref}[n+1]$ in the expression of the control law (14) makes it challenging to obtain this law when the reference is to be provided by a feedback loop. A possible solution is to use a predictive approach to get the value of the reference current from (36) one switching period ahead of time using (12). While this would work theoretically and in simulation, it would require increased computational resources in an experimental digital platform such as a DSP. A much simpler solution is to redefine the discrete-time sliding-mode surface as follows:

$$\sigma[n] = i_\text{ref}[n-1] - i_L[n] \tag{40}$$

and the resulting expression of the duty cycle becomes

$$d[n] = \text{sat}\left(\frac{L(i_{\text{ref}}[n] - i_L[n])}{Tv_o[n]} + \frac{v_o[n] - v_g[n]}{v_o[n]}\right) \tag{41}$$

The first term in the expression of $d[n]$ in (41) is non null only in the reaching phase. Once the sliding-mode regime is reached, this term becomes zero and only the second term forces the system to evolve toward the equilibrium point if the stability of the closed-loop system is ensured.

Figure 9 shows a block diagram of the large-signal model of the system with a two-loop control based on DSMC. The presence of a discrete-time integrator in the external voltage loop will impose that in steady state $V_o := v_o[\infty] = V_{\text{ref}}$. Furthermore, in steady-state one will have $Q := q[\infty] = i_L[\infty]$ and $I_L := i_L[\infty] = \frac{P}{V_g}$. Therefore, the coordinates of the equilibrium point are

$$I_L = \frac{P}{V_g}, \quad V_o = V_{\text{ref}}, \quad Q = I_L = \frac{P}{V_g} \tag{42}$$

where I_L, V_o and Q stand for the steady-state values of the state variables i_L, v_o and q respectively.

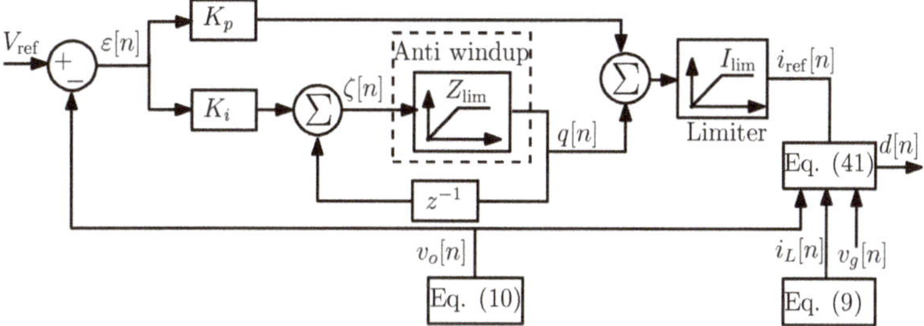

Figure 9. Block diagram of the large-signal model of the system with a double control loop based on the proposed DSMC.

6. Design of the Output Voltage Feedback Loop Using the Root-Locus Technique

The block diagram corresponding to (31) is depicted in Figure 10. The small-signal model can be used to design the feedback compensator to obtain a stable closed-loop system with a regulated output voltage.

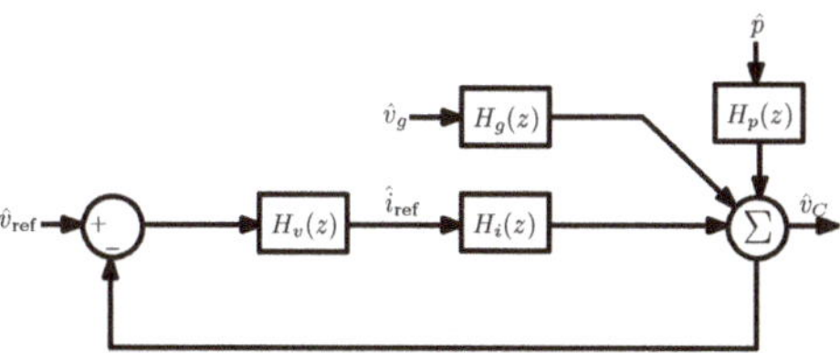

Figure 10. Block diagram of the $z-$domain small-signal model.

The focus in this section is on the design of the output voltage regulator. The time response characteristics are related to closed-loop pole locations. Hence, a design based on root-locus approach will be used. The aim is to design a controller such that the dominant closed-loop poles have a desired damping ratio ζ and a settling time t_s. According to (36) and (38), the transfer function for the outer digital PI voltage controller can be expressed as follows

$$H_v(z) = K_p + \frac{K_i}{z-1} \equiv K_p \frac{z - z_{pi}}{z-1} \tag{43}$$

where $z_{pi} = 1 - K_i/K_p$ is the zero introduced by the digital PI compensator. The loop gain of the system is

$$\mathcal{L}(z) = H_i(z)H_v(z) = -K_p R_i \frac{(z - z_{pi})(z - z_c)}{z(z-1)(z - z_p)} \tag{44}$$

Please note that because of the one cycle delay present in the current reference in (40), a pole at the origin is added to the loop gain.

Although (44) can be used to obtain numerically the suitable parameter values for the desired pole position, it is always more useful to have an explicit mathematical expression. For many applications, the feedback gain K_p is a design parameter that should be adjusted accordingly to the values of other parameters to get a system response with the desired performance. The purpose in this section is to perform an analytical study by carrying out a realistic approximation. Unfortunately, there is no universal procedure to approximate the loop gain. To simplify the design, the integral gain $K_i = K_p(1 - z_{pi})$ can be appropriately selected so that the zero of the PI controller is placed slightly smaller than 1. This means that the term K_i/K_p must be selected much smaller than 1 and the loop gain can be approximated by

$$\mathcal{L}(z) \approx -K_p R_i \frac{z - z_c}{z(z - z_p)}, \tag{45}$$

The approximate closed-loop characteristic polynomial equation can be expressed as follows

$$1 + \mathcal{L}(z) = 0 \equiv z^2 - (K_p R_i + z_p)z + K_p R_i z_c = 0 \tag{46}$$

The closed-loop poles can be selected at the break-away point z_{ba} on the real axis to correspond to a damping factor $\zeta = 1$ and a settling time $t_s = -4T/\ln|z_{ba}|$. For finding the break-away points, one must find the value of $z = z_{ba}$ that maximizes or minimizes the gain K_p [41] hence obtaining the following approximate expression for the values of the break-away points and the corresponding proportional gain of the PI controller

$$z_{ba} \approx z_c \pm \sqrt{z_c^2 - z_p z_c}, \quad K_{p,ba} \approx \frac{(z_{ba} - z_p)z_{ba}}{R_i(z_{ba} - z_c)} \tag{47}$$

The value of z_{ba} with positive sign is omitted because it corresponds to a break-in point outside the unit circle leading to an unstable system.

7. Numerical Simulations and Model Validation

7.1. System Startup and Steady-State Operation

The initial value of the duty cycle can be obtained from the initial values of the state variables. Usually, the obtained value at startup is saturated. With delay, the number of initial saturated cycles increases since the inductor is continuously charged during a few cycles leading to an increase of the inrush current.

7.1.1. Validation of the Closed-Loop Large-Signal Model and Guaranteeing the Sliding-Mode Operation

Let us consider the nominal values of the power stage parameters, the desired output voltage and the switching frequency depicted in Table 1. The steady state of the current reference is $I_{ref} = P/V_g = 5$ A. During the startup, the current reference i_{ref} and the integral variable q were limited to $I_{lim} = 10$ A and $Q = 10$ A, respectively. The PI zero is selected at $z_c = 0.95$. First, the root locus of the closed-loop system is obtained and the result is depicted in Figure 11. Let us select the closed-loop poles at the break-away point $z_{ba} \approx 0.62 + 0j$ which corresponds to a proportional gain $K_p \approx 0.82$, damping coefficient $\zeta = 1$.

The performance of the DSMC will be validated by means of numerical simulations from both the derived large-signal discrete-time model and from a detailed switched model implemented in PSIM$^©$ software.

Table 1. The used parameter values for the system.

L	C	P	v_g	V_{ref}	f_s
326 μH	20.8 μF	1 kW	200 V	380 V	100 kHz

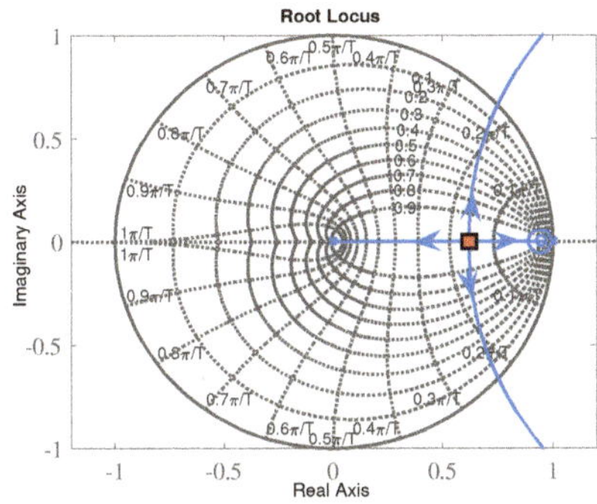

Figure 11. Root locus of the system. At the double pole position $z_{ba} \approx 0.62 + 0j$ marked by a square on the real axis, the gain $K_p \approx 0.82$ leading theoretically to damping coefficient $\zeta = 1$ and null overshoot.

Figure 12a shows the startup and the steady-state responses of the system from both models. It can be observed that during startup the inductor current reference $i_{\text{ref}} = I_{\text{lim}}$ remains constant due to the saturation hence limiting the inrush current. As soon as the capacitor voltage reaches the vicinity of the voltage reference V_{ref}, the PI controller comes to play and the current reference is no longer constant but time varying, state-dependent and provided by the PI compensator according to (36). The data from the discrete-time model are plotted together with the simulated data from the detailed switched model implemented in PSIM$^©$ software. It can be observed that the responses from the two models are very close. The voltage waveforms from the switched model and the discrete-time model cannot be distinguished from each other. Hence, the above simulations show that the large-signal model derived in this work can predict accurately the large-signal behavior of the system.

Remark 1. *Since during startup the average inductor current value is the regulated variable and it is supposed to reach the reference current in one cycle, the ripple of this variable can only exceed the limit I_{lim} by the switching ripple Δi_L given by*

$$\Delta i_L = \frac{T v_g (v_o - V_g)}{2 v_o L} \tag{48}$$

At the initial time, $v_o = V_g$ due to the presence of the auxiliary diode D_a and Δi_L should be zero as can be confirmed in Figure 12a. However, in a practical implementation of a digital controller, saturation of the duty cycle during the initial cycles and propagation delays always exist and it is expected that the average inductor current will still overpass the imposed current limit. The current ripple amplitude Δi_L from (48), superposed to the averaged current I_L is plotted in the bottom panel of Figure 12a together with the current waveforms obtained from the switched model. The agreement is remarkable both in the startup, in steady state and in the transient phase.

It is worth noting that one cycle delay inherently existing in the used commercial device has been eliminated by appropriately modifying the C code programming of the device. However, computation delay is unavoidable. By adding a computation delay $\tau_d = 5.5$ µs, the results depicted in Figure 12b are obtained where it can be observed that a small inrush current still exists in the current startup response. The value of the computation delay used is the one corresponding to the experimental prototype to be described later. The propagation delay makes higher the number of cycles during which the duty cycle is saturated making the sliding-mode condition not satisfied during these cycles which in turns lead to higher inrush current at startup. As a remedy of this problem, one can force the initial values of the duty cycle to lie within the interval (0,1) in a few cycles either by scaling down the value of the duty cycle obtained from the control law (41) or by limiting the rate of change of the reference current from zero to I_{lim} in startup. The last solution will be adopted later in the section related to the experimental validation of the results.

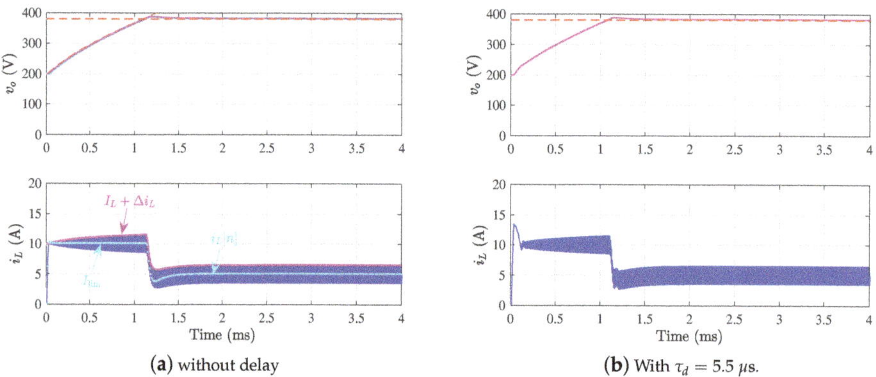

Figure 12. Startup and steady-state response from numerical simulations using a detailed switched model implemented in PSIM© and the discrete-time model.

By filtering out the high frequency component of the current reference at the abrupt change in startup, the inrush current can be suppressed. However, the presence of a filter also slows down the system response. Another way to limit the inrush current in the presence of computation delays without degrading the system response is by limiting the rate of change of the current reference during startup.

Remark 2. *The system is unstable when the saturation is taking place and the system is current-mode control with open voltage loop. During this phase, the output voltage tends to infinity although the average of the inductor current is theoretically well regulated to the maximum allowed current I_{lim}. During this phase, the system operates like the nonlinear first-order network shown in Figure 3. In the case of the boost converter, this type of instability only makes the output voltage to increase from the initial voltage V_g. This increase in the output voltage is desired since under this operation the system is approaching its desired operating point.*

7.1.2. The Importance of Operating in Sliding-Mode Regime for Inrush Current Limitation

To reveal the importance of different aspects in the controller, some cases are simulated below.

Figure 13 shows the startup of the system with the sliding-mode non-guaranteed at startup with two different cases. In the first case (Figure 13a), the auxiliary diode D_a is omitted and the system starts with zero initial condition. Without time delay, the system exhibits severe problems for starting-up. The inrush current in this case is very large even under current limitation. This is because since the sliding-mode operation is not guaranteed, the inductor current does not tightly track the limited current reference $I_{\lim}$. The mean reason is that since the output voltage is less than the input voltage, the driving signal is switched ON and it remains ON as long as $i_L[n] < I_{\lim}$ hence charging the inductor current. When $i_L[n] = I_{\lim}$ for the first time, the output voltage $v_o[n]$ is still smaller than the input voltage and sliding-mode condition is not yet satisfied. The output voltage increases, and the sliding-mode condition is satisfied as soon as the output voltage is equal to the input voltage. However, when this occurs, the inductor current has already reached a high value leading to an unacceptable inrush current. In the second case (Figure 13b), with the presence of the auxiliary diode D_a, the sliding-mode condition (28) is guaranteed at startup but the non-saturated reference generated by the outer voltage loop is very high and the inductor current well tracks it which is undesirable. In the simulation instead of not using a limit for the current reference, a high value ($I_{\lim} = 20$ A) is used.

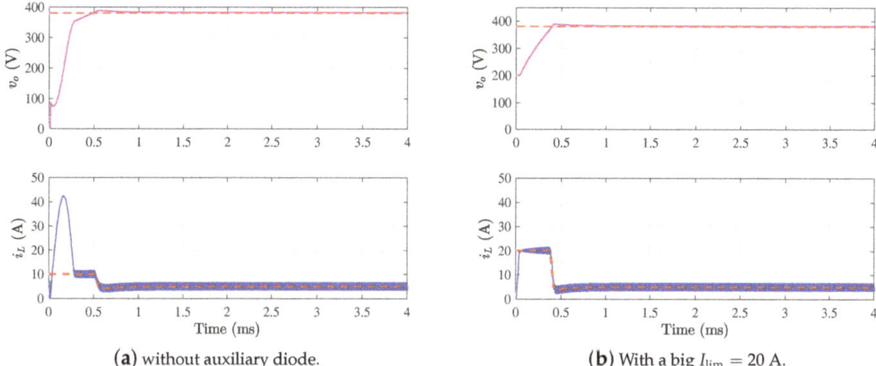

(**a**) without auxiliary diode.   (**b**) With a big $I_{\lim} = 20$ A.

Figure 13. Startup and steady-state response from numerical simulations using a detailed switched model implemented in PSIM© revealing the importance of the auxiliary diode D_a and the current limitation in guaranteeing the sliding-mode operation and hence the inrush current limitation.

It is worth noting that the DSMC derived in this study is based on a full-order representation of the system in the state space. In both DSMC and DPC, the control law is nonlinear since the output voltage appears in the denominator of the expression of the duty cycle. A discussion on approximating the samples of the output voltage by their constant steady-state values was presented in [42] concluding that it is reasonable to use the constant output voltage instead of its instantaneous samples. However, the startup and the transient response performance under this condition were not discussed and only steady-state behavior was evaluated. It should be noted that when the samples of the output voltage are substituted by their constant steady-state value, the controller becomes linear, and the computation effort is reduced. However, this occurs at the expense of losing the sliding-mode operation which in turns results on a high inrush current and unsatisfactory system performance during both startup and under transient response. In an experimental circuit, the high inrush current would even destroy the switching semi-conductor devices.

The expression of the duty cycle (equivalent control) contains the instantaneous values of the state variables and none of them is substituted by its constant steady-state value. This is indispensable for the system to work in sliding mode, which is crucial for inrush current limiting. This will change the system behavior during startup since the sliding-mode operation will be lost.

Figure 14 shows the startup response of the system when the samples of the output voltage are replaced by their steady-state value in (41) but using the instantaneous values of the output voltage in the expression of i_{ref} given by (37). As can be observed, the current reference is not well tracked during the startup phase and the system exhibits an unacceptable inrush current hence demonstrating the importance of operating under sliding-mode regime which is guaranteed by using the instantaneous values of the state variables not their steady-state values.

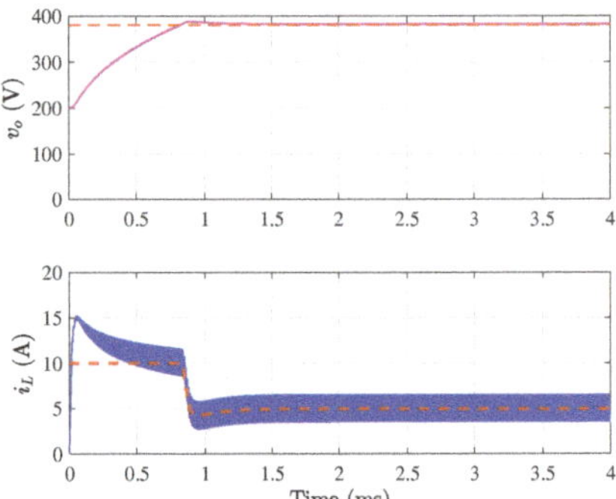

Figure 14. Startup and steady-state response from numerical simulations using a detailed switched model implemented in PSIM© when the samples of the output voltage are replaced by their steady-state value in (41) but using the instantaneous values of the output voltage in the expression of i_{ref} given by (37).

7.2. Small-Signal Response to Output Voltage Variation. Non-Minimum Phase Behavior

Figure 15 shows the response of the system to a ± 4 V step change in the reference voltage using a switched model. It can be observed that the system exhibits a small undershoot in the output voltage response immediately after the positive step change of voltage reference. The inductor current follows the reference current tightly as dictated by the DSMC.

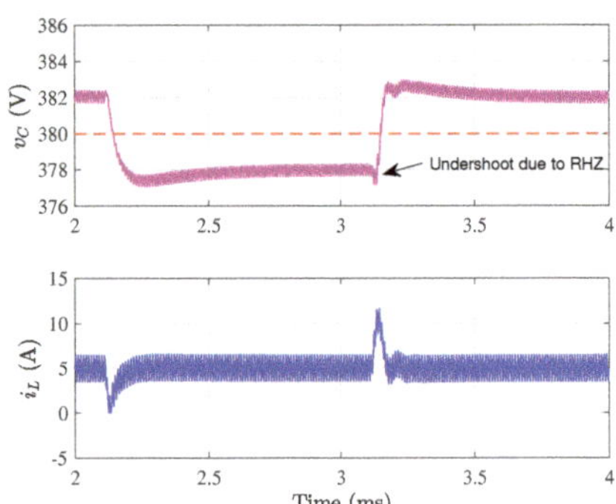

Figure 15. Small-signal transient response in front of a ± 4 V step change between 378 V and 382 V in the reference voltage from numerical simulations using a detailed switched model implemented in PSIM© and the discrete-time model showing a non-minimum phase behavior.

7.3. Small-Signal Response to Input Voltage Disturbance

Figure 16 shows the transient response in the presence of 38% step change in the input voltage from the detailed switched model implemented in PSIM© software. It can be observed that the output voltage is tightly regulated to its desired value. The steady-state average inductor current is $I_{ref} = P/V_g$ as predicted by the theoretical analysis. The inductor current and capacitor voltage show a fast recovery of the steady state after a disturbance takes place.

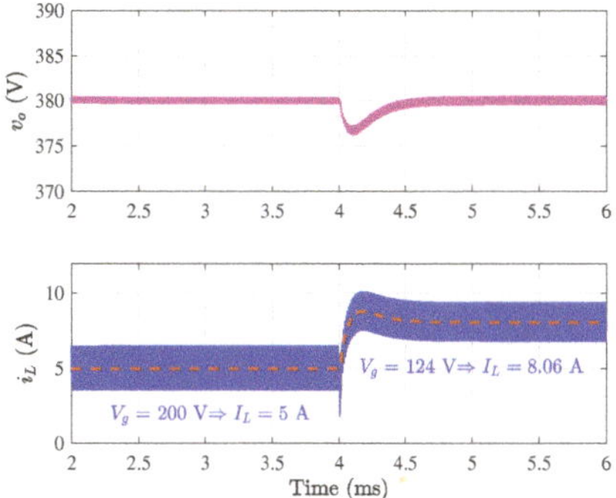

Figure 16. Small-signal transient response in front of a 38% step change in the input voltage from 200 V to 124 V using numerical simulations using a detailed switched model implemented in PSIM© and the discrete-time model.

7.4. Small-Signal Response to Power Disturbance

Figure 17 shows the transient response in the presence of 50% step change in the load power from the detailed switched model. A zero steady-state error in the output voltage can also be observed while the dynamic current reference i_{ref} is tracked by the inductor current i_L as imposed by the inner DSMC loop. The steady-state value of i_{ref} is $I_{ref} = P/V_g$ as predicted by the theoretical analysis. As before, both inductor current and capacitor voltage show a fast recovery of the steady state. As can be observed, in all the cases, both small-signal and large-signal behaviors show a similar behavior, confirming the validity of the model developed in the previous sections. Hence the large-signal model can be used for repeated simulations while the small-signal model can be used for control design and performance specifications.

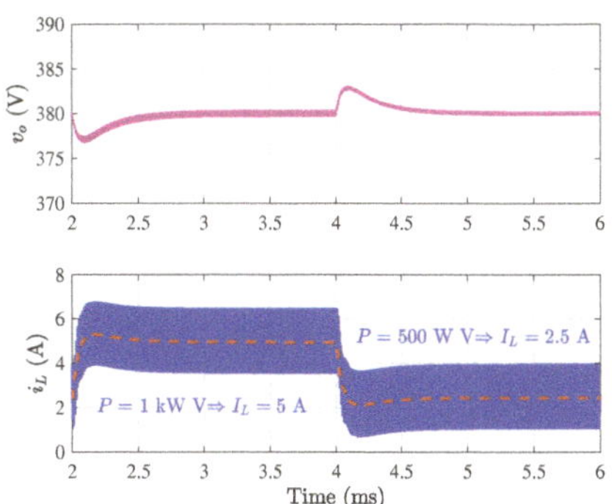

Figure 17. Small-signal transient response in front of 50% step change in the nominal power from numerical simulations using a detailed switched model implemented in PSIM© and the discrete-time model.

Remark 3. *Theoretically, according to the small-signal design procedure followed, the system should present no overshoot for a positive step change with the chosen values of parameters. However, a small overshoot can be still appreciated in the system response of Figure 15. The discrepancy is mainly due to the computation delay ($\tau_d \approx 5.5$ μs) not taken into account in the analysis.*

8. Experimental Results

8.1. Experimental Setup

A boost converter under DSMC was implemented to verify the validity of the control design approach proposed in the previous sections. A picture of the experimental benchmark and the implemented experimental prototype is depicted in Figure 18. The measured large-signal and small-signal response are compared with the theoretical predictions for the same set of parameter values and under the same conditions. The developed DSMC algorithm was programmed in the Digital Signal Processor (DSP) TMS320F28335 of TEXAS INSTRUMENTS. The samples of the state variables are captured and adapted to the voltage values supported by the DSP, connecting to a pin of the ADC module through an operational amplifier operating as a buffer to isolate the DSP. The signals are sampled at the switching frequency rate. The duty cycle was calculated according to (41) and processed in the PWM of the DSP which uses a symmetric triangular signal to generate the driving signal with a time delay of about $\tau_d = 5.5\ \mu s$. The CPL has been emulated by the electronic load 9000 EL-DE ELEKTRO-AUTOMATIK which has been programmed in constant power mode. The experimental waveforms shown below, have been measured by using the oscilloscope Tektronix TDS 754C and the probes TEKTRONIX TCP202 for illustrating the current waveforms.

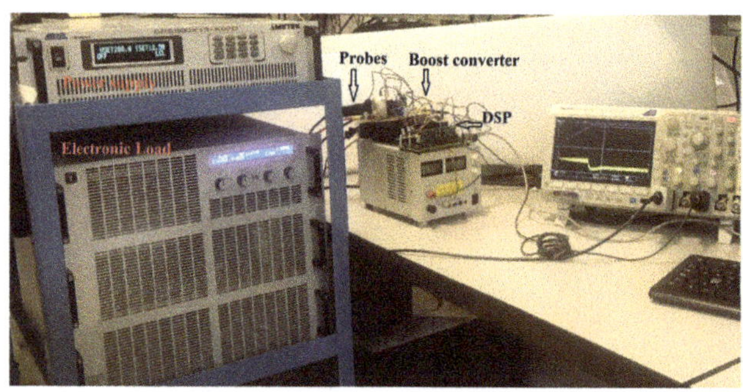

(**a**) Experimental setup (**b**) Experimental prototype and DSP

Figure 18. Picture of the benchmark and the implemented experimental prototype.

8.2. Results

Figure 19 shows the measured responses of the system during startup and in steady state. It can be observed that the measured responses shown in Figure 19 and the simulated responses presented in Figure 12 are very close. However, the inrush current exhibited in the experimental circuit is larger than in the numerical simulation. This is mainly due to the saturation of the inductor and the decrease of its inductance value at high current levels. To completely suppress the still remaining inrush current, a slope limiter is introduced in the current reference at the startup to guarantee the sliding-mode conditions given in (28). The effect of adding this slope limiter is shown Figure 20 where it can be observed that inrush current is completely suppressed thanks to the operation under sliding-mode regime. In all the cases, the output voltage regulation to 380 V in steady state is also well achieved.

Figure 21 depicts the measured response to a ± 4 V step change in the voltage reference. Note that the measured inductor current i_L tracks tightly and accurately the current reference i_{ref}. The output voltage is regulated to its desired reference. A small undershoot of the output voltage immediately after a positive step change can be observed.

The effect of a 50% step change on the power is shown in Figure 22. A good agreement can be observed between the measured and the simulated responses.

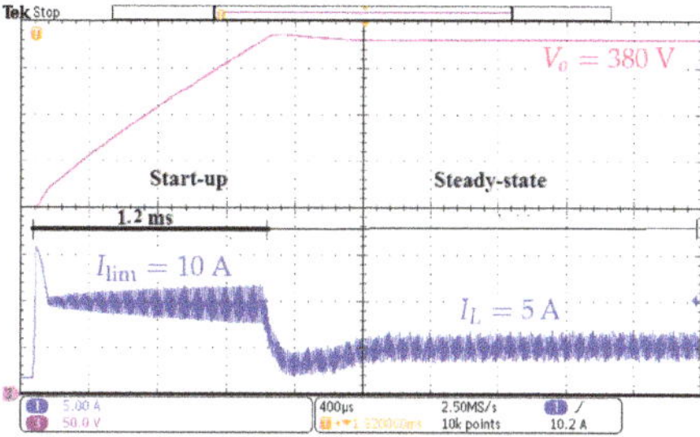

Figure 19. Startup and steady-state responses of the system from experimental measurements.

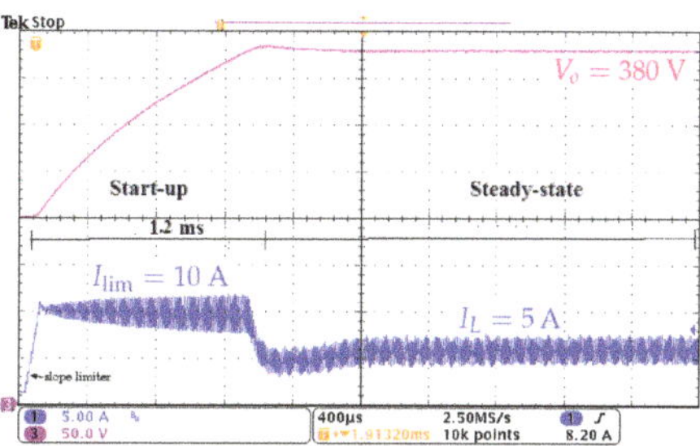

Figure 20. Startup and steady-state responses of the system from experimental measurements with slope limiter $\left.\dfrac{di_{\text{ref}}}{dt}\right|_{\text{lim}} = 100\ \text{kA/s}$.

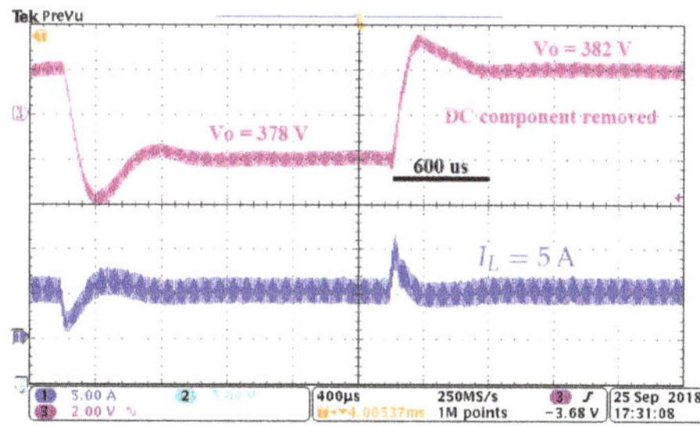

Figure 21. Response to a ± 4 V step change in the voltage reference from experimental measurements. DC component has been removed from the output voltage.

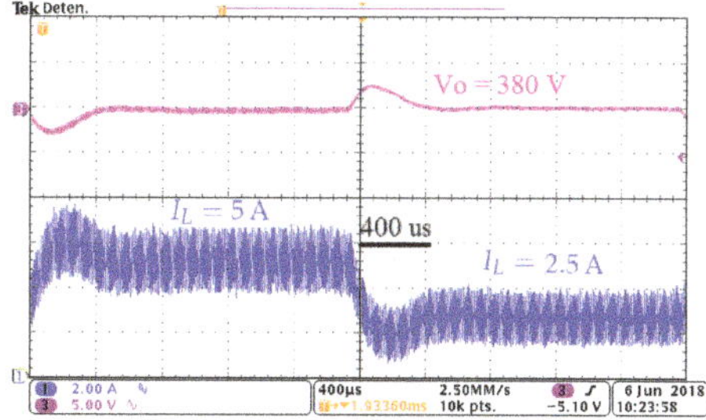

Figure 22. Response to a 50% step change in the power P from experimental measurements.

It should be noted that in all the experimental tests, the system was started with zero initial current. The initial voltage $v_O(0) = V_g$ is due to the presence of the auxiliary diode D_a.

9. Conclusions

DC-DC converters loaded by a CPL appears in many modern and emerging electrical energy conversion systems. This work has presented a digital sliding-mode approach for designing a two-loop controller for inrush current limiting and output voltage regulation in a DC-DC boost converter supplying a constant power load. The following features can be remarked

- The boost converter loaded with CPL is unstable in open loop.
- The boost converter loaded with CPL is unstable with digital sliding-mode current-mode control in a clear-cut contrast with the same converter with resistive load. With an appropriate choice of the outer voltage loop, the system can be stabilized to its desired operating point.
- The operation under sliding-mode regime helps in the inrush current limitation at startup
- The presence of propagation delay worsens the inrush current and the problem can be relieved by forcing a non-saturated value of the duty cycle within the few initial switching cycles. This can be accomplished either by scaling down the value of the duty cycle obtained by the control law or by limiting the rate of change of the current reference at startup.

The designed controller is based on an inner current control loop guaranteeing sliding-mode operation in discrete-time and an outer voltage control loop in the form of a digital PI compensator to stabilize the system and regulate its output voltage. The proposed control combines the advantages of analog controllers in terms of fast system response and the benefits of a digital implementation such as programmability and noise immunity while offering the additional profit of operating at constant switching frequency. The large-signal and the small-signal models of the system have been derived using a general procedure that can be applied to other converter topologies. The model of the system with voltage loop open has been expressed as a single nonlinear difference equation. The small-signal transfer functions have also been derived showing that the system is non-minimum phase, like for the case of resistive load, and that it is unstable with voltage loop open. A boost converter under the proposed control has been implemented to verify the large-signal and the small-signal models derived in the paper. The evaluation using numerical simulation from a detailed switched model and experimental validation suggests that this two-fold controller can effectively enhance the performance of a DC-DC boost converter with CPLs in a wide operating range and this has been proven showing the inrush current limitation during startup and a better voltage response during both startup and close to the steady-state operation. Problems related to delay effects and nonlinearity of the CPL have been addressed. The mitigation of these problems has allowed to design a digital sliding-mode approach for designing a two-loop controller with inrush current limiting capability. The measurements, the numerical simulations and the theoretical predictions have shown a very good agreement. The measurements, the numerical simulations and the theoretical predictions have shown a very good agreement. By comparing with the state-of-the art digital controllers, it has been found that the proposed digital sliding-mode controller exhibits better small- and large-signal responses.

Author Contributions: Conceptualization, A.E.A.; Methodology, A.E.A. and E.V.-I.; Software, A.E.A.; Validation, B.-A.M.T.; Formal Analysis, A.E.A. and E.V.-I.; Investigation, A.E.A.; Resources, A.C.-P.; Data Curation, E.V.-I.; Writing—Original Draft Preparation, A.E.A.; Writing—Review & Editing, A.E.A. and B.-A.M.T.; Visualization, B.-A.M.T.; Supervision, A.E.A.; Funding Acquisition, A.C.-P. and E.V.-I.

Funding: This work has been sponsored by the Spanish Agencia Estatal de Investigación (AEI) and the Fondo Europeo de Desarrollo Regional (FEDER) under Grants DPI2017-84572-C2-1-R and DPI2016-80491-R (AEI/FEDER, UE).

Conflicts of Interest: The authors declare no conflict of interest.

Abbreviations

The following abbreviations are used in this manuscript:

ADC	Analog-to-digital converter
CPL	Constant power load
CPS	Constant power source
DPC	Digital predictive control
DPWM	Digital pulse width modulation
DSMC	Digital sliding mode control
EV	Electric vehicle
KCL	Kirchhof current law
PWM	Pulse width modulation
SMC	Sliding mode control

References

1. Belkhayat, M.; Cooley, R.; Witulski, A. Large signal stability criteria for distributed systems with constant power loads. In Proceedings of the 26th Annual IEEE Power Electronics Specialists Conference, PESC '95 Record, Atlanta, GA, USA, 18–22 June 1995; Volume 2, pp. 1333–1338.
2. Emadi, A.; Fahimi, B.; Ehsani, M. On the Concept of Negative Impedance Instability in the More Electric Aircraft Power Systems with Constant Power Loads. *SAE Technol. Pap. 01-2545* **1999**. [CrossRef]
3. Choi, B.; Cho, B.; Hong, S.-S. Dynamics and control of dc-to-dc converters driving other converters downstream. *IEEE Trans. Circuit Syst. I Fund. Theory Appl.* **1999**, *46*, 1240–1248. [CrossRef]
4. Emadi, A.; Ehsani, A. Dynamics and control of multi-converter dc power electronic systems. In Proceedings of the 2001 IEEE 32nd Annual Power Electronics Specialists Conference, Vancouver, BC, Canada, 17–21 June 2001; Volume 1, pp. 248–253,
5. Khaligh, A. Realization of parasitics in stability of DC-DC converters loaded by constant power loads in advanced multiconverter automotive systems. *IEEE Trans. Ind. Electron.* **2008**, *55*, 2295–2305. [CrossRef]
6. Rivetta, C.; Emadi, A.; Williamson, G.; Jayabalan, R.; Fahimi, B. Analysis and control of a buck DC-DC converter operating with constant power load in sea and undersea vehicles. *IEEE Trans. Ind. Appl.* **2004**, *2*, 1146–1153.
7. Kwasinski, A.; Onwuchekwa, C. Dynamic Behavior and Stabilization of DC microgrids with instantaneous constant-power loads. *IEEE Trans. Power Electron.* **2011**, *26*, 822–834. [CrossRef]
8. Lu, X.; Sun, K.; Guerrero, J.M.; Vasquez, J.C.; Huang, L.; Wang, J. Stability enhancement based on virtual impedance for DC microgrids with constant power loads. *IEEE Trans. Smart Grid* **2015**, *6*, 2770–2783. [CrossRef]
9. Al-Nussairi, M.; Bayindir, R.; Padmanaban, S.; Siano, L.M.P.P. Constant Power Loads (CPL) with Microgrids: Problem Definition, Stability Analysis and Compensation Techniques. *Energies* **2017**, *10*, 1656. [CrossRef]
10. Singh, S.; Fulwani, D. Constant power loads: A solution using sliding-mode control. In Proceedings of the IECON 2014—40th Annual Conference of the IEEE Industrial Electronics Society, Dallas, TX, USA, 29 October–1 November 2014; pp. 1989–1995.
11. Emadi, A.; Khaligh, A.; Rivetta, C.; Williamson, G. Constant power loads and negative impedance instability in automotive systems: Definition, modeling, stability, and control of power electronic converters and motor drives. *IEEE Trans. Veh. Technol.* **2006**, *55*, 1112–1125. [CrossRef]
12. Smithson, S.; Williamson, S. Constant power loads in more electric vehicles—An overview. In Proceedings of the IECON 2012—38th Annual Conference on IEEE Industrial Electronics Society, Montreal, QC, Canada, 25–28 October 2012; pp. 2914–2922.
13. Cespedes, M.; Xing, L.; Sun, J. Constant-Power Load System Stabilization by Passive Damping. *IEEE Trans. Power Electron.* **2011**, *26*, 1832–1836. [CrossRef]
14. Rahimi, A.M.; Emadi, A. Active damping in DC-DC power electronic converters: A novel method to overcome the problems of constant power loads. *IEEE Trans. Ind. Electron.* **2009**, *56*, 1428–1439. [CrossRef]
15. Sulligoi, G.; Bosich, D.; Giadrossi, G.; Zhu, L.; Cupelli, M.; Monti, A. Multiconverter medium voltage DC power systems on ships, Constant power loads instability solution using linearization via state feedback control. *IEEE Trans. Smart Grids* **2014**, *5*, 2543–2552. [CrossRef]

16. Li, Y.; Vannorsdel, K.; Zirger, A.; Norris, M.; Maksimović, A. Current mode control for boost converters with constant power loads. *IEEE Trans. Circuit Syst. I Fund. Theory Appl.* **2012**, *59*, 19206. [CrossRef]

17. Rahimi, A.M.; Emadi, A.Discontinuous conduction mode DC-DC converters feeding constant power loads. *IEEE Trans. Ind. Electron.* **2010**, *57*, 1318–1329. [CrossRef]

18. Rodríguez-Licea, M.-A.; Pérez-Pinal, F.-J.; Nu nez-Perez, J.-C.; Herrera-Ramirez, C.A. Nonlinear Robust Control for Low Voltage Direct-Current Residential Microgrids with Constant Power Loads. *Energies* **2018**, *11*, 1130. [CrossRef]

19. Marcillo1, K.-E.L.; Guingla, D.-A.P.; Barra, W.; de Medeiros, R.L.P.; Rocha, E.-M.; Vaca-Benavides, D.A.; Nogueira, F.G. Interval Robust Controller to Minimize Oscillations Effects Caused by Constant Power Load in a DC Multi-Converter Buck-Buck System. *IEEE Access* **2018**, *6*. [CrossRef]

20. Hossain, E.; Perez, R.; Nasiri, A.; Padmanaban, S. A Comprehensive Review on Constant Power Loads Compensation Techniques. *IEEE Access* **2018**, *6*, 33285–33305. [CrossRef]

21. Utkin, V.; Guldner, J.; Shi, J. *Sliding Mode Control In Electro-Mechanical Systems, (Automation and Control Engineering)*; CRC Press: Boca Raton, FL, USA, 2009.

22. Utkin, V. Sliding mode control of dc/dc converters. *J. Frankl. Inst.* **2013**, *350*, 2146–2165. [CrossRef]

23. Venkataramanan, R. Sliding Mode Control of Power Converters. Ph.D. Thesis, California Institute of Technology Pasadena, Pasadena, CA, USA, 1986

24. Utkin, V.I. Sliding mode control design principles and applications to electric drives. *IEEE Trans. Ind. Electron.* **1993**, *40*, 23–36. [CrossRef]

25. Calvente, J.; El Aroudi, A.; Giral, R.; Cid-Pastor, A.; Vidal-Idiarte, E.; Martínez-Salamero, L. Design of Current Programmed Switching Converters Using Sliding-Mode Control Theory. *Energies* **2018**, *11*, 2034. [CrossRef]

26. Martinez-Salamero, L.; Cid-Pastor, A.; El Aroudi, A.; Giral, R.; Calvente, J. Sliding-mode control of DC-DC switching converters. *IFAC Proc. Vol.* **2011**, *44*, 1910–1916. [CrossRef]

27. Cid-Pastor, A.; Giral, R.; Calvente, J.; Utkin, V.; Martinez-Salamero, L. Interleaved converters based on sliding-mode control in a ring configuration. *IEEE Trans. Circuits Syst. I Regul. Pap.* **2011**, *58*, 2566–2577. [CrossRef]

28. Haroun, R.; El Aroudi, A.; Cid-Pastor, A.; Garcia, G.; Olalla, C.; Garcia, G. Impedance matching in photovoltaic systems using cascaded boost converters and sliding-mode control. *IEEE Trans. Power Electron.* **2015**, *30*, 3185–3199. [CrossRef]

29. Cid-Pastor, A.; Martinez-Salamero, L.; El Aroudi, A.; Giral, R.; Calvente, J.; Leyva, R. Synthesis of loss-free resistors based on sliding-mode control and its applications in power processing. *Control Eng. Pract.* **2013**, *21*, 689–699. [CrossRef]

30. Bodetto, M.; El Aroudi, A.; Cid-Pastor, A.; Calvente, J.; Martinez-Salamero, L. Design of AC-DC PFC high-order converters with regulated output current for low-power applications. *IEEE Trans. Power Electron.* **2016**, *31*, 2012–2025. [CrossRef]

31. Benadero, L.; Cristiano, R.; Pagano, D.; Ponce, E. Nonlinear analysis of interconnected power converters: A case study. *IEEE J. Emerg. Sel. Top. Circuits Syst.* **2015**, *5*, 326–335. [CrossRef]

32. Singh, S.; Fulwani, D.; Kumar, V. Robust sliding-mode control of dc/dc boost converter feeding a constant power load. *IET Power Electron.* **2015**, *8*, 1230–1237. [CrossRef]

33. Gao, W.; Wang, Y.; Homaifa, A. Discrete-time variable structure control systems. *IEEE Trans. Ind. Electron.* **1995**, *42*, 117–122.

34. Bartoszewicz, A. Discrete-time quasi-sliding-mode control strategies. *IEEE Trans. Ind. Electron.* **1998**, *45*, 63637. [CrossRef]

35. Schirone, L.; Celani, F.; Macellari, M. Discrete-time control for DC-AC converters based on sliding-mode design. *IET Power Electron.* **2012**, *5*, 833–840. [CrossRef]

36. Vidal-Idiarte, E.; Carrejo, C.E.; Calvente, J.; Martinez-Salamero, L. Two-loop digital sliding-mode control of DC-DC power converters based on predictive interpolation. *IEEE Trans. Ind. Electron.* **2011**, *58*, 2491–2501. [CrossRef]

37. Utkin, V.; Guldner, J.; Shi, J. *Sliding Mode Control in Electromechanical Systems*; Taylor & Francis: Boca Raton, FL, USA, 1999.

38. Vidal-Idiarte, E.; Marcos-Pastor, A.; Garcia, G.; Cid-Pastor, A.; Martinez-Salamero, L. Discrete-time sliding-mode -based digital pulse width modulation control of a boost converter. *IET Power Electron.* **2015**, *8*, 708–714. [CrossRef]
39. Vidal-Idiarte, E.; Marcos-Pastor, A.; Giral, R.; Calvente, J.; Martinez-Salamero, L. Direct digital design of a sliding mode-based control of a PWM synchronous buck converter. *IET Power Electron.* **2017**, *10*, 1714–1720. [CrossRef]
40. Marcos-Pastor, A.; Vidal-Idiarte, E.; Cid-Pastor, A. Interleaved digital power factor correction based on sliding-mode approach. *IEEE Trans. Power Electron.* **2016**, *31*, 4641–4653. [CrossRef]
41. Ogata, K. *Discrete-Time Control Systems*; Prentice-Hall, Inc.: Upper Saddle River, NJ, USA, 1987.
42. Chen, J.; Prodic, A.; Erickson, R.W.; Maksimovic, D. Predictive digital current programmed control. *IEEE Trans. Power Electron.* **2003**, *18*, 411–419. [CrossRef]

Article

Sliding Mode Control of the Isolated Bridgeless SEPIC High Power Factor Rectifier Interfacing an AC Source with a LVDC Distribution Bus

Oswaldo Lopez-Santos [1,*], Alejandro J. Cabeza-Cabeza [1], Germain Garcia [2] and Luis Martinez-Salamero [3]

[1]   Facultad de Ingeniería, Universidad de Ibagué, Carrera 22 Calle 69 Barrio Ambalá, 730001 Ibagué, Colombia
[2]   Laboratoire d'Analyse et d'Architecture des Systèmes, Centre Nationale de Recherche Scientifique (LAAS-CNRS), Institut National des Sciences Appliquées (INSA), 7 Avenue du Colonel Roche, 31077 Toulouse, France
[3]   Departament d'Enginyeria Electrònica, Elèctrica i Automàtica, Universitat Rovira i Virgili, Av. Paisos Catalans, No. 26, 43007 Tarragona, Spain
*   Correspondence: oswaldo.lopez@unibague.edu.co; Tel.: +57-8-2760010 (ext. 4007)

Received: 2 July 2019; Accepted: 6 September 2019; Published: 7 September 2019

Abstract: This paper deals with the analysis and design of a sliding mode-based controller to obtain high power factor (HPF) in the bridgeless isolated version of the single ended primary inductor converter (SEPIC) operating as a single-phase rectifier. In the work reported here, the converter is used as a unidirectional isolated interface between an AC source and a low voltage direct current (LVDC) distribution bus. The sliding-mode control is used to ensure the tracking of a high quality current reference at the input side, which is obtained from a sine waveform generator synchronized with the grid. The feasibility of the proposal is validated using simulation and experimental results, both of them confirming a reliable operation and showing good static and dynamic performances.

Keywords: sliding-mode control; isolated SEPIC converter; high power factor rectifier; isolated PFC rectifier; bridgeless rectifier; DC distribution bus

---

1. Introduction

The use of rectifiers operating with high power factor (HPF) is a mandatory issue today in AC-DC conversion [1]. This fact reduces the negative impact on the power quality of AC distribution networks caused by the increasing introduction of new energy processing technologies such as electric mobility and efficient lighting, among others. For example, hybrid electric vehicles need battery chargers, which must not only provide high power density and plug-and-play operation [2], but also meet the requirements of the power quality international standards. In the same context, it is possible to integrate the HPF rectification function in multi-mode converters, which can provide bidirectional power transfer capability or multiple power conversion types (AC-DC, DC-DC, and DC-AC), which eventually has an important impact on the power density of the converter in the electric vehicle [3]. At lower power levels, HPF rectifiers allow improving the input power quality and the general performance of LED lamps, which currently constitute the leading lighting technology in the market [4].

Utilization of HPF rectifiers is also fundamental in the newest applications related to feeding DC loads [5], DC distribution, and microgrids. On the basis of the high efficiency and flexibility of the low voltage direct current (LVDC) distribution systems [6], HPF rectifiers take part of the main DC source because they incorporate control systems that can help to meet the strict requirements of grid compatibility, safety [7], and power quality [8]. This type of power distribution is also currently integrated in microgrids, which constitutes a popular research topic in several industrial electronics areas [9]. Hybrid or AC-DC microgrids are extensively used and are supplied from renewable resources such as photovoltaic and wind, besides AC sources, namely, either the AC mains, electric machines, or fuel generators [10]. Although bidirectional power flow is a desired feature in many applications, unidirectional power flow is efficiently used in wind power integration, speed regulation, plug-in electric vehicles, and other two-quadrant applications [11]. In all cases, it is intended that the AC powered devices accomplish the international power quality standards in terms of total harmonic distortion (THD) and power factor [12].

The HPF rectifiers can be classified in a simple way as isolated and non-isolated. Non-isolated topologies, in turn, can be classified regarding the use of diode bridges. In particular, if a topology does not require one or more diode bridges, it is denominated bridgeless [13]. The bridgeless topologies generally exhibit a better performance because the number of semiconductor elements in a circulating current path is low. The conventional bridgeless single ended primary inductor converter (SEPIC) was presented by Ismail et al. [14], providing the basis for subsequent works developed by the same authors. This converter was studied in the work of [15] in discontinuous conduction mode (DCM), which resulted in a relatively simple control and a reduced size of the components at the expense of increased stress in semiconductors. More recently, in the works of [16,17], the bridgeless SEPIC rectifier topology was modified by adding multiplier cells in order to extend the operational input voltage range. Although the efficiency in the work of [16] was considerably improved (values above 98%), no increased performance was reported in the power quality indicators such as THD and power factor.

Although galvanic isolation between the AC source and DC distribution bus is not a mandatory issue in HPF rectifiers, it is clear that this feature can contribute to improving the system reliability, especially when the use of DC distribution buses implies the interconnection of multiple sources, different loads, and ancillary elements. For that reason, there is a particular interest in the development of HPF rectifiers with isolated topologies. For example, in the works of [18,19], an isolated rectifier topology is obtained by means of a special configuration of low-frequency transformers (Scott transformer), in which two separated bridge rectifiers based on either boost or buck converters are used to feed a split DC-bus. In another paper [20], isolation is obtained by integrating two isolated Cûk rectifiers configuring a bridgeless topology. A relevant feature of that configuration is the use of coupled inductors operating at high frequency, which eventually results in a significant improvement in the cost, size, and weight of the converter in comparison with the use of low-frequency transformers.

Different isolated architectures based on the SEPIC converter have been proposed in the literature for HPF rectifiers. An interesting alternative with interleaved configuration and a bridge-based SEPIC rectifier topology is presented in the work of [21], attaining unity power factor for a wide range of the output voltage. Nonetheless, the reported THD increases up to nearly 10% for some operation conditions. The isolation in that case is provided by a second conversion stage based on an LLC converter. Also, an active clamp topology operating in both continuous conduction mode (CCM) and DCM has been reported another paper [22], showing an acceptable performance at the expense of an additional input filter and high current stress in semiconductors. The three-phase architecture of isolated SEPIC rectifier presented in the work of [23] uses three high-frequency transformers and three bridge-based bidirectional switches, improving the performance of the single-phase bridgeless SEPIC rectifier. In the latter work, DCM operation is used to obtain a THD of 4% and unity power factor. However, the mentioned parameters were not evaluated in the entire range of operation rated for the converter. It is worth mentioning that the resulting efficiency in the aforementioned cases ranges from 75% to 90%.

The bridgeless configuration of the HPF SEPIC rectifier depicted in Figure 1 has been presented in the work of [24], working at constant switching frequency, which is imposed by a pulse width modulation (PWM) operation in the control loop. Besides the galvanic isolation, the authors of that work have highlighted the relatively low number of components and the low levels of resulting Electro Magnetic Interference (EMI) as the main advantages of the topology. The main features in the isolated version of the SEPIC converter are as follows: (i) the existence of a series inductor in the input port imposing a continuous behavior to the input current, (ii) the capability to either step-up or step-down the input voltage, and (iii) the enhancement of the input voltage range with respect to other topologies.

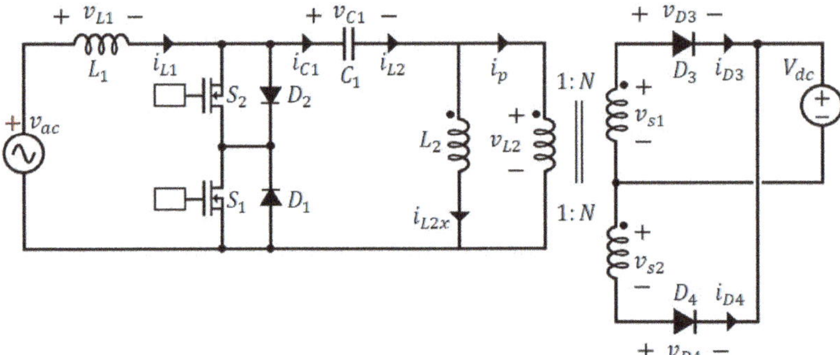

Figure 1. Circuit diagram of the isolated single ended primary inductor converter (SEPIC) rectifier.

The control of HPF rectifiers based on the SEPIC converter can be easily implemented using continuous-time linear methods and PWM. The wave-shaping, as defined by Tanitteerapan et al. in the literature [25,26], leads to low values of THD, while a good reference generation method allows achieving power factor correction. This is carried out by an inner control loop that normally processes the input current and is complemented by an outer loop regulating the output voltage in the case of rectifiers operating as pre-regulators. In addition, discrete-time control techniques such as repetitive control have been applied to control the SEPIC rectifier [27]. With this technique, the THD is high at low power levels, but enters in the permissive range for 50% of the nominal power, which suggests that there is still an important gap to improve the rectifier performance in terms of THD. Besides, in the work of [28], a linear control approach is compared with a feedback linearization technique, demonstrating a better performance of the linear technique in power factor and THD for nominal power. The nonlinear technique shows a slightly advantageous performance by regarding the THD distribution along the whole range of the converter power. In the same work, the robustness of the control system is improved by applying an adaptive passivity-based feedback linearization approach.

Sliding-mode control (SMC) has been also applied to control the SEPIC rectifier using a PWM implementation [29]. The power factor obtained with this technique in always higher than 0.97, while the current THD is only lower than 5% around the nominal power. It can be observed again that the decrease of the THD is an open problem in the SEPIC rectifier.

SMC using a hysteresis implementation results in variable switching frequency, but offers robustness, fast response, reliability, and simple implementation using either analog or digital electronics. It has been demonstrated that this technique is able to track periodic references, forcing loss-free resistor behavior. This implies resistive behavior at the input port and power source behavior at the output port of a power converter [30], which in fact transforms the set rectifier-load into a virtual resistance, as it is successfully developed for a semi-bridge-less pre-regulator in the work of [31] and a three-phase HPF Vienna rectifier in the work of [32]. As in the case of the grid-connected inverters, the control must track a reference, which can be directly provided by a measurement of the input voltage or indirectly by a synchronized reference generator [33]. On the other hand, as the output voltage of the rectifier (voltage of the DC bus) is not regulated, the power injected into the DC bus is given by the amplitude of the input current. This amplitude is in turn provided by an outer control loop that can be a part of a high-level layer of hierarchical control architectures [34].

The main goal of this work is to apply a hysteresis-based implementation of the sliding mode control approach to improve the performance of the isolated SEPIC converter, increasing the range of power in which the international standards are fulfilled.

Unlike the approach in the work of [24], the SEPIC converter analyzed in this paper feeds a non-resistive load and operates at a variable switching frequency. The first constraint has not been studied in any of the reported SEPIC-based rectifier topologies. More specifically, the main differences of the topology depicted in Figure 1 with respect to the conventional SEPIC are as follows:

-   The single controlled switch was changed by a bidirectional switch (S_1/D_1, S_2/D_2) providing control for voltage and current in both half-cycles of the grid voltage.
-   A transformer with three windings, the first one being the primary, replaces the second inductor. The secondary is split in two identical windings, which are interconnected through a central tap. The transformer ratio is 1:N.
-   No output capacitor is used because the converter is directly connected to a voltage regulated DC bus.

Moreover, this paper considers the sliding mode-based current control of a bridgeless isolated SEPIC rectifier tracking a sinusoidal reference and injecting power to an LVDC bus without considering additional control outer loops. The rest of the paper is organized as follows. Modeling and analysis of the sliding-mode current control are developed in Section 2. The implementation of the proposed control is described in Section 3. Simulation and experimental results are shown in Section 4. Finally, conclusions are given in Section 5.

2. Converter Model and Control

2.1. Detailed Description of the Converter

As can be observed in Figure 1, the converter is fed by the AC source v_{ac}, while the distribution bus is represented by the constant DC source V_{dc}. The converter is composed by controlled switches S_1 and S_2; diodes D_1, D_2, D_3, and D_4; capacitor C_1; and inductors L_1 and L_2. The latter element is a coupled inductor with three windings, one primary (n_1 turns) and two identical secondary windings ($n_2 = n_3$ turns), which are represented in the figure by inductance L_2 and an ideal transformer with ratio $N = n_2/n_1 = n_3/n_1$. The voltage at the primary side of the coupled inductor is denoted as v_{L2}, while the voltage at the secondary windings is defined as v_{s1} and v_{s2}, respectively.

The operation of the converter as a HPF rectifier is accomplished by means of the controlled switches S_1 and S_2 commutating at high frequency along each half-cycle of the AC source and the natural switching of diodes D_3 and D_4, which are connected to the secondary windings of the transformer at the output of the converter. To correctly ensure a safe operation of S_1 and S_2, D_1 and D_2 are activated and deactivated simultaneously with S_2 and S_1, respectively.

The isolated SEPIC rectifier in CCM [35] exhibits four possible configurations, that is, two for the positive half-cycle of v_{ac} and two for the negative one. Also, the proposed circuit can operate in an additional configuration during the zero crossings of the AC input. In that configuration, the input side of the converter can be represented by a single mesh with both inductors L_1 and L_2 in series with capacitor C_1, like in a series resonant converter. The control signal u takes the values 0 or 1 during the off and on states of the controlled switches, respectively. Variable a will allow us to distinguish operation in CCM ($a = 0$) and the operation in the above-mentioned additional configuration ($a = 1$).

During the positive half-cycle, in the on state (Figure 2a), inductor L_1 is directly connected to the source v_{ac} through the path composed by S_1 and D_2, while capacitor C_1 is directly connected to L_2, while the DC source V_{dc} is disconnected from the secondary because D_3 and D_4 are open. In the off-state during the positive half-cycle (Figure 2b), inductor L_1 is connected in series with C_1 and L_2, while the load is connected to the secondary v_{s2} through D_3. It has to be pointed out that L_2 operates primarily as a coupling inductor, charging energy during the on state and then discharging it during the off state. However, during the off state, L_2 operates as a transformer, directly transferring energy from the primary to one of the secondary windings depending on the half-cycle. Differential equations modeling the converter dynamic behavior in the on and off states during the positive half-cycle are listed in Table 1.

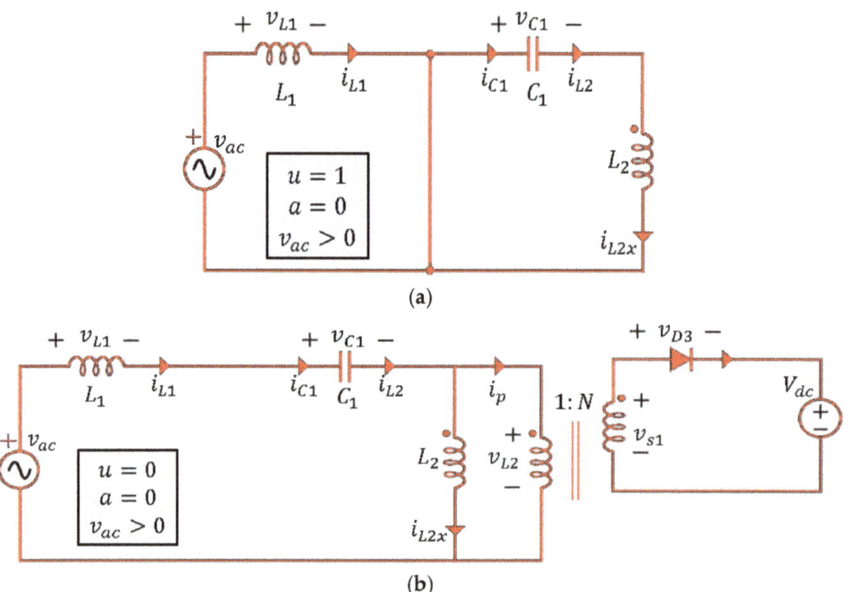

Figure 2. Circuit configurations of the isolated SEPIC rectifier: positive half-cycle.

Table 1. Converter dynamics during a positive half-cycle.

Differential Equations On-State (u=1)	Differential Equations Off-State (u=0)
$L_1 \frac{di_{L1}}{dt} = v_{ac}$	$L_1 \frac{di_{L1}}{dt} = v_{ac} - v_{C1} - \frac{V_{dc}}{N}$
$L_2 \frac{di_{L2x}}{dt} = -v_{C1}$	$L_2 \frac{di_{L2x}}{dt} = \frac{V_{dc}}{N}$
$C_1 \frac{dv_{C1}}{dt} = i_{L2x}$	$C_1 \frac{dv_{C1}}{dt} = i_{L1}$
$i_{D3} = i_{D4} = 0$	$i_{D3} = 0; i_{D4} > 0$
$v_{S1} = v_{S2} = -Nv_{C1}$	$v_{S1} = V_{dc}$

During the negative half-cycle, the on state of the controlled switch S_2 and the activation of the diode D_1 result in the configuration depicted in Figure 3a, which is equal to the one depicted in Figure 2a. Different to the case of the positive half-cycle, the off state leads to the circuit configuration in Figure 3b, because D_4 is forward biased. The differential equations modeling the dynamic behavior during the on and off states of the negative half-cycle are listed in Table 2.

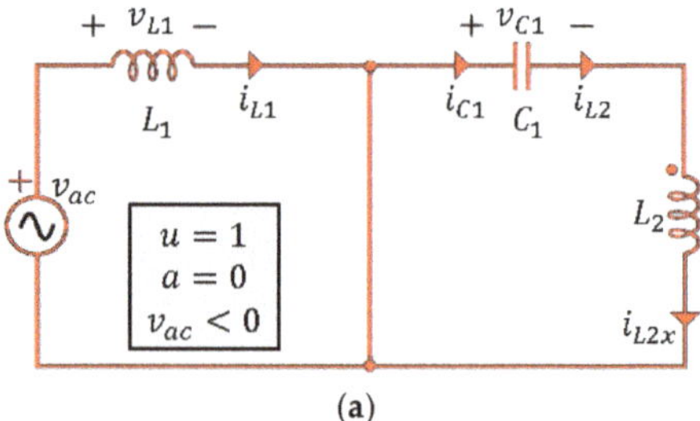

(a)

Figure 3. *Cont.*

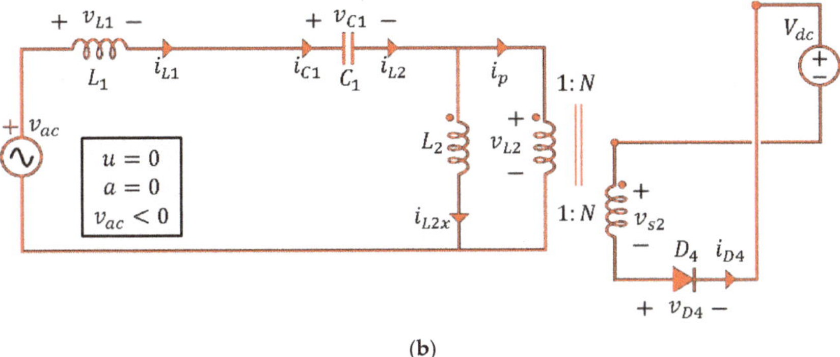

(b)

Figure 3. Circuit configurations of the isolated SEPIC rectifier: negative half-cycle.

Table 2. Converter dynamics during a negative half-cycle.

Differential Equations On-State (u=1)	Differential Equations Off-State (u=0)
$L_1\frac{di_{L1}}{dt} = v_{ac}$	$L_1\frac{di_{L1}}{dt} = v_{ac} - v_{C1} - \frac{V_{dc}}{N}$
$L_2\frac{di_{L2x}}{dt} = -v_{C1}$	$L_2\frac{di_{L2x}}{dt} = \frac{V_{dc}}{N}$
$C_1\frac{dv_{C1}}{dt} = i_{L2x}$	$C_1\frac{dv_{C1}}{dt} = i_{L1}$
$i_{D3} = i_{D4} = 0$	$i_{D3} > 0;\, i_{D4} = 0$
$v_{S1} = v_{S2} = -Nv_{C1}$	$v_{S2} = V_{dc}$

As will be explained later, the on and off states of the controlled switches are imposed by a hysteresis comparator, which in turn constrains the switching frequency of the converter to permissible values. The operation of the converter in CCM is ensured practically in almost the entire period of the input signal v_{ac} through the selection of the inductor L_1 (see details in Section 3). However, this mode is missed for short intervals denoted by $|V_{ac}| \leq \varepsilon$ for $\varepsilon \approx 0$, which correspond to the zero crossing points of the AC input. This feature results in two circuit configurations:

(a) The configuration depicted in Figures 2a and 3a at the start of both half-cycles when the initial condition of the current i_{L1} is zero. The differential equations modeling the dynamic behavior are the same as shown in columns describing the on state in Tables 1 and 2.

(b) The configuration shown in Figure 4, where L_2, L_1, and C_1 constitute de facto a series resonant circuit. The differential equations modeling this dynamic behavior are listed in Table 3.

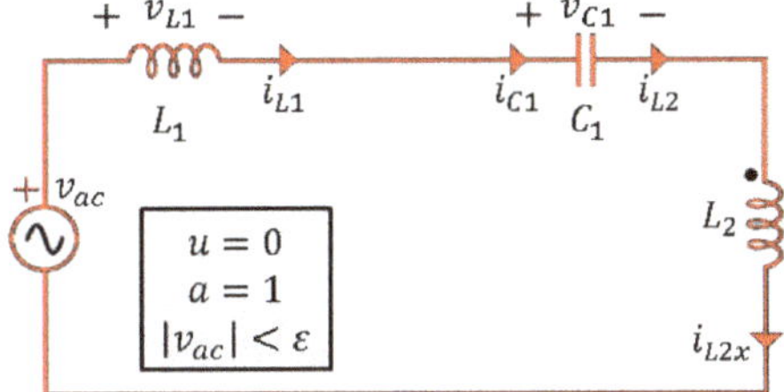

Figure 4. Configuration of the isolated SEPIC rectifier during the zero crossing of the AC input.

Table 3. Converter dynamics during zero crossings of the alternative current (AC) input signal.

Differential Equations ($u=0$)
$(L_1 + L_2)\frac{di_{L1}}{dt} = v_{ac} - v_{C1}$
$(L_1 + L_2)\frac{di_{L2x}}{dt} = v_{ac} - v_{C1}$
$C_1\frac{dv_{C1}}{dt} = i_{L2x} = i_{L1}$
$i_{D3} = i_{D4} = 0$
$v_{S1} = v_{S2} = 0$

The dynamic behavior of all circuit configurations can be expressed in compact form as follows:

$$(L_1 + aL_2)\frac{di_{L1}}{dt} = v_{ac} - v_{C1}(1 - u) - sgn(v_{ac})\frac{V_{dc}}{N}(1 - u)(1 - a),\tag{1}$$

$$(L_2 + aL_1)\frac{di_{L2x}}{dt} = v_{ac}a - v_{c1}(1 - u)a - v_{C1}u + sgn(v_{ac})\frac{V_{dc}}{N}(1 - u)(1 - a),\tag{2}$$

$$C_1\frac{dv_{C1}}{dt} = i_{L1}(1 - u) + i_{L2x}u.\tag{3}$$

The state variables are i_{L1}, v_{C1}, and i_{L2x}, with the latter representing the magnetizing inductance current, which is given by the following:

$$i_{L2x} = i_{L2} - i_p = i_{L2} - Ni_{D3} + Ni_{D4},\tag{4}$$

where i_p is the current of the primary winding of the transformer. Furthermore, currents of output diodes i_{D3} and i_{D4} can be expressed as follows:

$$i_{D3} = \frac{1 + sgn(v_{ac})}{2N}(1 - u)(i_{L1} - i_{L2x}),\tag{5}$$

$$i_{D4} = -\frac{1 - sgn(v_{ac})}{2N}(1 - u)(i_{L1} - i_{L2x}).\tag{6}$$

2.2. Existence Conditions of Sliding Modes for a Constant Reference

For this analysis, it is assumed that $a = 0$ (circuit structures in Figure 2a,b and Figure 3a,b), and v_{ac} and i_{ref} are considered as positive or negative constant values. First, consider the following sliding surface:

$$S(x) = i_{L1} - i_{ref}.\tag{7}$$

To demonstrate the existence of sliding-motions, it is assumed first that both v_{ac} and i_{ref} can be expressed by two constants K_{VI} and K_{CI}, respectively, of the same sign. Under such a hypothesis, $\frac{dS(x)}{dt} = \frac{di_{L1}(t)}{dt}$ and $S(x)\dot{S}(x) < 0$ is satisfied in the configurations depicted in Figure 2a–c, as is demonstrated next in the analysis of each topology.

2.2.1. Case 1

Particularizing (1) for $u = 1$, $a = 0$, and $v_{ac} = K_{VI} > 0$ (Figure 2a), the following is derived:

$$\frac{dS(x)}{dt} = \frac{K_{VI}}{L_1} > 0.\tag{8}$$

It can be observed that $u = 1$ implies $S(x) < 0$ or equivalently $i_{L1} - K_{CI} < 0$, which implies that the signs of the switching surface and its time-derivative are opposite.

2.2.2. Case 2

Similarly, for $u = 1$, $a = 0$, and $v_{ac} = -K_{VI} < 0$ (Figure 3a), Inequality (8) is derived again. Although current i_{L1} circulates now in the opposite sense, it also enters through the positive terminal of the inductor voltage, which exhibits the same voltage drop K_{VI}. Besides, it can be observed that $u = 1$ implies $S(x) < 0$, that is, the current circulating in the opposite sense is below the corresponding reference $-K_{CI} < 0$. Hence, the sliding-mode existence condition is again satisfied.

2.2.3. Case 3

For $u = 0$, $a = 0$, and $v_{ac} = K_{VI} > 0$ (Figure 2a), the following is obtained:

$$\frac{dS(x)}{dt} = \frac{K_{VI} - v_{C1} - \frac{V_{dc}}{N}}{L_1} < 0. \tag{9}$$

To justify the negative sign of Inequality (9), observe first that adding the state equations of i_{L1} and i_{L2x} (Equations (1) and (2)) for $a = 0$ results in

$$L_1 \frac{di_{L1}}{dt} + L_2 \frac{di_{L2x}}{dt} = v_{ac} - v_{C1}. \tag{10}$$

In the case of constant input voltage K_{VI} and constant reference K_{CI}, it can be expected that the coordinates of the equilibrium point, that is, the steady-state values of i_{L1}, i_{L2x}, and v_{c1}, are also constant provided that sliding-motions are obtained. In that case, it can be deduced from Equation (10) that the steady-state value of v_{c1} will be equal to that of v_{ac}, and thus equal to K_{VI}. Hence, Expression (9) becomes the following for values around the equilibrium point:

$$\frac{dS(x)}{dt} = \frac{-\frac{V_{dc}}{N}}{L_1} < 0. \tag{11}$$

It has to be pointed out that $u = 0$ corresponds to $S(x) > 0$ or equivalently $i_{L1} - K_{CI} > 0$, which implies that the signs of the switching surface and its time derivative are again opposite.

2.2.4. Case 4

Finally, for $u = 0$, $a = 0$, and $v_{ac} = -K_{VI} < 0$ (Figure 3b), Inequality (11) is derived again. Although current i_{L1} circulates now in the opposite sense, it also enters through the negative terminal of the inductor voltage, which exhibits the same voltage drop $\frac{V_{dc}}{N}$. Besides, it can be observed that $u = 0$ corresponds to $S(x) > 0$, that is, the current circulating in the opposite sense is above the corresponding reference $-K_{CI} < 0$. Therefore, the sliding-mode existence condition is again satisfied.

From the above analysis, the control law of the converter can be defined in a compact form as follows:

$$u = \begin{cases} for\ v_{ac} > 0,\ u = \begin{cases} 0\ if & S(x) > 0 \\ 1\ if & S(x) < 0 \end{cases} \\ for\ v_{ac} < 0,\ u = \begin{cases} 0\ if & S(x) < 0 \\ 1\ if & S(x) > 0 \end{cases} \end{cases}. \tag{12}$$

2.3. Analysis for Time-Varying Current References Using the Equivalent Control Method

We have demonstrated that there will be stable sliding motions for a constant input voltage K_{VI} and a constant reference K_{CI}. Now, we can expect that there will also be a sliding-mode regime in the case of time-varying functions such as $v_{ac} = V_m \sin \omega t$ and $i_{ref} = I_m \sin \omega t$ if the frequency of the sinusoidal signal is significantly smaller than the resulting switching frequency imposed by the sliding-mode operation. In that case, both v_{ac} and i_{ref} can be interpreted in terms of two periodic sequences of values of K_{VI} and K_{CI} at grid frequency, such that

$$K_{VI} \in \{-V_m, \ldots, -V_2, -V_1, 0, V_1, V_2, \ldots V_m\}, \tag{13}$$

$$K_{CI} \in \{-I_m, \ldots, -I_2, -I_1, 0, I_1, I_2, \ldots I_m\}, \tag{14}$$

$$\frac{V_m}{I_m} = \ldots = \frac{V_2}{I_2} = \frac{V_1}{I_1}. \tag{15}$$

The switching law will impose a sliding regime for each pair (K_{VI}, K_{CI}), so that the converter will eventually exhibit a sequence of equilibrium points $\left(I_{L1}^*, I_{L2x}^*, V_{C1}^*\right)$ whose coordinates will evolve in a periodic way at the grid frequency and, as consequence, the input inductor current will perfectly track its sinusoidal reference. A detailed analysis of this type of tracking can be found in the work of [36].

Consider now in Equations (1)–(3), $v_{ac} = V_m \sin \omega t$, where $\omega = 2\pi f$, f is the grid frequency, and V_m is the amplitude of the sinusoidal signal. Then, the input current is forced to track the reference $i_{ref} = I_m \sin \omega t$. If a sliding mode imposes $i_{L1} = I_m \sin \omega t$, then the equivalent control will be given by the following:

$$u_{eq} = 1 - \frac{V_m \sin \omega t - L_1 \frac{di_{ref}}{dt}}{v_{C1} + sgn(v_{ac}) \frac{V_{dc}}{N}}. \tag{16}$$

Consider now that $L_1 \frac{di_{ref}}{dt} \ll V_m$. Only for very low value of the input voltage v_{ac}, these two values are comparable, and then Equation (16) becomes the following:

$$u_{eq} \approx 1 - \frac{V_m \sin \omega t}{v_{C1} + sgn(V_m \sin \omega t) \frac{V_{dc}}{N}}. \tag{17}$$

Hence, by adding Equations (1) and (2), the following is obtained:

$$v_{C1} = v_{ac} - L_1 \frac{di_{L1}}{dt} + L_2 \frac{di_{L2}}{dt} \approx v_{ac} = V_m \sin \omega t. \tag{18}$$

From Equation (3), the following is deduced:

$$i_{L2x} = \frac{C_1 \frac{dv_{C1}}{dt} - i_{L1}\left(1 - u_{eq}\right)}{u_{eq}}. \tag{19}$$

By defining $K \triangleq \frac{V_{dc}}{N} sgn(\sin \omega t)$, Equation (19) leads to the following:

$$i_{L2x} = \frac{V_m}{K}\left[\frac{V_m C_1 \omega}{2} \sin 2\omega t - I_m \sin^2 2\omega t + K C_1 \omega \cos \omega t\right]. \tag{20}$$

As it can be observed in Figure 2, $i_p = i_{C1} - i_{L2x}$, then

$$i_p = -\frac{V_m}{K}\left[\frac{V_m C_1 \omega}{2} \sin 2\omega t - I_m \sin^2 2\omega t\right]. \tag{21}$$

Therefore, the current of output diodes D1 and D2 can be expressed as follows:

$$i_{D3} = \begin{cases} \frac{i_p}{N} & for\ 0 \le \omega t < \pi \\ 0 & for\ \pi \le \omega t < 2\pi \end{cases}, \tag{22}$$

$$i_{D4} = \begin{cases} 0 & for\ 0 \le \omega t < \pi \\ -\frac{i_p}{N} & for\ \pi \le \omega t < 2\pi \end{cases}. \tag{23}$$

The average current I_{dc} injected into the source V_{DC} can be computed as follows:

$$I_{dc} = \langle i_{dc} \rangle = \langle i_{D3} + i_{D4} \rangle = \frac{1}{2\pi} \int_0^{2\pi} (i_{D3} + i_{D4}) d\omega t,$$
$$I_{dc} = \langle i_{dc} \rangle = \langle i_{D3} + i_{D4} \rangle = \frac{1}{\pi} \int_0^{\pi} i_{D3} d\omega t, \tag{24}$$

which results in the following (see Appendix A):

$$I_{dc} = \frac{V_m I_m}{2V_{dc}}. \tag{25}$$

It can be observed from Equation (25) that the POPI nature of the converter, that is, DC output power equal to DC input power [30], operating as HPF rectifier is verified ($V_m I_m = 2I_{dc} V_{dc}$). Figure 5 depicts a simulation of the main variables of the converter for a grid frequency period in order to illustrate the sliding motion in the converter variables and the resulting high frequency components. The parameters used for the simulation are listed in Table 4 (see Section 4).

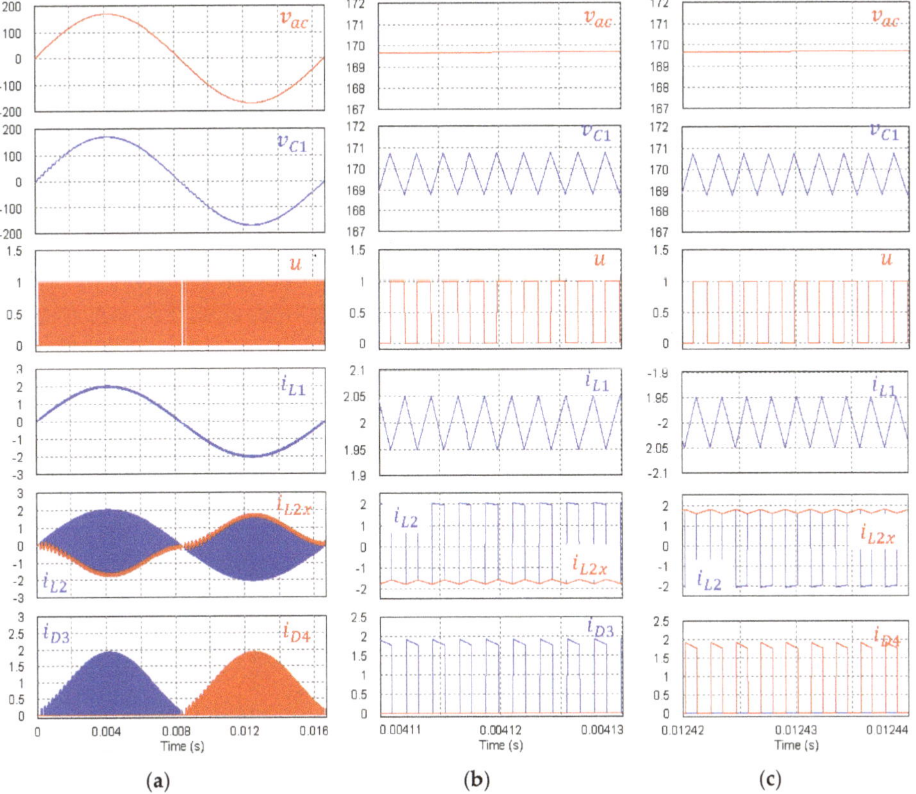

Figure 5. Simulated waveforms of v_{C1}, u, i_{L1}, i_{L2}, i_{L2x}, i_{D3}, and i_{D4} during a cycle of the AC input: (a) one-cycle waveforms; (b) zoom at positive peak of v_{ac}; and (c) zoom at negative peak of v_{ac}.

A detail of the equivalent control waveform is presented in Figure 6, showing that it is correctly constrained between zero and one along the entire period of the input signal.

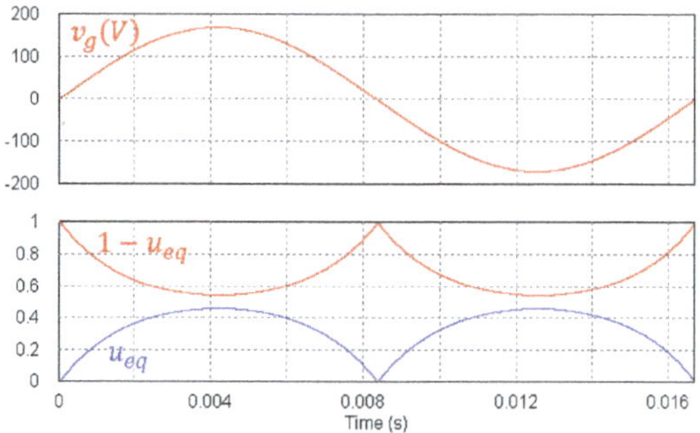

Figure 6. Simulated waveform of the equivalent control u_{eq} and $u'_{eq} = 1 - u_{eq}$ during a cycle of v_{ac}.

2.4. Behavior of the System at the Zero Crossing Points

2.4.1. Condition 1. Figure 4: $a = 1, i_{L1} = i_{L2}, u = 0$.

In this case, from Equations (1)–(3), the following is obtained:

$$\frac{di_{L1}}{dt} = \frac{v_{ac} - v_{c1}}{(L_1 + L_2)}, \frac{dv_{C1}}{dt} = \frac{i_{L1}}{C_1}. \tag{26}$$

Taking the time derivative of $\frac{di_{L1}}{dt}$ and replacing the expression of $\frac{dv_{C1}}{dt}$ leads to the following:

$$\frac{d^2 i_{L1}}{dt^2} + \omega_a{}^2 i_{L1} = \frac{V_m \omega \cos \omega t}{(L_1 + L_2)}, \tag{27}$$

where $\omega_a{}^2 = \frac{1}{C_1(L_1 + L_2)}$. Considering initial conditions equal to zero in Equation (27) and assuming that $\omega_a \gg \omega$, the current of the input inductor is given by the following:

$$i_{L1} \approx V_m \omega C_1 (1 - \cos \omega t). \tag{28}$$

Therefore, the sliding surface is reached when i_{L1} attains the sinusoidal current reference, and then the sliding motion is ensured from that moment. Figure 7a shows a simulated detail for condition 1 and the parameters are given in Table 4. The simulation shows the expected behavior of the current i_{L1} and how the sliding surface is reached. The behavior when finite switching frequency is imposed using a hysteresis comparator is also depicted (Figure 7b). In the latter case, it is possible to observe how the current remains inside the hysteresis band of width $\pm\Delta$ around i_{ref}.

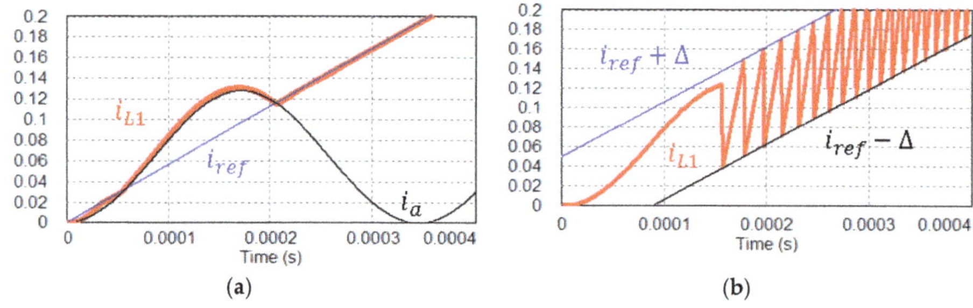

Figure 7. Detail of i_{L1} waveform at a zero crossing for condition 1: (**a**) using an ideal comparator and (**b**) using a hysteresis comparator.

2.4.2. Condition 2. Figures 2a and 3a: $a = 0, u = 1$

Irrespective of the behavior of the current i_{L2}, the following is derived from Equation (1):

$$\frac{di_{L1}}{dt} = \frac{v_{ac}}{L_1}. \tag{29}$$

By solving the differential equation for $\frac{di_{L1}}{dt}$, the current of the input inductor is defined by Equation (30).

$$i_{L1} = \frac{V_m \sin \omega t}{L_1} t. \tag{30}$$

Again, the sliding surface is reached when i_{L1} attains the reference i_{ref}. Figure 8a shows a simulated detail of the special case 2 and the parameters are given in Table 4. The behavior when a finite switching frequency is imposed using a hysteresis comparator with hysteresis band $\pm\Delta$ around i_{ref} is also depicted.

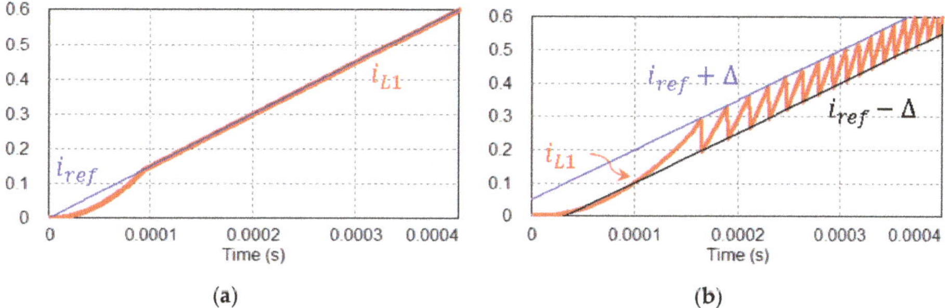

Figure 8. Detail of i_{L1} waveform at a zero crossing for condition 2: (**a**) using an ideal comparator and (**b**) using a hysteresis comparator.

3. Design Considerations and Control Implementation

3.1. Converter Design Considerations

The implementation of the proposed sliding mode control imposes a constant amplitude δ in the ripple content of the inductor current i_{L1}, which in turn forces the switching frequency of the converter to vary along the period of the AC input. As a consequence, the inductor L_1 can be selected to ensure the operation of the converter in CCM and constrain the maximum limit of the switching frequency. From Equations (1) and (18) for $a = 0$, it is possible to obtain the duration of the on and off intervals ($u = 1$ and $u = 0$, respectively) and compute the switching frequency as a function of the instantaneous angle in a period of the AC input voltage:

$$f_{s_{max}} = \frac{V_{dc}V_m}{2\delta L_1 (V_{dc} + NV_m)}.$$

Then, the value of the inductor L_1 can be derived from the expression:

$$L_1 = \frac{V_{dc}V_m}{2\delta f_{s_{max}}(V_{dc} + NV_m)}.$$

Similarly, the value of the inductor L_2 can be obtained from the expression:

$$L_2 = \frac{V_{dc}V_m}{\Delta_{iL2} f_{s_{max}}(V_{dc} + NV_m)},$$

where Δ_{iL2} can be defined as close to 2δ or higher to reduce the final value of L_2.

3.2. Control Implementation Scheme

As it is observed in Figure 9, the current reference i_{ref} is provided by a digitally implemented sine waveform generator, which uses an external phase looked loop (PLL). The PLL delivers a square signal with the same frequency f of the input voltage and a high frequency ($2^m f$) square signal, which is used as a clock signal to reproduce sample by sample a discrete sine waveform stored in a look-up table in a microcontroller. The low-frequency signal also ensures synchronization at zero phase. The value of m defines not only the resolution of the reference, but also the amount of memory required to store it. A value of $m = 11$ is used to ensure a THD lower than 1%, requiring 2^9 memory locations in order to store a quarter of cycle of the sine waveform, which is enough to easily reconstruct the complete sinusoidal signal. The continuous-time version of the reference can be produced by adding a serial digital to analog converter (DAC) [33], or alternatively by using the PWM modules of the microcontroller [37].

Theoretically, the sliding motion appears when the system switches at an infinite frequency. However, in the real implementation, this is not possible because of the limitation of semiconductor devices. Then, a hysteresis band is introduced to enforce the frequency into a finite range. The control law becomes the following:

$$
u = \begin{cases}
for\ v_{ac} > 0 & \begin{cases} 0 & if \quad S(x) > \delta \\ 1 & if \quad S(x) < -\delta \end{cases} \\
for\ v_{ac} < 0 & \begin{cases} 0 & if \quad S(x) < \delta \\ 1 & if \quad S(x) > -\delta \end{cases}
\end{cases}. \tag{31}
$$

A simple electronic implementation is obtained using two analog integrated comparators and one S-R type flip-flop, as is shown in Figure 9.

As is also depicted in Figure 9, the implementation of the proposed control requires the measurement of the input inductor current and the AC input voltage. The precision of the measurements directly compromises the power quality of the rectifier, which requires the use of specific sensors to provide preferably isolation and wide bandwidth.

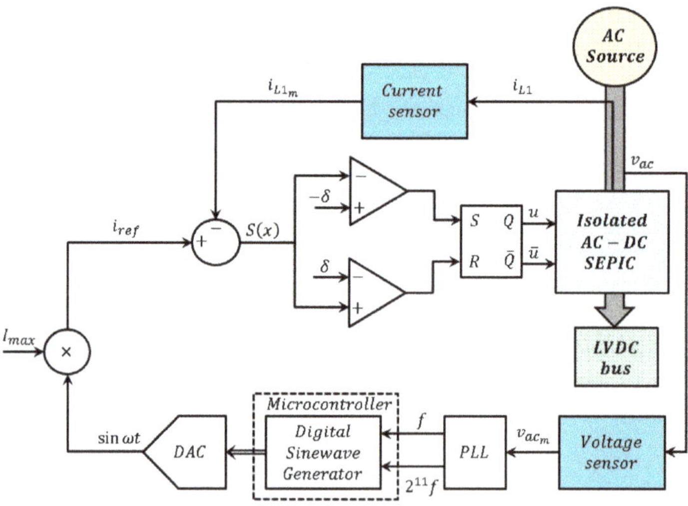

Figure 9. Control diagram of the isolated SEPIC high power factor (HPF) rectifier.

4. Experimental Validation

A simulation was developed in PSIM software obtaining current and voltage waveforms of the converter without considering parasitic elements was used along with the paper to complement the theoretical analysis. This section presents experimental results validating the theoretical predictions of Section 2. Besides, to assess the tracking capability of the sliding mode control in the current loop, a comparison of simulated and experimental results is presented, showing that the influence of parasitic resistances, inductances, and capacitances on the converter behavior is negligible. This means that the converter ideal model is sufficient to design the sliding-mode strategy, which eventually provides the insensitivity to the parasitic elements.

4.1. Converter Prototype and Experimental Setup

A 100 W prototype of the isolated HPF SEPIC rectifier was developed in the laboratory. Details on the converter parameters, passive components, and power semiconductors are listed in Table 4.

Table 4. Specification, parameters, and main components of the laboratory prototype.

Parameter	Symbol	Value	Units
Nominal input voltage	V_g	120	V
Nominal power	P	100	W
Input frequency	f_g	60	Hz
DC voltage	V_{dc}	400	VDC
SuperMesh Technology N-Channel MOSFET	S_1 and S_2	950	V
STD6N95K5		9	A
MOSFET on-resistance	R_{ds-on}	1.25	Ω
SiC Schottky diodes	D_1, D_2, D_3 and D_4	1700	V
GP2D005A170B		5	A
Input inductor (Bourns 1140-222-RC)	L_1	2	mH
Coupled inductor (TDK E 55/28/21 Core)	L_2	1	mH
Primary winding (20 AWG)	N_1	36	turns
Secondary windings (24 AWG)	N_2, N_3	78	turns
Intermediate capacitor (EPCOS Z115056959)	C_1	1	µF

The current reference generator is implemented using one microcontroller dsPIC30F4011, one DAC-SPI (digital to analog converter for serial peripheral interface) MCP4812 (Microchip, AZ, USA), and one analog multiplier AD633 (Analog Devices, MA, USA), and one. The amplitude of the reference is selected manually using a precision potentiometer 3590S-2-503L (Bourns, CA, USA) and an external signal. This signal is constrained between 1 and 5 V, representing the real amplitude of the input current between 0.1 and 1 A.

The sliding-mode controller is implemented using two comparators of the IC LM339 and a flip-flop of the IC CD4027. The MOSFETs are triggered using two MOSFET photo-drivers TLP350 (Toshiba International Corporation, Houston, TX, USA) fed by 24 V isolated power sources. The current measurement for control feedback is implemented using an isolated closed-loop Hall-effect transducer CAS 6-NP (LEM, Plan-les-Ouates, Switzerland). The AC source voltage measurement is implemented using one isolated closed-loop Hall-effect transducer LV-20P. Both sensor signals are conditioned by means of operational amplifiers LMV324. The IC CD4047 is used as a voltage controlled oscillator (VCO) to provide the high frequency signal of the PLL, while the IC CD4060 is used as a frequency divider to produce the low frequency signal. In the same prototype, an alternative way to obtain a square signal from the AC input voltage is included for comparison. The latter circuit includes two photocouplers of the IC MCT6, one comparator of the IC LM339, and one flip-flop of the IC CD4027.

A picture of the converter prototype is shown in Figure 10, wherein the passive components, power semiconductors, input and output terminals, microcontroller, and sensor can be observed.

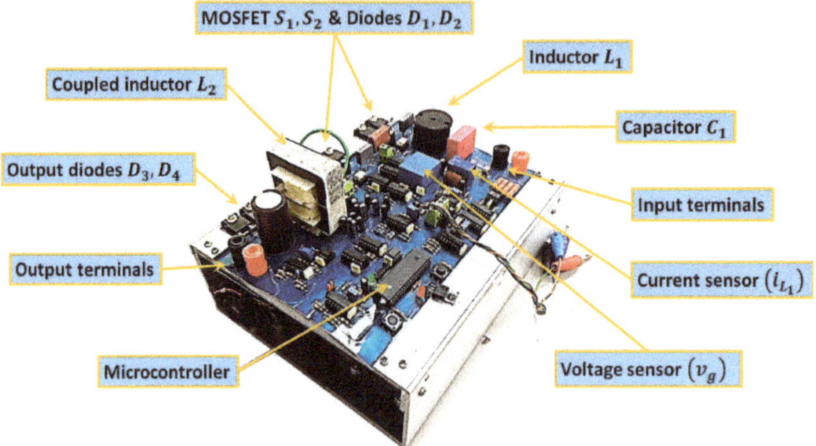

Figure 10. Converter prototype.

The workbench to obtain the experimental measurements depicted in Figure 11 is composed of the equipment set described in Table 5.

Figure 11. Experimental setup used in laboratory experiments.

Table 5. Equipment used in the experimental setup.

Equipment	Model	Capacity
Programmable DC load	IT8512B+	600 V/300 W
Oscilloscope	TDS2024C	50 MHz
Digital Multimeter	34401A	600 V
Power source for control circuits	GPC-3030D	30 V/3 A
Power quality analyzer	FLUKE 43B	600 V

4.2. Tracking of the Current Reference—Comparison of Simulated and Experimental Results

A comparison between simulated and experimental results is presented to validate the reproducibility and effectiveness of the proposed control in the tracking of the current reference. The harmonic content of the AC input voltage measured in the experiments was introduced in PSIM simulations (see Table 6). A fixed value of ±0.2 A was used to define the hysteresis band around the sliding surface.

Table 6. Frequency content of the input voltage (total harmonic distortion (THD) = 3.5%).

Frequency (Hz)	Amplitude (V)	Phase (°)
60	$120\,\sqrt{2}$	0
300	$3.4\,\sqrt{2}$	−144
420	$1.4\,\sqrt{2}$	20

Figure 12a shows the simulated waveforms of input voltage, input current, and input current reference when the converter works with a power level of 31 W. It can be observed that the tracking of the sinusoidal current reference, that is, the sliding motion, is accomplished along the entire cycle of the AC source. Figure 12b shows the experimental results for the same operating conditions. Namely, the same input voltage, the programmable load operating as constant voltage load at 400 V, and the current reference configured manually at the same value (0.35 A). The results are very similar, showing that the effect of parasitic components in the real circuit has a negligible effect on the behavior of the controlled circuit.

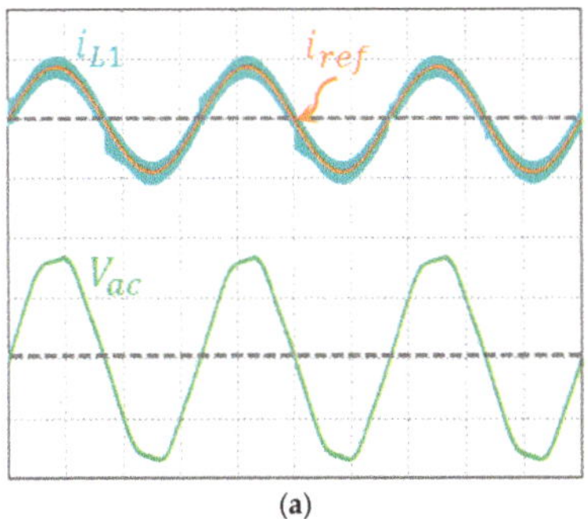

(a)

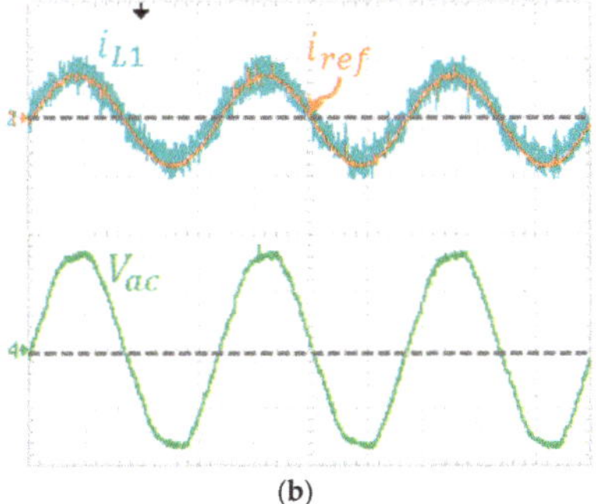

(b)

Figure 12. Current and voltage waveforms at the AC side of the converter during a 31 W test: (**a**) simulated results and (**b**) experimental results. Scales: 0.5 A/div and 100 V/div.

Similarly, Figure 13a shows the simulated waveforms of input voltage, input current, and input current reference when the converter works with a power level of 95 W. Again, it can be observed that the tracking of the sinusoidal current reference, that is, the sliding motion, is accomplished along the entire cycle of the AC source. Figure 13b shows the experimental results for the same operating conditions, which correspond to a current reference of 0.8 A. It can be observed that simulated and measured waveforms are very similar, showing that the effect of parasitic components in the real circuit is still negligible for high levels of power. This fact allows us to assert that the converter control has a good behavior along a wide power range.

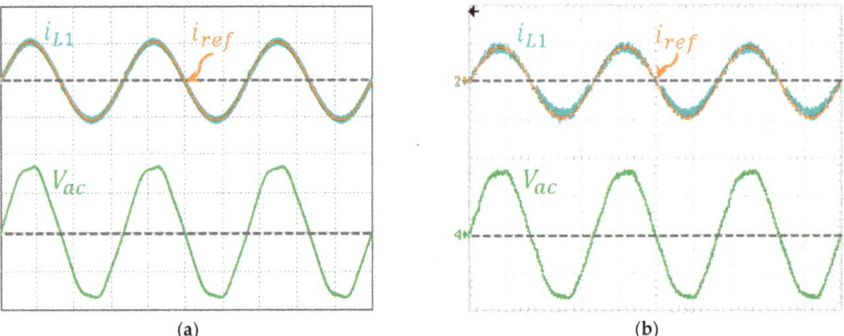

Figure 13. Current and voltage waveforms at the AC side of the converter during a 95 W test: (**a**) simulated results and (**b**) experimental results. Scales: 1 A/div and 100 V/div.

4.3. Experimental Power Quality Assessment

The power quality obtained with the proposed control was evaluated for different power levels, and it was found that the power factor (PF) is always higher than 0.95 when the converter operates between 10% and 120% of its nominal power (100 W), whereas the displacement power factor (DPF) is always 1.0. The THD always takes satisfactory values below the requirements of the international standards [12].

Figure 14 shows the measurements obtained with the power quality analyzer for two power levels, namely 31 W and 95 W. As can be observed for the low power level, the PF is low and the THD-R is high. On the contrary, for the high power level, the PF is higher and the THD is lower. It is worth highlighting the low value of THD-R obtained in the current at high power levels (1.6%), which illustrates the effectiveness of the proposed rectifier control.

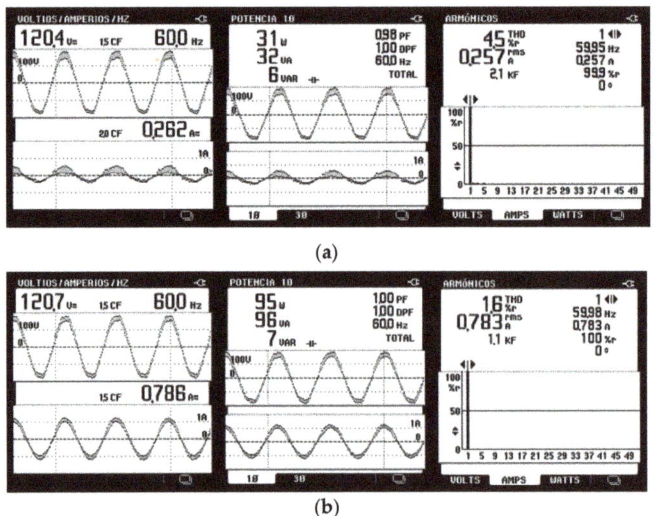

Figure 14. Power quality measurements: (**a**) test at 25% of the nominal power and (**b**) test at 80% of the nominal power.

4.4. Transient Behavior Assessment

The output port of the rectifier is connected to an LVDC bus whose behavior is emulated in experiments by using a programmable load configured as a constant voltage load. The input port of the converter is connected directly to the grid (120 V/60 Hz). Then, input and output voltages are imposed by external conditions. The transient behavior of the variables in the proposed HPF rectifier was evaluated by applying sudden changes in the current reference, which were introduced in the control circuit by replacing the precision potentiometer by a signal generator. As can be observed in Figure 15, the amplitude of the current reference follows a square periodic waveform of 2 Hz with a minimum level of 0.25 A and a maximum level of 0.75 A. As expected, the input current responds immediately to the induced changes, while the other variables preserve their stability.

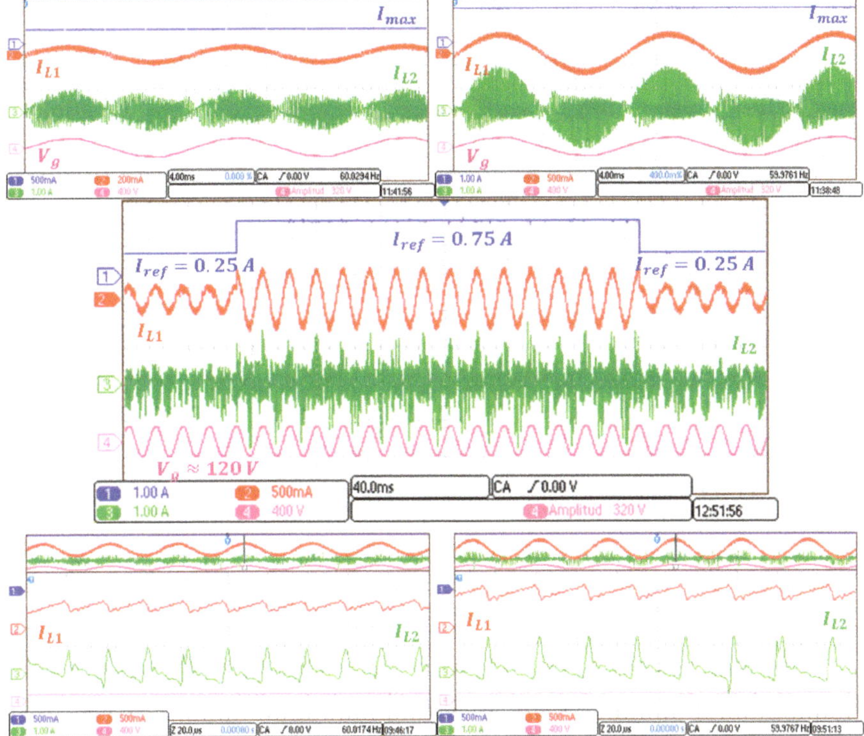

Figure 15. Oscilloscope captures during 400 ms showing dynamic behavior of the converter for changes from 0.25 A to 0.75 A, and vice versa. At the top, a 40 ms zoom details three cycles of the current waveform for each current level. At the bottom, a 200 μs zoom details the high frequency components.

4.5. Comparison with Previous Results

The isolated bridgeless SEPIC rectifier studied in this paper was controlled to ensure a high power factor interfacing an AC source with an LVDC distribution bus. The work reported here has taken into account some constraints that have not been considered before in the literature, namely the absence of output voltage regulation loops and the absence of the dynamic effect of the output capacitor in the converter dynamics. The benefits of applying sliding-mode control in the proposed converter with the mentioned constraints can be observed in terms of simple implementation, fast and robust response, and general improvement of quality indicators such as high power factor (>0.98) and reduced THD

(<5%). These quality indicators for the entire range of operation of the converter are above the quality levels reported before for the same topology or for modified versions of the basic circuit [14–17,21–28].

5. Conclusions

In this paper, the modeling and nonlinear control of the isolated bridgeless SEPIC rectifier interfacing an AC source with an LVDC bus were presented. The use of a hysteresis-based sliding mode control approach to ensure the tracking of a high quality current reference yielded very satisfactory results. A simple sliding surface allowed us to ensure high power quality and robustness in the isolated SEPIC rectifier with an easy electronic implementation. The correct operation of the proposed control was validated using several simulation and experimental results. It was demonstrated that the proposed solution exhibits adequate performance in power quality indicators such as THD always being lower than 3.5% (1.6% for the best case) and a PF higher than 0.95 (0.99 for the best case).

The simplicity of the implementation and the low levels of THD demonstrated that the proposed control method is comparable to the best strategies reported in the literature. Hence, the studied SEPIC converter with the proposed control is a promising alternative for the insertion of HPFs in emerging energy processing applications.

Moreover, the analysis carried out in the paper tackled—for the first time—the behavior of the isolated SEPIC circuit at zero crossing points. This additional mode cannot be considered as a trivial finding, because it differs from the known behavior of the conventional SEPIC topology used in DC–DC conversion. Nonetheless, in spite of the existence of this additional mode, the proposed sliding mode controller results to be immune to it because, after a brief dwell time in that mode, the sliding surface is quickly attained.

Our future work contemplates the study of the converter control when the coupled inductor operates in discontinuous conduction mode. This would introduce a substantial theoretical difference for the sliding motion, but it could result practically in important advantages in both the whole performance and converter power density.

Author Contributions: Conceptualization, O.L.-S., G.G., and L.M.-S.; Formal analysis, O.L.-S., G.G., and L.M.-S.; Funding acquisition, O.L.-S. and L.M.-S.; Investigation, O.L.-S., G.G., L.M.-S., and Á.J.C.-C.; Methodology, O.L.-S., G.G., and L.M.-S.; Project administration, O.L.-S.; Validation, O.L.-S. and A.J.C.-C.; Writing—original draft, O.L.-S.; Writing—review & editing, O.L.-S., G.G., and L.M.-S.

Funding: This research was developed with the partial support of Colciencias under contract 018-2016, the Gobernación del Tolima under Convenio de cooperación 1026-2013, the Universidad de Ibagué under project 16-435-SEM, and the Spanish Agencia Estatal de Investigación under grants DPI2015-67292-R (AEI/FEDER, UE) and DPI2016-80491-R (AEI/FEDER, UE).

Conflicts of Interest: The authors declare no conflict of interest.

Nomenclature

i_{L1}	Current through the inductor $L1$.
i_{L2}	Current through the inductor $L2$.
i_{L2x}	Current through the magnetizing inductance of $L2$.
I_m	Amplitude of the sinusoidal current reference i_{ref}.
I_{dc}	Average current through the load voltage source V_{dc}.
i_{ref}	Reference of the current control loop.
i_p	Input current in the primary winding of $L2$.
i_{C1}	Current through the intermediate capacitor.
i_{D3}	Current through the diode D_3.
i_{D4}	Current through the diode D_4.
v_{ac}	AC input voltage.
v_{L1}	Voltage across the inductor $L1$.
v_{L2}	Voltage across the inductor $L2$.
v_{C1}	Voltage across the intermediate capacitor.
V_{dc}	LVDC bus voltage.

V_m	Amplitude of the sinusoidal input voltage v_{ac}.
v_{D3}	Voltage across the diode D_3.
v_{D4}	Voltage across the diode D_4.
v_{s1}	Voltage across the secondary winding 1 of $L2$.
v_{s2}	Voltage across the secondary winding 2 of $L2$.
L_1	Input inductance.
L_2	Coupled inductance.
C_1	Intermediate capacitance.
S_1	Switch configuring the current path with D_2.
S_2	Switch configuring the current path with D_1.
N	Turns ratio of the coupled inductor $L2$.
n_1	Number of turns of the primary winding of $L2$.
n_2	Number of turns of the secondary winding 1 of $L2$.
n_3	Number of turns of the secondary winding 2 of $L2$.
D_2	Diode configuring the current path with S_2.
D_2	Diode configuring the current path with S_1.
D_3	Diode connected to the secondary winding 1.
D_4	Diode connected to the secondary winding 2.
ε	Time interval in which the sliding surface is attained after a zero crossing.
a	Variable modelling the special case of zero crossing when it takes a value equal to one.
u	Discrete control signal.
$S(x)$	Sliding surface.
K	Voltage value defined to simplify Expression (19).
K_{VI}	Constant value assigned to v_{ac} to verify the existence of sliding modes.
K_{CI}	Constant value assigned to i_{L1} to verify the existence of sliding modes.
u_{eq}	Equivalent control.
f	Grid frequency.
ω	Grid angular frequency.
ω_a	Resonance frequency of the series LC circuit that results when $a = 1$.
δ	Value defining the limits of the hysteresis band.
m	Exponent defining the frequency $2^m f$ of the high-frequency signal of a digital PLL.

Appendix A

By replacing K in expression (21), the following is obtained:

$$i_p = \begin{cases} -\frac{V_m}{K}\left[\frac{V_m C_1 \omega}{2}\sin 2\omega t - I_m \sin^2 2\omega t\right] & for\ 0 \le \omega t < \pi \\ -\frac{V_m}{K}\left[\frac{V_m C_1 \omega}{2}\sin 2\omega t - I_m \sin^2 2\omega t\right] & for\ \pi \le \omega t < 2\pi \end{cases} \tag{A1}$$

Then, Expression (22) can be rewritten as follows:

$$i_{D3} = \begin{cases} -\frac{V_m}{K}\left[\frac{V_m C_1 \omega}{2}\sin 2\omega t - I_m \sin^2 2\omega t\right] & for\ 0 \le \omega t < \pi \\ 0 & for\ \pi \le \omega t < 2\pi \end{cases} \tag{A2}$$

By replacing (A2), Expression (24) becomes the following:

$$I_{dc} = \frac{1}{\pi}\int_0^\pi i_{D3}d\omega t = \frac{1}{\pi}\left[-\frac{V_m^2 C_1 \omega}{4V_{dc}}\int_0^\pi \sin 2\omega t\, d\omega t + \frac{I_m V_m \pi}{2V_{dc}} - \frac{I_m V_m}{8V_{dc}}\int_0^\pi \cos 4\omega t\, d\omega t\right],$$

$$I_{dc} = \frac{1}{\pi}\left[\left.\frac{V_m^2 C_1 \omega}{4V_{dc}}\cos 2\omega t\right|_0^\pi + \frac{I_m V_m \pi}{2V_{dc}} - \left.\frac{I_m V_m}{8V_{dc}}\sin 4\omega t\right|_0^\pi\right], \tag{A3}$$

$$I_{dc} = \frac{I_m V_m}{2V_{dc}}.$$

References

1. Garcia, O.; Cobos, J.A.; Prieto, R.; Uceda, J. Single phase power factor correction: A survey. *IEEE Trans. Power Electron.* **2003**, *18*, 749–755. [CrossRef]
2. Sekar, A.; Raghavan, D. Implementation of Single Phase Soft Switched PFC Converter for Plug-in-Hybrid Electric Vehicles. *Energies* **2015**, *8*, 13096–13111. [CrossRef]
3. Feng, W.; Yutao, L. Modelling of a Power Converter with Multiple Operating Modes. *World Electr. Veh. J.* **2018**, *9*, 7–34. [CrossRef]
4. Cheng, C.-A.; Chang, C.-H.; Cheng, H.-L.; Tseng, C.-H.; Chung, T.-Y. A Single-Stage High-Power-Factor Light-Emitting Diode (LED) Driver with Coupled Inductors for Streetlight Applications. *Appl. Sci.* **2017**, *7*, 167. [CrossRef]
5. Tseng, S.-Y.; Huang, P.-J.; Wu, D.-H. Power Factor Corrector with Bridgeless Flyback Converter for DC Loads Applications. *Energies* **2018**, *11*, 3096. [CrossRef]
6. Han, J.; Oh, Y.-S.; Gwon, G.-H.; Kim, D.-U.; Noh, C.-H.; Jung, T.-H.; Lee, S.-J.; Kim, C.-H. Modeling and Analysis of a Low-Voltage DC Distribution System. *Resources* **2015**, *4*, 713–735. [CrossRef]
7. Rodriguez-Diaz, E.; Chen, F.; Vasquez, J.C.; Guerrero, J.M.; Burgos, R.; Boroyevich, D. Voltage-Level Selection of Future Two-Level LVDC Distribution Grids: A Compromise Between Grid Compatibiliy, Safety, and Efficiency. *IEEE Electrif. Mag.* **2016**, *4*, 20–28. [CrossRef]
8. Barros, J.; de Apráiz, M.; Diego, R.I. Power Quality in DC Distribution Networks. *Energies* **2019**, *12*, 848. [CrossRef]
9. Dragicevic, T.; Vasquez, J.C.; Guerrero, J.M.; Skrlec, D. Advanced LVDC Electrical Power Architectures and Microgrids: A step toward a new generation of power distribution networks. *IEEE Electrif. Mag.* **2014**, *2*, 54–65. [CrossRef]
10. Guerrero, J.M.; Loh, P.C.; Lee, T.L.; Chandorkar, M. Advanced Control Architectures for Intelligent Microgrids—Part II: Power Quality, Energy Storage, and AC/DC Microgrids. *IEEE Trans. Ind. Electron.* **2012**, *60*, 1263–1270. [CrossRef]
11. Liu, J.; Liu, Y.; Zhuang, Y.; Wang, C. Analysis to Input Current Zero Crossing Distortion of Bridgeless Rectifier Operating under Different Power Factors. *Energies* **2018**, *11*, 2447. [CrossRef]
12. IEC. *Limits for Harmonic Current Emissions (Equipment Input Current ≤ 16 A per Phase), EMC Part 3-2, IEC 61000-3-2*, 3rd ed.; International Electrotechnical Commission: Geneva, Switzerland, 2005.
13. Lin, X.; Wang, F.; Ho-Ching Iu, H. A New Bridgeless High Step-up Voltage Gain PFC Converter with Reduced Conduction Losses and Low Voltage Stress. *Energies* **2018**, *11*, 2640. [CrossRef]
14. Ismail, E.H. Bridgeless SEPIC Rectifier with Unity Power Factor and Reduced Conduction Losses. *IEEE Trans. Ind. Electron.* **2009**, *56*, 1147–1157. [CrossRef]
15. Sabzali, A.J.; Ismail, E.H.; Al-Saffar, M.A.; Fardoun, A.A. New Bridgeless DCM SEPIC and Cûk PFC Rectifiers with Low Conduction and Switching Losses. *IEEE Trans. Ind. Appl.* **2011**, *47*, 873–881. [CrossRef]
16. Al Gabri, A.M.; Fardoun, A.A.; Ismail, E.H. Bridgeless PFC-Modified SEPIC Rectifier with Extended Gain for Universal Input Voltage Applications. *IEEE Trans. Power Electron.* **2015**, *30*, 4272–4282. [CrossRef]
17. de Melo, P.F.; Gules, R.; Romaneli, E.F.R.; Annunziato, R.C. A Modified SEPIC Converter for High-Power-Factor Rectifier and Universal Input Voltage Applications. *IEEE Trans. Power Electron.* **2010**, *25*, 310–321. [CrossRef]
18. Badin, A.A.; Barbi, I. Unity Power Factor Isolated Three-Phase Rectifier with Two Single-Phase Buck Rectifiers Based on the Scott Transformer. *IEEE Trans. Power Electron.* **2011**, *26*, 2688–2696. [CrossRef]
19. Badin, A.A.; Barbi, I. Unity Power Factor Isolated Three-Phase Rectifier with Split DC-Bus Based on the Scott Transformer. *IEEE Trans. Power Electron.* **2008**, *23*, 1278–1287. [CrossRef]
20. Bist, V.; Singh, B. A Unity Power Factor Bridgeless Isolated Cûk Converter-Fed Brushless DC Motor Drive. *IEEE Trans. Ind. Electron.* **2015**, *62*, 4118–4129. [CrossRef]
21. Shi, C.; Khaligh, A.; Wang, H. Interleaved SEPIC Power Factor Preregulator Using Coupled Inductors in Discontinuous Conduction Mode with Wide Output Voltage. *IEEE Trans. Ind. Appl.* **2016**, *52*, 3461–3471. [CrossRef]
22. Chen, Y.; Mo, S. A bridgeless active-clamp power factor correction isolated SEPIC converter with mixed DCM/CCM operation. In Proceedings of the 1st International Future Energy Electronics Conference (IFEEC), Tainan, Taiwan, 3–6 November 2013; pp. 1–6.

23. Tibola, G.; Barbi, I. Isolated Three-Phase High Power Factor Rectifier Based on the SEPIC Converter Operating in Discontinuous Conduction Mode. *IEEE Trans. Power Electron.* **2013**, *28*, 4962–4969. [CrossRef]

24. Hou, D.; Zhang, Q.; Liu, X. An Isolated Bridgeless Power Factor Correction Rectifier Based on SEPIC Topology. *Inf. Techn. J.* **2011**, *10*, 2336–2342. [CrossRef]

25. Tanitteerapan, T.; Mori, S. Simplified input current waveshaping technique by using inductor voltage sensing for high power factor isolated SEPIC, Cûk and flyback rectifiers. In Proceedings of the Seventeenth Annual IEEE Applied Power Electronics Conference and Exposition (APEC.), Dallas, TX, USA, 10–14 March 2002; pp. 1208–1214.

26. Tanitteerapan, T. Analysis of power factor correction isolated SEPIC rectifiers using inductor detection technique. In Proceedings of the 47th Midwest Symposium on Circuits and Systems, Hiroshima, Japan, 25–28 July 2004.

27. Kim, J.; Han, S.; Cho, W.; Cho, Y.; Koh, H. Design and Analysis of a Repetitive Current Controller for a Single-Phase Bridgeless SEPIC PFC Converter. *Energies* **2019**, *12*, 131. [CrossRef]

28. Rosa, A.H.R.; de Souza, T.M.; Morais, L.M.F.; Seleme, S.I. Adaptive and Nonlinear Control Techniques Applied to SEPIC Converter in DC-DC, PFC, CCM and DCM Modes Using HIL Simulation. *Energies* **2018**, *11*, 602. [CrossRef]

29. Shieh, H.-J.; Chen, Y.-Z. A Sliding Surface-Regulated Current-Mode Pulse-Width Modulation Controller for a Digital Signal Processor-Based Single Ended Primary Inductor Converter-Type Power Factor Correction Rectifier. *Energies* **2017**, *10*, 1175. [CrossRef]

30. Singer, S.; Erickson, R.W. Canonical Modeling of Power Processing Circuits Based on the POPI Concept. *IEEE Trans. Power Electron.* **1992**, *7*, 37–43. [CrossRef]

31. Marcos-Pastor, A.; Vidal-Idiarte, E.; Cid-Pastor, A.; Martínez-Salamero, L. Loss-Free Resistor-Based Power Factor Correction Using a Semi-Bridgeless Boost Rectifier in Sliding-Mode Control. *IEEE Trans. Power Electron.* **2015**, *30*, 5842–5853. [CrossRef]

32. Flores-Bahamonde, F.; Valderrama-Blavi, H.; Martínez-Salamero, L.; Maixé-Altés, J.; García, G. Control of a three-phase AC/DC Vienna converter based on the sliding mode loss-free resistor approach. *IET Power Electron.* **2014**, *7*, 1073–1082. [CrossRef]

33. Lopez-Santos, O.; Garcia, G.; Avila-Martinez, J.C.; Gonzalez-Morales, D.F.; Toro-Zuluaga, C. A simple digital sinusoidal reference generator for grid-synchronized power electronics applications. In Proceedings of the IEEE Workshop on Power Electronics and Power Quality Applications (PEPQA), Bogota, Colombia, 2–4 June 2015.

34. Jin, C.; Wang, P.; Xiao, J.; Tang, Y.; Choo, F.H. Implementation of hierarchical control in DC microgrids. *IEEE Trans. Ind. Electron.* **2014**, *61*, 4032–4042. [CrossRef]

35. Erickson, R.W.; Maksimovic, D. *Fundamentals of Power Electronics*, 2nd ed.; Kluwer: Norwell, MA, USA, 2001.

36. Flores-Bahamonde, F.; Valderrama-Blavi, H.; Bosque –Moncusi, J.M.; Garcia, G.; Martinez-Salamero, L. Using the Sliding-Mode Control Approach for Analysis and Design of the Boost Inverter. *IET Power Electron.* **2016**, *9*, 1625–1634. [CrossRef]

37. Lopez-Santos, O.; Tilaguy-Lezama, S.; Garcia, G. Adaptive Sampling Frequency Synchronized Reference Generator for Grid Connected Power Converters. *Commun. Comput. Inf. Sci.* **2018**, *915*, 573–587.

MDPI
St. Alban-Anlage 66
4052 Basel
Switzerland
Tel. +41 61 683 77 34
Fax +41 61 302 89 18
www.mdpi.com

Energies Editorial Office
E-mail: energies@mdpi.com
www.mdpi.com/journal/energies

9 783039 280988